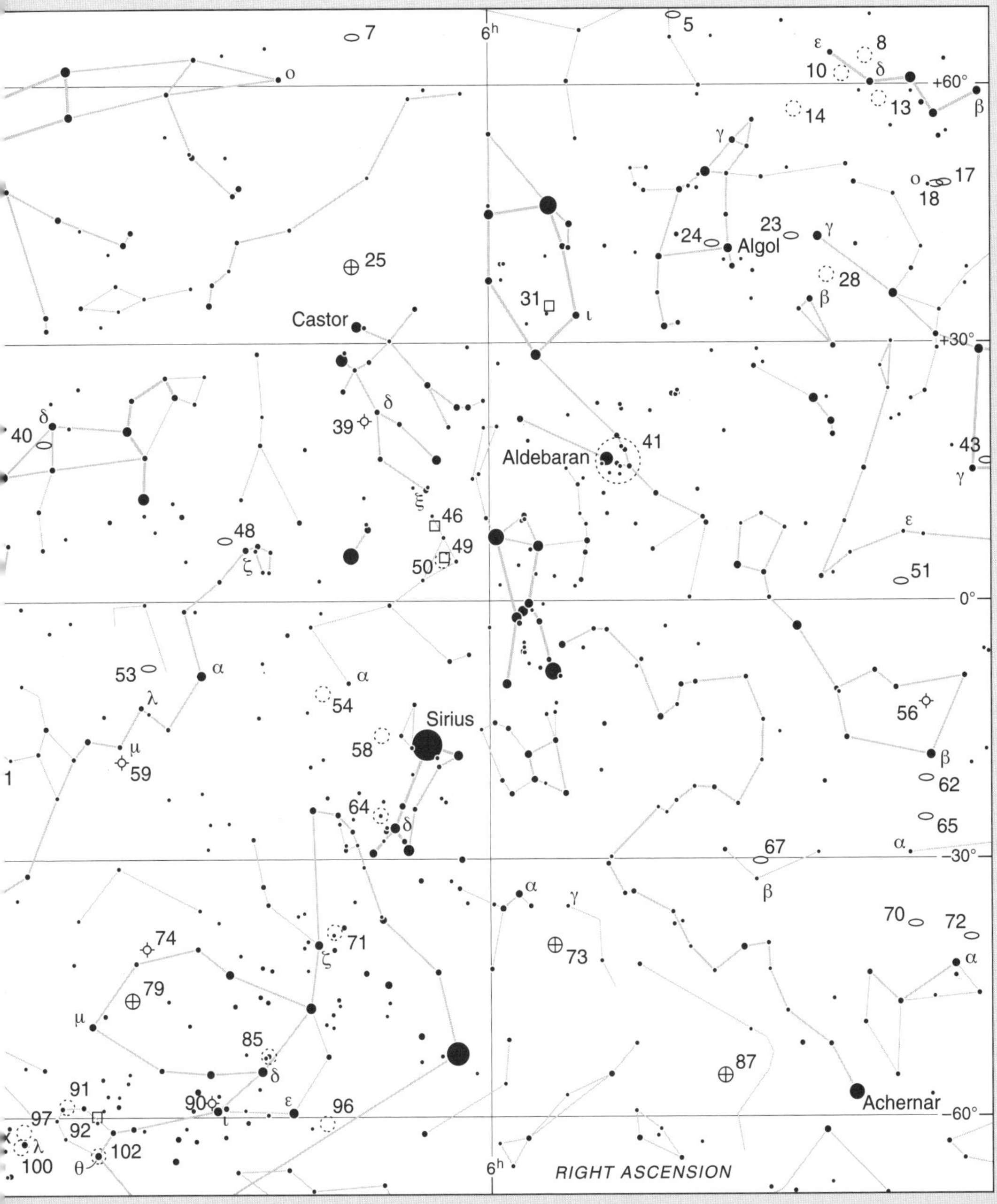

6h
+60°
+30°
0°
–30°
–60°
Castor
Algol
Aldebaran
Sirius
Achernar
5
7
8
10
13
14
17
18
23
24
25
28
31
39
40
41
43
46
48
49
50
51
53
54
56
58
59
62
64
65
67
70
71
72
73
74
79
85
87
90
91
92
96
97
100
102
6h
RIGHT ASCENSION

The Caldwell Objects

To Donna,

My Eternal Flame.

And in memory of Milky Way, Miranda-Pywackett, and Pele.

Your time never ceased in my heart;

I will always love you.

DEEP-SKY COMPANIONS

The Caldwell Objects

Stephen James O'Meara

With a Foreword by Patrick Moore

SKY & TELESCOPE
SkyandTelescope.com

CAMBRIDGE UNIVERSITY PRESS

Published by Sky Publishing Corporation, 49 Bay State Road,
Cambridge, MA 02138-1200, USA; http://SkyandTelescope.com

and by the Press Syndicate of the University of Cambridge
The Pitt Building, Trumpington Street, Cambridge, United Kingdom

Cambridge University Press
The Edinburgh Building, Cambridge, CB2 2RU, UK http://www.cambridge.org
40 West 20th Street, New York, NY 10011-4211, USA http://www.cup.org
477 Williamstown Road, Port Melbourne, VIC 3207, Australia
Ruiz de Alarcón 13, 28014 Madrid, Spain
Dock House, The Waterfront, Cape Town 8001, South Africa

Library of Congress Cataloging-in-Publication Data

O'Meara, Stephen James, 1956 –
The Caldwell objects / Stephen James O'Meara with a foreword by Patrick Moore.
p. cm. -- (Deep-sky companions)
Includes bibliographical references and index.
ISBN 0-933346-97-2 (alk. paper)
1. Astronomy -- Observers' manuals. 2. Astronomy -- Charts, diagrams, etc. 3. Galaxies -- Charts, diagrams, etc. 4. Stars -- Clusters -- Charts, diagrams, etc. 5. Nebulae -- Charts, diagrams, etc. I. Title. II. Series

QB64 .O64 2002
522 -- dc21

2002070831

A catalog record for this book is available from the British Library.

ISBN 0-933346-97-2 (Sky Publishing Corporation)
ISBN 0-521-82796-5 (Cambridge University Press)

Printed and bound in the United States of America

Contents

Foreword by Patrick Moore vii

Preface ix

Acknowledgments xiii

1 **About This Book** 1

2 **The Caldwell Objects** 19

3 **Twenty Spectacular Non-Caldwell Objects** 430

Appendix A

The Caldwell Catalog: Basic Data 444

Appendix B

Why Messier Did Not Include the Double Cluster in His Catalog 449

Appendix C

William Herschel: The Greatest Observer of All Time *Larry Mitchell* 454

Appendix D

Photo Credits 467

Index 469

Foreword

The story behind the Messier Catalog is familiar to many. In 1758 a French comet hunter named Charles Messier began compiling a list of star clusters and nebulae that would enable him to distinguish these objects from his beloved comets. His final catalog, published in the French almanac *Connaissance des Temps* for 1784, listed 103 objects. Since then, 7 more have been added, and 1, a duplicate entry, has been removed; the number of unique entries now totals 109. In themselves, the nebulae and clusters interested Messier not at all — they were merely "objects to avoid." Therefore, he omitted those that could not possibly lead to confusion or that he could not see rise above his Paris horizon.

In fact, from the seasoned observer's point of view, the catalog is incomplete. This fact struck me forcibly one evening, not so long ago, while I was out browsing around the sky. I looked casually at NGC 7293, the Helix Nebula in Aquarius, and wondered why it did not have a Messier number; it could easily be taken for a large comet. And there are many other objects of equal or greater interest than those with "M" designations. Many of these non-Messier sights are shamefully neglected.

I am not a reputed deep-sky observer. My subject is the Moon, and I like to joke that anything beyond the orbit of Neptune is rather too remote for me. But I do enjoy looking at the deep sky. Chasing Messier objects is most interesting, and there are countless Messier clubs. Almost every long-time deep-sky observer has made a round of the objects in the Messier Catalog. Why not draw up a catalog to list objects not identified by Messier? Why not extend that list so that it includes objects below the Paris horizon? The southern night sky has many spectacular sights. Why not list them as well? Amateurs are continually on the move today, traveling to star parties and expanding their observing horizons. The sizes of amateurs' telescopes are also expanding, allowing them to peer deeper into the night, to see things unimagined by Messier. Why not add a few of these challenging objects as well?

On the following morning I sat at my desk and began work. I decided to list 109 objects (the same number as in the Messier Catalog) but to arrange them in order of declination, beginning in the far north and ending in the far south — from Cepheus to Chamaeleon. The arrangement, of course, has utilitarian value. For example, from my home at Selsey, in South England (latitude 51° N), I could theoretically see to my colatitude (–39°), making the 72nd object in my list, the edge-on galaxy NGC 55 in Sculptor, theoretically within my grasp. A more reasonable object would be one, say, 5° above the horizon.

All the objects had to be accessible to amateurs using modern 4-inch and larger telescopes under a dark sky, but some would have to be more difficult than others, so that any observer setting out to survey them all would be faced with something of a challenge. I decided to include objects of all kinds: open star clusters, globular star clusters, galaxies, planetary nebulae, bright nebulae, and dark nebulae. Possibly the most elusive is the very last, C109 (NGC 3195), a decidedly dim planetary nebula in Chamaeleon. I had personally observed all these, and, as I say, I am a lunar cartographer first and foremost.

Next, the question of a name, like Messier, to go along with the numbering system. I could hardly use M (for Moore), because Messier begins with that letter. Fortunately my proper name is a double-barreled one: I am officially Patrick Alfred Caldwell-Moore. In official documents I have to use my full name, and you will find me listed under the C's. But from sheer laziness I generally sign myself simply "Patrick Moore." For the present purpose I would have to use the full name, so it became the Caldwell Catalog.

Having completed my list, I put it in an envelope and sent it off to *Sky & Telescope*. To be candid, I thought very little more about it. Creating the list had been an interesting exercise, but would anyone else be interested in it? I was absolutely taken aback by the reaction. *Sky & Telescope* gave it full space in its December 1995 issue; other observers took it up, and before long I realized that the Caldwell Catalog had really taken off. It became more or less official within a few months. Caldwell clubs, along the lines of Messier clubs, sprang up, and books were planned. Stephen James O'Meara's is particularly useful because he has taken the trouble to observe nearly all of the objects with the same telescope.

Will the Caldwell Catalog be really useful? I hope so, and the indications are that it will, because anyone who surveys all the objects will end up with a good knowledge of the nebular sky — and, as we know, amateurs can be very useful in observational branches of the science of astronomy such as supernova hunting. There is a certain wry humor in the fact that the catalog was drawn up more or less on the spur of the moment by an observer of the Moon, but at least it seems that my choice of objects has been approved. At any rate, I have so far had no adverse criticism.

So here is the Caldwell Catalog. I am grateful to Steve O'Meara for his hard work, and to *Sky & Telescope* for producing it so expertly. If you decide to set out and observe all 109 objects, I wish you well. One extra challenge: who will be the first to survey all the Messier and all the Caldwell objects in the fewest possible nights?

Patrick Moore
Selsey, Sussex
April 4, 2000

Preface

ISN'T IT IRONIC THAT A LUNAR EXPERT would draw up a catalog of 109 deep-sky objects that a planetary observer would then turn into a book? Actually it's not. In fact, it makes sense. Do not be fooled by Patrick Moore's humble confession of a passion for the Moon. Moore is his generation's most outstanding figure in astronomy. Born in Middlesex, England, in 1923, Moore was an astronomical child prodigy. By the age of 11 he had been elected a member of the British Astronomical Association; he became its president 50 years later. He is a Renaissance observer and has authored more than 100 books on astronomy, books whose titles range from *How to Make and Use a Telescope* to *Life on Mars*. His BBC television show, *The Sky at Night*, has been on the air now for more than four decades. It is among the most successful and long-running television shows in history. Moore is a consummate popularizer of astronomy, and he is a seasoned observer of both the solar system and the deep sky; he can address any astronomical topic with a sparkling air of authority. In 1988 Moore was made a Commander of the British Empire for his enduring service to astronomy, and in 2000 the Council of the Royal Astronomical Society bestowed a special Millennial Award on Moore in recognition of his unique contribution to the field.

The Eta Carinae Nebula (Caldwell 92)

In his "Star Trails" column for *Sky & Telescope* (May 1997), comet hunter David H. Levy called Moore the "ultimate amateur." Moore's been at it long enough to know what skywatchers like him desire. As Moore told Levy, "I have never pretended to be anything but an amateur. My only role now is spreading the word." The Caldwell Catalog is Moore's latest "word" — a stellar smorgasbord of celestial delights destined to whet our appetites and enrich our observing experiences.

I've known Patrick Moore since I was about 10 — not personally (until recently) but through his books. I grew up on a special diet of Moore: on the naked-eye sky, on satellite watching, on Moon and planet watching. Then, in 1970 — when I first seriously started hunting down the Messier objects and other deep-sky wonders with my newly acquired 4½-inch Tasco reflector — I used Moore's 1960 *Guide to the Stars* to acquaint me with the deep, dark night. In that book you will find many of the deep-sky objects that now appear in the Caldwell Catalog: wonders like the Double Cluster in Perseus, the spiral galaxy NGC 7331 in Pegasus, the R Coronae Australis Nebula, the Jewel Box Cluster, the dark Coalsack in Crux, and so on. The foundation for the Caldwell Catalog, then, was not laid down on a whim. It was the culmination in print of 40 years of thought and observation arising from the simple yet profound revelation that there's more to observe in the night sky than the Messier objects.

When *Sky & Telescope* introduced the Caldwell Catalog in its December 1995 issue, I had just completed the manuscript for *Deep-Sky Companions: The Messier Objects*, my own first foray away from the planets and into the fascinating realm of the deep sky. I suppose, like Moore, I too am hard to pigeonhole as an observer. But that's really nothing new. Two of history's most revered "deep-sky observers" were, well, solar-system observers. Messier was an experienced lunar and solar observer and, of course, a comet aficionado. In fact, Messier was not at all interested in the deep sky; the objects we now admire in his list were nothing but celestial mosquitoes to him — pests. And, as Larry Mitchell, a member of the Houston Astronomical Society, explains in his biography of William Herschel (Appendix C), Herschel was a "prolific observer of everything in the heavens." Herschel was fascinated by transient phenomena on the Moon, the polar caps and "luminous spots" of Mars, Saturn's rings, and the belts and flattened poles of Jupiter. It was Herschel, that great deep-sky observer, who discovered the planet Uranus, one of its moons, and two moons of Saturn. Herschel also cataloged double stars, tracked variable stars, and investigated stellar distances. So Moore and I are simply a part of an extended family of observers, a melting pot of people who love to look at all manner of heavenly wonders. What we like to observe is merely a reflection of what soothes our hearts and inflames our souls. (Besides, *what* you observe is not nearly as important as *how* you observe.)

Since the Caldwell Catalog was advertised as a list of 109 objects "beyond Messier," it seemed a natural candidate for the second book in my *Deep-Sky Companions* series. As soon as the Messier manuscript was mailed, I grabbed my 4-inch Tele Vue Genesis refractor and began another long but exhilarating visual journey across the night sky. Soon I learned that some observers believed the words "beyond Messier" were shorthand for "Here's a list of incredible deep-sky objects missed by Messier." Neither Moore nor *Sky & Telescope* ever intended to promote that view. A quick scan of the catalog's contents, which include irregular dwarf galaxies, prodigious sources of cosmic radio waves, and dim bubbles of gas blown off of Wolf-Rayet stars, proves otherwise, as does the simple fact that more than half of the Caldwell objects lie south of the celestial equator. In fact, I hope no one ever publishes a catalog of objects "missed" by Messier, because that would be a grave injustice to the famed ferret of comets. As I explain in *Deep-Sky Companions: The Messier Objects*, Charles Messier was first and foremost a comet hunter. His primary goal was to discover comets, not deep-sky objects. He did not systematically survey the night sky to discover every nebula and cluster in it, as did William Herschel. Messier explains as much in the *Connaissance des Temps* for 1801: "After me, the celebrated Herschel published a catalog of 2,000 [nebulae] which he has observed. This unveiling of the sky, made with instruments of great aperture, does not help in a perusal of the sky for faint comets. Thus my object[ive] is different from his, as I only need nebulae visible in a telescope of two feet [in length]." To fault Messier for excluding objects he would never have confused with comets — or that were beyond the visible limit of his modest telescopes — would be like faulting treasure hunters for not listing the names of birds hopping around the piles of gold and diamonds they discover. No, the Caldwell Catalog is a list of some of the most incredible non-Messier objects in the entire sky, *plus* a sprinkling of delights that will challenge small-telescope users or excite us with astrophysical intrigue.

The 109 Caldwell objects comprise 35 galaxies, 28 open star clusters, 18 globular star clusters, 13 planetary nebulae, 12 bright nebulae, two segments of a single supernova remnant, and one dark nebula. It includes many of our favorite non-Messier objects: Omega Centauri, 47 Tucanae, the Tarantula Nebula, the Eta Carinae Nebula, Centaurus A, the Helix Nebula, the Ghost of Jupiter Nebula, the Saturn Nebula, the Jewel Box, the North America Nebula, the Double Cluster, the Cat's Eye, and more. It includes the largest and most massive globular star clusters in our Milky Way; the shredded remains of a 15,000-year-old stellar explosion that blasted several solar masses of material across nearly 1,600 trillion kilometers of space; an enigmatic galaxy voraciously consuming a smaller spiral some 500 million years after the two collided; and an enormous star-forming region centered on the bloated body of a star that experienced the biggest explosion any star is known to have survived. And these are just a few of the species of celestial "wildlife" populating the Caldwell Catalog. In it you'll also find interacting sheets of cold, black molecular matter; flickering emission nebulae; planetary nebulae that "blink"; ancient globular clusters with young stars; black holes at the centers of galaxies; and open star clusters that may or may not exist. LINERs, FLIERs, Seyferts, Cepheids, and starbursts, they're all here. What's more, every Caldwell object can be seen in a 4-inch telescope under a dark sky. Most are visible with binoculars, and many can be seen with the unaided eye.

As with *Deep-Sky Companions: The Messier Objects*, the purpose of this book is to offer you a fresh perspective on the history and visual appearance of each Caldwell object; to help you find each object; and to summarize the latest research findings on each. In Chapter 1, "About This Book," I discuss the telescopes I used to observe the Caldwell objects, my observing sites and methods, helpful observing hints, and more. Since the history, astrophysics, and visual descriptions of many of these objects (especially the southernmost ones) have never been described at length in any other popular work, this chapter also explains my approach to presenting the information. The 109 Caldwell objects are detailed in Chapter 2. In many cases the essays describe recent observations from the world's largest ground-based telescopes, the Hubble Space Telescope, and an armada of spacecraft that have peered into the universe with X-ray- and infrared-sensitive "eyes." The essays are flush with historical anecdotes, including solutions to many outstanding mysteries; observational challenges; and descriptions of other interesting objects that happen to lie nearby on the dome of the sky.

Chapter 3 briefly describes 20 spectacular non-Caldwell (and non-Messier) objects — yet more proof that the heavens are replete with wonders beyond the realm of any single catalog of reasonable proportions. In fact, if you take the 109 Caldwell objects, add my 20 extras, and string them together with the other objects mentioned in this book, you'll find that this book contains perhaps 300 interesting telescopic targets. Add those to the roughly 250 objects mentioned in my book *Deep-Sky Companions: The Messier Objects*, and you'll realize that the two books introduce you to half a thousand celestial wonders — potentially enough for a lifetime of observing.

The appendixes round off the work. Appendix A tabulates each Caldwell object's position, type, angular size, and apparent magnitude. Appendix B explores why Charles Messier did not include the Double Cluster in Perseus (Caldwell 14) in his catalog; though everyone talks about it, I'm not aware of any detailed discussion of this "omission" in the amateur literature. And since William Herschel discovered more than half of the objects in this catalog — and his sister Caroline and son John found several of the others — you'll find a new and informative biography of the great 18th-century astronomer in Appendix C. The esteemed deep-sky observer Larry Mitchell wrote it. The most apparent living successor to William Herschel, Mitchell owns a portable 36-inch reflector; has observed every object that William Herschel discovered; and has compiled the positions of about 100,000 otherwise anonymous galaxies, which have not been published anywhere. This list has since become known as "Mitchell's Anonymous Catalogue," and the positions have been included in the *MegaStar* software package. Appendix D credits the talented astrophotographers whose work graces the book's pages.

As always, my desire in creating the *Deep-Sky Companions* books is to help you progress as an observer; to expand your visual envelope; and to make observing a fun and memorable experience, one that will inspire you to share the love, joy, and romance of the night sky with others. For this is what has made Patrick Moore so special in the eyes of millions. As David Williams, a former president of the Royal Astronomical Society, said about Moore receiving the Millennial Award: "Patrick Moore has been the foremost popularizer of astronomy [in the United Kingdom] for more than 40 years, and has served as an exemplary ambassador for our science to the British public and around the world. He was responsible for first sparking the interest in many of us who went on to become astronomers and he has always encouraged young people, giving generously of his time and expertise." The accolades have not ended, for as the year 2000 drew to a close, the ever-humble Moore was "astounded" to learn that he had been knighted for his services to science and broadcasting. And thanks to Sir Patrick Moore's Caldwell Catalog, we can now go out under the stars and, in a systematic manner not unlike William Herschel's, behold what beauty the night has to offer beyond the Messier objects.

Acknowledgments

The Milky Way

FIRST AND FOREMOST I would like to thank Patrick Caldwell-Moore for originating the concept of a Caldwell Catalog and the editors of *Sky & Telescope* for having the foresight to publish it in the magazine — otherwise, this book might not have happened.

Starting to work on *Deep-Sky Companions: The Caldwell Objects* was a bit like starting my first job at *Sky & Telescope*; I first had to create my office, which meant putting up sheet rock, plastering the walls, painting, laying the carpet, and moving in the furniture. Only once everything was in order, and the typewriter (yes, a typewriter) was on the desk, could I begin the long process of thinking about how to put a story together. The point is, before I could begin writing about each Caldwell object, I had to first observe and draw it, research the literature on it, gather observations of it, and sort out all kinds of conflicting details. Only then could I start to think about what I was going to say, especially about the more perplexing cases involving historical mysteries.

That said, I cannot bow low enough to thank Brent Archinal for invaluable research and insights into nearly every Caldwell mystery. Had it not been for his keen knowledge of astronomical history, or for his access to many old volumes of forgotten astronomical lore, my essays on these objects would not have been nearly as interesting or as complete. Barbara Wilson, Brian Skiff, and Hal Corwin also were instrumental in helping me solve many mysteries. A big "thank you" also goes to my long-time friend and observing companion Daniel W. E. Green, who helped me find historical passages that appeared in books 6,000 miles away from me but a few hundred yards away from him at the Harvard College Observatory library. I also have to thank Barbara for supplying me with William Herschel's original observing notes, the original description of Herschel's classification system, and the object descriptions listed in the 1864 *General Catalogue of Nebulae*. I also am very indebted to Larry Mitchell, who created the outstanding history of William Herschel for this book — a work that has not been published anywhere before. (Larry, Sir William would have been proud.) Barbara and Larry also were responsible for hog-tying me at the Texas Star Party years ago and turning me into a Wild Child of the Deep Sky. Another TSPer, Jay McNeil, was a great help in sorting out some of the planetary-nebula observations and mysteries.

To me, amateur astronomy is a global family of friends who love to look at the night sky and share its splendors. There are too many faces (sometimes "faceless" ones in the night) to thank, but I appreciate all of you for showing me your favorite deep-sky wonders or inspiring me along the way. That being said, I would like to call out special attention to Graham Blow, Rob and Leslie Hall, Grant Christie, and Gordon Herdman for opening their homes to me during my stays in New Zealand; they also were instrumental in opening the doors to the Carter and Auckland observatories. A warm *aloha* and *mahalo* also go out to Ian Cooper, New Zealand's finest deep-sky observer, and to my good friends and extended families in Florida, Texas, Guatemala, El Salvador, Panama, and Honduras: thank you, thank you, thank you — you make observing fun!

Helping me to keep all the facts straight — so that, in the end, each essay made sense — was the job of *Sky & Telescope*'s Joshua Roth. Josh is a rare bird; he is a professionally trained research astronomer with the heart and soul of an amateur. For this project, he was a godsend. As he edited this work, Josh not only massaged or corrected the astrophysics and the text; he also took his telescope outside to look at as many Caldwell objects as he could (in part to inspect my instructions, in part to enjoy the view). Josh also checked every fact in this book, and I say that in complete sincerity. It is almost inconceivable how many facts appear here: celestial directions, coordinates, sizes, object types, magnitudes, dimensions, distances, historical quotes, catalog entries, dates, translations, names, places, and reams of astrophysical data gleaned from professional journals and other sources. Thanks to Josh's monumental fact-checking efforts, any errors that had the potential to appear in this book all but vanished. Of course, nothing in life is perfect, and I take full responsibility for any *faux pas* that might remain.

The production team at Sky Publishing, headed by Derek Corson and managing editor Bud Sadler, was a tour de force. Foremost I must thank Sally MacGillivray, who helped to organize the project from its inception. Her substantial experience in book publishing was invaluable to the project, and she was a tremendous help to the production crew. Imelda Joson, my "little sister," did extensive research in *Sky & Telescope*'s photo archives and, with the help of veteran observer and jack-of-all-trades Tal Mentall, hunted down all the Caldwell photographs that eluded detection. Imelda and Tal selected photos that would best benefit a visual observer behind the telescope. On that note, I must send a global "thank you" to all the highly talented photographers listed in Appendix D, some of whom went out of their way to shoot particular Caldwell objects by request. I deeply thank all of you for making such a valuable contribution to this work. *Sky & Telescope* staff photographer Craig Michael Utter scanned all the photos, as well as my drawings, which was no easy task, especially since the faintest details in my drawings are a challenge to reproduce; yet, he did an incredible job in preserving the finest details.

Roger Sinnott wrote the software that created all the star charts; he also created the wide-field and polar star maps of the Caldwell objects that appear as endpapers. Thank you. And thank you, Tal, for slogging through the copy, tabulating all the objects, checking object orientations, and generating the charts, which illustrator Gregg Dinderman then transformed into works of art. Nina Barron proofread the manuscript. *Deep-Sky Companions: The Caldwell Objects* is the fourth book she has copy-edited or proofed for me. I admire her sensitive eye, which never seems to falter.

Likewise, thank you, Sue Bryant, for proofing Larry Mitchell's Herschel history. Special thanks go to Linda Kenny Sloan and Joshua Roth for putting together the labor-intensive index. And thank you, Lynn Sternbergh, for your most elegant design — the book looks beautiful.

Thank you, Al Nagler, for creating such a wonderful and versatile telescope. Thank you, Kilauea volcano, for continuing to erupt, because as long as you do so, the skies on my side of the island will remain dark. And thank you, Donna, for your love and support during this long journey. You are the brightest star in my eyes. And thank you, Pele, Milky Way, and Mandy, for your constant companionship in the dungeon of my office.

Finally, thank you, Rick Fienberg (Sky Publishing) and Simon Mitton (Cambridge University Press) for believing in this project and publishing the work. I also must thank you for your understanding, patience, and support, as this project sailed on through an eternal night. But look — the voyage has ended. We are somewhere over the rainbow, and the skies are blue!

CHAPTER 1
About This Book

Arcturus rising

ON ONE SPRING EVENING IN THE 1980s I was working late at *Sky & Telescope*, which is produced in three old yellow houses in Cambridge, Massachusetts. My office was on the second floor of one of those buildings, and my desk faced a window that looked out to the south. Everyone else had gone home, the Sun had set, and the stars were just beginning to shine. I love the subtle tones of twilight and soon found myself gazing out the window, transfixed. That's when the phone rang. Usually I wouldn't take a call after hours, but this time I did. On the other end of the line the faint, slightly strained voice of an elderly man asked, "Could you tell me the name of the yellow star shining outside my window?" At first I thought the call might be a prank, but the tone of the man's voice told me otherwise; it had an almost dire quality to it, like that of a sick child asking for comfort. "I remember seeing the star when I was a child," he explained, "but I never bothered to learn its name." He paused a moment to catch his breath. "You see, I'm dying."

An electric shock raced up my spine. For a moment I didn't know how to react. I wanted to say something sympathetic; instead, I asked him where he was looking. The star was low in the east, he replied. Without looking, I knew his star. Still, I had to be certain. He deserved that much. I asked the man to hold on, then ran downstairs and dashed outside. The air was cool and still, and it smelled sweet. Looking to the east I saw a stand of trees bursting with fresh spring leaves. How ironic, I thought, as I spied his star above them. And for the first time I really saw that star and realized its significance. I returned to the phone and, with a voice that was both excited and solemn, said, "It's Arcturus." I paused, then added, "or Hoku-lea, the Hawaiian star of gladness, the star that leads great voyagers home." I heard a sniff on the other end of the line, followed by a whispered "Thank you." The phone clicked. I never learned the man's name.

I'll never forget that call. It made me realize how important the stars are to humanity, how comforting they can be in times of distress. But why, I wondered, had this man never bothered to learn that star's name or to pursue his interest in the night sky? There's so much in the heavens to delight our eyes and warm our hearts. By nature we are an inquisitive species. When we smell a rose, the fragrance makes us giddy, and immediately we ask, "What kind of rose is that?" And we learn its name — say, a Lady X. The same happens when we see a flash of red feathers in our yard. "What bird is that?" we wonder. "That's a cardinal," we're told. Pick up a pure green gem, and our mind queries, "Emerald?" And we seek the answer. But how many millions of people look to the sky at night and simply say "Wow," leaving it at that? To many the sky is overwhelming, mind-boggling, too metaphysically deep.

That's why if you're interested in learning the night sky, chances are you'll be alone in your pursuit. You'll be one of many "backyard" astronomers, isolated lovers united under one sky. But we are not really alone. We have friends, we have companions, and they are the ever-loyal stars — and even more, entire clusters and galaxies of them. (What's a cluster anyway, but a family of stars traveling through space?) We have above us a cityscape of starlight. We have within our grasp all the natural wonders in the heavens. And each has a story to tell. To "hear" those stories all we have to do is look, "listen," and read. That's why I've created the *Deep-Sky Companions* series — to help introduce you to some new companions beyond the friendly veneer of stars. It's also why I use a conversational tone in the text. I want you to feel my presence as I describe what it is you're seeing. I want to make your nights less lonely, more complete. The late Leslie C. Peltier opens the sixth chapter of his endearing book *Starlight Nights* with a drawing of a child out under the stars. That child — Peltier himself — holds open before him a book, Martha Evans Martin's classic, *The Friendly Stars*, and you can sense his enamored skyward gaze. When I look at that drawing I can almost hear the book speaking to him, helping him become, as Robert Frost said, "acquainted with the night." Similarly, I try to describe and guide you to each Caldwell object in a friendly and fun manner.

If you're a beginner and have not used *Deep-Sky Companions: The Messier Objects*, I encourage you to do so, because it is . . . well, a companion to this work, the first in what promises to become a family of titles. More important, its second chapter is designed to help newcomers get started in astronomy. That section is not repeated in this work. But do not despair. This book can stand on its own. Its charts and text will enable you to locate each Caldwell object with ease. But they do not eliminate the need for newcomers to learn how to navigate the sky or to become familiar with astronomical terms. If you live in the Southern Hemisphere, please note that the book is written with a definite Northern Hemisphere bias. For instance, when I refer to the object's location above the horizon, it's generally as seen through the eyes of someone living north of the equator. But the star charts, photographs, sketches, and descriptive essays can be used by anyone, anywhere.

FIELD WORK

I live on the Big Island of Hawaii, the same island that supports the fleet of monster telescopes atop the nearly 14,000-foot-high summit of Mauna Kea. My house, however, is in a little town called Volcano at an altitude of 3,600 feet on the burning back of Kilauea volcano. I generally observe from Kilauea's 4,200-foot-summit in the Hawaii Volcanoes National Park. There I get unobstructed views of declinations (celestial "latitudes") as far south as –70°. That means only four Caldwell objects are below my horizon, though that number increases to seven if I want to get a decent view a few degrees above the horizon.

In the photograph on the next page you can see one problem I sometimes have when observing near the horizon of my unique observing site. The dark band skirting the distant horizon is vog, or volcanic smog, from a cone erupting on Kilauea's southeastern flank. When this photograph was taken the vog was concentrated to the east, toward the erupting vent; on other days the vog can drift farther to the south. It's not always a menace, but sometimes vog can back up and cause a smelly whiteout in Volcano. On average, Kilauea releases anywhere from 350 to 1,850 tons of sulfur dioxide per day; about 90 to 260 tons of that is released at the summit, where I observe. When sulfur dioxide is released at the summit it reacts chemically with moist air to form sulfuric acid or, with water vapor, acid rain. With a pH as low as 2.0, this acidic precipitation keeps the landscape around the crater nearly devoid of vegetation, which is why my horizons are free of obstructions.

I'm not trying to give you a geochemistry lesson. I just thought you might like to know that after I completed the observations for *Deep-Sky Companions: The Messier Objects*, I discovered a problem. Four years of exposure to the acidic Kilauea environment had all but obliterated the coatings on the objective lens of my 4-inch Tele Vue Genesis refractor. The glass elements also had some strange semitransparent blemishes, like alien bacteria magnified a million times over, and hairy growths had taken hold around the objective's elements, on the telescope's diagonal mirror, and . . . well, all over. Some of those abstract creations might have been mold nurtured by the tropical downpours of La Niña. I suppose I could have sacrificed the scope to the volcano and then gone out to buy a new one. But I'm a sentimental observer. Not only do I love my Genesis, I love my old, beat-up, sulfur-eaten Genesis. I wanted to observe all the Caldwell objects with it. So I called Tele Vue Optics founder Al Nagler, a true friend to the amateur community, and asked if his company could clean up an old friend. "Of course," Al said. So the telescope went on a 6,000-mile journey to New York. A week later, Al called. "Stephen!" he cried (not really), "what did you do to this telescope?" After I told him, I think I heard him faint and hit the floor.

Anyway, I got my Genesis back with a card from the optician, who said Al was still recuperating. I unpacked the telescope and looked at the lens — perfect. I had back my old 4-inch f/5 (500-mm) Genesis refractor. (Actually, Al recently looked up this telescope's original "specs" and told me that it has a focal length of 504.8 millimeters.) And with this new lease on life, my Genesis turned its tiny 4-inch eye out over the volcanic landscape and began hunting down, one by one, the 102 Caldwell objects I planned on observing from Hawaii. To my surprise I was also able to use my restored Genesis to make an additional, dramatic sighting: I spotted NGC 4833 (Caldwell 105) just 13′ above the southern horizon, making it the most southerly Caldwell object visible from the United States.

For this five-year-long study I used three eyepieces (also made by Tele Vue): a 22-mm

Panoptic, a 7-mm Nagler, and a 4.8-mm Nagler. On the Genesis these eyepieces provide magnifications of 23×, 72×, and 105×, respectively. A 1.8× Barlow lens gave me additional magnifications of 41×, 130×, and 189×, while a 3× Barlow gave me high powers of 216× and 315×. On the Genesis 216× is equal to a textbook 54× per inch of aperture, while 315× gives nearly 80× per inch of aperture — perfect for those nights when the atmosphere is extremely steady, or for high-power observations of planetary nebulae. Occasionally I would pop in some of my old Fecker eyepieces, which I used for planetary observing a quarter century ago with the 9-inch and 15-inch refractors at Harvard College Observatory. These eyepieces aren't manufactured anymore, though they are superb two-element eyepieces (albeit with little eye relief). Observing with them is sort of like looking through the eye of a sewing needle, but the star images are tack-sharp. It is with these eyepieces that I made pre-Voyager 1 observations of the spokes in Saturn's rings and determined the rotation period of Uranus's cloud tops (pre-Voyager 2) among other things. Unfortunately, because of their minimal eye relief, only a handful of today's observers would feel comfortable using these eyepieces. On this project I employed these veterans only when I needed to search for, say, a planetary nebula's central star, or to try to resolve a troubling knot in a galaxy's arm. As a "finderscope" I now use a Tele Vue Qwik Point (it's like a laser pointer). The Genesis offers me a field of view nearly 3° wide when I use the 22-mm Panoptic, and I can't remember a time when I had trouble locating an object with that eyepiece and the Qwik Point.

The weather, as anywhere, can be weird in Volcano. Clouds can suddenly appear out of nowhere and force me to move my telescope. But that's another reason why I like the Genesis and its sturdy Gibraltar mount; in two minutes I can pack it up and move on. (I hope Al never sees the mount, though, because one leg is being held together by duct tape, and it has all manner of love taps and bruises.)

In the five years it took to make the observations in this book, I watched my night sky brighten, not from light pollution but from airglow. When I made the observations for *Deep-Sky Companions: The Messier Objects* in the mid-1990s, the sky over Hawaii was the darkest it had been in 10 years, with sunspot minimum occurring in 1995. But when I began observing the Caldwell objects in 1996, solar activity had begun to rise, and sunspot data indicate that solar maximum occurred in May 2000, just as I was wrapping up. The increase in natural sky brightness over the years was subtle yet remarkable. For instance, in June 2000, shortly after astronomical twilight, the base of the zodiacal light stretched across the northern horizon, where I saw an extremely pale green glow at a height of about 10°. This latter luminescence was not artificial light pollution but airglow, a sort of permanent aurora that intensifies during heightened solar activity. Overhead, the effect dims starlight, perhaps by a magnitude. It's an incredible phenomenon.

By emphasizing natural, not manmade, light pollution, I do not want to give you the impression that everything is rosy in Volcano. In the seven years that I've lived here I have seen unnatural changes in the night sky as well. I had been traveling to the island of Hawaii for 15 years before I moved here in 1994. Now, for the first time in about 20 years, I can see skyglow emanating from Hilo, 25 miles to the northeast, and from some closer subdivisions. It's not much compared to what others have to contend with at dark sites on the U.S. mainland, and the artificial glow only appears when there is moisture in the air. But it's there. I've also seen our village grow and more streetlights go up. An armed-forces recreational facility in the national park was recently renovated, and this has added light to the summit environment. The saddest experience I've had, though, occurred one night as I traveled to the summit of Mauna Kea and noticed a strange dim reflection on the west side of the Keck Observatory domes. Light from Kona, 14,000 feet below the observatory, was defiling these temples to the stars. The eastern horizon had an unhealthy pallor caused by Hilo's sodium-vapor lamps, whose brightness seemed to have intensified. And lights from the neighboring island of Maui, which I had never noticed before, burned through low clouds to the northwest. "But there are lighting ordinances and low-pressure sodium-vapor lamps on the island, aren't there?" you ask. Well, yes. That's true. But take a drive around the island. Most of the lights are unshielded, and extended-tube lighting is becoming more widespread. Many of the new lamps are overpoweringly strong, so much so that the roads are difficult to see in a rainstorm. And so it's happening: one of the last untouched astronomical frontiers is being peeled away in strips. I suppose that as long as the major observatories have their filters, they're all set, unlucky as we may be down below. But I have not sat idle. I have met with senators, local entrepreneurs, and park officials. I've given lectures on light pollution and voiced my concerns at meetings. Some village businesses such as the Kilauea Lodge and the Steam Vent Cafe took immediate action to shield or remove troublesome lights. So light pollution can be mitigated. It just takes a little time and faith in the human condition.

My first Hawaii observation for *Deep-Sky Companions: The Caldwell Objects* took place in August 1996, and my last in August 2001. Between those dates I made two trips to New Zealand (one successful, one dampened by clouds), and one to South Africa, to observe the southernmost Caldwell objects. In August 1997 I was a guest at the Auckland Observatory (in Auckland), at Carter Observatory (in Wellington), and of amateurs Rob and Lesley Hall in Wellington. Auckland Observatory has a fantastic 20-inch f/13.5 Zeiss reflector with a 4½-inch refractor as a finderscope. Despite the temptation to use only the 20-inch, I primarily observed the globular star clusters Caldwell 105 (the one I saw 13′ above my Hawaii horizon), Caldwell 107, and Caldwell 108 with the 4½-inch finder, whose aperture was comparable to that of the Genesis — though I did spy the objects through the main scope as well. In fact, my drawings of these objects are composites based on these views. While the main structures and shapes represent what was seen through the 4½-inch at 19×, 75×, and 150×, I added stars seen through the 20-inch. I did this because I was observing under the light of a bright gibbous Moon; my moonlit views through the 4½-inch alone were not representative of what patient observers could hope to see under darker skies with a small telescope. And though I also observed Caldwell 109 (a planetary nebula) with the 4½-inch, my drawing of it is based solely on my view through the 20-inch. This planetary is easily visible in any small telescope, but small apertures reveal only its swollen disk.

My observations of Caldwell 103, 104, and 106 (and Caldwell 107, again) were made with a Celestron 8-inch Schmidt-Cassegrain in the Halls' Wellington backyard. As I explained in *Deep-Sky Companions: The Messier Objects*, the 4-inch Genesis under a dark sky shows as much detail in an object as an 8-inch Schmidt-Cassegrain does from a suburban sky. Therefore I did not stray from my mission.

Why didn't I bring the Genesis to New Zealand? That's simple. I would have had to check it as luggage, and the airline was unwilling to reimburse me if the telescope were lost en route. Why New Zealand? It's beautiful and I have friends there — and that's half the observing experience.

I have been traveling to the Southern Hemisphere since 1982, and over the years I have accumulated impressions of some of the more prominent Caldwell objects through a variety of instruments. I observed some of them through Carter Observatory's impressive 9-inch refractor in 1982. Four years later I observed some from the Australian outback. I've spied a few during visits to Central America; and I've seen others from the altiplano in Bolivia and game reserves in South Africa. But for this book, I've focused on how 96 percent of these objects look through the same 4-inch telescope from my home in Hawaii.

HOW TO USE THIS BOOK

To locate a Caldwell object, first look up its celestial coordinates (right ascension and declination) in its section of Chapter 2, or in the table in Appendix A. Next, flip to the all-sky maps located on the endpapers and use the coordinate grid to find the object in the sky. Take note of the brightest stars (those with Greek letters or Flamsteed numbers) that lie within a few degrees of your Caldwell pick. Next, locate those stars on the detailed finder chart that accompanies the object's photograph and text in Chapter 2. (The detailed finder charts in Chapter 2 are oriented with north up and west to the right.) Finally, find the part of the text that describes how to locate the object and simply follow the directions.

Each object's entry in Chapter 2 opens with a photograph (oriented with north up and west to the right, unless otherwise noted) and a list of essential data: Caldwell number; common name(s), if any; NGC, IC, or other catalog number, if any; object type; constellation; equinox 2000.0 coordinates; apparent magnitude; angular size or dimensions; surface brightness in magnitudes per square arcminute (for galaxies); distance; and the object's discoverer and discovery date. The text includes a history of the object's discovery; recent research findings; naked-eye or binocular impressions; the object's appearance through the 4-inch Genesis refractor at various magnifications; descriptions by other observers using larger instruments; a visual challenge or two; and brief notes on any interesting objects in the same patch of sky.

A drawing also accompanies the text, so you can compare your view of any Caldwell object with my own. The views may be very dissimilar, but that's okay; seeing things differently is a human condition.

You'll find William Herschel's original published description of the object or, if William did not observe the object, his son John's. Deep-sky aficionado Barbara Wilson (like Larry Mitchell, a member of the Houston Astronomical Society, and a modern incarnation of William Herschel) supplied me with William's original notes, which she drew from his original catalogs as they appeared in the *Philosophical Transactions* of the Royal Society of England.

John Herschel's quotations have been gleaned from a variety of secondary sources. For objects in the northern skies I relied on *Burnham's Celestial Handbook* and Walter Scott Houston's "Deep-Sky Wonders." For the southern objects I used Herschel quotes as given by the "Deepsky Observer's Companion" (www.fortunecity.com/roswell/borley/49/), which was created by the Astronomical Society of South Africa to promote its Deepsky Observing Section. The quotes are from John

Herschel's original observations, published in 1847 as "Results of Astronomical Observations made during the years 1834, 5, 6, 7, [and] 8, at the Cape of Good Hope; Being the completion of a telescopic survey of the whole surface of the visible heavens, commenced in 1825." John Herschel resided in South Africa from 1834 to 1838, during which time he cataloged more than 1,700 clusters and nebulae. During his stay John Herschel often made several observations of each object. The quotes used in this book's tables, however, refer only to his first observation; a date is given only if the junior Herschel discovered the object.

Incidentally, when Barbara and I began to compare William's notes with the descriptions in the 1864 *General Catalogue of Nebulae* (or "*GC*," as it is commonly called), we noted that the *GC*'s descriptions of a given object do not always match William Herschel's. For example, compare William Herschel's description of NGC 7023 (Caldwell 4), a reflection nebula in Cepheus, with that in the *GC*:

W. HERSCHEL: [Observed 18 October 1794] A star of 7th magnitude. Very much affected with nebulosity, which more than fills the field. It seems to extend to at least a degree all around; fainter stars, such as 9 or 10 magnitude, of which there are many, are perfectly free from this appearance. (H IV-74)

GC: Extremely faint, star of 7th magnitude in nebula (?).

John Herschel compiled the data for the *GC*, and he based the object descriptions largely on his observations and his father's. Thus I have included *GC* descriptions for Caldwell objects, when available. The *GC* descriptions that don't match William's original notes probably reflect observations made by John Herschel and not by his father.

An *NGC* description (whenever available) follows each Caldwell object's *GC* description. In 1888 Johann Louis Emil Dreyer published a modified version of the *General Catalogue*; he merged the *GC* and all prior catalogs, as well as any new discoveries available to him, into a new catalog: the *New General Catalogue*, or *NGC*. The *NGC* repeated most of the original *GC* object descriptions unaltered. But every now and again we find slight, and sometimes major, variations. Compare the following *NGC* description of NGC 7023 with Herschel's original description and that from the *GC*, given above:

NGC: A star of 7th magnitude in an extremely faint, extremely large, nebulosity.

The Caldwell Catalog also includes descriptions from the *Index Catalogues*, supplements to the *NGC*. Caldwell 5, for instance, is the 342nd object in the Index Catalogues, and thus goes by the name IC 342. And Caldwell 9's moniker, Sh2-155, identifies that nebula as the 155th object in Stewart Sharpless's *Catalogue of H II Regions*, which appeared in the December 1959 *Astrophysical Journal Supplement*.

At the end of each Herschel description is a code contained in parentheses ("H IV-74," for instance, or "h 34"). This code dates to a classification system created and used by the Herschels. "H" stands for the elder Herschel and "h" for his son.

Nothing is ever simple when dealing with historical observations, so I'll make this point brief. William Herschel had his own numbering system for his objects. So did John. When John Herschel combined his observations with his father's to create the *General Catalogue*, he created an entirely new numbering system. Of course, when Dreyer modified the *GC* to create the *NGC*, he did away with the *GC* numbering system and assigned to the objects the *NGC* numbers we use today. Here's an example of the possible confusion: Caldwell 2, a planetary nebula in Cepheus, is also H IV-58, h 8, GC 20, and NGC 40.

Now to explain the Roman numerals and numbers in William Herschel's system. The Roman numeral identifies the class into which Herschel placed each object:

I. Bright nebulae
II. Faint nebulae
III. Very faint nebulae
IV. Planetary nebulae: Stars with burs, with milky chevelure, with short rays, remarkable shapes, etc.
V. Very large nebulae
VI. Very compressed and rich clusters of stars.
VII. Pretty much compressed clusters of large or small stars.
VIII. Coarsely scattered clusters of stars.

The Arabic numeral that follows is simply the order in which that object appears in that class. So H IV-58 is the 58th object in Herschel's Class IV (planetary nebulae).

Herschel and *GC* designations no longer see much use. But I give them here for two reasons. William Herschel's designations tell us something about the 18th-century perception of these objects' very natures. Remember that Herschel did not know about galaxies as we think of them today, and he believed that planetary nebulae were solar systems in formation. In fact, Herschel believed that *all* nebulae could be resolved into stars provided the observer used a telescope of sufficient aperture and magnification. Indeed, some of his "clusters" are actually galaxies peppered with foreground stars, while many of his "nebulae" are in fact galaxies — island universes like our Milky Way, endowed with billions or trillions of stars.

These classical descriptions have also helped me, and other astronomers, solve some historical mysteries attending several of the Caldwell objects. The histories of many of these objects have never been published in any detail in the popular literature, and this book required extensive research. In doing the research, I have found some perplexing puzzles. Two of the most dramatic examples of this kind of historical detective work are provided by the open cluster NGC 6885 (Caldwell 37) in Vulpecula and the nebula-and-cluster complex that was listed as IC 2944 (Caldwell 100) in the original Caldwell Catalog. The findings are too complex to discuss here, but I encourage you to read the fascinating accounts given in Chapter 2.

As these accounts make clear, I did not solve the puzzles alone. Barbara Wilson and Brent Archinal (U.S. Naval Observatory) were my main archive companions, and in some cases we had to call upon the assistance of other sleuths, all of whom are credited in the essays in the next chapter. The historical mysteries were rather vast. Roughly half of the objects in the Caldwell Catalog were subjects of some historical intrigue. For instance, are you aware that many observers (including professional astronomers) do not know which star in the Kappa Crucis Cluster (Caldwell 94) is Kappa Crucis? Do you know the proper designation of the Cocoon Nebula (Caldwell 19)? (According to Archinal, the "Caldwell Catalog has it right and everyone else has it wrong.") And did you know that Eta Carinae (in Caldwell 92) may have pulsed in and out of view since the dawn of humanity? These are but a few of the many findings that follow in the individual Caldwell accounts.

MAGNITUDES THEN AND NOW

One important, albeit technical, point of historical contention involves the apparent magnitudes of stars and deep-sky objects. In *Deep-Sky Companions: The Messier Objects* I thoroughly discuss limiting magnitudes and the origins of the magnitude formula we use today. But what about the magnitudes we see in the historical records? How do they compare to those that we use today and to one another? I was always fascinated by the following entry in the log of the 19th-century Harvard astronomers William Cranch Bond and his son, George. On October 11, 1847, these men used the Harvard College Observatory's 15-inch refractor to probe the depths of the Trapezium star cluster in the Orion Nebula. While doing so they found that "a star of the *fifteenth* magnitude is seen with this Telescope within, or less than, twenty minutes of sunrise, and this without any particular effort, a good evidence of its tenacious grasp of light" (emphasis mine). How could this be, I long wondered — seeing a star more than a full magnitude fainter than Pluto in bright twilight?

Now consider the brightness of the central star in NGC 40 (Caldwell 2), a planetary nebula in Cepheus. William Herschel estimated the star to be of magnitude 9, the *GC* records it as magnitude 10, and the *NGC* (as well as many modern references) lists it as 12th magnitude. Has the star faded over the years? It's certainly possible. Many central stars of planetary nebulae vary in brightness. Does this string of historical data prove that this particular star did so? Not necessarily.

In the *Monthly Notices* of the Royal Astronomical Society (Vol. XVII, 1857), Rev. Norman R. Pogson, the father of our modern magnitude formula, points out that, though many "great catalogue-makers" adopted his brightness ratio of 2.512 between stars that were one magnitude apart (so a difference of five magnitudes would correspond exactly to a brightness ratio of 100 to 1), some great observers of his day adopted *different* ratios — among them the Struves, William Dawes, the Bonds, John Herschel, and Adm. William Henry Smyth. The implications are extraordinary.

Pogson explained that the Struves had 6 magnitudes spanning the entire range between the limit of naked-eye vision (namely, 6th magnitude) and the visual limit of his 15-inch refractor (namely, 12th magnitude). This implied that each successive step on the apparent-magnitude scale corresponded to a brightness ratio of 4. By contrast, Pogson's ratio suggests that the visual limit for that telescope was magnitude 15.1, more than 9 magnitudes fainter than the dimmest stars visible to the naked eye. Dawes adopted Struve's ratio. The Bonds, on the other hand, chose to designate the visual limits of 1-inch and 15-inch telescopes as magnitudes of 9 and 20, respectively, while Herschel divided such stars between magnitudes of 6 and 18. And Smyth "took 16th magnitude as the limit for [an instrument with an] aperture of 5.9 inches" while Pogson's formula suggests a limiting magnitude of 13 for a telescope of that size.

Pogson's table, reproduced on the next page, allows us to get a feeling for the visual limiting magnitudes that Struve, Bond, Herschel, and Smyth would have attributed to instruments of various apertures. It gives, in the first column, a "standard" magnitude, with each step corresponding to a decrease in stellar brightness by a factor of 2.512 (Pogson's ratio). The second column gives the "apertures required by an *average sight* to reduce such stars to a limit of vision" (emphasis mine) — in other words, the aperture required to make such stars visible at the eyepiece (assuming that the fully dark-adapted eye has a limiting magnitude of 6.0 and a pupil diameter of 0.23 inch, or 5.8 mm). And the remaining columns show the corresponding limiting magnitudes of Struve, Bond, Herschel, and Smyth.

The table's bottom line tells the story by itself. Through a 14.4-inch telescope Pogson's "average" observer would see stars as faint as 15th magnitude, while Struve would deem the limit 12th magnitude; Bond, 20th magnitude;

STANDARD MAGNITUDES	APERTURES IN INCHES	CORRESPONDING MAGNITUDES OF			
		STRUVE	BOND	HERSCHEL	SMYTH
6	0.229	6.0	6.0	6.0	6.0
7	0.363	6.6	7.0	7.3	7.0
8	0.575	7.3	8.0	8.6	8.0
9	0.912	8.0	9.0	10.0	9.6
10	1.44	8.6	10.0	11.3	11.2
11	2.29	9.3	11.6	12.6	12.8
12	3.63	10.0	12.9	13.9	14.4
13	5.75	10.6	15.2	15.2	16.0
14	9.12	11.3	17.5	16.6	—
15	14.4	11.9	19.8	17.9	—

and Herschel, 18th magnitude. So it's clear we have to take into account the magnitude scale used by each observer before saying anything definitive about an object's apparent brightness — much less any brightness changes — on the basis of their logs. A discussion of *your* telescope's visual limits can be found in *Deep-Sky Companions: The Messier Objects.*

DATA SOURCES

The data in this book were drawn from a variety of modern sources. Many of these sources were used in *Deep-Sky Companions: The Messier Objects,* so you can compare the properties of the respective catalogs' objects with confidence. Generally speaking, recent research findings on the physical nature of these objects were gleaned from the *Astronomical Journal* or the *Astrophysical Journal;* citations are frequently given. From each object's apparent diameter and distance I calculated its physical dimensions using the formulas that appear on page 35 of *Deep-Sky Companions: The Messier Objects.* Other information, such as constellation lore; properties of stars; and objects' positions, apparent magnitudes, angular sizes, and surface brightnesses, come from the following excellent sources (primary sources are listed first):

Star Names, Constellations, and Mythology

Allen, Richard Hinckley. *Star Names: Their Lore and Meaning.* New York: Dover Publications, 1963.

Staal, Julius D. W. *The New Patterns in the Sky: Myths and Legends of the Stars.* Blacksburg, VA: McDonald and Woodward, 1988.

Motz, Lloyd and Carol Nathanson. *The Constellations: An Enthusiast's Guide to the Night Sky.* New York: Doubleday, 1988.

Stellar Magnitudes and Spectra

Hirshfeld, Alan, Roger W. Sinnott, and François Ochsenbein, eds. *Sky Catalogue 2000.0,* Vol. 1, 2nd ed. Cambridge, England: Cambridge University Press, and Cambridge, MA: Sky Publishing Corp., 1991.

SIMBAD (*S*et of *I*dentifications, *M*easurements, and *B*ibliography for *A*stronomical *D*ata). Strasbourg, France: Centre de Données Astronomiques de Strasbourg. http://simbad.u-strasbg.fr/ and http://simbad.harvard.edu/.

Stellar Distances

The Hipparcos and Tycho Catalogues. Noordwijk, The Netherlands: European Space Agency, 1997.

Double Stars

The Washington Double Star Catalog. Washington, DC: Astrometry Department, U.S. Naval Observatory. http://ad.usno.navy.mil/ad/wds/wds.html.

Couteau, Paul. *Observing Visual Double Stars.* Cambridge, MA: MIT Press, 1981.

Hirshfeld, Alan, and Roger W. Sinnott, eds. *Sky Catalogue 2000.0,* Vol. 2. Cambridge, MA: Sky Publishing Corp.; Cambridge, England: Cambridge University Press, 1985.

Luginbuhl, Christian B., and Brian A. Skiff. *Observing Handbook and Catalogue of Deep-Sky Objects.* Cambridge, England: Cambridge University Press, 1989.

Variable Stars

Sky Catalogue 2000.0, Vol. 2.

Levy, David H. *Observing Variable Stars: A Guide for the Beginner.* Cambridge, England: Cambridge University Press, 1998.

Payne-Gaposchkin, Cecilia. *Variable Stars and Galactic Structure.* London, England: Athelone Press, 1954.

Open Star Clusters

Hynes, Steven J. and Brent A. Archinal. *Star Clusters.* Richmond, VA: Willmann-Bell, Inc., publication pending. (A draft was made available in advance of publication.)

Globular Star Clusters

Harris, William E. *Catalog of Parameters for Milky Way Globular Clusters.* Hamilton, ON: McMaster University. http://physun.mcmaster.ca/~harris/mwgc.dat.

Skiff, Brian A. "Observational Data for Galactic Globular Clusters." Webb Society *Quarterly Journal* 99:7 (1995); updated May 2, 1999, at www.ngcic.com/papers/gcdata.htm. (most diameters)

Planetary Nebulae

Skiff, Brian A. "Precise Positions for the NGC/IC Planetary Nebulae." Webb Society *Quarterly Journal* 105:15 (1996). (positions)

Observing Handbook and Catalogue of Deep-Sky Objects. (dimensions)

Cragin, Murray, James Lucyk, and Barry Rappaport. *The Deep Sky Field Guide to Uranometria 2000.0,* 1st ed. Richmond, VA: Willmann-Bell, Inc., 1993. (apparent magnitudes and dimensions of far-southern planetary nebulae)

Acker, Agnès, and Francis Gleizes. *Catalogue of the Central Stars of True and Possible Planetary Nebulae.* Strasbourg, France: Observatory of Strasbourg, 1982.

Hynes, Steven J. *Planetary Nebulae: A Practical Guide and Handbook for Amateur Astronomers.* Richmond, VA: Willmann-Bell, 1991.

Planetary-nebula distances generally were gleaned from the professional research literature or from the World Wide Web page of the Space Telescope Science Institute (www.stsci.edu).

Diffuse Nebulae

The Deep Sky Field Guide to Uranometria 2000.0.

Sky Catalogue 2000.0, Vol. 2.

Galaxies

Tully, R. Brent. *Nearby Galaxies Catalog.* Cambridge, England: Cambridge University Press, 1988. (types, angular sizes, distances, masses, and luminosities)

The Deep Sky Field Guide to Uranometria 2000.0. (apparent magnitudes and surface brightnesses)

NASA Extragalactic Database. Pasadena, CA: Infrared Processing and Analysis Center. http://nedwww.ipac.caltech.edu/. (positions and, when not available from the above references, other parameters)

Extragalactic Supernovae

"List of Supernovae." Cambridge, MA: Central Bureau for Astronomical Telegrams. http://cfa-www.harvard.edu/iau/lists/Supernovae.html.

Note that World Wide Web Uniform Resource Locators, or URLs, are subject to change. The dimensions, magnitudes, and positions of non-Caldwell objects generally were taken from *The Deep Sky Field Guide to Uranometria 2000.0.* Incidentally, the two-volume *Uranometria 2000.0* atlas was radically revised as this book went to press, as was its *Deep Sky Field Guide* companion; descriptions of *Uranometria 2000.0* charts refer to previous incarnations.

Finally, in Chapter 2 and Appendix A the Caldwell objects' celestial coordinates are stated to varying degrees of precision. In each case the decision was based on the object's angular size and degree of concentration or symmetry (tiny planetary nebulae merit arcsecond precision, for example, while loose, 1°-wide open clusters are listed to the nearest arcminute). The coordinates should get each Caldwell object into the field of view of a computer-driven telescope or an equatorially mounted one being aimed with the help of setting circles.

Incidentally, you'll note that I do not refer to objects in this book as "C" objects, as in "C101." I spell out the name "Caldwell." For the same reason that Moore decided not to use the prefix "M" (because that would conflict with the Messier, or "M," objects), I decided not to call his picks "C" objects either. As Archinal points out, the "C" prefix has already been used for other astronomical catalogs and is widely recognized as a prefix designating star clusters. "The Dictionary of Nomenclature of Celestial Objects . . . shows 13 catalogs already designated by the 'C' name," he says. "In fact the Task Group on Astronomical Designations of IAU Commission 5 specifically recommends . . . that among other things, 'an acronym shall consist of at least three characters.'" Thus for this book I have adopted Archinal's recommendation, and use the full Caldwell name. ("C" designations do appear on the detailed finder charts and some photos for brevity, however.)

SOME ORIGINAL SINS

This work corrects several unfortunate errors that appeared in the December 1995 issue of *Sky & Telescope,* where the Caldwell Catalog first was published. In several cases the tabular information (magnitudes, sizes, etc.) in this book also differs significantly from that found in the original *Sky & Telescope* presentation or on early editions of the popular Caldwell Card. Here are but a few of the most obvious errors that have been corrected in this book.

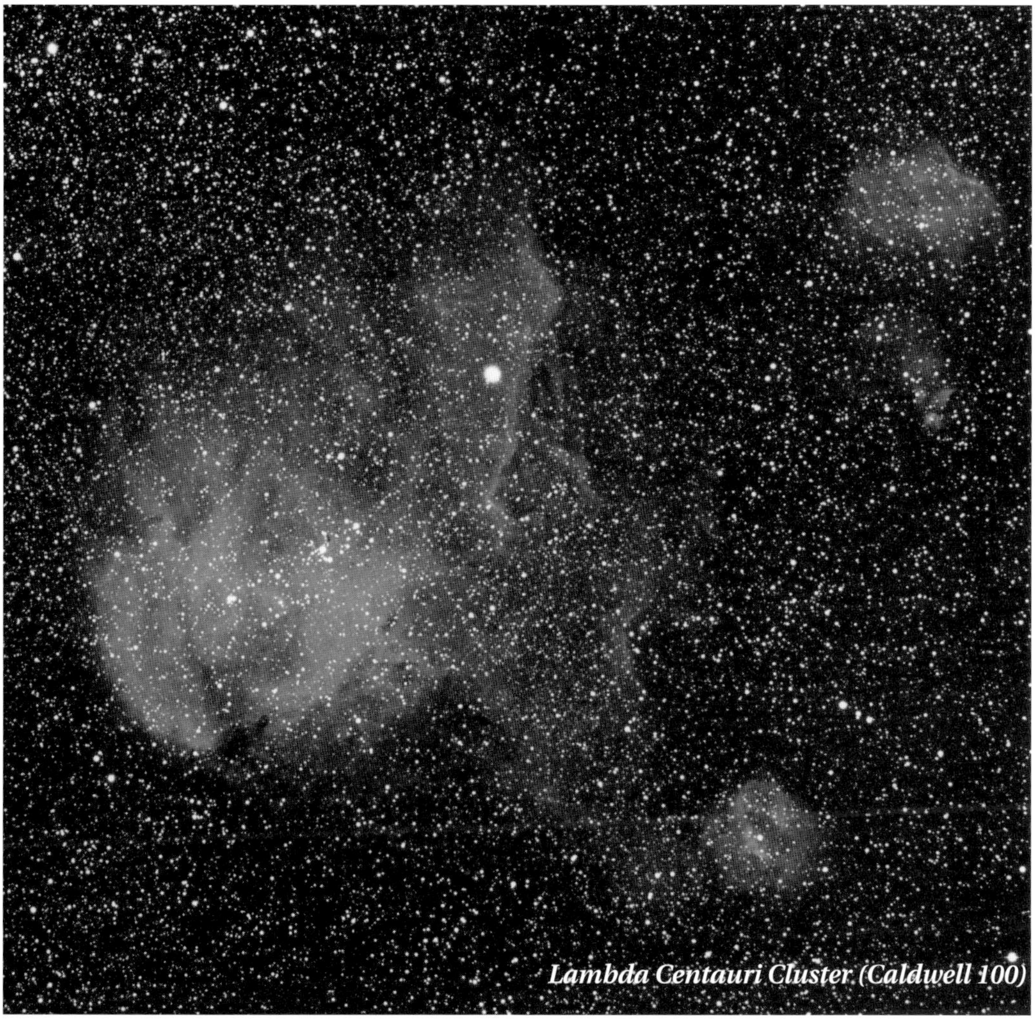

Lambda Centauri Cluster (Caldwell 100)

Caldwell 26. The objects in the Caldwell Catalog were meant to be listed in order of descending declination. Therefore Caldwell 26, or NGC 4244, a 10th-magnitude spiral galaxy with a declination of +37° 48′, should not be between Caldwell 25 (NGC 2419, a globular cluster at declination +38° 53′) and Caldwell 27 (NGC 6888, a circumstellar nebula at declination +38° 21′). It is out of place. NGC 6888 should be Caldwell 26 and NGC 4244 should be Caldwell 27. But what's done is done, and there's no need to create further confusion. Just be aware of this fact.

Caldwell 37. The open cluster associated with this Caldwell number has a significantly different position and size than given in the original Caldwell Catalog. The differences attest to a major historical enigma, details of which are given in Chapter 2.

Caldwell 41. Like Caldwell 26, the Hyades are

out of order with respect to declination; the cluster should be listed between NGC 7814 (Caldwell 43) and NGC 7479 (Caldwell 44).

Caldwell 49. The designation of the Rosette Nebula as NGC 2237-9 is incorrect; the Rosette comprises NGC 2237, NGC 2238, and NGC 2246.

Caldwell 89. The S Normae Cluster is not NGC 6067, as originally published in *Sky & Telescope,* but NGC 6087.

Caldwell 100. This entry was originally labeled "IC 2944" and "Gamma Centauri Cluster." But this identification is quite garbled. First, there is no Gamma Centauri Cluster; the name should have read "Lambda Centauri Cluster." But that alone doesn't fix the problem, for in the *Index Catalogues* IC 2944 is a nebula, not a star cluster. (Steven Lee and David Ratledge came close to solving the puzzle of Caldwell 100's designation in the June 2000 issue of *Sky & Telescope,* page 12, but even their informative discussion didn't close the case on Caldwell 100. See Chapter 2.)

That's just the tip of the research iceberg. Once I started accumulating historical and observational information on other Caldwell objects, more problems materialized; their solutions are detailed in the appropriate sections of the next chapter.

One more concern. Throughout the book I will refer to several star atlases: Wil Tirion's *Sky Atlas 2000.0* (both the first and second editions), *Uranometria 2000.0,* the *Millennium Star Atlas,* and more. Why? Just as there are inconsistencies within the historical record and the amateur literature, so too are there problems with these star charts (omissions, misplottings, etc.). Since I have no idea which star charts you own, I often take the time to tell you which objects appear on which atlases, and what was omitted or misplotted. You shouldn't have any problems with the star charts in this book, of course, and they alone should enable you to locate all 109 Caldwell objects without having to resort to other atlases.

THE FINDER CHARTS

As noted above, finding a Caldwell object that you've never seen before is a three-step process. First look up its celestial coordinates (its right ascension and declination) alongside the object's description in Chapter 2, or in Appendix A. Next, use those coordinates to find the object on the all-sky star charts located on the endpapers. The all-sky charts depict the constellations much as a planisphere would, and they should tell you at a glance the general direction in which to look (assuming your chosen object is above the horizon). Finally, return to Chapter 2 and look at the object's detailed finder chart. That chart will have at least one Bayer (Greek-lettered) or Flamsteed (numbered) star in common with the all-sky chart. Beginning with that star, which should be visible to the naked eye from all but the most light-polluted sites, you can star-hop to the Caldwell object, either by following the route described in Chapter 2 or by blazing your own trail.

The detailed charts in Chapter 2 are outwardly similar to Wil Tirion and Roger W. Sinnott's *Sky Atlas 2000.0,* second edition. Like the Field and Desk editions of that work, they have a scale of 3.55° (3° 33′) per inch (or 7.1 mm per degree), and they depict stars as faint as magnitude 8.5, providing a generous density of stellar "stepping-stones" for a wide-field telescope. North is up and west is to the right throughout (though for objects that lie near one of the celestial poles there is some distortion near the edges of the charts). Generally, one symbol size is used for all open and globular clusters, planetary nebulae, and galaxies, regardless of the objects' apparent magnitudes or sizes. Galaxies are represented by circles if face on (or nearly so) and by ellipses if significantly elongated; in the latter case the ellipse is oriented to represent the galaxy's position angle on the sky. Large diffuse nebulae have been traced to reflect their borders on astrophotographs, while small ones are represented by open squares. A key appears here. Extensive field testing has shown that a moderately experienced observer with a rich-field telescope at a suburban site can readily locate Caldwell objects using these charts.

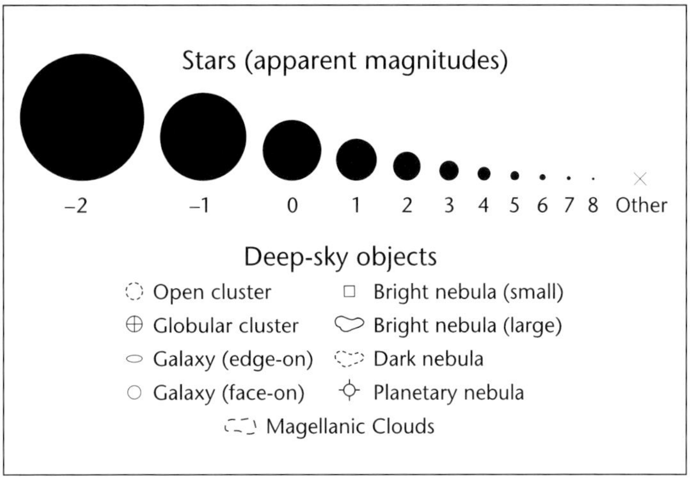

Perhaps the most noteworthy aspect of these elegant and unique charts — the fruit of many hours of painstaking work by Gregg Dinderman, E. Talmadge Mentall, Joshua Roth, and Roger Sinnott at *Sky & Telescope* — is their intimate correspondence to each Caldwell object's description in Chapter 2. Nearly all the deep-sky objects mentioned during each Caldwell "tour" are shown, as are named or numbered stars. (Sometimes, though, such companion objects are labeled on the photographs rather than on the charts, especially when the field is crowded, as is the case with the galaxy clusters inhabited by Caldwell 24 and Caldwell 35.) However, even if two or more Caldwell objects have charts that frame the same part of the sky, their *non-Caldwell* companions are plotted only once, in the section of Chapter 2 where they are discussed. A book of bicycle tours through the Great Southwest might outline each tour on a map of the region, but it wouldn't chart every scenic wonder in the Southwest when discussing a trip to the Grand Canyon. Likewise, we haven't plotted the star clusters embedded within the Eta Carinae Nebula (Caldwell 92) on the chart for the Jewel Box Cluster (Caldwell 94), even though the Eta Carina Nebula itself appears in the Jewel Box chart. Of course, each chart shows all the Caldwell (and Messier) objects that lie within its borders.

THE PHOTOGRAPHS

As in my *Deep-Sky Companions: The Messier Objects,* all the photographs in this book are reproduced in black and white. The decision was not made merely to save money. While it's possible to see hints of color in some of the brighter star clusters, nebulae, and galaxies, such objects typically manifest themselves in shades of gray at the eyepiece. Color photographs of deep-sky objects may mislead beginners into thinking that they can expect to see vivid reds, greens, and blues, when they will not.

Imagine you're a bird watcher who has never seen a raven. You want to see the bird, because the photograph in your guidebook has set your heart on fire. The raven, it appears, is a most striking specimen. Its beak is yellow, its head a rose-tinted aquamarine, its body olive, its wings vibrant indigo, and its tail feathers squid-ink black. You must see it! Now imagine the disappointment when you go birding with a friend, who points out the black raven. Shocked, you show your friend the photograph in your guidebook. Your friend laughs and tells you that that's a heavily enhanced image that exaggerates the iridescence of the bird's feathers.

The point is, why risk disappointing new observers when we can try to help them discover the beauty of reality? Furthermore, you most likely will compare the photographs to your eyepiece views with the help of a dim red flashlight, and under such illumination a color photograph actually would be a handicap.

Whenever possible, photographs were chosen to reflect the kinds of detail an observer might see with his or her own eyes. For that reason a few photographs may appear underexposed compared to more famous views you may have seen elsewhere.

That we were able to reproduce such excellent photographs of the Caldwell objects is largely a result of the generosity and perseverance of many talented astrophotographers. These dedicated students of the night sky work much harder than a purely visual observer might imagine in order to capture their revealing images.

While we attempted to obtain high-quality astrophotos of every Caldwell object from the worldwide community of amateur astronomers, in several cases we have resorted to reproducing digitized photographs taken by enormous Schmidt telescopes in both hemispheres. These photos have been made available to astronomers and scientists worldwide by the visionary architects of the Digitized Sky Survey (DSS), which can be perused on the World Wide Web at http://archive.stsci.edu/dss/. (The copyright for the DSS photos used in this book rests with the Anglo-Australian Observatory Board, the United Kingdom Particle Physics and Astronomy Research Council, the California Institute of Technology, and the Associated Universities for Research in Astronomy; they are used here with permission.) Detailed credits appear in Appendix D.

In a few cases we have also reproduced images from the Hubble Space Telescope and large telescopes like those on Mount Wilson and Palomar Mountain. For the most part, these images are included to illustrate scientific points discussed in the accompanying essays.

THE DRAWINGS

All the sketches in Chapter 2 are composites of field drawings I made at various magnifications. They are shown approximately with north up and west to the right, unless otherwise noted. Whenever possible, the orientation matches that of the corresponding photograph. Scale bars are included to help you size up each object in your own telescope. (Note that the scale of a drawing might differ markedly from that of the accompanying photograph; if so, a scale bar will appear on both the photograph and the drawing.) The composites show details visible at low, medium, and high magnifications. For instance, on the one hand, I might have seen a galaxy's faint halo readily at low power but not at high power. On the other hand, a small knot in that galaxy's arm might have been obvious at high power but inconspicuous at low power. Both the halo and the knot will appear in my sketch.

The drawings also show details seen in fits and starts over several hours observing from a dark-sky site. Some objects — like the Eta Carinae Nebula, the Coalsack, and the Bubble Nebula, among others — required several nights of observing. Unless otherwise noted, the drawings all represent views obtained from the 4,200-foot-high summit of Kilauea volcano, under one of the finest night skies on Earth — thanks to the frequent eruptions that help keep people (and lights) away. In *Deep-Sky Companions: The Messier Objects,* I claimed that a 4-inch Genesis refractor at an elevation of 4,200 feet will perform as well as an 8- to 10-inch Schmidt-Cassegrain telescope will from a dark, sea-level site. Since then, I have further convinced myself that this is true at various star parties over the years. This means that the details I portray in my drawings should be within the visual grasp of everyone using the large reflectors and catadioptric telescopes that are so popular today.

Having been a planetary observer before becoming a deep-sky observer has had its advantages. I am often asked how is it that I can see so much detail with so little aperture. My response has not changed. It's patience, and the accumulation of experience. Planetary observers are notorious for staring through the eyepiece for hours on end, waiting for those fleeting moments when the atmosphere suddenly freezes and the view sharpens to crystal clarity. That moment may last a second, maybe less. But it sends a strong visual signal to the brain, a signal that says, "Did you see that?" The analytical brain's response might be "Yes," or it might be "I think so, but I need to see that feature again in the same place and with the same orientation to be sure." And so the planetary observer sits and waits. The clock ticks away the minutes. Suddenly it happens again: snap! Another moment of fantastic seeing. The brain has already registered one quantum of visual information, so it's especially alert, ready to compare its impressions with the next photon "hit." Each positive sighting helps to validate the reality of a specific feature. And that's how I observe behind the eyepiece, ever watchful for those ephemeral moments of perfect clarity. That's why it takes me so long to make my observations. And what are nebulae but gaseous filaments, not unlike the swirling clouds on Jupiter, or its feathery festoons? Star clusters are but thousands upon thousands of double stars crammed into Moon-sized disks. And what about "planetary" nebulae — well, aren't they like Uranus and Neptune with a stellar twist?

I couple this "serious" kind of study with playful imaginings. In other words, I like to have fun when I observe. Why do it any other way? If you find it difficult to "see" creatively and would like to learn how, take a stroll to a park one day, lie down, and do nothing but look at the clouds for a half hour. When you first look up, you'll probably see, say, a cumulous cloud, all puffed up and white. With time you'll then start to see that this cloud has a particular shape or form. If you have a particularly creative bent, you may make out some recognizable pattern — a familiar face (Abraham Lincoln's, say) or a dragon. As William Sheehan and I explain in our book, *Mars: The Lure of the Red Planet,* our distant ancestors made sense of the stars by looking for similar figures among the chaos of the night sky. "The sense organs," explains Cambridge University psychologist Richard L. Gregory, "receive patterns of energy, but we seldom see patterns: we see objects. A pattern is a relatively meaningless arrangement of marks, but objects have a host of characteristics beyond their sensory features. They have pasts and futures; they change and influence each other, and have hidden aspects which emerge under different conditions." It's normal human nature to look at a cluster of stars and see Donald Duck. It's part of our genetic code. Why fight it? Let loose. Be free. Have fun.

Let's return to our cloud and *look* at it. Select one cauliflower head or bright rim and study it carefully. What do you see? You should see more than you did with your first glance, perhaps some delicate shadowing effects. You are starting to define texture, a layer of complexity. Take your eye away from the cloud. Relax for a moment, then return to it anew. This time, do not look at the cauliflower head or bright rim in its entirety, but look hard at one tiny section within it. You should start to see even more details — and details within details. If someone were to ask you now to draw that cloud, you'd have to decide how much detail you want to show, for you wouldn't be able to show everything you'd seen. You'd have to be selective and try to create an accurate, though representative, drawing of what you'd seen after such prolonged immersion. It is this kind of visual voyage that you want to take through your telescope.

In this book I've striven to make the drawings as accurate as possible, to represent what I saw through the telescope. I did not attempt to draw all the stars in open or globular clusters, and some nebulae might tend to show a stylized flair. You see, as with the clouds mentioned in the paragraph above, what I'm trying to capture is the essence of what I see — not only what my eye sees but also what my heart feels. That too is why I like turning some clusters into whimsical creatures. That adds a level of fun to the observing experience. But don't worry; if it's precision you seek, you've got the photographs of each Caldwell object to help you out. Don't worry if you don't see exactly what I see at the eyepiece. No one can do that — just as I can't see exactly what you see, even if we were to observe side by side with identical telescopes. We're all different. And that's what makes this hobby exciting. This is how we share. Sit seven artists in front of a tree and ask them to draw it, and not one rendition of that tree will be the same. In fact, each artist would probably buy the other artists' works, finding them more beautiful than his or her own. It's no different with drawing deep-sky objects.

So go out, I say, into the night with the eyes of a child and look upon these objects with unrestrained innocence. Put to paper not only what you see but also what you feel. And relive the joy of discovery.

1

NGC 188
Type: Open Cluster
Con: Cepheus

RA: $00^h\ 47.5^m$
Dec: +85° 14.5′
Mag: 8.1 (cataloged);
7.1 (O'Meara)
Diam: 15′
Dist: 4,800 light-years
Disc: John Herschel, 1831

J. HERSCHEL: [Observed 3 November 1831] Cluster, very large, pretty rich, 150 to 200 stars of 10th to 18th magnitude; more than fills the field. (h 34)

GC/NGC: Cluster, very large, round, 150 to 200 stars of magnitude 10 to 18.

UNLIKE THE MESSIER CATALOG, WHICH opens with a bang — the Crab Nebula, a supernova remnant — the Caldwell Catalog opens with a whisper. Open star cluster NGC 188 lies in a celestial graveyard about 4° from Polaris. The region is largely devoid of bright stars and avoided by most amateurs whose telescopes ride on equatorial mounts; to point anywhere near the pole with such a setup requires a truckload of patience. For these reasons NGC 188 could be considered one of the most forgotten clusters of the night sky. John Herschel discovered the cluster in 1831 and listed it as number 34 in his 1833 catalog of nebulae and clusters of stars. His description of it as "very large" and "pretty rich" is appropriate, but, as the late *Sky & Telescope* "Deep-Sky Wonders" columnist Walter Scott Houston wrote, "it is virtually invisible with poor transparency, or in too feeble a telescope." The cluster is by no means a stunning sight.

NGC 188 in Cepheus is a timeworn cluster. While most open clusters populate our galaxy's spiral arms, NGC 188 lies about 1,800 light-years above the galactic plane. We tend to think of open clusters as collections of hot, young stars. But NGC 188 holds a surprise. Its hottest main-sequence star has a relatively tepid spectral type of *F*2, and its 10 brightest stars are evolved yellow giants with spectral types ranging from *G*8 to *K*4. Indeed, this open cluster is one of the oldest known. With an estimated age of 5 billion years NGC 188 is about 1 billion years older than the aged open cluster M67 in Cancer and about 2 billion years younger than open cluster NGC 6791 in Lyra — currently the oldest known open cluster in our Milky Way galaxy.

NGC 188 may have formed when our solar system did, but it will not last as long. Both internal and external forces tend to rob open clusters of their stars. Massive binary stars within a cluster can cast out solitary members passing near them. Stars grazing a cluster's outskirts can free members from the cluster's weak gravitational grip. And open clusters are subjected to the tidal shear of our galaxy, which is constantly working against the gravitational forces that hold the clusters together. Over the next few billion years we can expect NGC 188's stars to slowly drift apart, each following a random path into space, until the cluster fully dissipates and its constituents become part of the scintillating backdrop of the night sky.

Northern Hemisphere observers can find the loose and aged NGC 188 year-round. The cluster is conveniently placed 1° south-southwest of a 4th-magnitude star in Cepheus. The first edition of *Sky Atlas 2000.0* identifies this star as 2 Ursae Minoris, but you will not find this designation on other star charts. Walter Scott Houston notes that the star was engulfed by Cepheus when the constellation boundaries were redefined by Belgian astronomer Eugene Delporte, then adopted by international agreement in 1930; the Flamsteed designation has since been dropped to avoid confusion. Brent Archinal of the U.S. Naval Observatory adds that NGC 188 lies so close to the pole that the cardinal directions here will change noticeably in several decades because of precession.

The cluster's cataloged magnitude of 8.1 may be deceiving. Using 7×35 binoculars I estimated the cluster to be a full magnitude brighter. I could not detect the object with the unaided eye, however; its dim glow is spread across nearly ¼° of sky, and it does not stand out from the bland background sky. A low-power glimpse through the 4-inch Genesis reveals what appears to be the spirit of a deceased open cluster, one seemingly fading from view. Houston himself called NGC 188 a "ghostly glow" as seen through his 4-inch Clark refractor. (He also noted that "since the cluster is always at nearly the same altitude, amateurs will find it a good object for checking observing conditions" — a classic "Scotty" remark.)

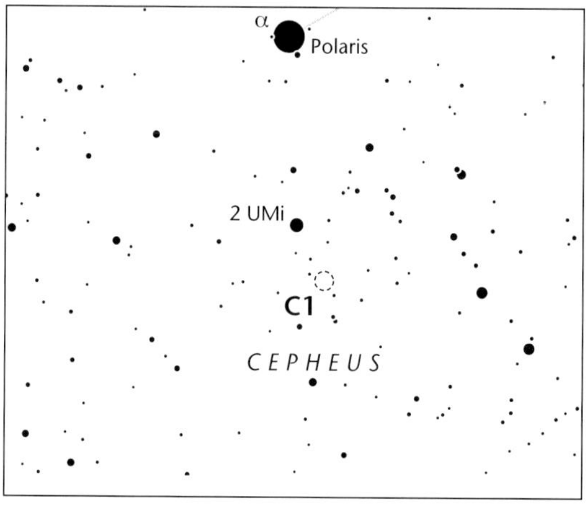

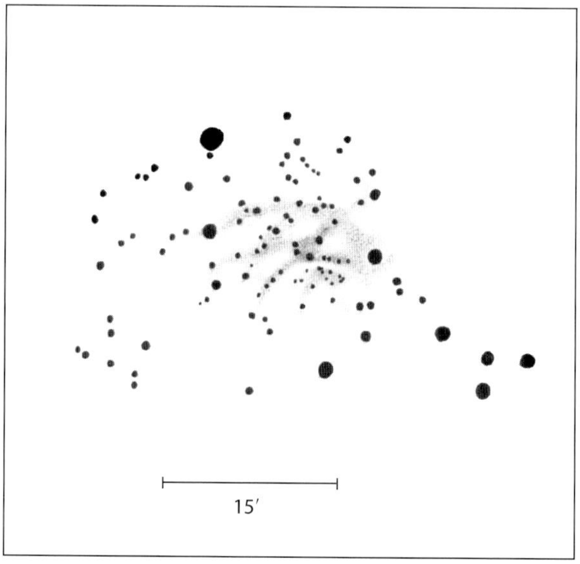

Because of its location far from the star-clotted plane of the Milky Way, NGC 188 is among the most secluded objects of its class. Its loneliness is magnified from the Big Island of Hawaii, where it can climb no more than 25° above the horizon. Still, I have been able to see it in 7×35 binoculars as a shy haze that seems to hide behind several 8th- to 10th-magnitude field stars. Binoculars also reveal five pairs of stars, all within 1° of the cluster, that look like markers in this circumpolar celestial graveyard.

Through the 4-inch at 23× the cluster resolves into an elongated scattering of dim stars centered on a more spherical, though largely ill-defined, background glow. Increased magnification does not better the overall view. High power shows the cluster as a very loose and dim gathering of distant suns. At 189× it's almost hard to believe that this is a cluster and not a chance association of stars. Moderate power (72×), however, does bring out a dark lane, trending east-west, in the cluster's northern half; the lane curves southward on either end between two sinewy rows of similarly bright stars. The suns to the north of these star chains are fainter than those to the south. The background sky seems to gradually consume the stars farthest from the cluster's fuzzy center. With a relaxed gaze, and a little imagination, the stars to the south appear to overflow the stellar banks of that dark stream, forming rivulets of unresolved suns separated by thin wisps of darkness. A more poetic impression is that of a mimosa blossom, or the feathery red lehua blossom that adorns the stark branches of Hawaii's ohia tree. A darker view would be that of a skeletal wrist and fingers, which seems somewhat more appropriate for this ancient cluster.

2

Bow-Tie Nebula
NGC 40
Type: Planetary Nebula
Con: Cepheus

RA: $00^h\ 13^m\ 01^s$
Dec: +72° 31′ 19″
Mag: 12.3 (nebula);
11.6 (central star)
Dim: 38″ × 35″
Dist: ~3,500 light-years
Disc: William Herschel, 1787

W. HERSCHEL: [Observed 25 November 1787] A star [of] 9th magnitude surrounded with very faint . . . much nebulosity. The star is either double, or not round. Less than 1′ diameter. (H IV-58)

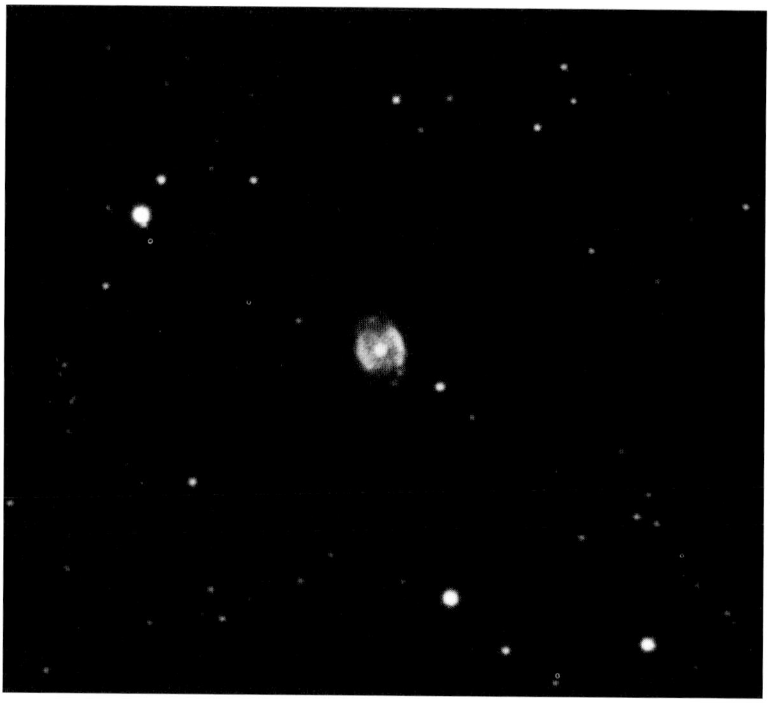

GC: Very faint, very small, round, very suddenly much brighter in the middle to a star of 10th magnitude, a star of 12th magnitude toward position angle 241.4°, diameter 25″.

NGC: Faint, very small in angular size, round, very suddenly much brighter in the middle, a star of 12th magnitude south preceding [to the southwest].

THE PATCHY FAR-NORTHERN REACHES OF THE Cepheus Milky Way all but hide the peculiar 12th-magnitude planetary nebula known as NGC 40. In 1918 Heber D. Curtis of Lick Observatory first noted the object's peculiar nature on photographic plates. Curtis saw an irregular shell of gas that resembled a "truncated ring, from the ends of which extend much fainter wisps." Photographs and images taken with modern research telescopes have revealed that Curtis's "wisps" curve gently away from the bright truncated ring, forming a series of loop-like structures reminiscent of a bow tie. The wisp extending from the nebula's northern end breaks sharply to the east, looking like a snapped twig. Photographs taken in red light reveal a weak, irregular outer envelope.

How do planetary nebulae, which originate from the largely spherical envelopes of red-giant stars, develop such weird shapes? The answer lies, in part, in the interaction between the central star's fast wind of particles and material cast off previously, when the central star was a bloated giant. NGC 40's progenitor, which has the spectral characteristics of a Wolf-Rayet star — a very luminous, very hot star with an expanding atmosphere — contains about 70 percent of our Sun's mass. Its extremely high surface temperature (about 90,000° Kelvin) should suffice to strongly ionize the surrounding nebula, causing it to glow. Yet observations show that NGC 40 isn't quite as strongly energized as would be expected. This fact suggests that some of the ultraviolet radiation from the very hot central star is being absorbed by a shield of dust between the star and the glowing nebula. The dust could have been created when the central star's fast wind (racing at speeds of about 1,800 km per second) collided with the much more slowly expanding planetary-nebula shell (which crawls along at a mere 40 km per second). Earlier mass-ejection events could be responsible for the nebula's outer shell and other peripheral features.

Seen through a small backyard telescope, NGC 40 is a visually unimposing sight. The nebula's compact disk can be found about 5½° southeast of magnitude-3.2 Gamma (γ) Cephei, and midway between, and slightly west of, two roughly 9.5-magnitude stars. Through the 4-inch at 23×, NGC 40 appears virtually stellar; this is not surprising, considering that the magnitude-11.6 central star is nearly twice as bright as the surrounding nebula. Increasing the magnification to 72× helps separate the diminutive nebula from the star; the dark cavity between the central star and its shell is barely discernible. With averted vision and prolonged effort on an observer's part, the nebula looks distinctly oval, oriented southeast-northwest, and a 13.5-magnitude star pops into view just 1¼′ to the nebula's southwest. High power brings out a beaded nature to the ring, which has two prominent knots: one to the southeast, the other to the northwest. (The other beads are rather amorphous.) The positions of the two brightest beads correspond well to the luminous arcs that show up in photographs of the nebula's truncated ring. Magnifications in excess of 200× should make these arcs stand out more clearly.

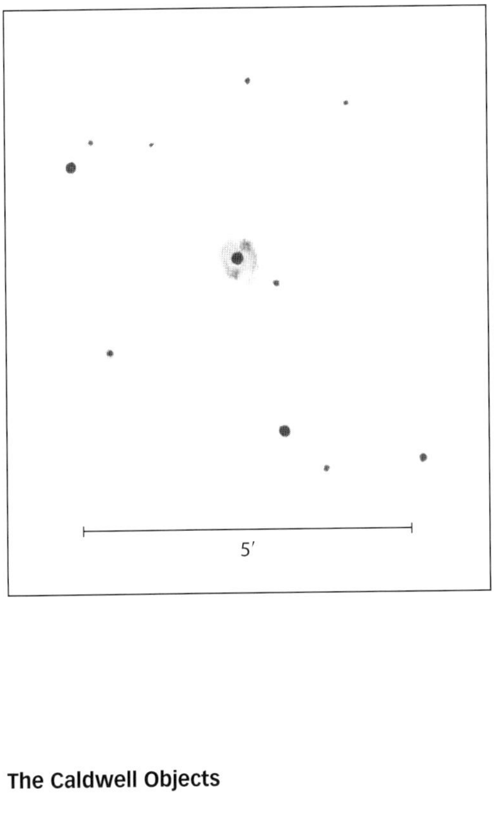

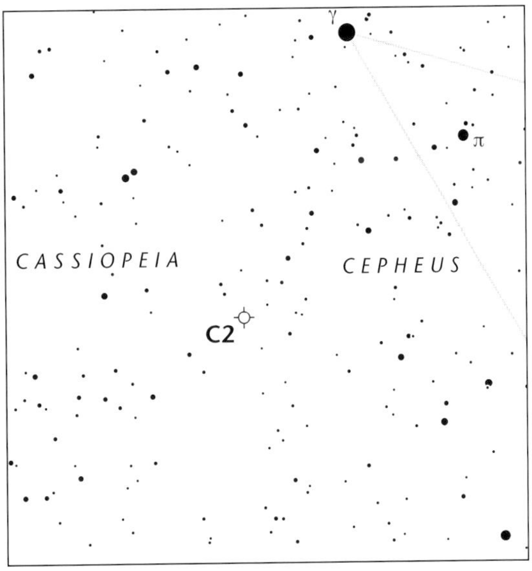

3

NGC 4236
Type: Barred Spiral Galaxy (SBdm)
Con: Draco

RA: $12^h\ 16.7^m$
Dec: +69° 27.8′
Mag: 9.6
Dim: 19.6′ × 7.6′
SB: 15.0
Dist: 7 million light-years
Disc: William Herschel, 1793

W. HERSCHEL: [Observed 6 April 1793] Very faint, much extended toward position angle 70° north preceding to south following [north-northwest to south-southeast]. About 25′ long, and losing itself imperceptibly, about 6′ or 7′ broad. (H V-51)

GC/NGC: Very faint, extremely large, much extended toward position angle about 160°, very gradually brighter in the middle.

IN LONG-EXPOSURE PHOTOGRAPHS TAKEN with large telescopes, the barred spiral galaxy NGC 4236 is a fantastic object, almost a visual enigma. A glance at the photograph above can give one the false impression that the galaxy's gossamer arms have escaped the pull of gravity and are fleeing wildly in all directions. Blotchy patches of star formation pepper the arms, which appear almost orphaned from the long, diffuse bar that constitutes the galaxy's brightest portion (the most prominent patches are 5.5′ southeast of the galaxy's center). Indeed, the "SBdm" classification means that this galaxy is a loose, clumpy, ragged system with a prominent bar but an otherwise rather faint disk.

About one-third of all galaxies have bars to some degree. A bar may provide a galaxy's stars a way to settle into stable orbits. Another mechanism for forming galactic structure is suggested by the patchiness of NGC 4236. According to some theorists, shock waves from supernova explosions could plow into the galaxy's interstellar medium, causing gas to collapse into stars. The most massive of those stars would become supernovae in turn, and a sporadic wave of star birth would pass through the galaxy's disk. In a 1994 *Astrophysical Journal* paper, Jean L. Turner (UCLA) and Paul T. P. Ho (Harvard-Smithsonian Center for Astrophysics) studied the relatively strong radio waves from star formation in the cores of nearby spiral galaxies, NGC 4236 included. They found that recent star formation appears to dominate both the infrared and radio spectra of these galaxies, pointing to a possible large population of young supernova remnants, and hence to copious star formation in the recent past.

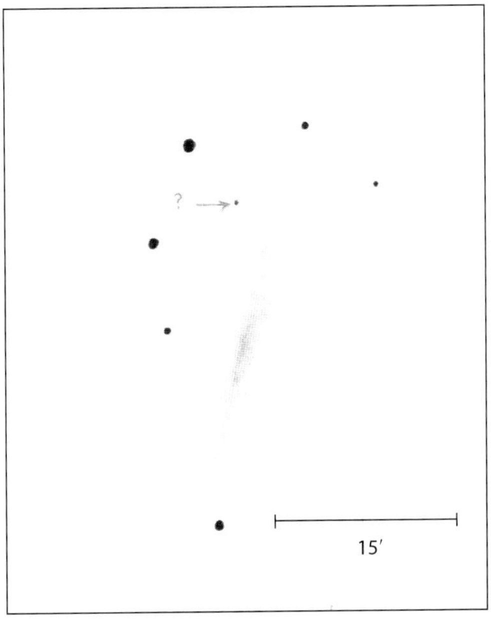

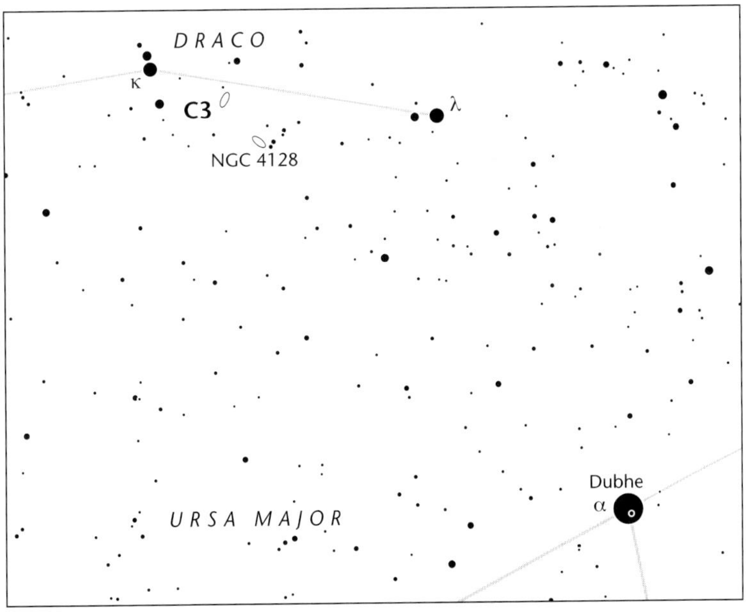

Unfortunately, few of the details visible in a long-exposure photograph of NGC 4236 can be seen at the eyepiece of a modest amateur telescope. Through my 4-inch the galaxy is extremely ill-defined. In fact, I've never seen a galaxy look so supernaturally faint. With dimensions of 19.6′ × 7.6′, the galaxy is quite large and elongated, but averted vision and low power are required to see it against the sky background, even under the pristine dark skies of Volcano, Hawaii. At 23× the galaxy virtually vanishes with direct vision, and it is all but lost with higher magnifications. Christian B. Luginbuhl and Brian A. Skiff note that the galaxy is difficult to view even in a 12-inch telescope. In his *Visual Astronomy of the Deep Sky*, Roger N. Clark calculates that 16× is the optimum magnification for detecting the galaxy with a 4-inch telescope and 27× for doing so with an 8-inch.

To find NGC 4236, first locate 2nd-magnitude Beta (β) Ursae Minoris (Kochab) and Alpha (α) Ursae Majoris (Dubhe), then look for 4th-magnitude Kappa (κ) Draconis roughly halfway between them; Kappa is the brightest star in a tight chain of three. The galaxy is 1½° to the west-southwest of Kappa Draconis. Light pollution will greatly diminish your chances of seeing this dim but intriguing object. If you do locate it, keep in mind that the ghostly glow you are beholding is a vast island universe measuring 34,000 light-years in diameter, with a total mass of 6.6 billion Suns. NGC 4236 is receding from the Milky Way with a cosmologically negligible velocity of only 2 km per second. This means that local galaxy dynamics and the Sun's motion within the Milky Way have negated the redshift that one would expect the universe's expansion to impart to a galaxy 7 million light-years distant.

If you're looking for more of a challenge, try spying the tiny 12th-magnitude galaxy NGC 4128, only 1° southwest of NGC 4236. NGC 4128 has a bright stellar core surrounded by a dim, slightly elliptical halo.

southern binocular highlight | by les dalrymple

Nearby Galaxy NGC 55

SITUATED JUST NORTH of the border separating the constellations Phoenix and Sculptor lies NGC 55 (also known as Caldwell 72), one of the brightest galaxies in the sky. It is also a member of the Sculptor Group of galaxies and the South Galactic Pole Group, which includes NGC 247, 253, 300, and 7793.

NGC 55 itself is perhaps only 4.2 million light-years away — just a little farther than the Andromeda Galaxy and M33. Due to its proximity, NGC 55 appears very large and bright (magnitude 7.5), but in reality it is quite a small galaxy compared to our Milky Way. It is more akin in size and structure to the Large Magellanic Cloud — a barred irregular galaxy.

Locating NGC 55's position with binoculars is simple. Start by centering the 2.4-magnitude star Ankaa (α Phoenicis), then move the star just outside the southeastern edge of the binocular field to center NGC 55. Bear in mind that while this is one of the brightest galaxies in the sky, its angular size is very large. Accordingly, it has very low surface brightness — its light is spread out over a very large area. A fairly dark sky is required to show it well. In 10 × 50 binoculars, NGC 55 is a 15′ hazy, cigar-shaped smudge, elongated southeast to northwest. A sizable, slightly brighter zone is offset from center toward the northwest.

Looking toward Phoenix, we are peering straight out of the plane of the Milky Way, and, accordingly, the field contains a few 7th- and 8th-magnitude stars. While you're gazing at NGC 55, consider what the view might look like in reverse. From there, our Milky Way Galaxy would appear face on and span a degree or more. It would be bright enough to be seen with the naked eye and would unquestionably make a wonderful binocular highlight!

Using the Star Chart

WHEN

Late November	Dusk
Early November	9 p.m.
Late October	10 p.m.
Early October	11 p.m.
Late September	Midnight

These are standard times.

HOW

Go outside within an hour or so of a time listed above. Hold the map out in front of you and turn it around so the label for the direction you're facing (such as west or northeast) is right-side up. The curved edge represents the horizon, and the stars above it on the map now match the stars in front of you in the sky. The center of the map is the zenith, the point in the sky directly overhead.

Example: Turn the map around so the label "Facing SE" is right-side up. About a third of the way from there to the map's center is the bright star Canopus. Go out and look southeast about a third of the way from horizontal to straight up. There's Canopus! (Mercury is positioned for mid-November.)

Note: The map is plotted for 35° south latitude (for example, Sydney, Buenos Aires, Cape Town). If you're far north of there, stars in the northern part of the sky will be higher and stars in the south lower. Far south of 35° the reverse is true.

ONLINE

You can get a sky chart customized for your location at any time at **SkyandTelescope.com/ observing/skychart**

SKY & TELESCOPE | southern hemisphere's sky

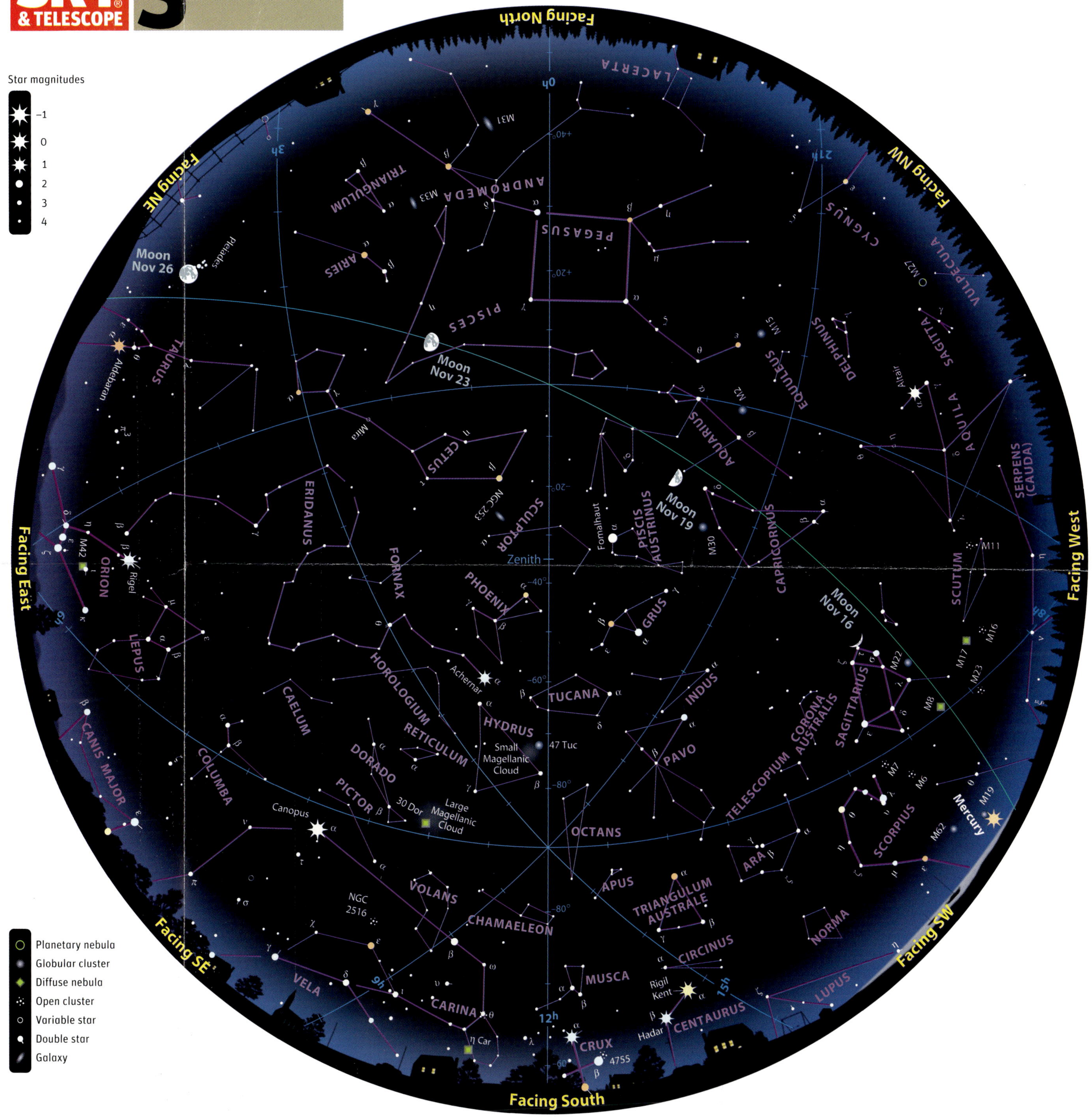

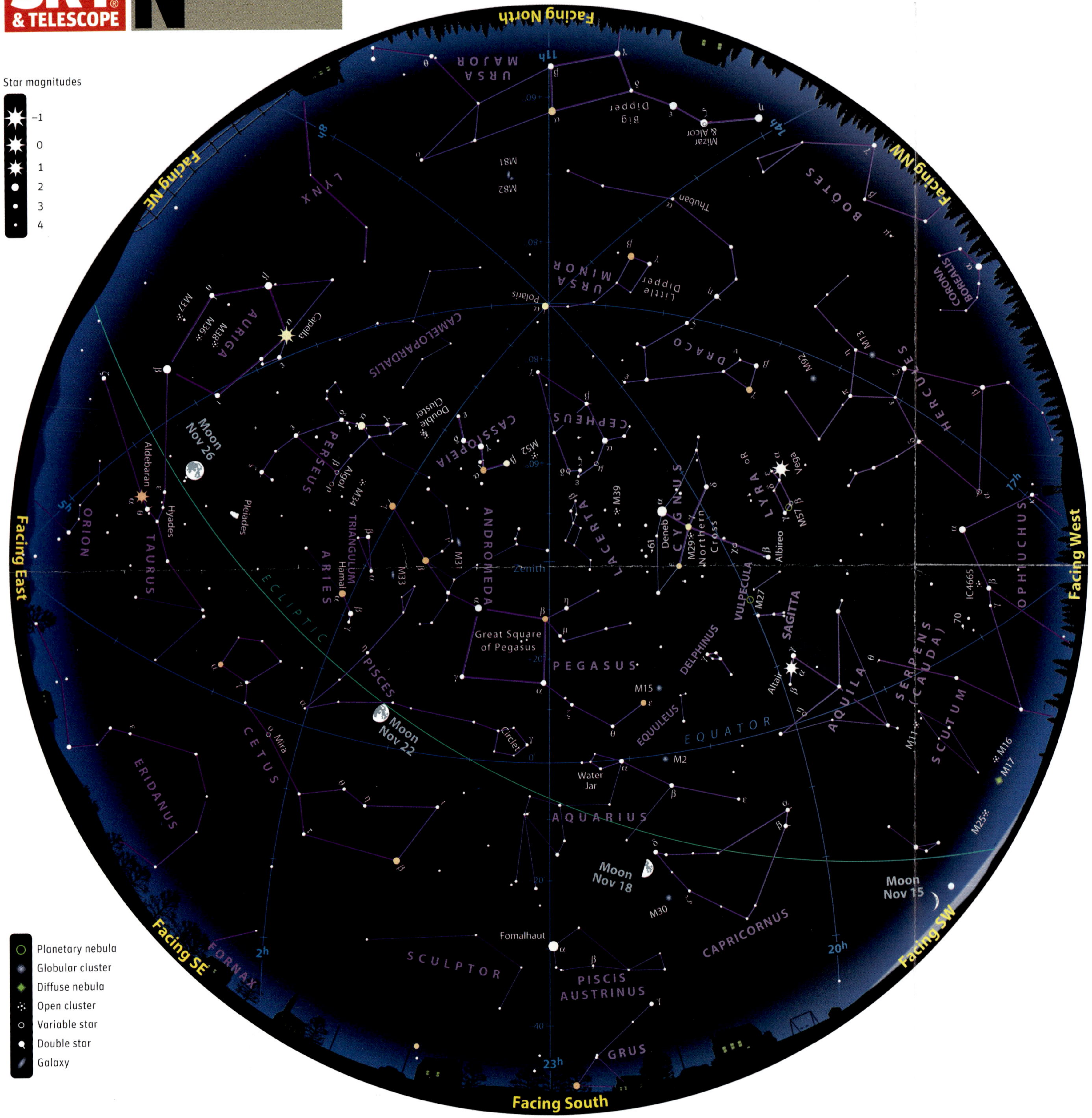

Using the Star Chart

WHEN

Late November	7 p.m.
Early November	8 p.m.
Late October	10 p.m.*
Early October	11 p.m.*
Late September	Midnight*

*Daylight-saving time

HOW

Go outside within an hour or so of a time listed above. Hold the map out in front of you and turn it around so the label for the direction you're facing (such as west or northeast) is right-side up. The curved edge represents the horizon, and the stars above it on the map now match the stars in front of you in the sky. The center of the map is the zenith, the point in the sky directly overhead.

Example: Turn the map around so the map has "Facing NE" right-side up. About a third of the way from there to the map's center (the zenith) is the bright yellow star Capella. Go out, face northeast, and look about a third of the way from horizontal to straight up. There's Capella! (No planets are easy to see in the evening sky this November.)

Note: The map is plotted for 40° north latitude (for example, Denver, New York, Madrid). If you're far south of there, stars in the southern part of the sky will be higher and stars in the north lower. Far north of 40° the reverse is true.

ONLINE

You can get a sky chart customized for your location at any time at **SkyandTelescope.com/observing/skychart**

northern binocular highlight | by gary seronik

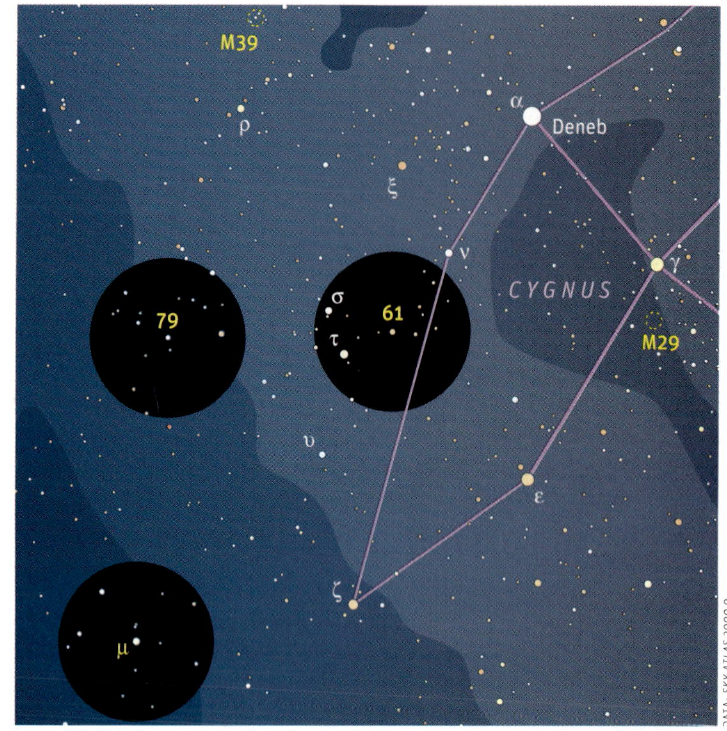

Three Cygnus Doubles

TUCKED AWAY IN eastern Cygnus is a trio of binocular double stars that neatly demonstrate the challenge and the pleasure of observing stellar pairings. As a rule of thumb, the closest pair of equally bright stars you can split with steadily held binoculars can be estimated by dividing 300 arcseconds by your binoculars' magnification. For example, 10× binos should just be able to resolve stars separated by 30 arcseconds (300/10). But do actual observations bear this out?

Our first Cygnus double is Mu (μ) Cygni. Its stars (magnitudes 4.4 and 7.0) are separated by a generous 198″ and, as expected, they are an easy split in my 10 × 30 stabilized binoculars. The duo is set in an attractive field, with Mu forming a nearly equilateral triangle with two other 5th-magnitude stars lying a couple of degrees to the east. While you're there, look a half degree due north of Mu to see a slightly wider double consisting of twin 7th-magnitude stars.

About 9° north lies 79 Cygni. This pair of 5.7- and 7.0-magnitude stars are closer together (150 arcseconds) yet not much more difficult to split than Mu. This is because there is only 1½-magnitude difference between the component stars compared to 2½ magnitudes for Mu.

The last double on our list is 61 Cygni — a pair of 5.2- and 6.0-magnitude stars separated by 31″. Our rule of thumb suggests that this will be a tough split for 10× binoculars, and indeed it is. I was able to separate the components only when my binoculars were perfectly focused. However, my 15 × 45 stabilized binoculars easily resolved the pair. And an attractive sight it is, too — stellar neighbors of the Sun side by side on a rich Milky Way background!

4

NGC 7023
Type: Reflection Nebula
Con: Cepheus

RA: $21^h\ 01.6^m$
Dec: +68° 10′
Mag: 6.8, star;
7.7, nebula (O'Meara)
Dim: 10′ × 8′
Dist: ~1,400 light-years
Disc: William Herschel, 1794

W. HERSCHEL: [Observed 18 October 1794] A star of 7th magnitude. Very much affected with nebulosity, which more than fills the field. It seems to extend to at least a degree all around; [fainter] stars, such as 9th or 10th magnitude, of which there are many, are perfectly free from this appearance. (H IV-74)

GC: Extremely faint, star of 7th magnitude in nebula (?).

NGC: A star of 7th magnitude in an extremely faint, extremely large, nebulosity.

OF THE HALF DOZEN OR SO NEBULOUS COMplexes that adorn the boundaries of Cepheus, Caldwell 4 (NGC 7023) is the smallest but most rewarding visually. It is also an object of some confusion. Brent Archinal, a coauthor with Steven J. Hynes of *Star Clusters*, has this to say about it: "I was surprised to find NGC 7023 listed in my catalog as a star cluster. I assumed immediately the Caldwell Catalog was in error, but further checking showed I was wrong! The Caldwell Catalog may be the only modern catalog to get the object type correctly!"

William Herschel and the *NGC* describe NGC 7023 as a large nebulosity around a 7th-magnitude star. Archinal notes that though Herschel mentioned the existence of many stars associated with the nebulosity, he did not indicate a star cluster here; nor does the *NGC*. Archinal says the confusion occurred when Per Collinder erroneously listed NGC 7023 as a star cluster in 1931, though it is not clear whether Collinder discovered it. In any case Collinder erroneously equated this cluster with NGC 7023. Subsequent references repeat this error. The SIMBAD database currently indicates "Cl vdB 139" as an alias for this cluster. "This name should be denigrated," Archinal says, "as van den Bergh (1966) [in the *Astronomical Journal*] gives no indication there is a star cluster around his reflection nebula vdB 139 [, which he equates with] NGC 7023. NGC 7023 is indeed only a nebula, and the associated cluster's proper name is Collinder 429. Thus, all modern sources examined to date erroneously label NGC 7023 as a star cluster. NGC 7023 is a reflection nebula, with the star cluster Collinder 429 involved."

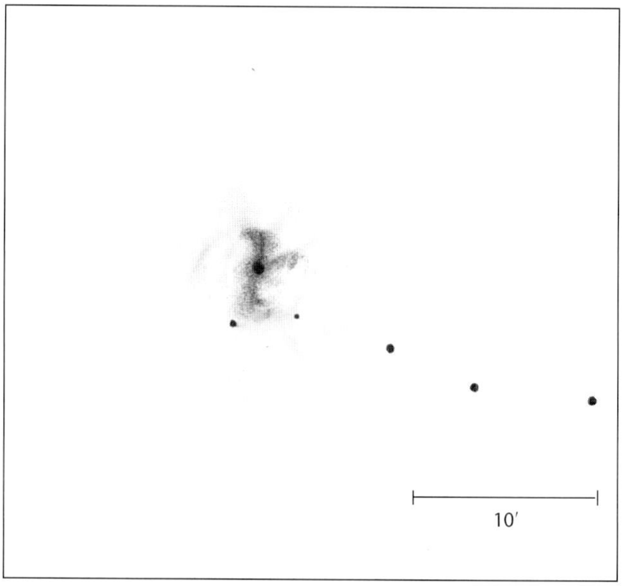

NGC 7023 lies about 3½° southwest of Beta (β) Cephei, or Alfirk, a quasi-Cepheid variable star that pulses between magnitude 3.2 and 3.3 about every 4½ hours. (Alfirk is also a nice double star, first noted by F. G. W. Struve in 1832, with an 8th-magnitude companion 13″ distant.) The nebula, whose brightness I estimated at magnitude 7.7, surrounds a 7th-magnitude star (HD 200775), and both can easily be detected in binoculars. (Use averted vision to note the extended size of that star compared to others in the field.) In the 4-inch at 23× the star and nebula are quite distinct, looking like a lamp seen through a window moist with breath. With direct vision the nebula measures a few arcminutes in diameter and looks irregularly round. Under averted vision, however, the nebula swells magnificently to a diameter of 10′. A quick glance out of the corner of one's eye makes the nebula look like a chiffon skirt suddenly puffed out by a gust of wind. (Try the same with any other similarly bright star in the field, and they will all remain skirtless.) By the way, estimating the nebula's brightness required some mental gymnastics, namely psychologically removing the 7th-magnitude star from view to concentrate solely on the glow emanating from the nebula.

At 72× NGC 7023 displays some arcing bands of nebulosity, knots (or faint stars embedded in the glow), and sharp streamers. Most noteworthy is a bar of light (oriented roughly north-south) that slices through HD 200775. The ends of the bar curve away in opposite directions, so the material there seems to fan out like the spray from a two-pronged lawn sprinkler. Another weaker streamer jets off the star to the north-northwest. The full glory of this object is revealed at higher powers and with averted vision. At 189× HD 200775 looks like a topaz stone that has been hastily woven into a spider's gossamer web. The nebula's densest regions (the spiral-like arms) have a feathery quality to them, while the remainder of the cloud has a stringy texture reminiscent of the fibers in high-quality paper. Prolonged study will bring out the nebula's faint, hollow lobes extending to the west and east.

Although reality dictates that we are seeing the passive reflection of starlight by interstellar dust within the nebula, it is not difficult to imagine a more exciting, though fanciful, drama unfolding before our eyes. Given NGC 7023's central star, bright spiral twists, and ballooning lobes, one could imagine that the star had experienced a violent novalike explosion that tossed out arcs of matter while shock waves plowed through the surrounding interstellar medium to form the lobes. But the prosaic truth is that NGC 7023 is a prototypical reflection nebula — one of the most extensively studied. It was one of the first regions where

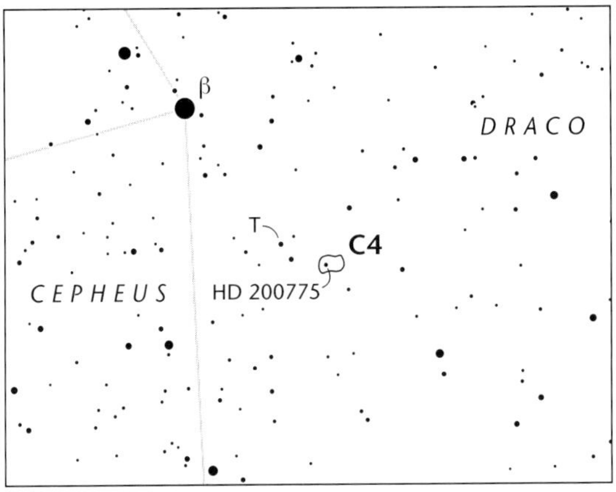

astronomers found direct evidence of dust grains in molecular clouds. The bright nebula we see, in fact, is light being reflected off dust particles that survived the star's birth some 6,400 years ago. If we accept 1,400 light-years as the nebula's distance, the vast cloudscape we see surrounding HD 200775 is actually about 4 light-years across, or 38 trillion km — a distance comparable to that between the Sun and Proxima Centauri.

Max Wolf noted in 1907 that NGC 7023 is surrounded by a "lacuna absolutely empty of faint stars." This void also was imaged by Isaac Roberts, who wrote that the "nebula appears in a region almost void of stars." Wolf said the "round cavity" encircling the nebula has a diameter of about ½°. Using your widest possible field, see if you cannot spy this feature, which, to me, looks like a dark four-leaf clover.

Before moving on to the next Caldwell object, note the nebula's proximity to the bright reddish Mira-type variable star T Cephei, roughly 1° to the east-northeast. T Cephei fluctuates between magnitude 5.2 and 11.3 every 388 days. Thus sometimes it is visible to the naked eye, while at other times an observer needs a small telescope to see it.

5

5

IC 342
Type: Mixed Spiral Galaxy (SABcd)
Con: Camelopardalis

RA: 03^h 46.8^m
Dec: +68° 06′
Mag: 8.4;
7.9 (O'Meara)
Dim: 16.1′ × 15.5′
SB: 15.0
Dist: 13 million light-years
Disc: William F. Denning, ~1890

HERSCHEL: None.

GC/NGC: None.

IC: Pretty bright, very small in angular size, a star of 12th magnitude close to the north.

ALTHOUGH MANY OF THE BIGGEST AND brightest deep-sky objects visible from the Northern Hemisphere — such as the Great Orion Nebula (M42), the Andromeda Nebula (M31), and the Pleiades (M45) — can be found in the Messier Catalog, some seemingly obvious telescopic objects are conspicuous by their absence. Of course, the Double Cluster in Perseus is the most glaring omission (see Appendix B). Another is the 8th-magnitude spiral galaxy IC 342 in Camelopardalis — one of many surprises you'll find in the Caldwell Catalog.

IC 342 is a moderately open, clumpy, low-surface-brightness spiral whose classification suggests a barred galaxy in transition. William F. Denning first spied the galaxy's sharp central glow and amorphous halo around 1890. But the spiral nature of the "nebula" remained unnoticed until Edwin P. Hubble and Milton L. Humason noted, in 1934, the galaxy's beautiful pinwheel pattern on plates taken from California's Mount Wilson.

In shape and form IC 342 looks similar to the open-faced spirals M74 in Pisces and M101 in Ursa Major. Quantitatively, too, the three galaxies share comparable visual characteristics. IC 342's apparent diameter of 16′ is about midway between that of M74 (10′ across) and M101 (29′). IC 342 is a full magnitude brighter than M74 and half a magnitude fainter than M101. And IC 342's surface brightness (15.0 magnitudes per square arcminute) nearly matches those of M74 (14.4) and M101 (14.8). These data clearly suggest that IC 342 should have been within the visual range of Messier

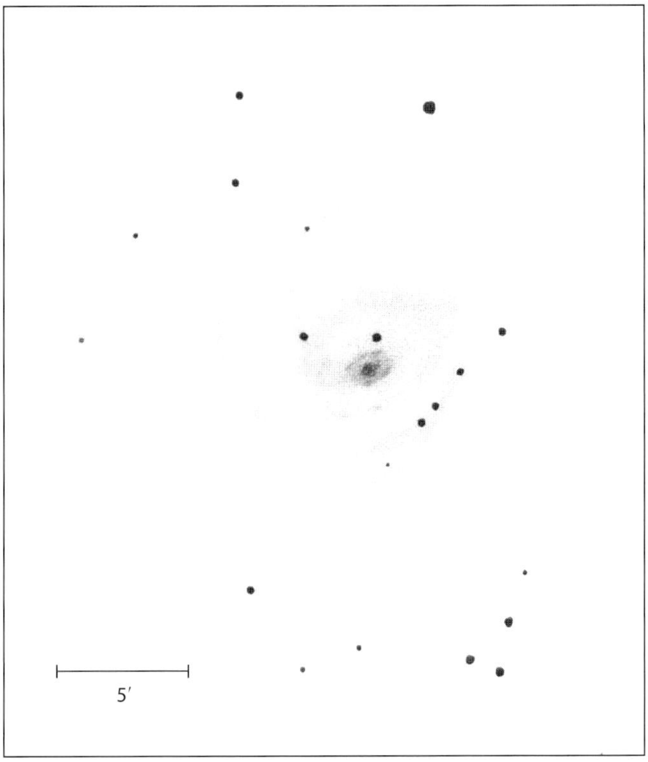

nucleus surrounded by a very large and faint hazy glow." This may be accurate for a view through an 8-inch telescope at moderately high power, which would diminish the view of such a large, low-surface-brightness object. But the beauty of IC 342 can be appreciated in a rich-field telescope with a low-power eyepiece. Roger Clark suggests that optimum magnifications for detecting IC 342 are 14× with a 4-inch telescope and 27× with an 8-inch.

Under a very dark sky, IC 342 is easily seen in 7×35 binoculars. Just look 3¼° south of 4.6-magnitude Gamma (γ) Camelopardalis for a large, diffuse glow. Nothing else of that nature is within several degrees of your target. The galaxy looks so obvious in binoculars that I have tried to detect it with the unaided eye but without success. I would not be surprised, however, if someone with younger eyes does see it without optical aid. I'm also suspicious of IC 342's listed magnitude of 8.4 and believe that it may be at least a half magnitude brighter. The galaxy's position against the outer reaches of the northern Milky Way adds to the visual splendor; unlike M74, which is a large and dim glow in a star-poor region of the sky, IC 342 is seen against a sky background littered with starlight.

At 23× the galaxy's core looks like a 12th-magnitude star trapped in a tight cocoon of nebulosity just south of a pair of similarly bright stars. Even at that low power there are immediate suggestions of appendages indicative of spiral structure. By scanning the field while sitting in a relaxing position, I was able to trace out three major spiral arms to nearly

and his contemporaries. Why then was it not noticed until the late 19th century? Perhaps its anonymity resulted from the combination of IC 342's far-northern declination (+68°, an unlikely region for comet hunting), low surface brightness, and isolation (being in a region devoid of any especially bright stars or conspicuous asterisms).

Unfortunately for us, IC 342 lies only 11° above the plane of our galaxy, where dense gas and dust dim the extragalactic spiral's light by nearly 3 magnitudes. Does that mean the galaxy is not worth investigating through small instruments? In his famous *Celestial Handbook*, Robert Burnham Jr. certainly gives that impression. He writes that IC 342 is "difficult for small telescopes" and while it "may be observed in an 8-inch glass under good conditions," it "appears as a small fuzzy 12th-magnitude

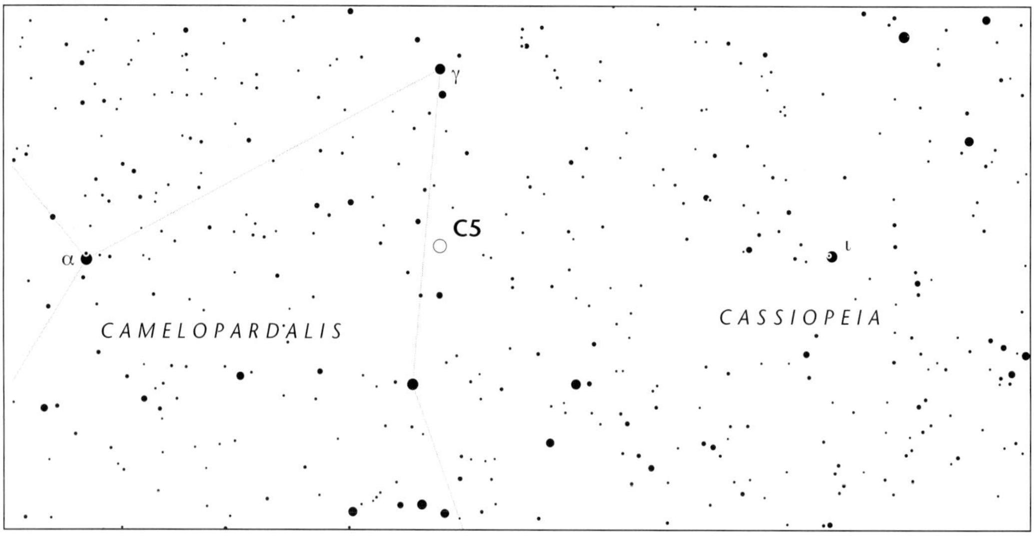

their full extent as seen in photographs. Hints of tangential spiral filaments could also be gleaned, though the reality of those features remains uncertain; the eye tends to follow chains of faint stars, tricking observers into believing they are seeing coherent extragalactic structures. Although faint, the spiral structure in IC 342 is much easier to trace than that in either M74 or M101.

IC 342 is a close neighbor of the Milky Way, cosmologically speaking; it resides only 13 million light-years away — 19 million light-years closer than M74 and 4.5 million light-years closer than M101. IC 342 is inclined at an angle of about 20° from face on. It measures 61,000 light-years in diameter and is receding from the solar system at a cosmologically negligible speed of 32 km per second. The galaxy belongs to the Coma-Sculptor Cloud and travels through space with NGC 2403 (Caldwell 7), NGC 6946 (Caldwell 12), and NGC 147 and NGC 185 (Caldwell 17 and 18, respectively).

IC 342 is a starburst galaxy of sorts. Its nucleus contains a cluster of hot, 10-million-year-young supergiant stars, each with a mass between 12 to 18 Suns. This cluster is surrounded by a gaseous star-forming ring about 230 light-years in diameter. Intense star formation is happening in two giant H II regions roughly 4″ to either side (east and west) of the nucleus. These regions are similar to "superclusters" like 30 Doradus in the Large Magellanic Cloud (see Caldwell 103) and are only about 5 million years old. The sweeping motion of an inner bar structure could be triggering these starbursts as it moves through molecular clouds populating the inner regions of IC 342.

6

Cat's Eye Nebula
NGC 6543
Type: Planetary Nebula
Con: Draco

RA: $17^h\ 58^m\ 33.4^s$
Dec: +66° 37′ 59″
Mag: 8.1
Dim: 23″ × 17″
Dist: ~3,000 light-years
Disc: William Herschel, 1786

W. Herschel: [Observed 15 February 1786] A planetary nebula. Very bright. Has a disk of about 35″ diameter, very ill-defined edge. With long attention a very bright well defined round center becomes visible. (H IV-37)

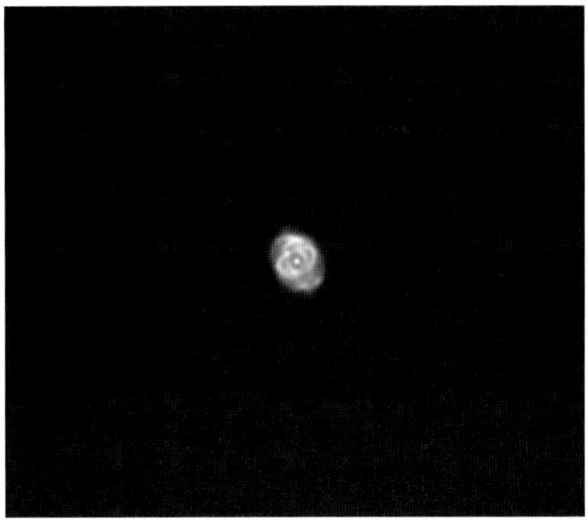

GC / NGC: Planetary nebula, very bright, pretty small in angular size, suddenly brighter in the middle to a very small nucleus.

By area, Draco ranks eighth among the 88 constellations, containing 1,083 square degrees of sky within its long, zigzagging borders. Despite its generous size, Draco possesses only one noteworthy planetary nebula: NGC 6543, commonly called the "Cat's Eye." Jay McNeil, a planetary-nebula aficionado from Kingwood, Texas, rates NGC 6543 as one of the best of the night sky's bright planetaries.

William Herschel is often credited with naming these curious objects "planetary nebulae" in 1785 because their round forms resembled that of the pale green planet (Uranus) he discovered four years earlier. But French astronomer Antoine Darquier de Pellepoix also deserves credit; in 1779 Darquier de Pellepoix discovered the famous Ring Nebula (M57) in Lyra and likened its appearance to a "fading planet." At the time, most observers believed these objects to be masses of unresolved stars rather than gaseous nebulae. The first clue to their true nature came on August 29, 1864, when English amateur astronomer William Huggins turned a spectroscope to NGC 6543; it was the first nebula of its kind to be studied this way. Huggins immediately recognized that the spectrum he saw was that of a luminous gas, not a haze of unresolved suns. Today we know planetary nebulae are indeed thick shells of gas ejected by, and moving out from, the extremely hot cores of dying red-giant stars. Such activity portends the fate of our Sun.

On January 11, 1995, the Space Telescope Science Institute introduced the Cat's Eye to the public eye. Magazines, newspapers, and calendars soon carried stunning HST images of NGC 6543. Ground-based images, like that above, merely show the planetary's bright central star seemingly tied in a loose knot of overlapping gas shells. But HST's powerful eye added depth to the scene, revealing a three-dimensional tangle of loops, bubbles, and jets. Apparently the nebula's innermost structure is an oblong bubble of gas blown out at supersonic speeds from what is now suspected to be a binary central star; the ends of this bubble seem to have burst open. The elongated central bubble resides within two larger lobes of gas blown off the star during an earlier ejection phase. These outer lobes are "pinched" by a clumpy ring of dense gas that presumably was ejected along the orbital plane of the central binary star. The HST images (one of which is reproduced to the lower right) also reveal jets of gas shooting out of opposite ends of the nebula at right angles to the equatorial ring. Those jets are fleeing the region at speeds of 25 km per second or more, colliding with the gas ahead of them and creating the bright arcs visible near the outer edges of the lobes. The jets themselves may be wobbling and turning on and off episodically, adding further complexity to these interpretations.

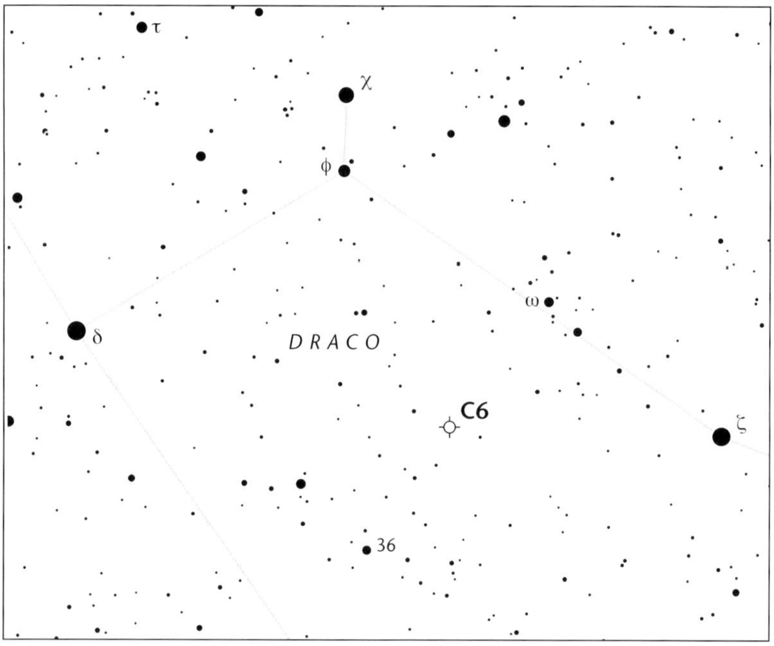

At the June 2000 meeting of the American Astronomical Society, Bruce Balick (University of Washington) described how he and graduate student Darren Reed had analyzed HST images of the nebula and noticed a series of very faint features surrounding it. These wisps, they realized, were a series of at least nine rings ejected earlier at regular 1,500-year intervals. Apparently the star puffed these rings out as part of its preparations for death. Each puff contains thousands of Earth masses of stellar material. "Each outburst would have been a spectacular sight from any nearby planet," Balick said. "The star would have swelled to about the size of the Earth's orbit." The excitement offered by HST's images of the Cat's Eye should help you imagine the ongoing chaos of that system as you squint at the nebula's delicate glow through your telescope. That is the beauty of being an amateur in the 21st century: we are more knowledgeable than ever about the "mystery" clouds that have haunted astronomers for centuries.

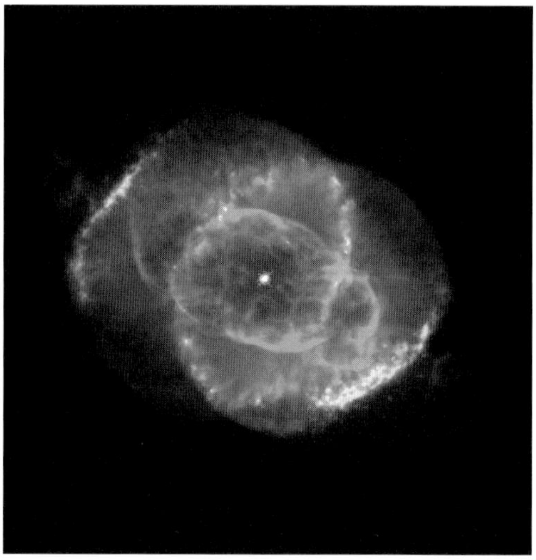

Eighth-magnitude NGC 6543 is located about halfway between the 3rd-magnitude stars Delta (δ) and Zeta (ζ) Draconis, a mere 10′ northwest of the north ecliptic pole. If you live under dark skies, you can further refine its location by placing your finderscope's cross

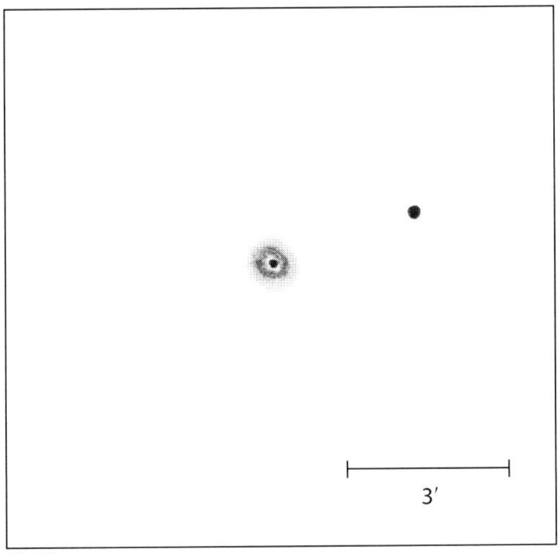

hairs halfway between the 5th-magnitude stars Omega (ω) and 36 Draconis. Then, using low power, look for a slightly bloated "star" just 3′ east-southeast of an 8th-magnitude star. At 23× through the Genesis refractor the planetary looks stellar with direct vision, creating the illusion of a nice 8th-magnitude double star. With averted vision, though, the planetary swells (the Cat's Eye winks!), giving away its identity. The nebula shines with a pale green color even at low power. As with all planetaries, high magnifications give the best views. In fact, because of their small sizes and high surface brightnesses, most planetaries can handle magnifications well beyond those you might normally consider using on deep-sky objects. I would recommend using at least 75× per inch of aperture on an average night.

Alas, little detail can be seen in the Cat's Eye in the 4-inch, even at high power. The planetary's sharp 11th-magnitude central star is closely bordered by a luminous ring. A tenuous outer shell of nebulous matter surrounds these features, giving the planetary a slightly elongated appearance. Using 450× and a 12-inch telescope, Christian Luginbuhl and Brian Skiff note that the "southern end is diffuse while the northern side has a broken bright edge on the east and west sides." Of course, detecting fine structure in any nebula requires excellent seeing. If you can't resolve much detail in your telescope, console yourself with this: looking through the famous 60-inch reflector on Mount Wilson in California, David A. Allen recorded NGC 6543 as merely a "pale green uniform nebula" with a central star. His words magnify the importance of the Hubble Space Telescope.

The familiar and most conspicuous features of the Cat's Eye are young, only about 1,000 years old. In part it is this youthful aspect of planetary nebulae that excites many amateurs like McNeil. "In a hobby literally filled with such ages as 4.5 billion years and more," says he, "planetary nebulae such as NGC 6543 represent events that are measured on time scales easily grasped relative to human standards." McNeil also notes something quite interesting (albeit little known) about the Cat's Eye. "Part of its 'outer halo' is actually visible in amateur scopes," McNeil says. "The object known as IC 4677 is often miscataloged as a 15.7 magnitude background galaxy residing less than 2′ due west of NGC 6543. However, further research has overwhelmingly shown that it is merely a brighter fragment of the outer halo of the nearby planetary nebula. Such semi-detached shells of planetary nebulae are generally extremely faint and difficult to visually observe. In fact, most exhibit a surface brightness 1,000 to 10,000 times lower than that of the planetary's bright inner disk. On the contrary, this one small piece was ironically obvious enough to receive its own designation as a nearly edge-on spiral galaxy in at least four different catalogs. It can even be glimpsed with a good 10-inch scope under nice, dark skies."

So there you have it: your Cat's Eye Challenge. Go to it!

7

NGC 2403
Type: Mixed Spiral Galaxy (SABcd)
Con: Camelopardalis

RA: 07^h 36.9^m
Dec: +65° 36′
Mag: 8.4; 7.3 (O'Meara)
Dim: 23.8′ × 12.8′
SB: 14.6
Dist: 14 million light-years
Disc: William Herschel, 1788

W. HERSCHEL: [Observed 1 November 1788] Considerably bright. Round. Very gradually brighter in the middle. Bright nucleus. 6′ or 7′ diameter, with a faint branch extending a great way to the north preceding [northwestern] side; not less than ½°, and to the north or north following [northeast] the nebulosity [is] diffused over a space, I am pretty sure, not less than a whole degree. (H V-44)

GC: Remarkable, considerably bright, extremely large, very much extended, very gradually much brighter in the middle to a 7′ nucleus.

NGC: Remarkable, considerably bright, extremely large, very much extended, very gradually much brighter in the middle, to a nucleus.

OF THE SKY'S 88 CONSTELLATIONS, Camelopardalis is not only one of the largest but also one of the faintest. In size the celestial Giraffe is significant; its jagged boundaries outline 757 square degrees of sky, making it the 18th-largest constellation on the celestial sphere. But Camelopardalis contains no stars brighter than 4th magnitude, and even though the Giraffe never sets when viewed from mid-northern latitudes, it remains an inconspicuous form under dark skies and a virtual washout from many urban locations. Yet it is in this wilderness of dim suns, tucked into the constellation's southeastern quadrant (actually within sniffing distance of the Great Bear's nose), that Camelopardalis hides a precious extragalactic treasure: the fertile spiral galaxy NGC 2403.

NGC 2403 belongs to the Coma-Sculptor Cloud of galaxies — a peppering of gravitationally bound systems that includes our Local Group and extends more than 10 million light-years beyond its fringe. Having the mass of some 47 billion Suns and spanning 98,000 light-years of space, NGC 2403 is one of the Cloud's more physically impressive members. From our perspective, we see NGC 2403's spiral form inclined 28° from edge on. The galaxy displays two tiers of arms. Those close to the nucleus are tightly wound; farther out the arms sweep wide and free.

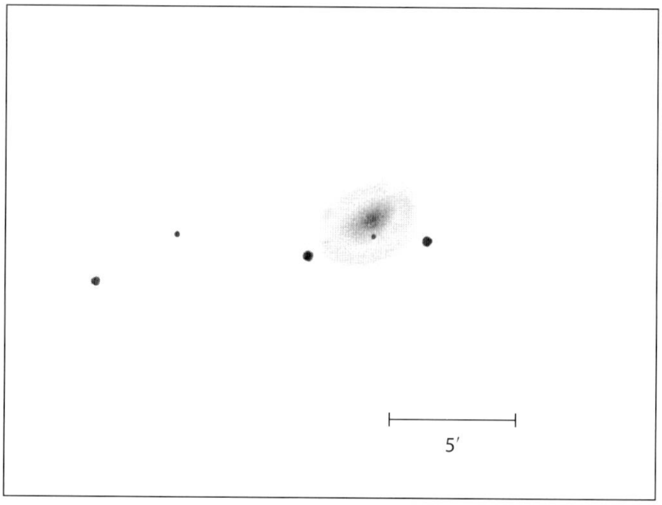

Any photograph will show the galaxy's face "exploding" with clumps of starlight; some of these are foreground stars, any one of which could mislead an unsuspecting observer to think he or she has stumbled upon an extragalactic supernova. The galaxy's clumpy appearance, however, is not due to foreground starlight alone. After surveying the galaxy on plates taken with the Palomar Mountain 200-inch reflector during the 1950s, Edwin Hubble detected more than 100 H II regions lining the galaxy's spiral arms. The largest of these nebulous star-forming regions, NGC 2404, lies 6′ north-northeast of the nucleus; its 20″ apparent diameter corresponds to a true diameter of 1,400 light-years at the galaxy's measured distance of 14 million light-years. This H II region is larger than any known in our Milky Way. The galaxy also contains many *OB* associations, whose hot, blue stars hint at copious star formation in the galaxy's recent past. The distribution and number of NGC 2403's H II regions resemble those found in other galaxies, such as M33, that are rife with gas and hot, young stars. Hubble also identified 27 variable stars in the system; as of 1960, 10 of them had been certified as Cepheid variable stars with periods ranging from 18 to 54 days. In fact, NGC 2403 is the first system beyond the Local Group in which Cepheids — whose period-luminosity relationship is an important tool for measuring extragalactic distances — were identified with the 200-inch.

Near-infrared images of NGC 2403 made in the mid-1990s reveal a weak two-arm spiral structure extending over as much as 180° in azimuth to a radius of 6,500 light-years. Other studies have determined that NGC 2403's most active star-forming regions are not in its nucleus, which appears quiescent, but along the outer ridges of its spiral structure, where, according to theory, the gas density is high enough to trigger star formation.

Locating NGC 2403 is really not much of a problem at all. It lies about 7½° northwest of Omicron (ο) Ursae Majoris (Muscida), the Great Bear's well-known nose, and 1° west of 6th-magnitude 51 Camelopardalis. Under dark skies the galaxy is immediately obvious in 10×50 binoculars, looking much like a tailless comet trying to squeeze through a close pair of 11th-magnitude suns. In his *Celestial Handbook* Robert Burnham Jr. reminds us that the galaxy is frequently picked up by comet

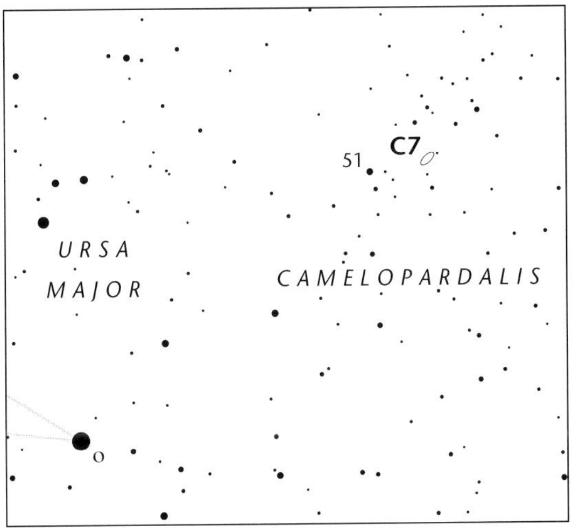

hunters; reading between the lines, one infers that the galaxy is often mistaken for a comet. Understandably so. Certainly the galaxy's obscure location and cometlike appearance have caused more than a few hearts to flutter. Actually, NGC 2403 may have sent many comet hunters' hearts aflutter more than once, because once they learn that they have stumbled upon a galaxy, how could they not suddenly wonder if one of the many stars across its face might be a supernova?

Burnham also points out an obvious discrepancy in the galaxy's apparent magnitude. The Shapley-Ames catalog lists NGC 2403's magnitude as 10.2, which, Burnham says, "seems definitely too faint, and should be corrected to magnitude 8.8." Indeed, several sources now list the galaxy's magnitude as 8.8; others list it as bright as magnitude 8.4. But I find even these "new" magnitudes demonstrably too faint. On the evening of January 26, 1998 (Hawaiian Standard Time), I estimated the brightness of Comet Tempel-Tuttle at magnitude 7.3. I then set out to find NGC 2403. The galaxy's brightness, it turned out, looked remarkably similar to the comet's that night. Interestingly, my entirely independent estimate for the galaxy's brightness turned out to be magnitude 7.3; I even used different stars to gauge the brightnesses of the comet and the galaxy.

The wide range of values for NGC 2403's brightness may be a result of the method used to estimate the galaxy's magnitude. Noted comet observer Charles Morris and I have independently developed an unconventional method for estimating the brightnesses of comets, or of diffuse objects like galaxies. Now known as the "Modified-Out Method," the technique is described in *Deep-Sky Companions: The Messier Objects*. (An excellent, comprehensive review of *all* methods used in making magnitude estimates of comets appears in the *Guide to Observing Comets*, a special issue of the *International Comet Quarterly* that was published in January 1997). The Modified-Out Method is most effective when a diffuse object displays a bright inner region surrounded by an equally large but fainter outer halo — as NGC 2403 does. Standard magnitude-estimation methods tend to diminish the contribution of light from the outer halos of such objects, making them seem fainter than they really are.

At 23× in the 4-inch, the galaxy is an extremely well-defined elliptical glow punctuated by two 11th-magnitude stars on either side of its core. The core itself is faintly lens-shaped with no discernible nucleus, though there is a gradual brightening toward the center. A definite patchiness or mottled quality becomes apparent with a prolonged gaze and averted vision. Two arcs of light hug the core, one to the north of it, the other to the south; each is separated from the core by a thin channel of darkness. Together, the core and these arcs look rather like an eye. An increase in power to 72× reveals a faint, roughly 13th-magnitude sun next to the hazy concentration of light marking the galaxy's nucleus. (It certainly had me checking photographs for a possible supernova.) The arms are broken into several distinct petals of light, all of which display delicate irregularities. When Barbara Wilson viewed NGC 2403 through her 20-inch f/4 reflector at the 1990 Texas Star Party, she likened it to a delicate flower. I'd call it a rose, bedashed with dew. The galaxy's surface brightness is too low to handle higher power well with the 4-inch.

The number of stars one sees superimposed on the galaxy increases with aperture: through their 10-inch telescope, Christian Luginbuhl and Brian Skiff recorded at least three stars in the galaxy's core; the number increased to eight with a 12-inch. Through her 20-inch and a UHC filter Wilson detected 13 stars and two H II regions on the galaxy's face. The illusion that the foreground stars belong to NGC 2403 can be dissolved in your mind's eye, however, by mentally placing the galaxy 14 million light-years distant and whisking it away from us at a speed of 760,000 km per hour (its measured recession velocity).

Finally, I'd like to share a historical treasure relating to Camelopardalis, which can be found in *The Constellations* by Lloyd Motz and Carol Nathanson. They write that the Chinese saw the stars of Camelopardalis as several smaller figures, including the "usual assortment of ministers and guards, and a symbolic personage named *Yin Tih,* or 'Unostentatious Virtue': an excellent description of Camelopardalis' qualities." Not only are NGC 2403's qualities equally befitting of the name *Yin Tih,* but the galaxy appears to be flying through space with an assortment of ministers (the foreground stars) and guards (the pair of 11th-magnitude suns).

8

Ghost's Goblet
NGC 559
Type: Open Cluster
Con: Cassiopeia

RA: 01^h 29.5^m
Dec: +63° 18′
Mag: 9.5
Diam: 7.0′
Dist: 3,700 light-years
Disc: William Herschel, 1787

W. HERSCHEL: [Observed 9 November 1787] A compressed cluster of some pretty [bright] and many very [faint] stars. Irregularly round. 6′ or 7′ diameter. (H VII-48)

GC: Cluster, bright, pretty large, pretty rich, stars of mixed magnitudes.

NGC: Cluster, bright, pretty large, pretty rich.

FOR STARWATCHERS IN THE NORTHERN Hemisphere, late autumn signals a time of change. Hot and hazy nights are replaced with brisk and transparent evenings. The powerful glow of the summer Milky Way, with its awesome central bulge, has all but slipped into the twilight. High in the north, a more delicate river of stars arches over the celestial pole. In its midst five bright stars trace a wide M (or W) that dominates the view, looking somewhat like a newborn deer extending its legs in an attempt to stand for the first time. These stars, which belong to the constellation Cassiopeia, lie in the direction of the Milky Way's outer fringes (though nowhere near as far).

The density of stars we see surrounding Cassiopeia increases sharply around the famous M asterism because that asterism lies on the galactic equator — the dense midplane of the galactic disk. It is a region rich in galactic, or open, star clusters. If you draw a 4°-wide circle around the midpoint between the 3rd-magnitude stars Delta (δ) and Epsilon (ε) Cassiopeiae, you will find at least eight such clusters within the reach of binoculars or a small telescope. The westernmost of these clusters is NGC 559 (Caldwell 8).

Most open clusters, which are loose packs of physically associated stars, will usually reward viewers no matter what size instrument is used to see them. But their visual beauty can be subtle. Take NGC 559, for example. The cluster shines meekly at magnitude 9.5. About 120 stars populate the cluster, but only about 40 of them are of magnitude 12 or brighter; the brightest cluster member is a paltry magnitude 9.0. The cluster's surface brightness (13.7 magnitudes per square arcminute) further

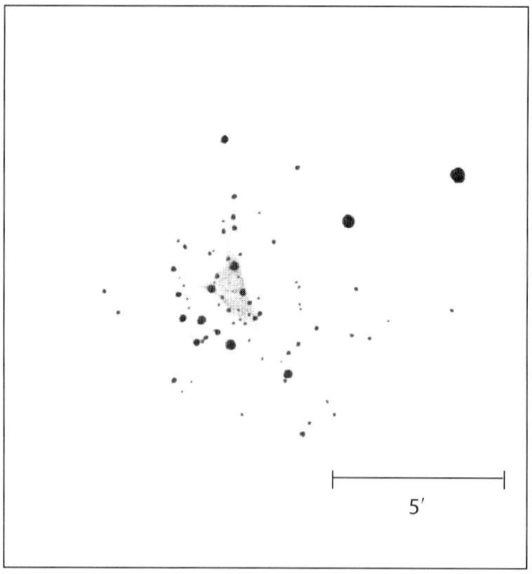

weakens its impact. On paper Caldwell 8 is by no means a showpiece. Yet, under a dark sky the cluster can be spied in 7×35 binoculars. In fact, it is quite noticeable as a 7′ fuzzy patch due west of a tight trapezoid of 7th-magnitude stars, the northernmost two of which are nice binocular doubles. Surprisingly, the cluster has more visual punch than the data suggest, and it appears even more attractive when coupled with the neighboring asterism rich in double stars.

Through the 4-inch, NGC 559 is a tight cluster with a strong core of stars of similar brightness (roughly 12th magnitude). At 23× the dense core appears immersed in a faint mist of unresolved starlight, with a trickle of faint suns extending toward the north. The best view comes at 72×, which reveals a wedge of suns south of the cluster's center. The view doesn't change much in larger apertures. Describing her views through a 20-inch reflector in Columbus, Texas, Barbara Wilson called NGC 559 a "nice cluster, block-shaped, with two chains of stars radiating from the box, like a little dipper." At 150× Wilson says the cluster is "pretty rich with 50 to 60 stars seen against the glow of many unresolved suns, the dimmest being around 14th magnitude."

Perhaps it is an illusion, but, when seen through the 4-inch at an altitude of 4,200 feet, the cluster appears to contain many colorful stars that glisten like jewels. The wedge of jewel-like suns mentioned above is contained in an arc of largely unresolved star-light in the shape of a cup (Wilson's "dipper," no doubt). The string of stars extending northward curves to the west, then south, where fainter suns gradually fade into what can only be called a rim of starlight. This faint rim connects to the south-western tip of the wedge. Another arc of stars to the east forms the base of the cup, giving it a seemingly three-dimensional form. The faint glow of unresolved starlight within the cup seems to be a brew of fleeting vapors. With imagination this central core takes on an eerie quality, that of a still photograph of a ghostly jeweled goblet in the act of vanishing. In a way, this metaphor reflects the cluster's advanced age (1.8 billion years).

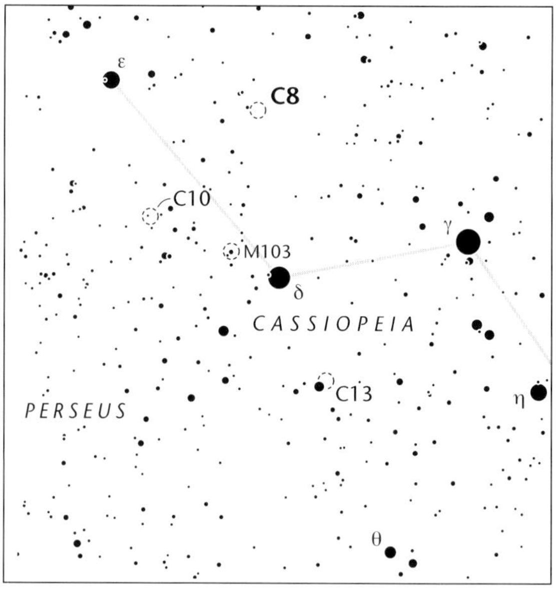

One other thing adds to the cluster's specterlike quality. Radio and optical observations of the cluster have revealed a 45′-wide supernova remnant (G127.1+0.5 R5) centered on a bright radio source at right ascension 01^h 25^m 00^s and declination +62° 55′. Radio observations show the remnant to have a distinct shell. According to *A Catalogue of Galactic Supernova Remnants* by David A. Green (Mullard Radio Astronomy Observatory, England, 1998), the relationship between the supernova remnant and NGC 559 is still under investigation. If the two are associated, the remnant would span nearly 50 light-years of space, which is five times larger than the cluster's true physical diameter. At face value, however, any association seems highly implausible, since the remnant appears to have been expanding for only 18,000 years — making it fully 100,000 times younger than the cluster.

9

Cave Nebula
Sharpless 2-155
Type: Emission Nebula
Con: Cepheus

9

RA: $22^h\ 56.8^m$
Dec: +62° 37′
Mag: 10.0 (O'Meara)
Dim: 50′ × 30′
Dist: ~2,800 light-years

Disc: noted by Stewart Sharpless on photographic plates and included in his *Catalogue of Emission Nebulae* (1953)
HERSCHEL: None.
GC/NGC: None.

WHEN CLOCKS STRIKE THE MIDNIGHT HOUR on August 11th, a look to the northern sky will reveal King Cepheus executing his highest reign over that celestial realm. That's also a propitious time to start watching the King get pelted by stones — meteors whipped across the Milky Way from that seemingly angry hero Perseus, rising in the northeast. If one of these meteors were to fly from Perseus to Cepheus along the northern edge of the dense Milky Way, it would skirt the famous variable star Delta (δ) Cephei. There, the Milky Way suddenly turns back on itself and forms a hollow, a cave of sorts. It's as if the imaginary meteor had suddenly bored a hole into our galaxy's star-studded disk. And it's at the mouth of this "cave" where telescopic adventurers can seek out a highly elusive visual "monster" — the emission nebula Sharpless 2-155 (Caldwell 9), ironically known as the Cave Nebula for its own hollowed-out shape.

A word of clarification. Although Stewart Sharpless included the Cave Nebula in his 1953 *Catalogue of Emission Nebulae,* he did not call it the Cave Nebula; he cataloged it simply as a large emission nebula. In fact, calling Sharpless 2-155 *the* Cave Nebula is a historical no-no. I refer you to the December 1908 *Monthly Notices of the Royal Astronomical Society,* where, in a paper titled "A New 'Cave-Nebula' in Cepheus," Max Wolf reports:

> I have the pleasure to announce the discovery of a new interesting nebula in the midst of the constellation Cepheus.
>
> The nebula was found by Dr. [August] Kopff on a plate taken by him with my Bruce-telescope on the night of October 21st, 1908.

The "Cave-Nebula in Cepheus" that Wolf goes on to describe is not Sharpless 2-155 but a nebula that now bears the designation vdB 152; it is located 9° northwest of Sharpless 2-155 and 3½° east of Beta (β) Cephei. What's key is that Wolf wrote that this new nebula "forms an important addition to this class of nebulae." And what class of nebulae might that be? "This nebula," Wolf continues, "is a good example of the singular phenomenon of cave-formation amongst Milky Way stars. In some respects it shows the general characteristics of *other* cave nebulae . . ." (emphasis mine).

Sharpless 2-155 is but one of several Caves in the heavens.

But let's follow modern custom and apply the Cave Nebula moniker to Sharpless 2-155. The Cave Nebula, it turns out, is part of an extensive region of ionized hydrogen gas — an H II region — associated with the Cepheus B giant molecular cloud. Part of the cloud is illuminated by a pack of hot, young stars known as the Cepheus *OB*3 association. Energetic photons from these stars strip the surrounding hydrogen atoms of their electrons, causing the gas in the visible nebula to glow. The now-defunct Rosat satellite has revealed some 40 point sources of X-rays in the association, the majority of which are T Tauri stars — hot young suns still partially swaddled in dusty blankets of gas. Radio and near-infrared observations also have revealed a small near-infrared star cluster coincident with a bright knot of hydrogen-alpha emission near the southern edge of the *OB* association (itself several degrees across). This cluster's location is tantalizing, for it lies near a sharp spike of ionized gas that delineates the interface between the H II region and the molecular cloud. The cluster may be the result of star formation that was triggered by a front of ionized gas, which pushed out from the *OB* association into the molecular cloud.

Employing your telescope, you will find this "invisible" action occurring halfway between the 5th-magnitude star 30 Cephei and the Bubble Nebula (Caldwell 11). Sharpless 2-155 also can be found by voyaging a little more than 3° due west and a tad north of the famous open star cluster M52 in Cassiopeia. By placing M52 just out of my Genesis's nearly 3° field, I can make the Sharpless 2-155 field come into the eyepiece from the west. Another tiny nudge northwestward brings the 6.5-magnitude star near the brightest part of the nebula into the center of my field of view. Notice I was careful to state that the star — *not* the nebula — is what appears in the eyepiece. Sharpless objects are for the most part notoriously faint. Indeed, no visual description of Sharpless 2-155 is readily available to the casual observer. The 50′ × 30′ object does not appear in the first edition of either *Sky Atlas 2000.0* or *Uranometria 2000.0*. Stewart Sharpless noticed this nebula on *National Geographic Society – Palomar Observatory Sky Survey* (POSS) plates, and he recorded it in his 1959 *Catalogue of H II Regions*. Observing guides like *The Deep Sky Field Guide to Uranometria 2000.0* and *Sky Catalogue 2000.0* do offer photographic descriptions, informing us that the object is obvious in red-light photographs, with the northern half of the crescent-shaped nebula (which has both amorphous and filamentary structure) being most noticeable. Indeed, the Cave Nebula is occasionally photographed by amateur astronomers and showcased in magazines like *Sky & Telescope*. But what visual impression does it make through a small telescope, if any?

In a cursory literary search I did learn that amateurs in Helsinki, Finland, found the Cave visible in a 14-inch telescope with the help of an ultra-high-contrast (UHC) filter. It can be seen with smaller, unfiltered apertures — but just barely. Through the 4-inch Genesis the

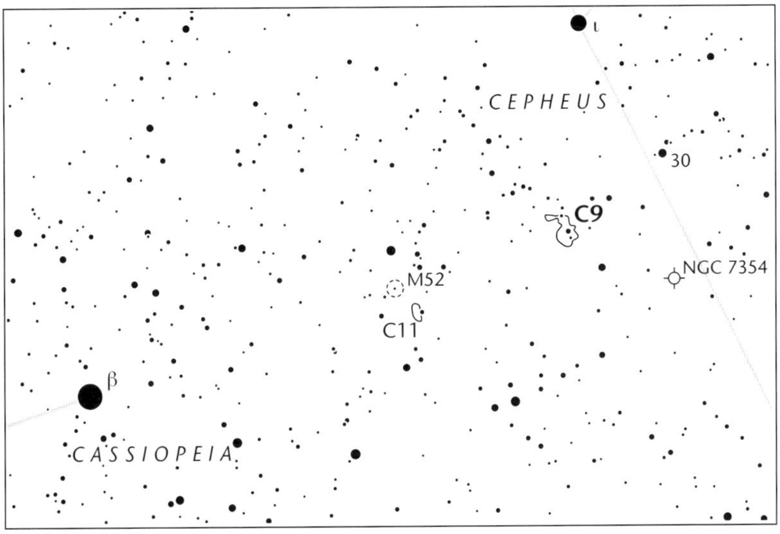

nebula is a near impossibility, even when viewed at an altitude of 4,200 feet under transparent Hawaiian skies. At 23× the Cave looks like an ill-defined haze playing with an arc of roughly 7th-magnitude suns. The weakly glowing gas is peppered with many faint stars, making it almost indistinguishable from the patchwork Milky Way star fields permeating the region. I find that the best eyepiece combination for glimpsing it is a 22-mm Panoptic and a 1.8× Barlow lens, which together yield 41× through the Genesis.

Of course, finding the right magnification to bring out the nebula turns out to be a delicate balance between power and contrast. While the nebula is best seen at 41× it virtually vanishes at 72×, though a filamentary "spike," which is a part of the Cave, does dimly stand out against the darkness. (This is the aforementioned ionization front.) Sweeping the telescope back and forth over the area will help the eye pick out the nebula's brightest (though low-contrast) glow. I find it at the center of the nebula, just west of the darkest part of the cavity, near a triangle of 9th-magnitude suns. It is this 10′-wide area that I estimated to shine at an apparent magnitude of 10.0. A patch of dark nebulosity can be seen east of the triangle. Also, brief looks into the eyepiece will help more than a prolonged stare. But beware! It's easy for one's imagination to creep into the mind's eye. There are many faint glows in the field, some of which I'm suspicious of—even though I included them in my drawing—because they tend to follow streams or clumps of faint stars rather closely. Just as some double stars (like M40) appear fuzzy at low power, so too do lines of stars and patches of the Milky Way. The use of a UHC filter should boost the contrast and bring out the nebula, which covers an area roughly 50 light-years across. If you only suspect seeing the Cave Nebula, do not give up on it. Atmospheric clarity varies, so try on many nights. Once spotted, the nebula actually becomes easier to see on subsequent nights.

Unfortunately, I was not able to see the entire dark mouth of the Cave in the 4-inch. Your challenge, then, is to determine the smallest aperture required to see this elusive feature without filters. While you're in the area, check out the 12th-magnitude planetary nebula NGC 7354 about 2½° to the southwest. Don't let the planetary's magnitude fool you. NGC 7354 is visible at 23×, and its 22″ × 18″ disk takes magnification very well.

10

The Horseshoe Cluster
NGC 663
Type: Open Cluster
Con: Cassiopeia

RA: 01^h 46.3^m
Dec: +61° 13′
Mag: 7.1;
6.7 (O'Meara)
Diam: 15.0′
Dist: 7,200 light-years
Disc: William Herschel, 1787

W. HERSCHEL: [Observed 3 November 1787] A beautiful cluster of pretty [bright] stars near 15′ diameter. Considerably rich. (H VI-31)

GC / NGC: Cluster, bright, large, extremely rich, stars pretty [bright].

TO THE TELESCOPIC TRAVELER, the open clusters that adorn the Cassiopeia Milky Way are like pale wildflowers along a country road. Anyone with modest equipment can find some five dozens of them planted in the constellation's 598 square degrees of sky. Most congregate around that busy highway of stars, the galactic equator. One particularly rich section lies only 2½° east-northeast of 3rd-magnitude Delta (δ) Cassiopeiae in the constellation's famous M (or W) asterism. Here you will find five dramatically different examples of these loosely packed stellar agglomerations. They are, in order of equinox 2000.0 right ascension: M103, Trumpler 1, NGC 654, NGC 659, and NGC 663. When seen through binoculars or a rich-field telescope, these clusters form an incomplete 2°-wide circle around a diamond of roughly 7th-magnitude stars.

Glimmering at 7th magnitude, NGC 663 (Caldwell 10) is brighter, but not more famous, than M103, itself a Christmas-tree-shaped cluster glistening with stellar ornaments about 1½° to the west-southwest. Messier objects have a powerful hold on observers. Like magnets, they immediately attract our gaze, pulling it away from nearby objects that might be equally interesting. So unless an observer feels adventurous, he or she will probably not seek out NGC 663 in lieu of M103 without reason (like this suggestion that you do so). I've been guilty of this oversight myself.

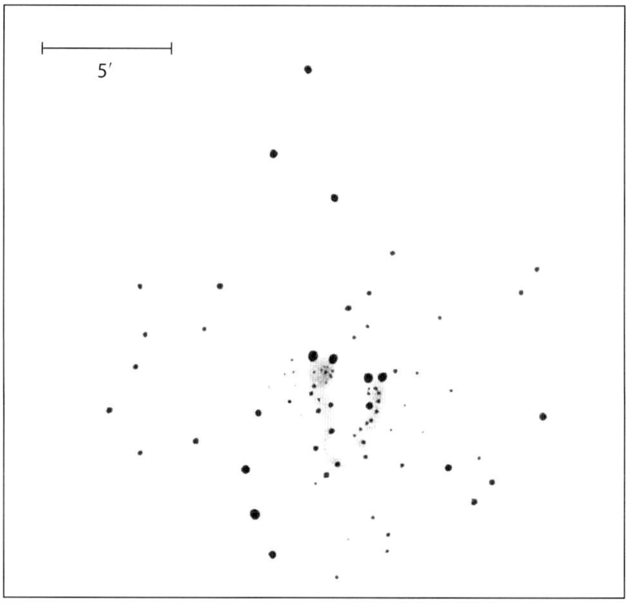

NGC 663, as it turns out, is a wonderful object, being at least 0.3 magnitude (32 percent) brighter, 1,000 light-years closer, and 2½ times larger than M103. The cluster's 108 measured members (roughly 8th magnitude and fainter) fill an area of sky measuring ¼°, or 31 light-years, across. By comparison, M103 packs some 172 suns in an area of sky only 6′ across, which, at its distance of 8,100 light-years, is equal to 14 light-years of space. (This high concentration of stars in such a small area of sky is, of course, why M103 packs such a visual punch.)

From a dark sky NGC 663 can be seen with the naked eye, but you have to take the time to resolve it from the two roughly 6.5-magnitude stars about ½° west and south-southwest of it. While comparing the apparent brightnesses of these three objects, I realized that NGC 663's listed magnitude (7.1) is too faint. Using 7×35 binoculars, I estimated the cluster's apparent visual magnitude to be 6.7. Barbara Wilson independently deduced a value of 6.5. So NGC 663 is arguably at least half a magnitude brighter than commonly believed. (Wilson also wonders why William Herschel's description in his original *Philosophical Transactions* catalog differs from that in the *General Catalogue*, since John Herschel did not observe the cluster and William did so only once.)

Viewed with the 4-inch at 23× the nearly 3° field centered on NGC 663 comes alive with clusters, double stars, patches of Milky Way, and streams of dark nebulosity. NGC 663 itself is immediately very well resolved. In fact, several of its brightest (8th-magnitude) members can be glimpsed in 7×35 binoculars. My visual impression with the 4-inch is one of an elliptical orb with snaking chains of 8th- and 9th-magnitude stars, all of which are projected against a lens-shaped backdrop of fainter suns. Similarly, Jere Kahanpää (Jyväskylä, Finland) used a 6-inch refractor at 52× to see a "beautiful cluster" with its major axis oriented east-west. In the Genesis at 23× the star chains radiating from the cluster's core seem to gravitate to the south; one especially long arm ends in a bright pair of stars that all but pinches NGC 659 ½° to the southwest.

At 72× the cluster's core is fragmentary, being composed of three stellar groupings shaped somewhat like a horseshoe. The horseshoe asterism is the result of a prominent north-south-trending dark lane that rips through the

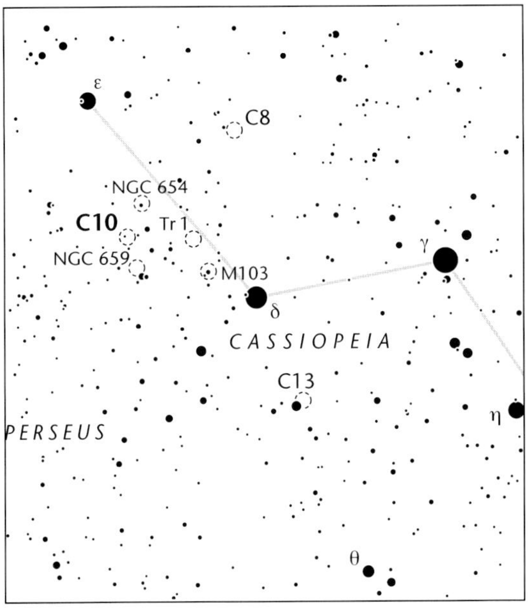

heart of the cluster, dividing it into two V-shaped stellar cities before tapering off into a triangular island of stars. Looking through a 20-inch f/4 reflector at 72×, Wilson notes that the horseshoe's dark lane is "very dark and devoid of stars, making the cluster appear [weakly] concentrated." Lining this dark lane like poorly shielded streetlights are several bright double stars. Christian Luginbuhl and Brian Skiff note that many of these doubles are faintly colored. Two distinct pairs punctuate the ends of the horseshoe asterism. Higher power in the 4-inch also brings out an attractive concentration of stars just south of the eastern pair; the scene looks like a sparkling mist trapped in a circlet of suns. And the eastern member of each wide pair is a telescopic double star with a separation just under 10″.

Within 15′ of NGC 663's fractured core are 24 irregular variable stars of spectral class *B*e. These special stars display bright hydrogen emission lines superimposed on the normal absorption spectrum of a hot, *B*-type star. *B*e stars are believed to rotate so rapidly that they shed mass, forming an expanding shell and a disk around each star's equator. One of NGC 663's *B*e stars is already known as an X-ray source and may be associated with a neutron star, the result of a supernova explosion. Future X-ray observations may enable astronomers to detect other compact objects in the system. Stepping beyond the corridors of reality, one can imagine that the great rift tearing through the cluster's core was the result of an extraordinary supernova explosion, one whose supersonic winds pushed through the crust of stars lining the rift like some violent volcanic event.

Finally, observers looking for a challenge might try spotting NGC 654 and NGC 659 in 7×35 binoculars. I did this with difficulty. NGC 654 is the most difficult, for it is but a breath of haze slightly offset from a yellowish 7th-magnitude star. NGC 659 is also difficult and is distinguishable from the background Milky Way only with averted vision. Telescopically, 7.9-magnitude NGC 659 is a small pack of stars — a C-shaped asterism —within a 5′-wide glow of partially resolved suns. NGC 654, on the other hand, is outstanding. The magnitude-6.5 cluster is tapered, and starlight drips off its tip like water off a needle. It's such an unusual gathering of irregularly bright suns that it stands out in the 4-inch.

11

Bubble Nebula
NGC 7635
Type: Emission Nebula
Con: Cassiopeia

RA: 23^h 20.7^m
Dec: +61° 12′
Mag: 10.0 (O'Meara)
Dim: 15′ × 8′ (nebula); 3′ × 3′ (bubble)
Dist: 7,100 light-years
Disc: William Herschel, 1787

W. HERSCHEL: [Observed 3 November 1787] A star of 9th magnitude with very faint nebulosity of small extent about it. (H IV-52)

GC: Very faint, a star of 9th magnitude involved a little excentric.

NGC: Very faint, a star of 8th magnitude involved a little excentric.

SAILING CLOSE TO THE CASSIOPEIA-Cepheus border, about 6° west-northwest of Beta (β) Cassiopeiae (Caph), is a whimsical pairing of deep-sky objects: the illustrious 7th-magnitude open cluster M52 and, about ½° to its southwest, the dauntingly dim emission nebula NGC 7635 (Caldwell 11). Once included in the *Catalogue of Galactic Planetary Nebulae* by Luboš Perek and Luboš Kohoutek, the Bubble Nebula has since been reclassified as an emission nebula. It found its way into the first edition of *Sky Atlas 2000.0* as a planetary nebula, but more modern atlases, such as the *Millennium Star Atlas, Uranometria 2000.0*, and the second edition of *Sky Atlas 2000.0*, portray it correctly as a diffuse emission nebula.

Detailed photographs of the Bubble help us understand the confusion surrounding this object. Red-light-sensitive plates taken with the Palomar Mountain 200-inch telescope show a magnificent loop of gas centered on, and seemingly blowing away from, a nondescript central star. But the Bubble is not a planetary nebula; it is a hollow cavity in the H II region known as NGC 7635, and it was formed by radiation pressure streaming out from a hot blue central star (9th-magnitude SAO 20575, with a spectral type of O6.5). Hubble Space Telescope images of the Bubble suggest a celestial ballet of windswept veils dancing playfully around that 9th-magnitude sun.

A closer look at the images reveals dense knots embedded within the nebula's central cavity. The surfaces of these knots are being boiled away by the region's flood of intense ultraviolet radiation. They also have been shocked by the central star's 2,000-km-per-second wind. By contrast, high-density knots outside the sharp northern rim of the Bubble show no shock-wind effects. The rim thus

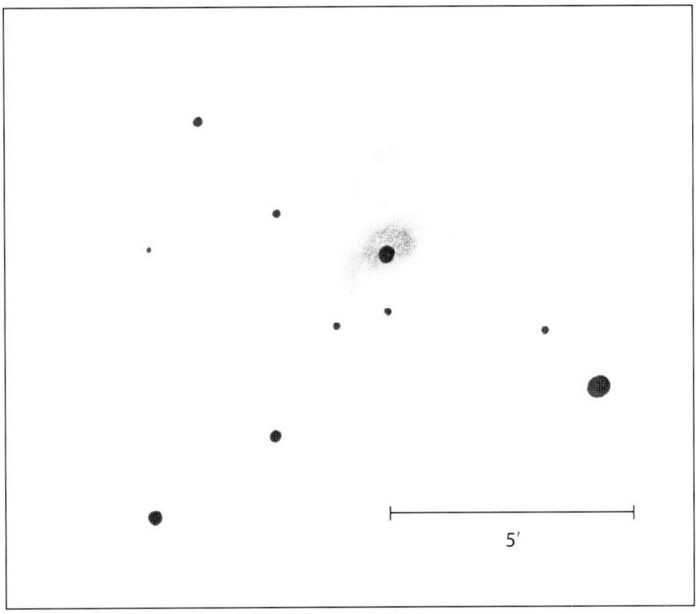

appears to be the edge of a rapidly expanding shock wave plowing through the surrounding H II region. Using one's imagination, these shock waves appear to be washing onto the nebula's shores like tidal waves intercepting a coastal plain. EGGs (evaporating gaseous globules), similar to those found in the Eagle Nebula (M16), have also been detected. But those in NGC 7635 are larger, in both angular and physical terms, than those in the Eagle. EGGs are solar-system-sized globs of matter in which conditions seem ripe for star birth and, possibly, planet formation.

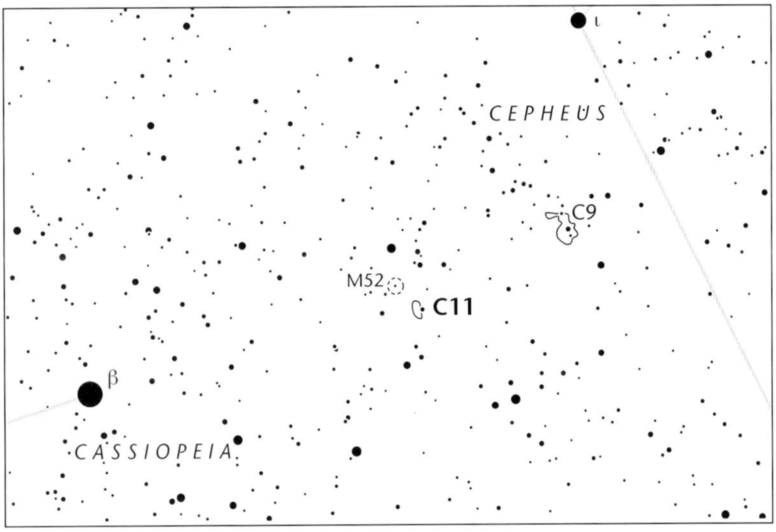

Undoubtedly anyone who has pointed a wide-field telescope at M52 has also received weak photon hits from NGC 7635. But far fewer have seen the Bubble itself. Many sources identify the Bubble with NGC 7635 while failing to clarify a very important point: the Bubble is only a *part* of NGC 7635. The Bubble is a dim 3′ ring of vapors surrounding (though not centered on) SAO 20575, which punctuates the south-central rim of the somewhat California-shaped NGC 7635; its brightest segment is an insignificant arc about 1′ in length. By contrast, the fully elongated form of NGC 7635 measures 15′ × 8′ and extends from the southeast to the northwest. The distinction between NGC 7635 and the Bubble is critical to visual observers, especially those with modest telescopes. Without this knowledge the 15′ × 8′ dimensions of NGC 7635 might easily be mistaken for those of the Bubble, leading one to search fruitlessly for something five times larger than it really is. Be warned: a roughly circular glow lies just north-northeast of the illuminating 9th-magnitude star. With a glance in the Genesis, I initially mistook this circle of light for the Bubble, especially since a patch of dark nebulosity occupies its center. But it is not the Bubble. (It is, however, the portion of NGC 7635 that shines at 10th magnitude. Also note that I have exaggerated the Bubble's northern rim in my drawing; otherwise it might not have been reproduced.) The Bubble, in fact, is the smallest of *three* bubbles associated with the central star.

Unfortunately the Bubble is not a feature to be glimpsed casually through small telescopes. It is not visible in its entirety in the 4-inch Genesis (though plenty of the surrounding nebulosity is). Christian Luginbuhl and Brian Skiff could not see it through a 6-inch telescope, saying that even NGC 7635 was "very faint and barely discernible." No, it takes large apertures (and perhaps a nebula filter) to see the Bubble well. Using a 13.1-inch f/4.5 reflector at 47× from Columbus, Texas, Barbara Wilson not only discerned the Bubble, finding it "lovely and delicate," she also saw a bright greenish tint to it — *without* a filter. Of course, nothing can beat a view of the Bubble through a *really* large telescope, like the view I enjoyed during one Winter Star Party in the Florida Keys through Tom and Jeannie Clark's 36-inch Yard Scope (now owned by Astronomy To Go). That night the full ghostly loop, supported by the amorphous greenish glow of NGC 7635 stood out sharply against the scintillating backdrop of the Milky Way. In fact, the Bubble and its surrounding nebulosity looked rather like a wad of bubble gum stuck to the underside of a table.

12

NGC 6946
Type: Mixed Spiral Galaxy (SABcd)
Con: Cepheus

RA: $20^h 34.9^m$
Dec: +60° 09′
Mag: 8.8
Dim: 11.2′ × 8.8′
SB: 14.2
Dist: 18 million light-years
Disc: William Herschel, 1798

W. HERSCHEL: [Observed 9 September 1798] Considerably faint, very large, of an irregular figure, a sort of bright nucleus in the middle. The nebulosity [spans] 6′ or 7′. The nucleus seems to consist of stars, the nebulosity is of the milky kind. It is a pretty object. (H IV-76)

GC/NGC: Very faint, very large, very gradually, then very suddenly brighter in the middle, partially resolved, some stars seen.

NGC 6946 IS A SPIRAL GALAXY WITH A PAIR of branching arms that have a puzzling propensity for supernova explosions. In the 20th century terrestrial astronomers witnessed six supernovae blazing forth in that galaxy (in 1917, 1939, 1948, 1968, 1969, and 1980). The longest duration between these events was 22 years, suggesting that, as of this writing, another seems about due. Caldwell 12 could be the ticket to fame for a dedicated amateur who decides to survey that galaxy whenever possible. As seen from Earth, the 1980 eruption was the galaxy's brightest. The new star rose to a peak visual magnitude of 11.4 and outshone any foreground star near it. Designated Supernova 1980K, the "guest star" was classified as a Type II supernova, one that occurs when a very massive supergiant star (one with 8 or more times the mass of our Sun) collapses abruptly after nuclear fusion grinds to a halt in its core. Astronomers now believe that the core of a Type II supernova progenitor collapses, then rebounds slightly, transferring prodigious amounts of energy to the star's envelope. The resulting shock wave blows much of that envelope into outer space. Since they are spawned by short-lived supergiant stars, Type II supernovae generally occur in the spiral arms of disk galaxies, where stars are continuously born.

Type II supernovae tend to show more diversity in their light curves and absolute magnitudes than do Type Ia supernovae, which make for much better "standard candles" — even though *their* cause remains unknown. Indeed, Supernova 1980K evinced such an

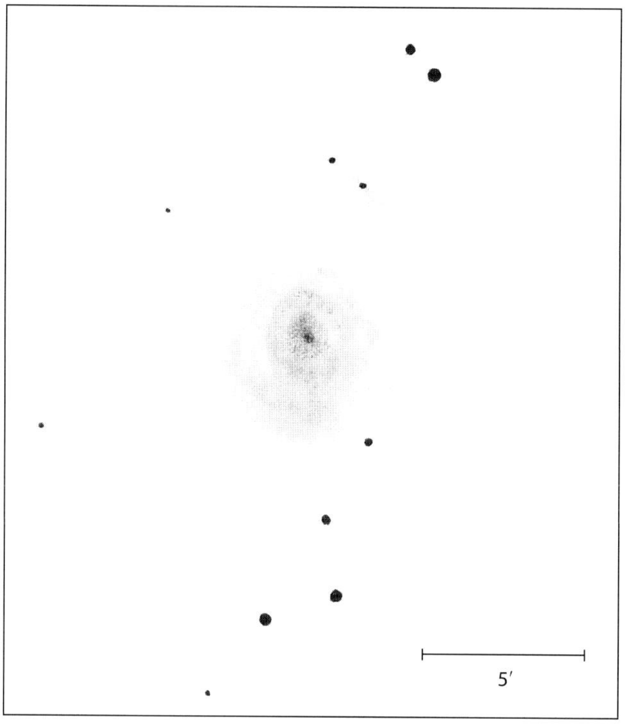

anomalously slow decline that it was rediscovered years after its initial apparition. Its slow decay is now attributed to interactions between the supernova ejecta and particles that had been cast off earlier by the progenitor star. X-ray observations with the now-defunct Rosat satellite support this theory.

Like IC 342 (Caldwell 5) and NGC 2403 (Caldwell 7), NGC 6946 is a member of the Coma-Sculptor Cloud of galaxies. Shining at magnitude 8.8, NGC 6946 lies only 11° from the central plane of the Milky Way, whose dust and gas dim its light by 1.6 magnitudes. Images taken through large telescopes reveal the system to be quite large. Its estimated linear diameter is 58,000 light-years, and it shines with a total luminosity of 30 billion Suns.

We see NGC 6946 inclined 42° from face-on; its clumpy spiral arms are flung wildly open. Recent observations of these arms have caused some confusion. In theory astronomers expect to find a spiral's strongest magnetic fields at the inner edges of the visible spiral arms. A 1990s map of NGC 6946's polarized radio emissions did indeed trace out two well-defined magnetic arms. But these magnetic arms lay between the visible ones. This finding defied then-prevailing theories. However, University of Chicago theorists Zuhui Fan and Yu-Qing Lou have since proposed that these out-of-phase magnetic arms are caused by a second, rather sluggish set of density waves that peak between the visible arms.

It's a shame that this astrophysical dynamo is not equally electrifying visually. Robert Burnham Jr. has little to say about NGC 6946's visual impression, except that "owing to the low surface brightness, the small central nucleus is the only detail which appears clearly to the visual observer." Surprisingly, Christian Luginbuhl and Brian Skiff do not include NGC 6946 in their *Observing Handbook and Catalogue of Deep-Sky Objects*, and, though it is listed in Roger Clark's *Visual Astronomy of the Deep Sky*, the galaxy's magnitude is misrepresented there as 11.1, a full 2.3 magnitudes fainter than the currently accepted value.

Certainly NGC 6946 is an object whose faint features can be fully appreciated only under a dark sky. The galaxy is nicely situated only 2° southwest of Eta (η) Cephei, just ⅔° from the brighter, more visually appealing open cluster NGC 6939 (magnitude 7.8). Seen through the 4-inch at 23× these objects make a magnificent pair: the cluster starts to be resolvable at low power and is cleanly sliced through the middle by a dramatic dark lane bordered

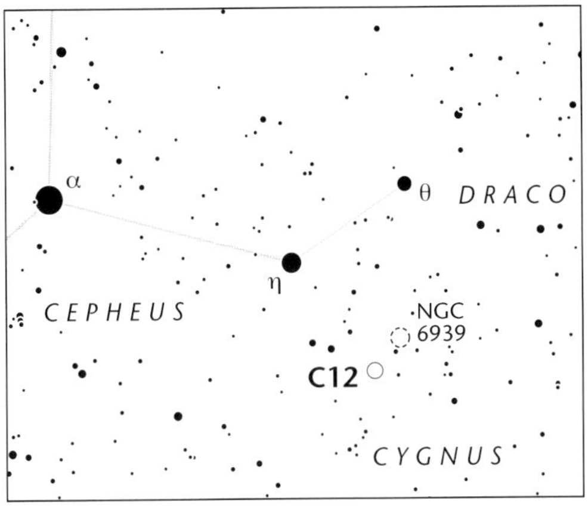

to the west by a row of similarly bright suns. The galaxy looks somewhat like a ghost image of the cluster, making me wonder why NGC 6939 is not the Caldwell object.

But give the galaxy time — a lot of time. Slowly the seemingly uniform circular haze takes on a more elliptical form, oriented northeast to southwest (like the cluster), with a sharp core. Now relax for a moment away from the eyepiece. Next, focus your attention on the galaxy's heart. In time, the core should appear to be surrounded by a tight, broken ring of light. That ring, in turn, is surrounded by yet another concentric glow whose jagged edges mark the beginnings of spiral structure — the positions of those ghostly density waves rolling through the galaxy like ripples in a lake. (Imagine the sharp core as the stone dropped in the water to create the ripples.) You can now use your mind's eye and transform the innermost ring into a dust-shrouded circle of star birth that astronomers have discovered close to the galaxy's nucleus. There, a fantastic episode of intense star formation has been occurring for the last 20 million years; probably it was triggered by a rotating bar that has been detected at infrared wavelengths.

NGC 6946 virtually vanishes at 72×, so users of small or moderate-size telescopes will have to be satisfied with what they can make out at low power. Unfortunately, this requires a lot of patience. But the effort will be rewarding as faint details briefly float in and out of view. On the other hand, if you own a large telescope, NGC 6946 should appear quite dramatic. Through her 20″ f/4 reflector at 220× Barbara Wilson saw several H II regions in the four spiral arms. The galaxy's 1′-wide core was very bright, and its outer arms were apparent, especially to the east. Had I stumbled upon this object in the 19th century with a telescope as good as the Genesis refractor — a very unlikely circumstance — I would have likened it to a globular cluster at the limit of resolution. William Herschel apparently alluded to this aspect of NGC 6946 when he wrote that "the nucleus seems to consist of stars."

13

13

E. T. Cluster;
Phi Cas Cluster
NGC 457
Type: Open Cluster
Con: Cassiopeia

RA: 01^h 19.5^m
Dec: +58° 17′
Mag: 6.4
Diam: 20′
Dist: indeterminate (see text)
Disc: William Herschel, 1787

W. HERSCHEL: [Observed 18 October 1787] A brilliant cluster of [bright] and very [faint] stars, considerably rich. (H VII-42)

GC/NGC: Cluster, bright, large, pretty rich, stars of magnitude 7, 8, and 10.

OF THE DOZENS OF OPEN CLUSTERS IN Cassiopeia, none inspires as many visual flights of fancy as NGC 457. Christian Luginbuhl and Brian Skiff rate it as "perhaps the most impressive of all the clusters in Cassiopeia," and *Sky & Telescope*'s Alan M. MacRobert calls it "one of my favorites in the whole sky." Certainly NGC 457 is fast becoming one of the most popular showpieces at amateur gatherings. In his book *Star-Hopping for Backyard Astronomers*, MacRobert explains why: "This is one of those clusters that especially provoke the human tendency to see animals, faces, and monsters in random patterns." In photographs this roughly T-shaped pack of glittering suns looks like the stick figure of a man ready to greet you: two prominent stars at the cluster's southeastern end mark the figure's gaping eyes; two chains of stars on the northeastern and southwestern sides of the cluster's tapered body extend outward like wide-open arms; and two isolated suns at the cluster's northwestern end mark the figure's comfortably spaced feet. The cluster looks like it is not only happy to see you but wants to give you a hug.

Shining at magnitude 6.4, NGC 457 is Cassiopeia's third-brightest open cluster, following magnitude-4.4 Stock 2 (a little-known 1°-wide cluster about 2° north-northwest of the Double Cluster's NGC 869) and Stock 5, whose brightest star is magnitude 4.9. Yet despite its prominence, NGC 457 is yet another one of those visual gems that tends to be lost in the shadow of fame surrounding a nearby Messier object, in this case M103. Both M103 and NGC 457 are in the Milky Way's Perseus Arm and can be found near Delta (δ) Cassiopeiae. M103 lies less than 1° northeast of Delta Cassiopeiae, while NGC 457 is 2° to

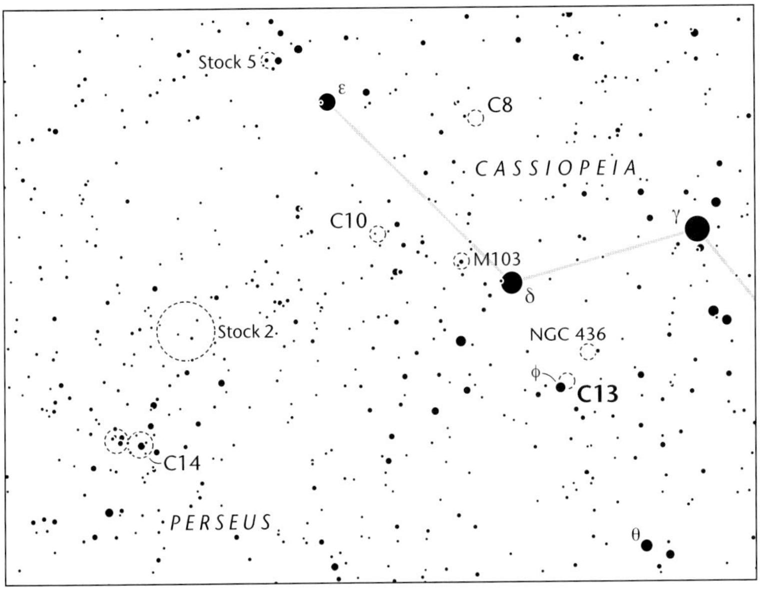

telescopic targets at star parties across North America and, undoubtedly, in other parts of the world. I was first introduced to the cluster as recently as 1994 while attending StarFest, an annual summer gathering of amateur astronomers outside Toronto, Ontario. The sudden sensation can be traced back to 1987, though, when Tom Lorenzin and Tim Sechler compared NGC 457 with Spielberg's charismatic alien hobbit in their book *1000+: The Amateur Astronomer's Field Guide to Deep Sky Observing.* Of NGC 457 they write, "ET waves his arms at you, and winks!" A search of the pre-1987 literature reveals no widely recognized nickname other than "Phi Cas" for the cluster.

the south-southwest. NGC 457, however, hides in the glare of 5th-magnitude Phi (φ) Cassiopeiae. Not surprisingly, amateurs tend to point their telescopes more often to M103. But the celestial winds are fast changing.

Classically, NGC 457 has been referred to as the Phi Cas Cluster. But this humdrum moniker fails to capture the cluster's spirit or to fire the observer's imagination. So, despite the fact that Robert Burnham Jr. spotlights NGC 457 in his *Celestial Handbook,* and despite repeated invitations from the late Walter Scott Houston to seek out the Phi Cas Cluster, interest in NGC 457 fell short of making it a national or international showpiece — until something magical happened.

In 1982 Steven Spielberg introduced *E. T. The Extra-Terrestrial* to the big screen. Five years later, NGC 457 mysteriously exploded onto the celestial stage. Amateurs began introducing NGC 457 to their peers, first by word of mouth, then through their telescopes set up under the shimmering Milky Way. By 1990 NGC 457 had become one of the preferred

As E. T. faded from the public scene, amateurs began personalizing the cluster. Over the years I have heard different telescope owners refer to NGC 457 as the Owl, Airplane, Stick Man, and Worry Doll Cluster. Some of these names may have been around a lot longer than "E. T.," but none of them had its universal charm or staying power.

Ironically, Lorenzin and Sechler's popularization of the E. T. Cluster may be doing science a favor. The bright double star Phi Cassiopeiae (magnitudes 5.0 and 7.0; 134″ separation) that marks E. T.'s eyes may not belong to the cluster. Brent Archinal notes that a 1961 article titled "Photometry of Stars in Galactic Cluster Fields" contains some of the first good combined photoelectric and photographic brightness measurements of stars in open clusters. "From the accompanying photograph and color-magnitude diagram," Archinal writes, "it looks fairly obvious that the two brightest stars in NGC 457 [Phi[1] and Phi [2] Cassiopeiae] are non-members. They are much brighter than the other (cluster) stars, and much higher on the color-magnitude diagram plot. There's, of course, no way to tell for sure from just this information whether they are members, but it seems unlikely." *Sky & Telescope*'s Roger W. Sinnott adds that parallax data from the Hipparcos satellite suggest a probable distance for Phi Cassiopeiae of 2,000 light-years, though uncertainties in the data could position the double star anywhere from 1,000 to 4,000 light-years out. Sinnott remarks that this distance is better than anything we've had before, despite the uncertainty.

Ambiguities also exist in determinations of NGC 457's distance. Gösta Lynga's value of 9,965 light-years, published in his 1983 *Catalogue of Open Cluster Data*, is probably the most reliable pre-Hipparcos estimate. Hipparcos based a provisional distance of 2,100 light-years on four bright stars that presumably belong to the cluster. Other estimates recently published in professional journals are 9,400, 10,400, and 8,150 light-years. Archinal says that "either the ground-based estimates have some common large error, the brightest stars are not members, or there is some large error common to the four Hipparcos stars used in this most recent study." On the one hand, if Phi Cassiopeiae is not related to the cluster, then NGC 457's brightest member shines at magnitude 8.6. On the other hand, if new data one day reveal unequivocally that Phi Cassiopeiae *is* a true member, the cluster's total apparent magnitude would certainly have to jump to a higher value.

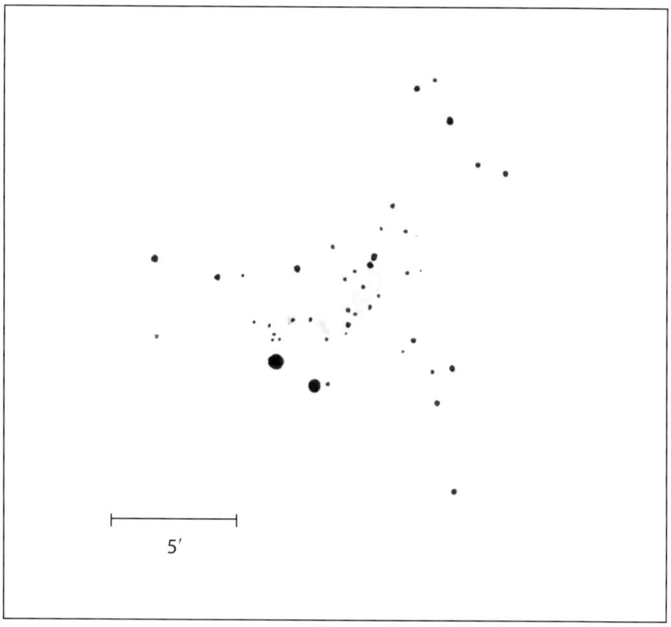

Phi Cassiopeiae is troublesome in another way. NGC 457 should be near the limit of naked-eye visibility for a diffuse object, but glare from Phi Cassiopeiae renders it invisible (at least I couldn't convince myself of its visibility). But the jury is still out. So here is a decent visual challenge for observers (especially young ones) who live under dark skies: see if you can pick up the cluster unambiguously without optical aid. Some guidance may come from descriptions of NGC 457's appearance in binoculars. In my 7×35s I see the tapered tail of a fragmenting comet. More poetically, MacRobert likens his binocular view to a "wisp of a candle flame blown from the two stars by the wind." One should wonder why Messier failed to notice this beautiful comet-like object.

Through the 4-inch at 23× the cluster has a more sinister appearance than it does in photographs. The cluster's bright "eyes" seem to pierce the night with the fiery gaze of a specter emerging from the dusty, cobwebbed corridors of space. The ghost's clothes hang in tatters from its skeletal limbs. The dusty cobwebs, of course, belong to the network of dark material that permeates the Cassiopeia Milky Way. The clothes, in contrast, belong to the combined glow of a hundred or more faint suns shimmering near or beyond the limit of resolution, like smoke rising from countless dying embers. Of the 200 cluster members within a circle 20′ (58 light-years) across, Luginbuhl and Skiff find 100 visible in an 8-inch telescope. Higher powers do little to enhance the view.

By the way, Phi[1] Cassiopeiae itself is a double star, with a magnitude-12.2 companion 49″ away at a position angle of 208°. And if you bump your telescope 40′ to the northwest, you will encounter the compact glow of the magnitude-8.8 open cluster NGC 436, whose brightest stars trace out three sharp arcs akin to the digits in a crow's foot.

14

14

Double Cluster
NGC 869 and NGC 884
Type: Pair of Open Clusters
Con: Perseus

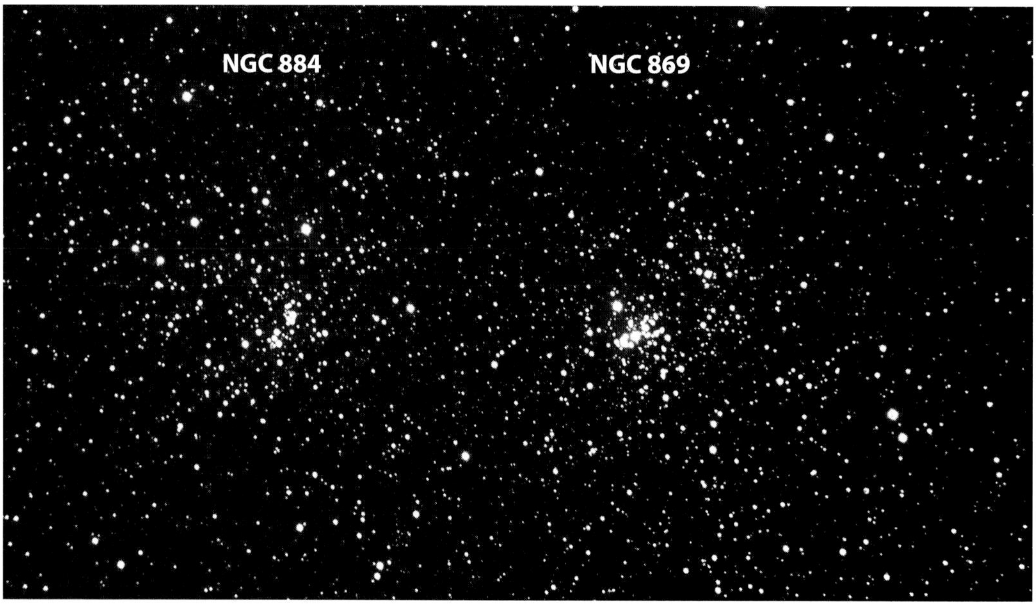

NGC 869
RA: 02^h 19.0^m
Dec: +57° 08′
Mag: 5.3; 4.5 (O'Meara)
Diam: 18′
Dist: 7,300 light-years
Disc. known since antiquity

W. HERSCHEL: [Observed 1 November 1788] A very beautiful and brilliant cluster of [bright] stars, very rich. The middle contains a vacancy. (H VI-33)

GC/NGC: Remarkable cluster, very large, very rich, stars ranging from magnitude 7 to 14.

NGC 884
RA: 02^h 22.0^m
Dec: +57° 08′
Mag: 6.1; 5.7 (O'Meara)
Diam: 18′
Dist: 7,300 light-years
Disc. known since antiquity

W. HERSCHEL: [Observed 1 November 1788] A very beautiful, brilliant cluster of [bright] stars, irregularly round. Very rich. Near ½° in diameter. (H VI-34)

GC/NGC: Remarkable cluster, very large, very rich, ruby star in the middle.

THERE'S A PARTY GOING ON IN THE SKY EVERY Christmas Eve, and it focuses on the Double Cluster in Perseus. This cheerful gathering of an imaginary celestial family is one you won't find mentioned in any previous book on star lore; it's a celestial play I created at the age of 10. (It sounds silly now, but it might make a good under-the-stars skit for children.) The play begins at 8:00 p.m., when my eyes knock on the door of the northwestern horizon. If the night is clear, the door opens. And there, standing with perfect posture, is the matronly form of the Northern Cross. Dressed in the evening gown of the Milky Way, she welcomes me with open arms: "Why, Stephen, what a pleasant surprise! Come and join us!" Orion and his dogs suddenly burst through the back door of the southeastern horizon; winds lap at me like wet tongues, sending shivers up and down my spine. I look to the north and see three more friends, Cepheus, Cassiopeia, and Perseus, parading over the snowcapped hill of the celestial pole. My attention is drawn to Perseus because he carries a special gift wrapped in mist. The moment has come. I use my telescope, and the magic of magnification, to open the gift. The mist evaporates to reveal two crystal spheres. I look away from the telescope and see Perseus hanging them at the apex of the heavens. "If enough people wish for snow on Christmas day," he tells me, "their thoughts will shatter the crystal and the pieces will fall and turn to snow." As you can imagine, if snow did fall the next day I could see in it pieces of this shattered crystal glinting sharply in the morning light.

It's not surprising that the Double Cluster could ignite a fire of imagination in a young skywatcher. Around the globe the great Double Cluster in Perseus is hailed as one of the most impressive telescopic sights in the heavens. Under dark skies the glow from this pair of stellar islands stretches across two Moon diameters and looks like a 4th-magnitude knot in the ruffled folds of the Milky Way, about halfway between Delta (δ) Cassiopeiae and Gamma (γ) Persei.

Although Herschel is credited with discovering these objects' star-cluster nature, the combined glow of NGC 869 and NGC 884 has been known since antiquity. Hipparchus recorded them as a "cloudy spot." In his book *Star Names: Their Lore and Meaning,* Richard Hinckley Allen says that Ptolemy noted them as a "dense mass," and subsequent pretelescopic astronomers saw them as nebulous entities. Allen also says that the twin clusters "seem strangely to have escaped the notice of astrologers, who, as a rule, devoted much attention to clusters as harmful objects which portended accidents to sight and blindness." Without the aid of a telescope, however, it's doubtful astrologers would have associated this "cloud" with an obvious cluster, such as the Pleiades. Today NGC 869 and 884 are simply known as the Double Cluster in the Sword Hand of Perseus.

One long-lived mystery concerning the Double Cluster is why the 18th-Century French comet hunter Charles Messier did not include it in his famous catalog. (Actually, Messier probably had a good reason not to do so; see Appendix B for details). An equally compelling mystery is that, while Hipparchus and other early skywatchers listed NGC 869 and NGC 884 as two objects, few of today's amateurs seem to believe the components can be resolved with the unaided eye. Most popular astronomy books and articles state that the Double Cluster can be seen with the unaided eye only as a hazy patch of light, which becomes two beautiful objects with the aid of binoculars or a rich-field telescope. With a sweep of the hand that statement is true. But I wonder how many observers have missed seeing two distinct components with the unaided eye (not to mention the brightest stars superposed on them) because no one told them to look.

Together the two clusters occupy a full degree of sky, and their bright cores are separated by 25′ — nearly a full Moon diameter. Why, then, have these distinct cores escaped the notice of the unaided eye? The problem may be one of contrast. With a casual glance the faintly luminous background of the Milky Way tends to fill the gap between the cores, making the split less distinct. Atmospheric clarity also may be a critical factor. When the Double Cluster is seen from nearly 14,000 feet atop Hawaii's Mauna Kea, a distinct lane of darkness separates the two components, even with direct vision. At an altitude of 4,200 feet, however, resolving the components requires a slight effort. The odds seem to worsen at sea level. You can avoid the issue by looking for the Double Cluster at the start of astronomical twilight, about 45 minutes after sunset; it emerges at about the same time as the brightest stars in Cassiopeia. Before the sky darkens completely the two cores pop clearly into view for several minutes, without any interference from the Milky Way.

It is also at this time that one can make an accurate estimate of each component's brightness. Published magnitudes for the individual clusters are wide-ranging. The *Observing Handbook and Catalogue of Deep-Sky Objects* lists the components' visual magnitudes as 3.5 (NGC 869) and 3.6 (NGC 884). *Visual Astronomy of the Deep Sky* pegs them at 4.4 and 4.7, respectively. And both the *Deep Sky Field Guide to Uranometria 2000.0* and *Star Clusters* list values of 5.3 for NGC 869 and 6.1 for NGC 884. Most sources determine a cluster's total apparent

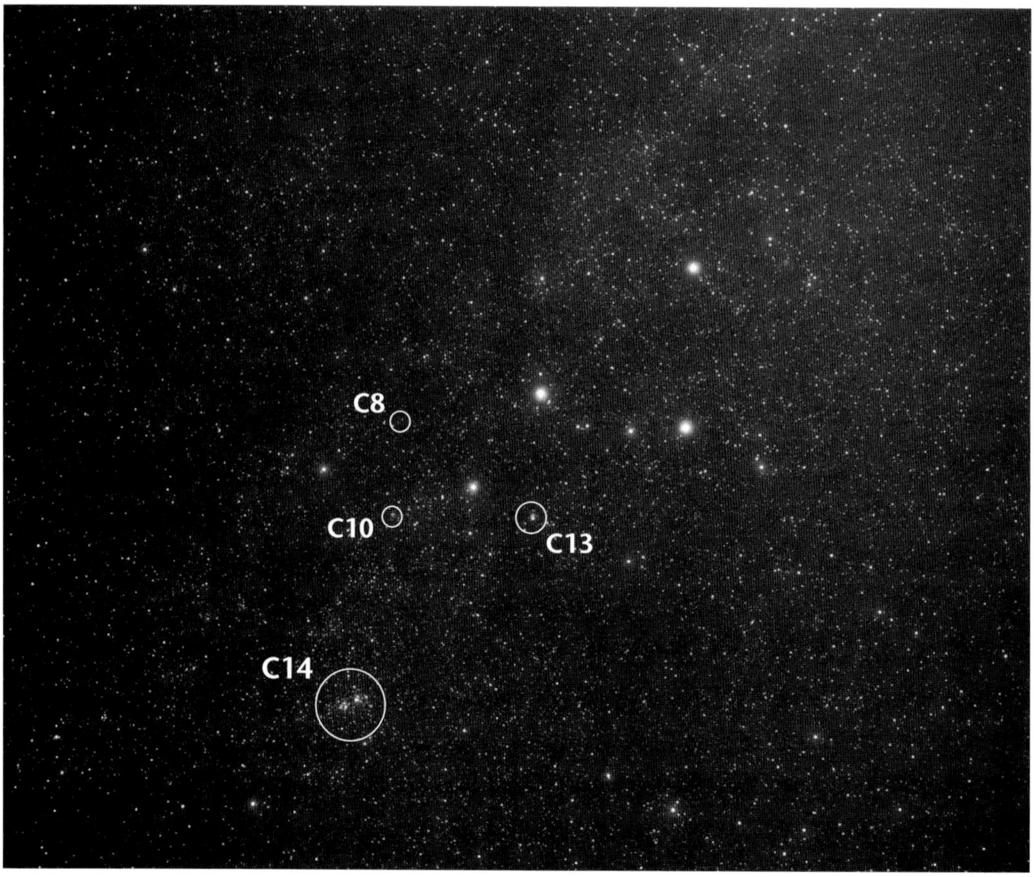

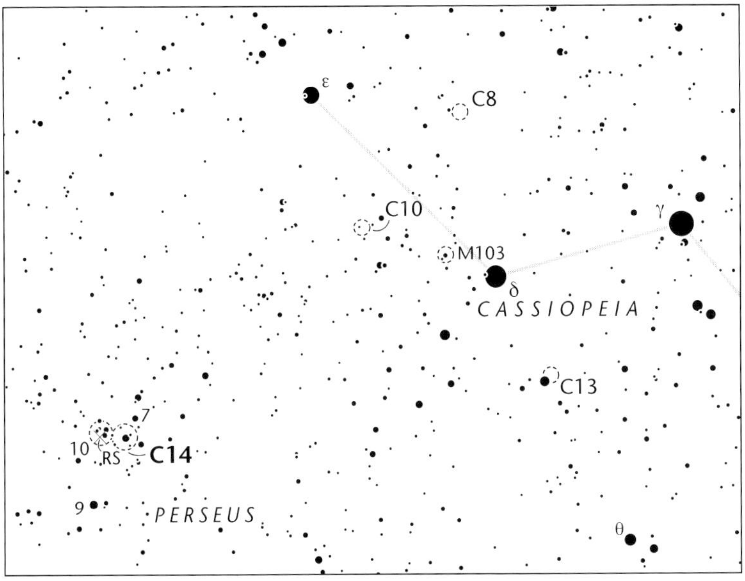

brightness by summing the contributions of the stars in each cluster as recorded on photographic plates. The differences among published values may reflect (1) differences in the numbers of stars used, and (2) whether the total magnitude includes any bright stars superimposed on the clusters. (On photographic plates the overexposed images of bright stars will lead to spuriously bright total magnitude estimates.) Motivated by these discrepancies, in February 1998 I estimated the brightness of the individual clusters with the naked eye, both in deep twilight and under a dark sky, and got an average visual magnitude of 4.5 for NGC 869 and 5.7 for NGC 884. Brent Archinal suspects that these visual estimates are close to what one would get by including the stars that are too bright to be measured on photographs.

The Double Cluster resides in a Milky Way spiral arm (the Perseus Arm) that lies farther from the galactic center than the one containing our solar system (the Orion Arm). A recent study places the clusters 7,300 light-years from Earth. These clusters probably formed from a single cloud of dust and gas, perhaps 13 million years ago. We see the Double Cluster through thick clouds of interstellar dust that line the plane of the Milky Way. Were it not for this dust, we would see the Double Cluster shining 1.6 magnitudes (4.4 times) brighter. The obscuration is unfortunate because, of the 300 or so suns that populate each of the two clusters, many are hot, blue-white supergiants; yet interstellar dimming lets even the brightest of these appear no brighter than magnitude 6.5. Imagine the grandeur of the Double Cluster if it were at the distance of the Pleiades (407 light-years). It would recall the starscape described in Isaac Asimov's story, "Nightfall." The twin mounds of blue supergiants would each span 5 Moon diameters and shine 300 times brighter than they actually do in our skies. One-quarter of the northern sky would be filled with the concentrated splendor of 600 suns, the brightest of which would shine with the brilliance of Vega. The earliest skywatchers would have succumbed to the Double Cluster's presence, looked to it for meaning, and planned their affairs in step with its eternal rhythms. Poets and artists would have looked upon these twin celestial cities to fuel their passions, or perhaps simply to dream.

But we have to settle for the more subtle view. To the naked eye, each cluster has a central condensation and a mottled outer halo. Try resolving some of the stars in the Double

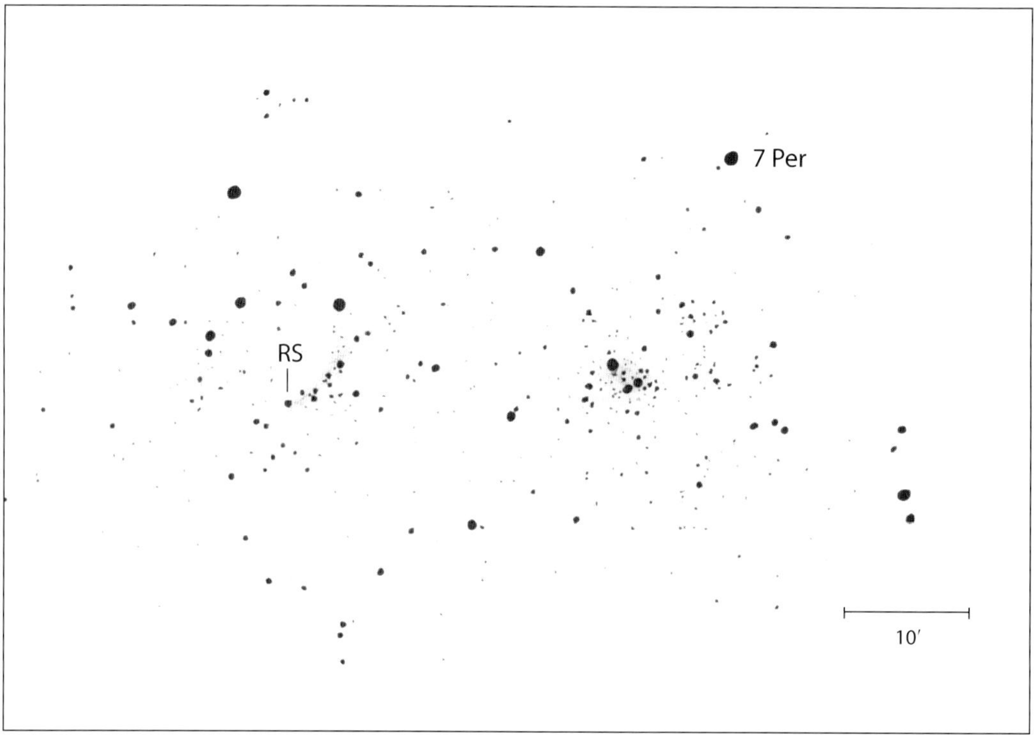

Cluster without optical aid. After looking for 30 minutes I was able to detect and chart four naked-eye stars with magnitudes ranging from 6.4 to 7.0. The positions of these suns and several other nearby ones make each cluster appear falsely elongated (NGC 869 to the east and west and NGC 884, north and south). Binoculars will help you verify the presence of any stars whose presence you suspect with the naked eye.

In 10×50 binoculars the core of NGC 884 looks like a translucent diamond surrounded by a cloud of white smoke, as if it has just fallen into a heap of lime. To its west the core of NGC 869 is punctuated by a pair of stars that form the center of a cross-shaped asterism. A stream of 8th- to 10th-magnitude stars divides the two cores; flowing in from the north, the stream curves eastward around the southern periphery of NGC 884. A second river of stars starts at the wide double star 7 Persei and follows the western fringe of NGC 869 before it meanders toward 9 Persei, more than 1° south of NGC 884. When viewed with east "up," the "weight" of NGC 869 seems to press down on this western river of stars, causing it to sag. Sweeping the binoculars to the south, then east toward 10 Persei should bring into view two rings of stars, each roughly 1° in diameter. The Double Cluster lies on the northern edge of the overlapping rings, each of which appears to have been set with a cluster of semiprecious stone. One could imagine the stars of the Double Cluster having been swept out of these rings into two piles. The view brings to mind two frozen ponds cleared of snow for skating. Skaters' tracks can be seen faintly etched in starlight within each circle.

At 23× in the Genesis the Double Cluster is transformed into an abundance of scintillating jewels. Here are sapphires, rubies, topaz, and diamonds bursting out of treasure chests mired in the black sands of a volcanic beach. Few of the heavens' sights are better. A closer look with higher magnification reveals a dark Y-shaped rift at the heart of NGC 869, just east of the cluster's elongated core. In comparison the core of NGC 884 resembles a hollow rib cage of stars surrounding a ruby heart — the pulsating semiregular variable star RS Persei. (The star's pulse is slow, its magnitude varying between 7.8 and 10.0 about every 224 days.) The *Millennium Star Atlas* shows a half dozen other variables in and around the Double Cluster.

Finally, near the Double Cluster lies the radiant of the Perseid meteor shower, which peaks in the predawn morning hours of August 12th. Allen notes that Dante may have made reference to the Perseid meteors in the *Purgatorio*:

> Vapors enkindled saw I ne'er so swiftly
> At early nightfall cleave the air serene,
> Nor, at the set of sun, the clouds of August;

"In the Middle Ages," Allen continues, the Perseids "were known as the Larmes de Saint Laurent, Saint Laurence's Tears, his martyrdom upon the red-hot gridiron having taken place on the 10th of August, 258." Thus together with the upwellings of starlight in the Double Cluster we have tears of fire and ice.

15

Blinking Planetary
NGC 6826
Type: Planetary Nebula
Con: Cygnus

RA: $19^h\ 44^m\ 48.2^s$
Dec: +50° 31′ 30″
Mag: 8.8; 8.5 (O'Meara)
Dim: 27″ × 24″
Dist: ~2,200 light-years
Disc: William Herschel, 1793

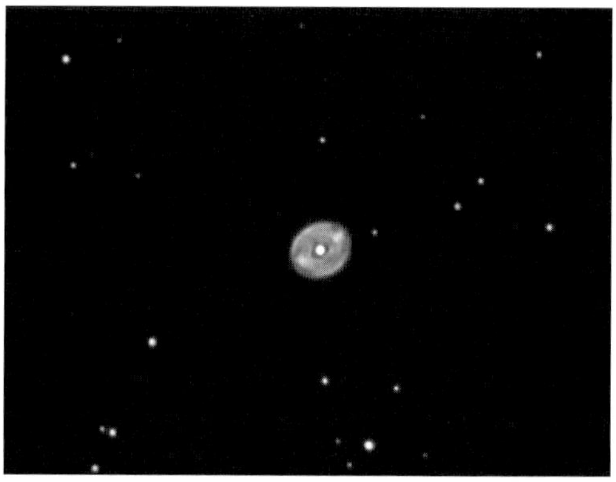

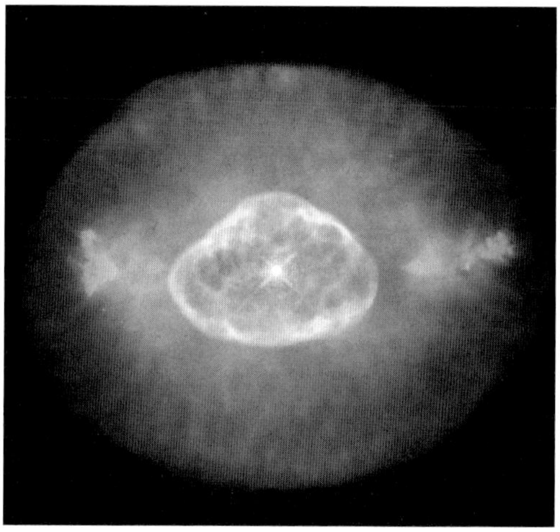

W. Herschel: [Observed 6 September 1793] A bright point, a little extended, like two points close to one another; as bright as a star of the 8.9 magnitude, surrounded by a very bright milky nebulosity suddenly terminated, having the appearance of a planetary nebula with a lucid centre; the border however is not very well defined. It is perfectly round, and I suppose about half a minute in diameter. It is of a middle species, between the planetary nebulae and nebulous stars, and is a beautiful phenomenon. (H IV-73)

GC/NGC: Planetary, bright, pretty large, round, a star of magnitude 11 in the middle.

While he was working at the Harvard-Smithsonian Center for Astrophysics in Cambridge, Massachusetts, in the late 1970s, Peter Collins took me into the modest dome of the 9-inch Clark refractor nestled on the building's roof and pointed the telescope to the northern wing of Cygnus, the Swan. "Have you ever seen the Blinking Planetary?" he asked. The answer was "no," but his words set my mind racing. "How can a planetary blink?" I wondered. The object he showed me, of course, was NGC 6826. The reason that planetaries of this type blink is simply a matter of where the object's telescopically harvested light hits the observer's eye.

Most planetary nebulae have a dense inner shell surrounded by a fainter outer halo. When you look straight at a planetary, the fainter outer envelope disappears because its light falls on the eye's *fovea centralis*. The fovea is populated largely by cone cells. Cone cells are responsible for our visual acuity, but they provide little light sensitivity. To see dim objects we rely upon the rod cells lining the surrounding retina. There are rod cells near the fovea centralis, and they will pick up relatively bright objects like stars and the condensed regions of planetary nebulae. But dim objects will not register with them. However, when you use indirect, or averted, vision, light from a planetary's dim

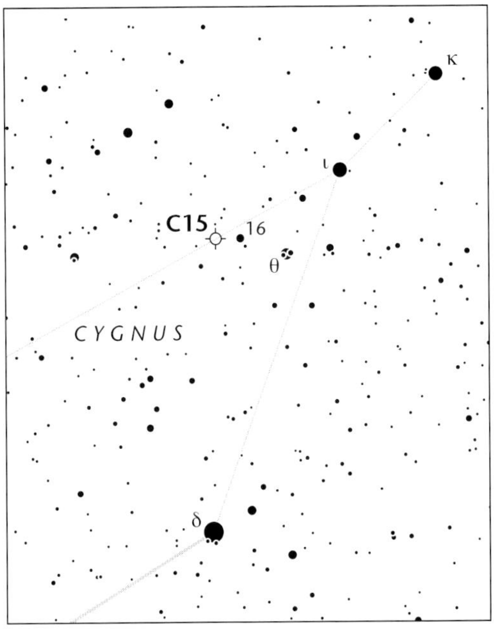

outer shell falls upon the retina's rod-lined periphery, which sends electrical signals to the brain, making the outer shell suddenly pop into view. Alternating between direct and indirect vision makes a double-shelled planetary appear to blink.

The blinking phenomenon is a rather good metaphor for the brief lives of these phantom gas clouds. A planetary nebula lasts but for a "blink of an eye" in astronomical terms. A Sun-like star can live for 10 billion years, but the shell we see around NGC 6826 will last for no more than some 10,000 years. Hubble Space Telescope images of the Blinking Planetary cause us to blink yet another way: in surprise. For in them we see mysterious knots of dense gas that astronomers have dubbed FLIERs (*f*ast *l*ow-*i*onization *e*mission *r*egions). FLIERs cause astronomers to blink because they challenge their understanding of planetary nebulae. No model of stellar evolution presently can explain their existence. In one sense, says Bruce Balick (University of Washington), some of their characteristics suggest that FLIERs are like sparks flung outward from the central star about 1,000 years ago. Yet their shapes, as revealed by HST, suggest that they are stationary, and that material ejected from the star flows past them, scraping gas from their surfaces.

It's best to approach this curious object in four simple steps. First locate 3rd-magnitude Delta (δ) Cygni, which marks the western arm of the Northern Cross asterism. Seven degrees north-northwest of Delta Cygni is 4th-magnitude Iota (ι) Cygni. Next locate the beautiful double star 16 Cygni 2¼° degrees to the southeast; the sight of its 6th-magnitude Sun-like stars, separated by 40″, equals that of the planetary itself. (What's more, we now know that the southeastern member of the pair is orbited every 2.2 years by a planet or substellar brown dwarf that may have as little as 1.5 times the mass of Jupiter.) NGC 6826 is about ½° east-southeast of the double. In fact, you can look for a "double star" at the planetary's position, for it forms a pair with a much dimmer 12th-magnitude star 1.6′ to the south. See if you can detect the planetary with 7×35 binoculars; the task should be easy.

At 23× through the 4-inch NGC 6826 is virtually stellar, though averted vision and a few minutes of staring shows its slightly swollen form. Increasing the magnification to 72× reveals the planetary as a small disk with a soft aqua hue reminiscent of the colors one sees when peering at Uranus or Neptune through a telescope. Steven Hynes, the author of *Planetary Nebulae*, noted a blue-green hue to NGC 6826 through his 10-inch reflector at 155×. At 189× in the 4-inch, the nebula's 10.6-magnitude central star stands out as a strong punctuation mark at the center of a large, diffuse shell of uniform light, though there is a hint of brightening toward the central star. Viewing through a 16½-inch telescope, Webb Society member

Malcolm J. Thomson of Santa Barbara, California, describes the planetary as having a curved dark area between the central star and the following (eastward) edge. Through a 12-inch reflector, Christian Luginbuhl and Brian Skiff also viewed this dark region as well as an uneven edge on the nebula's western limb.

Images taken through large telescopes and various filters reveal greater details in the Blinking Planetary. Especially noteworthy is the presence of a dense elliptical inner disk and knots at either end of that ellipse embedded in an outer halo. Would an Oxygen-III filter help amateurs bring out further details in NGC 6826 with their moderate-sized telescopes? I have not seen any reports of this. In 1995 Austrian astronomer Gerald Handler (Vienna University) announced that several relatively cool central stars of planetary nebulae, including those of NGC 6826 and NGC 6543 (Caldwell 6), show brightness variations of up to 0.2 magnitude on time scales of several hours to days. Amateurs with access to large telescopes and photometers might want to try their hands at recording these fluctuations.

With Mira, his 16-inch Dobsonian, Jay McNeil has recently made one of the most detailed visual studies ever performed on the Blinking Planetary. He spent a total of about 2½ hours studying and drawing this object on two separate nights. On both nights the seeing was better than 1″ and the transparency quite good, and the most pleasant and revealing view came at 586×. A power of 855× was also used at times to glimpse even smaller, more intricate details. According to McNeil, the nebula's brightest part was an elliptical oval (roughly 30″ × 25″), oriented east-southeast and west-northwest, around the central star. This oval ring was mottled near the tips of its major axis and, during moments of excellent seeing, McNeil detected a sharp inner ring about 10″ in diameter. The area between the inner ring and the central star was darker than any other part of the nebula but not as dark as the background sky. McNeil also sighted the southeastern FLIER complex, which appeared as a very slightly nonstellar nebulous "knot" less than 3″ in extent at the

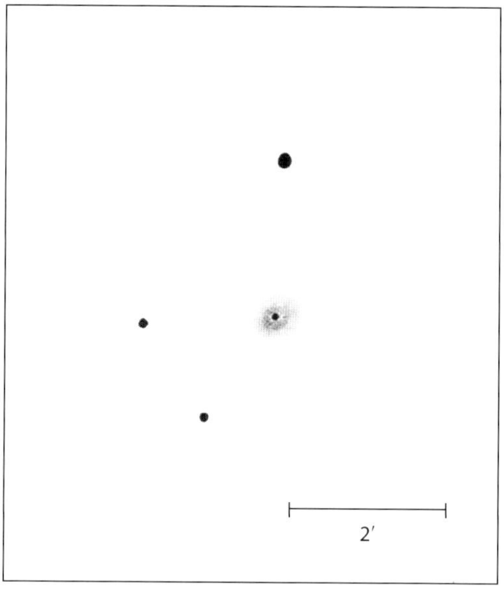

very edge of the outer envelope's easternmost point. "With a Lumicon Deep-Sky filter applied at 586×," he says, "this [feature] became nearly as apparent as the nebula's filtered nucleus at times." He also sighted the much fainter northwestern FLIER complex, which looked like a diffuse knot. At 855× shell structure within the nebula seemed quite complex, and the view was reminiscent of NGC 7009, the Saturn Nebula (Caldwell 55). At times the FLIERs extended very slightly beyond the nebula's bright outer envelope "as if the very edges of the major axis consisted of small 'nipples.'" With extreme averted vision a low-surface-brightness halo surrounded the entire structure out to 35″. The individual FLIERs or “knots” have also been noted by expert observer Alister Ling in an 8-inch f/6 telescope at 500×. So they are definitely within the reach of deep-sky observers. The key is the proper use of extremely high magnification — and patience.

16

16

NGC 7243
Type: Open Cluster
Con: Lacerta

RA: 22^h 15.0^m
Dec: +49° 54′
Mag: 6.4
Diam: 30′
Dist: ~2,500 light years
Disc: William Herschel, 1788

W. HERSCHEL: [Observed 26 September 1788] A cluster of coarsely scattered [bright] stars, a little extended, 16′ in diameter. (H VIII-75)

GC/NGC: Cluster, large, poor, little compressed, stars very [bright].

LACERTA, THE LIZARD, IS ONE OF THE TINIEST creatures inhabiting the far-northern celestial sphere. With only 201 square degrees of sky inside its roughly rectangular borders, it is also among the smallest constellations in the entire heavens. Johannes Hevelius formed the constellation in 1687, robbing stars from Cygnus, the Swan, and Andromeda, the Chained Maiden, to do so. He crafted the Lizard to fit within the narrow space between those previously established constellations. Today the Lizard slinks around the sky virtually unnoticed. Alpha (α) Lacertae, the constellation's brightest star, shines at a meager magnitude 3.8 (though it does make a fine double, with a 12th-magnitude companion 36″ away). Some beginners' observing guides and star charts fail even to recognize the constellation.

The Lizard seemingly leaps off the northern forefoot of the winged horse Pegasus toward the Swan's tail. In doing so it jumps straight into a rich swath of Milky Way. Even so, of the eight open clusters, three possible clusters, one asterism, three planetary nebulae, and scores of galaxies and nebulae in Lacerta, only two objects are worth spending time on with a small telescope: the open cluster NGC 7209 (briefly described in the Caldwell 19 installment) and the open cluster NGC 7243 (Caldwell 16).

To the naked eye NGC 7243 is a weak magnitude-6.4 glow 2½° west-southwest of Alpha (α) Lacertae. Seeing the cluster with the unaided eye is difficult because of its proximity to a

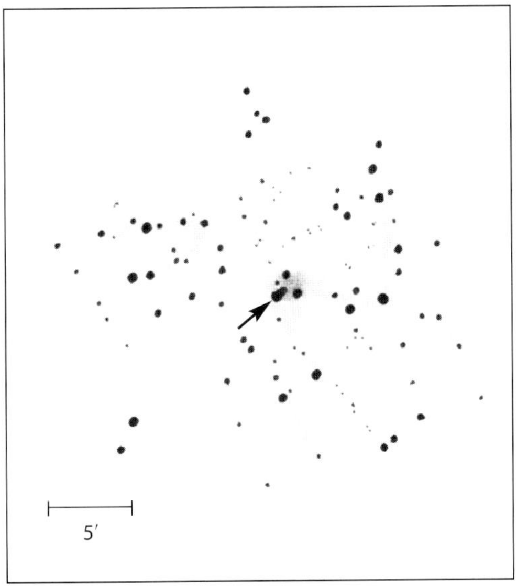

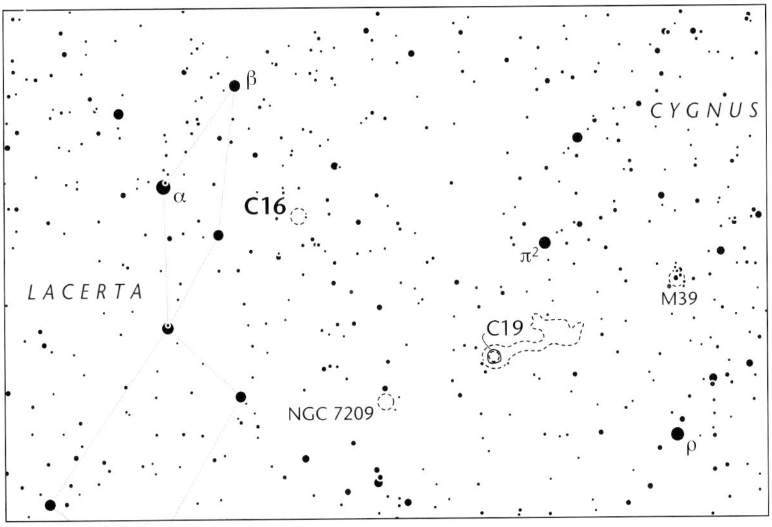

tight gathering of a half-dozen roughly 7th-magnitude stars just 30′ to the southwest. This grouping cannot be resolved by the naked eye into individual stars. Instead, it looks like an extended fuzzy glow, and its collected light appears greater than that of NGC 7243. The challenge, then, is to separate the cluster from this asterism. Lie on a blanket in late summer, when the Lizard climbs high in the evening sky, and plan to spend some time at this task under the darkest conditions possible.

In 7×35 binoculars the cluster's brightest members stand out from the starry background as a long and slender arrowhead of partially resolved stars pointing to the northeast. At 23× in the 4-inch NGC 7243 is a loose aggregation of some 40 reasonably bright suns spread across 20′ of sky, though there are fainter members that extend the cluster's dimensions another 10′. A north-south-trending dark rift winds through the cluster's midsection. The members lining the northern side of the arrow are brighter than those to the south. Most amusing is that the two halves of the cluster seem to mimic each other in shape and structure. The core of the southwestern half is V-shaped, with long looping arms of stars that look like the flapping wings of a seagull. The northeastern half of the cluster is a similarly shaped "bird." Using one's imagination, both birds appear to be flying away from the observer and into the cluster's background of unresolved suns.

The scene changes quite dramatically at 72×, when the fainter stars forming the haze start to sparkle into view. At this magnification the cluster's core appears to be centered on a tight triangle of stars at the apex of the southwestern V. The easternmost star (arrowed in my sketch is a fine double for small apertures. Known as Struve 2890 (Σ2890), it consists of a magnitude-9.3 star with a magnitude-9.7 companion 9.2″ away. Extending outward from that core is an obvious "spiral arm": a bright string of stars juts first to the west, then to the north; a weaker extension of dim suns continues to the east and back to the triangle. Paralleling that eastward-trending arm to the north is an isolated fishhook of equally dim suns. Opposite the fishhook, in turn, is an isolated triangle of faint suns in the cluster's northernmost reaches. The dark lane separating the two halves of the cluster abuts a north-south-trending string of stars that curves to the east. The cluster's northeastern half also has a triangular array of stars, with angular strings of stars emanating from it. Higher powers do not add to the view of the cluster, but do use them on the double star.

Once you've gone through this visual exercise, consider this: in a 1999 paper W. G. Braun (Bonn University, Germany) and two colleagues investigated whether NGC 7243 is a single open cluster, a double cluster, or a group of bright stars without any physical connection to one another. After analyzing two possible cluster concentrations on *National Geographic Society–Palomar Observatory Sky Survey* prints, Braun's team determined that the number of stars in the putative cluster does not exceed the expected background count. Furthermore, the researchers plotted color-magnitude diagrams for the two possible cluster regions; again, no evidence for a star cluster was found. Proper motions were calculated using catalogs dating from 1902 to 1996 along with the Hipparcos satellite database. This yielded only a hint of a small open cluster. Finally, the scientists examined the proper motions of stars within radii of 10′ and 20′ around the two putative cluster centers. No significant cluster signature emerged. The scientists conclude that if NGC 7243 is really a star cluster, it consists of only a few members. But no final determination has been made. The cluster is 76 million years young — similar in age to the Pleiades.

17

NGC 147
Type: Elliptical Galaxy (E5 pec)
Con: Cassiopeia

RA: $00^h\ 33.2^m$
Dec: +48° 30′
Mag: 9.5
Dim: 17.8′ × 11.0′
SB: 14.7
Dist: 2.3 million light-years
Disc: John Herschel, 1829

J. HERSCHEL: [Observed 8 September 1829] Very faint, very large, irregularly round, 4′ to 5′ diameter, loses itself insensibly, has a magnitude 11 star in centre. (h 29)

GC/NGC: Very faint, very large, irregularly round, gradually then suddenly much brighter in the middle to a star of magnitude 11.

18

NGC 185
Type: Elliptical Galaxy (E3 pec)
Con: Cassiopeia

RA: $00^h\ 39.0^m$
Dec: +48° 20′
Mag: 9.2
Dim: 17.0′ × 14.3′
SB: 14.7
Dist: 2.3 million light-years
Disc: William Herschel, 1787

W. HERSCHEL: [Observed 30 November 1787] Pretty bright, very large, irregularly round, very gradually much brighter in the middle, resolvable, 5′ or 6′ diameter. (H II-707)

GC/NGC: Pretty bright, very large, irregularly round, very gradually much brighter in the middle, resolvable.

ALTHOUGH THE DWARF ELLIPTICAL GALAXIES NGC 147 and NGC 185 lie within the boundaries of Cassiopeia, they are in fact companions of, and gravitationally bound to, the great Andromeda spiral galaxy M31, which looms some 7° to the south. (Here is a classic example of the arbitrary nature of constellation boundaries.) Both dwarf galaxies are essentially at the same distance as M31. NGC 147 is approaching us at a speed of 163 km per second and NGC 185 is doing so at 211 km per second. By comparison, M31 is rushing toward our solar system at 298 km per second. These velocities suggest that the two dwarfs are in a similar phase in their orbit about M31, either approaching the grand spiral from our side or heading away from it on the spiral's far side. Most astronomers would take the dwarfs'

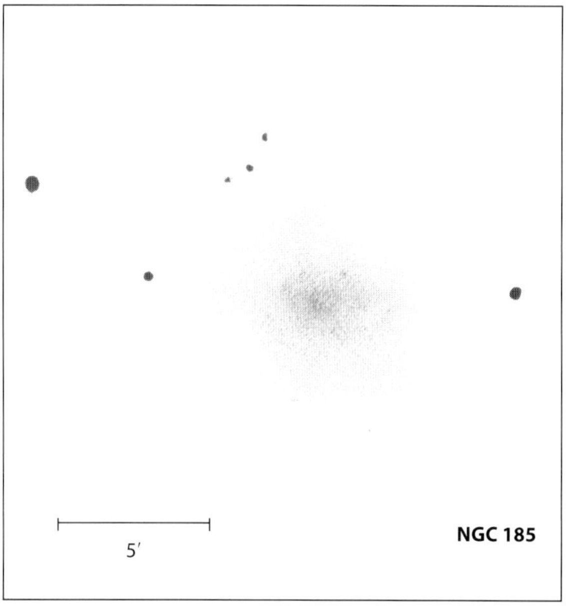

NGC 185

lower approach speeds as mere coincidence, but in a 1994 *Astrophysical Journal* paper, Halton C. Arp (Max Planck Institute for Astrophysics, Germany) announced that 22 major companions to Local Group galaxies — his entire sample — have recession velocities, or redshifts, that are positive with respect to their respective "parent" galaxies, as is the case with NGC 147, NGC 185, and M31. The chance that this can be an accidental configuration of velocities, Arp contends, is only 1 in 4 million, implying, to critics of the Big Bang, that galaxy redshifts have nothing to do with the universe's expansion.

Until I took on this project, I had never considered looking for the "other" companions to M31. Now I'm glad I did. To my surprise NGC 147 and NGC 185 both are visible in the same 7×35 binocular field (with effort) from a dark sky. The 9th-magnitude galaxies are separated by about 1°, with NGC 147 being the more difficult to see. The clarity of the night has a profound effect on NGC 147's visibility; even the slightest drop in transparency will steal the galaxy from view. Because of its slightly brighter apparent magnitude and its more concentrated light profile, NGC 185 has a greater chance of withstanding the atmospheric whims.

Locating these challenging objects is simple. Look for a north-south-oriented chain of three roughly 5th-magnitude stars 6° to the north of M31. The brightest of these stars, Omicron (ο) Cassiopeiae, marks the northern end of the chain. NGC 185 is about 1° west of Omicron, and NGC 147 is 1° farther to the west-northwest. As mentioned above, NGC 147 is the more challenging target. On one seemingly clear night I could not detect NGC 147 with binoculars, though I could see its companion quite clearly. The following night NGC 147 suddenly appeared to stand out. Light pollution will certainly kill either object.

Through the Genesis both galaxies can be viewed in the same low-power field. Two bright field stars bracket NGC 185 to the west and northeast. The galaxy is surprisingly easy to see at 23×, appearing as a slightly elliptical uniform glow that gradually brightens toward

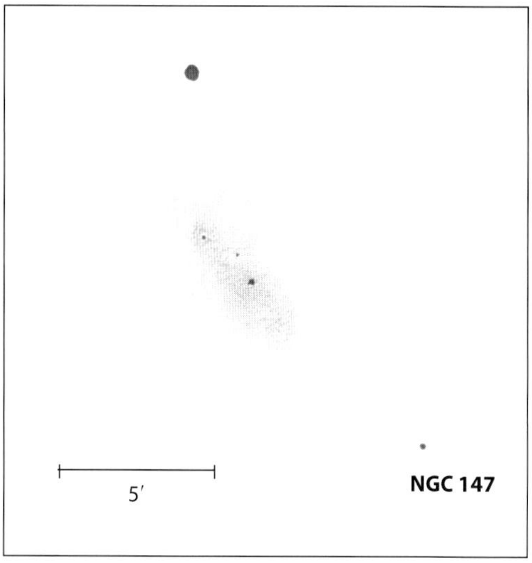

NGC 147

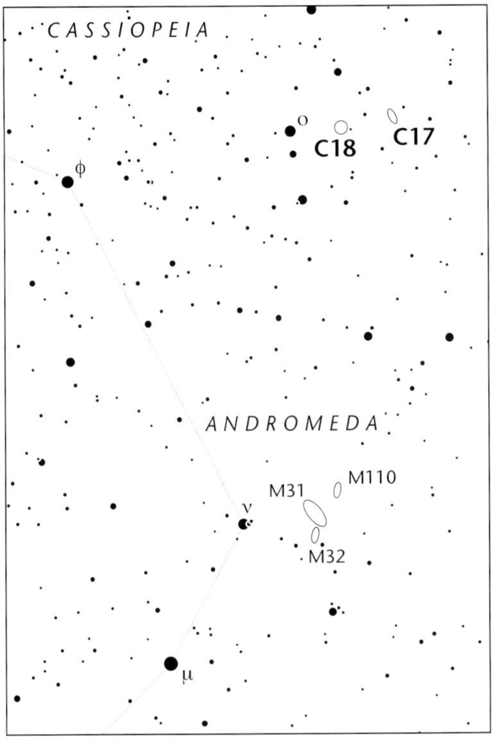

the center. A faint outer halo is also suggested. At 72× the galaxy's core looks tight, though not stellar. The 1′-wide core also seems elongated and mottled. Christian Luginbuhl and Brian Skiff found that through a 6-inch telescope the eastern edge of the galaxy's halo has a "granular" texture and with a 10-inch the galaxy "looks like an unresolved globular cluster."

NGC 147, however, is a demon at 23×. The region contains several clumps of field stars that mimic hazes, and only after careful plotting did I recognize NGC 147's true glow. (Do rely on both the chart and the photograph provided here to pinpoint it.) Dark skies are an absolute must for this dim galaxy. NGC 147 clearly looks more oval than NGC 185. Long and careful study of NGC 147 at 23× revealed hints of an elliptical center surrounded by a large, albeit faint, lens of light. Many times my eye caught the lens-shaped core of this elliptical system quite clearly; at other times, a larger, extended halo appears to round it out. The outer halo is contained within a sharp diamond-shaped pattern of bright field stars. At 72× the galaxy seems to scintillate with minute sparkles of light. Perhaps this is an illusion created by stars superposed on the galaxy. The more nights I spent observing this galaxy, the easier it became to pick out its faint glow on successive nights. This underscores how experience can make us better — and more efficient — observers.

M31 and its four brightest companions (NGC 147, NGC 185, M32, and M110) are all members of the larger Coma-Sculptor Cloud of galaxies. In a 1998 *Astronomical Journal* article Sidney van den Bergh (Dominion Astrophysical Observatory, Canada) stated that NGC 147 and NGC 185 probably form a stable binary system, orbiting M31 together like affectionate siblings. NGC 147's linear diameter is 12,000 light-years, while NGC 185 measures 11,000 light-years across; you could fit fully 10 galaxies the size of either dwarf across the length of the Great Andromeda spiral. In 1995 astronomer Paul Hodge (University of Washington) announced spectral evidence for some ongoing star formation in NGC 185. But in 1999 a trio led by David Martínez-Delgado (Astrophysical Institute of the Canaries, Spain) found that most of NGC 185's stars were formed at an early stage in the galaxy's evolution. "After that," these researchers write, "star formation proceeded at a low rate until the recent past, the age of the most recent traces of star formation activity detected in the galaxy being some 100 million years." They found no evidence for stars younger than a billion years beyond the galaxy's innermost few hundred light-years. "Since the larger concentration of stars of any age lies in the central part of a galaxy," Martínez-Delgado and his colleagues conclude, "it could be the case that the youngest stars originate from material ejected from dying stars and that this process would only be efficient enough in the center of the galaxy."

Interestingly, NGC 147 does not seem to have shared in this recent star-forming epoch. Hubble Space Telescope data have shown that star formation ceased there about 1 billion years ago. No evidence for star-forming regions could be found, nor any evidence for star-forming dust. The galaxy's youngest stars are concentrated toward the center, as is the case in NGC 185.

After scrutinizing these galaxies at the highest workable magnification, return to low power and try glimpsing both of them again in the same field of view. Seen this way, NGC 147 and NGC 185 remind me of M31's two brightest companions (M32 and M110) minus the great spiral itself. I can easily imagine NGC 185 taking the place of the round elliptical galaxy M32 and the more elongated NGC 147 representing M110. Try comparing all these systems on the same night to gain an appreciation for this remarkable collection of dwarf galaxies.

19

Cocoon Nebula
IC 5146 (nebula)
Collinder 470 (associated cluster)
Type: Emission Nebula
Con: Cygnus

RA: $21^h\ 53.4^m$
Dec: +47° 16′
Mag: 7.2, cluster
Mag: 9.3, nebula (O'Meara)
Diam: 20′, cluster
Dim: 10′ × 10′, nebula
Dist: ~3,300 light-years
Disc: Thomas E. Espin, 1899

HERSCHEL: None.

GC/NGC: None.

IC: Pretty bright, very large, irregularly faint, magnitude 9.5 star in the middle.

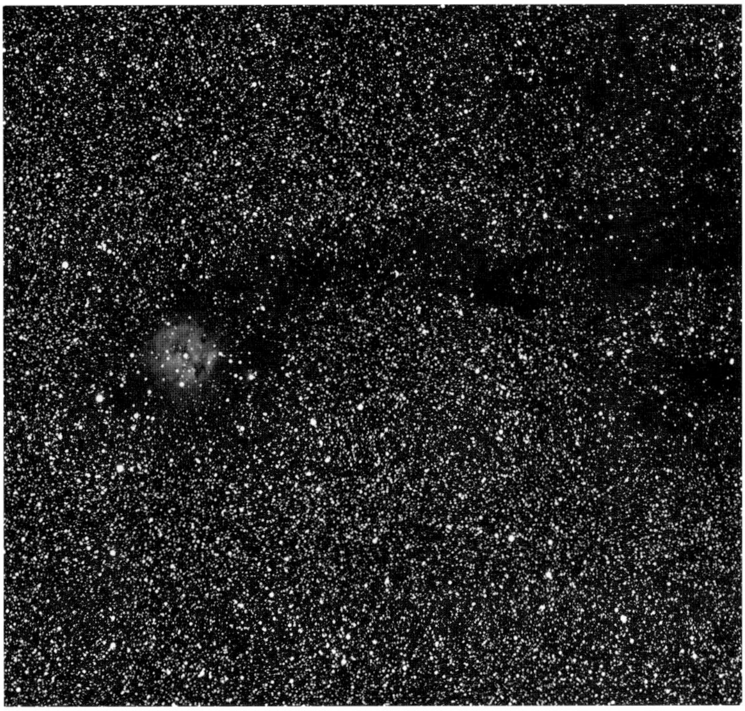

CALDWELL 19 SHOULD BE A PARADOX FOR visual observers, especially for those using small-aperture telescopes. Normally one would expect a bright nebula to be more conspicuous than a dark one, but this is not the case with the Cocoon and its neighboring field of dark nebulosity. Had I made the Caldwell Catalog, I would have mentioned IC 5146 in passing while showcasing the region's pool of dark nebulae about 1½° to the northwest. Under a dark sky the dark pool and its associated stream — which flows toward, then ponds around, IC 5146 — is one of the northern sky's most visually stunning dark nebulae, whether for the naked eye, binoculars, or low-power telescopes. (Known as Barnard 168, or B168, this dark nebula is outlined nicely in the *Millennium Star Atlas* and in *Sky Atlas 2000.0*, second edition, as well as on the finder chart on page 81.)

In 1899 the British clergyman and noted double-star observer Thomas E. Espin, whose observatory at Tow Law had 18- and 24-inch telescopes, first recorded the existence of the Cocoon Nebula on August 13th; he made a follow-up observation two nights later. In *Astronomische Nachrichten* for March 31, 1900, he refers to the nebula as a "faint glow about 8′, well seen each night." That same year Max Wolf took the first photographs of the Cocoon. In a 1904 installment of *Monthly Notices* of the Royal Astronomical Society, Wolf wrote extensively about this "Remarkable Nebula in Cygnus connected with Starless Regions." He described the Cocoon as being somewhat round and having a complicated structure resembling that of the Trifid Nebula (M20) in Sagittarius:

> It is placed centrally in a very fine lacuna, void of faint stars, which surrounds the luminous cloud like a trench. The most striking feature with regard to this object is that the star-void halo encircling the nebula forms the end of a long channel, running eastward from the western nebulous clouds and their lacunae to a length of more than two degrees.

Like William Herschel before him, Wolf believed in the possibility that the location within these stellar "voids" of the Cocoon and other extended nebulae — such as the Rho Ophiuchi and Eta Carinae nebulae — was not coincidental:

> . . . though partially or wholly surrounded by void zones, the nebulae are generally placed at the end of a longer extended lacuna, so that we are led to the impression that we here see the result of some cosmic movement, the end of the lacuna showing the place where this unknown event began.
>
> In regarding this nebula we are led to speculation. We might suppose the nebula were detached from the great western nebulous cloud, and as if it, or the cosmic process connected with its origin, had swept the long channel through the star-crowds of the Milky Way. Or is there a dark mass following the path of the nebula, absorbing the light of the fainter stars? We are far from knowing enough to settle these questions.

Wolf also noticed the existence of some brighter suns within the confines of the Cocoon. Today we know of some 110 stars as faint as 17th magnitude inside the nebula. These constitute a very young cluster, as do the stars at the heart of the great Orion Nebula (M42), and many authorities have equated IC

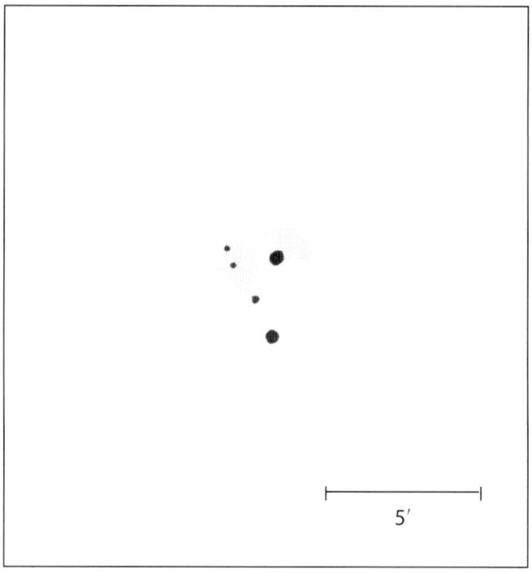

5146 with this young cluster. But Brent Archinal of the U.S. Naval Observatory notes that the Cocoon is "another case where the Caldwell Catalog has it right and everyone else has it wrong. IC 5146 (the Cocoon Nebula) is the nebula, while Collinder 470 (not IC 5146) is the correct name of the cluster. Espin and Wolf independently discovered the nebula and are credited in the *Index Catalogues*. I've looked at the discovery references and they make no report of a cluster here. Collinder was apparently the first to notice the cluster."

To track down the Cocoon, first locate the 4th-magnitude star Pi^2 (π^2) Cygni some 2½° east-northeast of the stunning open cluster M39. Now simply cast your gaze south of that star until the brightly glowing star clouds of the Cygnus Milky Way gradually fade into a vortex of dark material. The dark vortex is but a concentrated section of a larger, slightly elliptical web of dark concentric veins draping the region. The entire web measures about 2° in diameter; its westernmost edge (not plotted here) nearly abuts M39, while its eastern edge stops just shy of IC 5146. It is out of this black cobweb that a dark, narrow stream emerges to

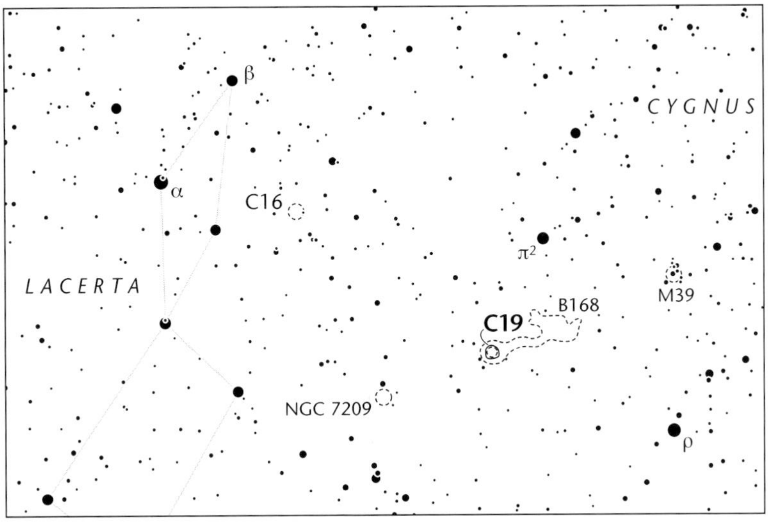

trickle into the pool surrounding the Cocoon. It's a challenge to see this 1°-long stream with the naked eye, but try sweeping your gaze up and down and left to right for a few minutes and see if you don't catch glimpses of it. (The darkest of skies are required for this exercise, of course.)

Through binoculars, a 7th-magnitude star shines at the heart of the dark web, and it appears to be the brightest star trapped there. The dark stream is stunning in binoculars, where it appears sharp and needlelike. At 23× in the 4-inch the stream meanders out of the large pool to its northwest, then straightens toward the Cocoon. Rivulets of dark nebulosity parallel it on either side. Follow the stream eastward from the large pool until it rounds out near an obvious pair of 7th-magnitude stars, then stop. Here is where you can start looking for the Cocoon in earnest.

As noted above, in long-exposure photographs the Cocoon, with its round form and dark lanes, resembles the Trifid Nebula (M20). Unfortunately large telescopes are required for an observer to see these features at the eyepiece. Through the 4-inch Genesis at 23× the Cocoon is a 9th-magnitude circular glow, easily overpowered by the pair of roughly 9.5-magnitude stars within it. Higher magnifications only diminish the view. This makes me wonder about amateurs' claims to have seen the nebula with binoculars. As most comet hunters will attest, close pairs of stars often look fuzzy when seen at low magnifications. In fact, history offers a striking parallel. In 1660 Johannes Hevelius cataloged a nebula in Ursa Major, which Charles Messier later discovered was only a pair of 9th-magnitude stars (the double star since dubbed Winnecke 4). Although he was not fooled by the double, Messier still deemed it worth including in his catalog, perhaps to alert other comet hunters to the double's deceitful nature. Today we know Winnecke 4 by its more popular designation, M40.

While on the subject of popular designations: just who named IC 5146 the Cocoon? Neither Espin nor Wolf makes specific references to a cocoon in their original discovery notes. One could infer that its origin is rooted in Wolf's theorem — namely, that the nebulosity could have been woven out of the faint stellar fabric of the Milky Way through which it moved at the time of its creation.

Before moving on to the next Caldwell object, be sure to check out the large, scattered open cluster NGC 7209 in Lacerta. (Be careful not to confuse NGC 7209 with NGC 7092, which is M39.) NGC 7209 lies 2° east-south-east of the Cocoon, shines at magnitude 6.7, and spans 25′ — nearly a full Moon diameter. Although the NGC 7209 field looks rich with stars, only about 25 of them are true cluster members, with the brightest shining between 9th and 10th magnitude, making the cluster an impressive view in a 4-inch telescope. Look for a loose and lazy zigzagging stream of stars, most of which lie near the cluster's center.

20

North America Nebula
NGC 7000
Type: Emission Nebula
Con: Cygnus

20

RA: 20^h 58.8^m
Dec: +44° 20′
Mag: 5.0; 3.8 (O'Meara)
Dim: 100′ × 60′
Dist: ~1,800 light-years
Disc: William Herschel, 1786

W. HERSCHEL: [Observed 24 October 1786] Very large diffused nebulosity, brighter in the middle. 7′ or 8′ long. 6′ broad and losing itself very gradually and imperceptibly. (H V-37)

GC / NGC: Faint, most extremely large, diffused nebulosity.

WHEN SEEN FOR THE FIRST TIME THROUGH A telescope, few deep-sky objects in the heavens live up to the poetic names pinned on them by their discoverers or by later observers. For their shapes to be seen with the mind's eye, many nebulae — like the Lagoon (M8), the Crab (M1), and the Owl (M97) — require either an impressive photograph or a lot of imagination at the eyepiece. But not the North America Nebula in Cygnus. Here is one fuzzy glow that lives up to its name even upon a casual glance through binoculars or a rich-field telescope.

Max Wolf first photographed NGC 7000 in 1890 and called it the North America Nebula for its uncanny resemblance to that continent. Look at the photograph on page 83 and you will see why. Among other "geographical" features, the nebula has a sharply defined eastern seaboard, a projecting Florida Peninsula, a dark Gulf of Mexico rimmed by Texas and Central America, and a West Coast running from Mexico to Canada. In his book *365 Starry Nights* Chet Raymo writes, "As I look at a photograph of the North America Nebula, I am reminded of the explorers who, on horseback and in birchbark canoe, mapped North America. Our own small planet has now been thoroughly explored. We have reached the heart of every dark continent. We now turn our curiosity to new 'dark continents' in space and the telescope will be our 'birchbark' canoe."

Your own adventure can start with the 1st-magnitude star Alpha (α) Cygni (Deneb), one of the most luminous blue supergiants in the night sky. Deneb marks the tail of Cygnus, the Swan, and the top of the Northern Cross. It is also the northernmost gem in the Summer Triangle, whose trio of stars hangs high overhead on late-summer evenings. The North America Nebula resides only 3° east of Deneb and 1° west of 4th-magnitude Xi (ξ) Cygni. Immediately west of the North America Nebula is the famous (though harder-to-see) Pelican Nebula (IC 5067 and IC 5070), named for its uncanny photographic resemblance to that fish-eating bird. Together these remarkable nebulae constitute the visible portions of a single large H II region; a dark dust lane running from northwest to southeast separates the landmass from the bird. Observations of this dust lane by Vytantas Straizys (Institute of Theoretical Physics and Astronomy, Lithuania) and colleagues suggest a distance of about 1,800 light-years for the entire complex. For a while astronomers believed the nebula was illuminated by Deneb, but their focus soon shifted to the hot (spectral type *O*6) 6th-magnitude star HD 199579 (20^h 56^m 34.8^s +44° 55′ 29″) in the northern quadrant of the H II region. Apparently that star emits enough energy to play a major role in the excitation of the surrounding nebula.

Some doubt has been shed on HD 199579's association with the nebula complex, for the star's distance was made out to be 3,300 light-years in 1965. But that distance is probably in error because it would make the H II

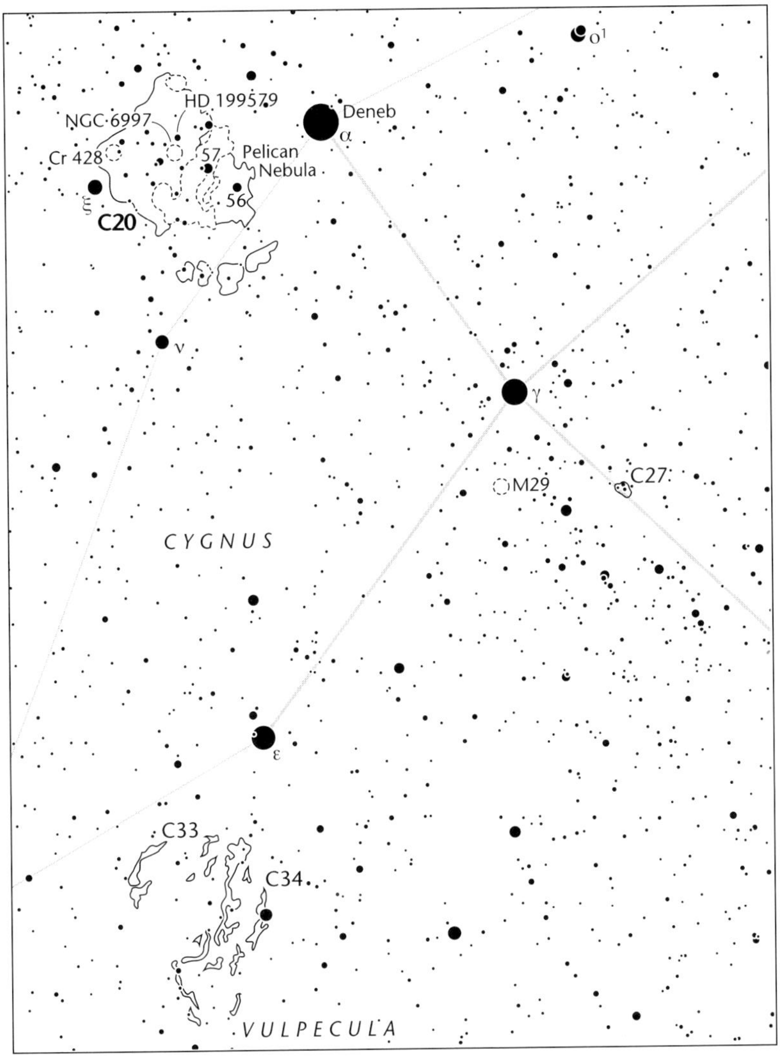

cluded that this star is a major source of illumination for both nebulae. Other ionizing sources have been found since; they appear to be pre-main-sequence stars with a wide range of masses.

To the naked eye the North America and Pelican complex stands out as an enhanced region in the Cygnus Milky Way. The North America Nebula itself spans more than 1½°. If we accept a distance of 1,800 light-years, the nebula spans nearly 50 light-years of space. Its glow is not powerful enough for us to see the entire continental outline without optical aid; the gaseous regions representing Mexico and Florida are the nebula's weakest parts by far. As I see it, the naked-eye view is more like that of a box that has been kicked by a mischievous child. But any pair of small binoculars will bring out the missing features. At 23× the 4-inch Genesis is the perfect instrument to show the North America Nebula's full grandeur, as its nearly 3° field of view fully encompasses the object with room to spare. And at 23× the nebula is a most rewarding sight. Countless suns are scattered across a pale green glow, looking like scintillating gems emerging from a sheet of melting ice.

region's size and mass extreme when compared with those of other H II regions. HD 199579 is still a contender for an energy source — but it's not the only one. In a 1980 *Astronomy and Astrophysics* report Thorsten Neckel (Max Planck Institute for Astronomy, Germany) and his colleagues announced their discovery, at near-infrared wavelengths, of an extremely reddened star at the center of the dust lane that divides the North America — Pelican Nebula complex. The scientists con-

Now let's tour the North America Nebula in some detail. Start in Central America. The bright strip of nebulosity running along the isthmus's eastern (on the sky, that is!) coast is reminiscent of the great volcanic spine that stretches the length of that region on Earth; this "mountain range" connects to the apex of a ghostly triangle representing Mexico. The central region of Mexico, in turn, is dimmer than its coasts. Likewise, the central portion of the United States is patchy, with a dim, grayish Y stretched across it. The 10th-magnitude open cluster NGC 6997 roughly marks the location of Ohio. Through the Genesis at 23× it's an elliptical gathering of about two dozen suns sprinkled across 15′ of sky; the number of stars doubles at 72×, though you have to strain to see the faintest ones. The cluster's major axis runs east-west. Look for a tail of stars that extends eastward from the cluster's eastern end before curving north. Other faint arms of stars branch out from the cluster's core to the north and south, while its western end sports two short and stubby legs. Seen in this way, the cluster looks somewhat like a human stick figure. The cluster's distance is 2,000 light-years, while its age (roughly 100 million years) excludes the possibility of an evolutionary relationship with the nebula complex.

Incidentally, confusion between NGC 6997 and NGC 6996 — supposedly also a cluster — is rife in both the amateur and the professional literature. The *New General Catalogue* lists both objects as open star clusters, and Brent Archinal has verified that the positions given therein are correct. NGC 6996 lies nearly 1° to the north of NGC 6997. So why the confusion? As Archinal explains in his 1993 monograph, *The "Non-Existent" Star Clusters of the RNGC*, the fault may lie with the 1975 *Revised New General Catalogue of Nonstellar Astronomical Objects*. This work deemed NGC 6997 a "non-existent" cluster and NGC 6996 a bona fide one. Yet a casual glance at the region through a telescope makes it clear that there is an obvious cluster at NGC 6997's position and (at best) a subtle star grouping at NGC 6996's. However, perhaps because of the *Revised New General Catalogue*'s error, *Uranometria 2000.0* put a cluster at NGC 6997's position but labeled it NGC 6996, while the *Millennium Star Atlas* plotted NGC 6996 at its correct location but omitted NGC 6997. The second edition of *Sky Atlas 2000.0* avoided the issue entirely, omitting both clusters (though the first edition correctly identifies and plots NGC

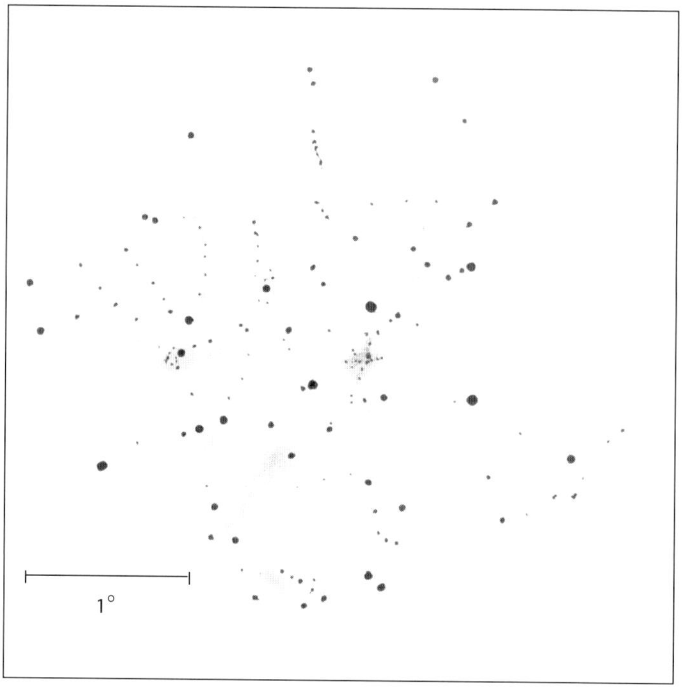

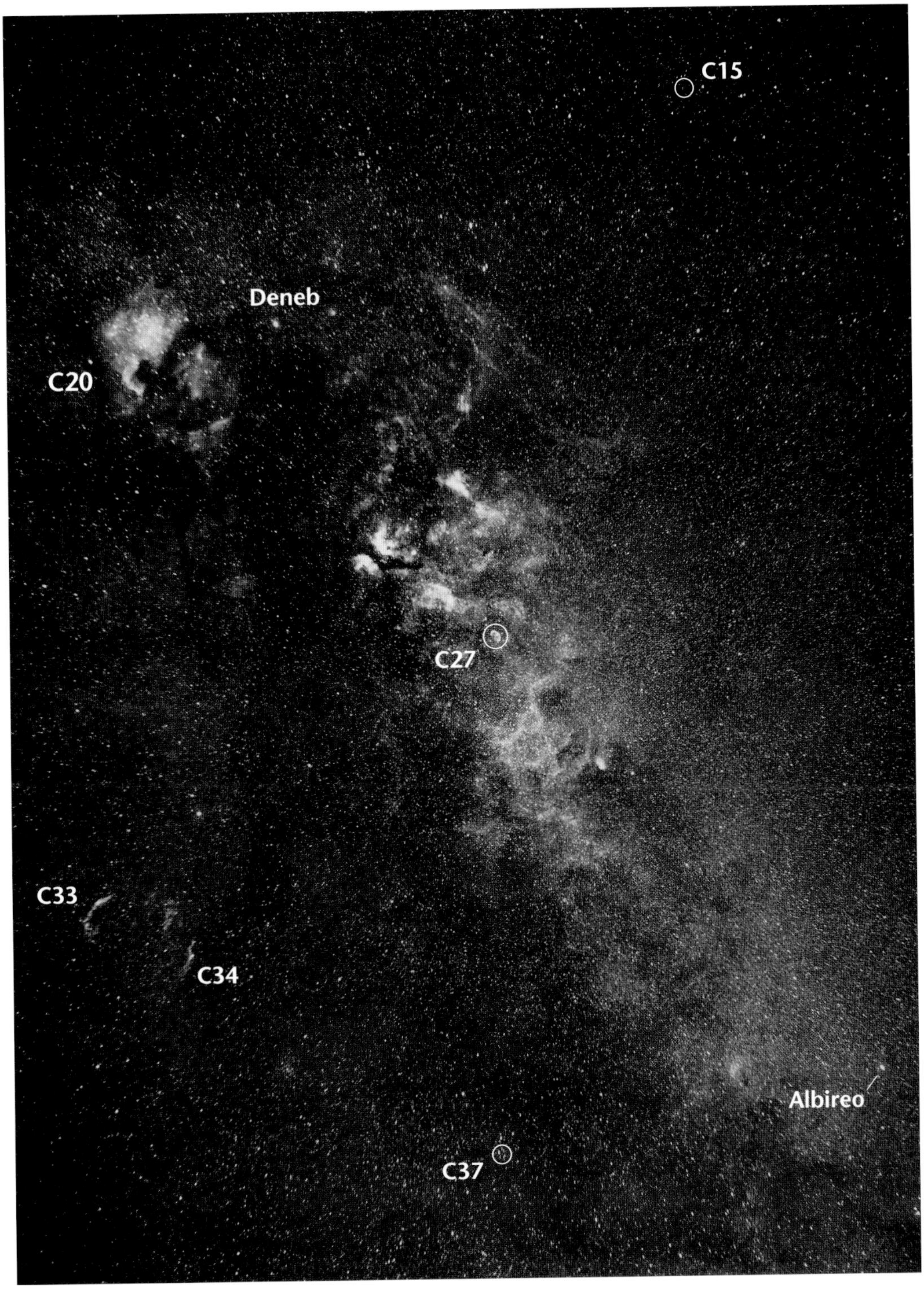
C15
Deneb
C20
C27
C33
C34
Albireo
C37

6997, as we do on page 85). The *Observing Handbook and Catalogue of Deep-Sky Objects* misidentifies NGC 6997 as NGC 6996, as does the second edition of *Sky Catalogue 2000.0*. Even the pros have gotten caught up in the confusion; a 1990 *Astrophysics and Space Sciences* paper titled "Photometry of the open cluster NGC 6996 in the North America Nebula" actually describes NGC 6997.

On the North America Nebula's "Pacific Northwest" coastline, a weak explosion of a dozen or so stars occupies a 10′-wide area at about the position of Washington State. This is the frequently overlooked 8th-magnitude cluster Collinder 428. I like to imagine its glow as the illuminated dust thrown into the atmosphere by the 1980 eruption of Mount St. Helens. The cluster's stars are arranged in fairly neat parallel lines trending northeast to southwest on the sky. There are also some perpendicular strings of stars to the east and south of the cluster's core. At 72× the pattern resembles a symbol in a Japanese newspaper. It might be fun to grab the famous "North America at Night" satellite image, which shows the wasteful glow of countless cities casting light into space, and try to match up a star in the North America Nebula with your hometown. Then, during summer star parties, you can show "your city" to guests.

If you're up for a challenge, try to trace the outline of the neighboring Pelican Nebula. If you can see the weak glow of "Florida" under your skies, you should be able to capture the Pelican. But it will take some time, because the Pelican is a faint and fairly uniform glow with numerous patches of dark nebulosity. The star 56 Cygni perches on the Pelican's back, while 57 Cygni sits at eye level on the bird's drooping beak. Raymo likes to think of the Pelican as "lost Atlantis risen from the sea."

Finally, while scanning the Milky Way surrounding the North America Nebula, I encountered a larger "North America" that I find easier to make out because of its size. This "object" shares many of NGC 7000's characteristics: eastern and western coasts, a dark Gulf of Mexico, and a Florida Peninsula. My "new" North America spans 5° on the sky. To see it, start at Deneb, which marks the position of Maine. The entire western edge of NGC 7000 serves as the southern half of the new continent's eastern (terrestrial) seaboard, with the southernmost portion of NGC 7000 representing Florida rather than Central America. The Gulf of Mexico is a large circular bay of darkness that loops around Xi (ξ) Cygni, while the star 68 Cygni, 2° to the east, marks the position of Mexico. The West Coast runs 4° to the northwest, where the star 63 Cygni lies on the border of Washington State and British Columbia. Now look about 4° farther east to the bright open cluster M39. Its glowing stars are placed just about where you would expect to find the Hawaiian Islands.

21

Box Galaxy
NGC 4449
Type: Irregular Barred Dwarf Galaxy (IBm)
Con: Canes Venatici

RA: 12^h 28.2^m
Dec: +44° 06′
Mag: 9.6
Dim: 5.4′ × 4.2′
SB: 12.8
Dist: 10 million light-years
Disc: William Herschel, 1788

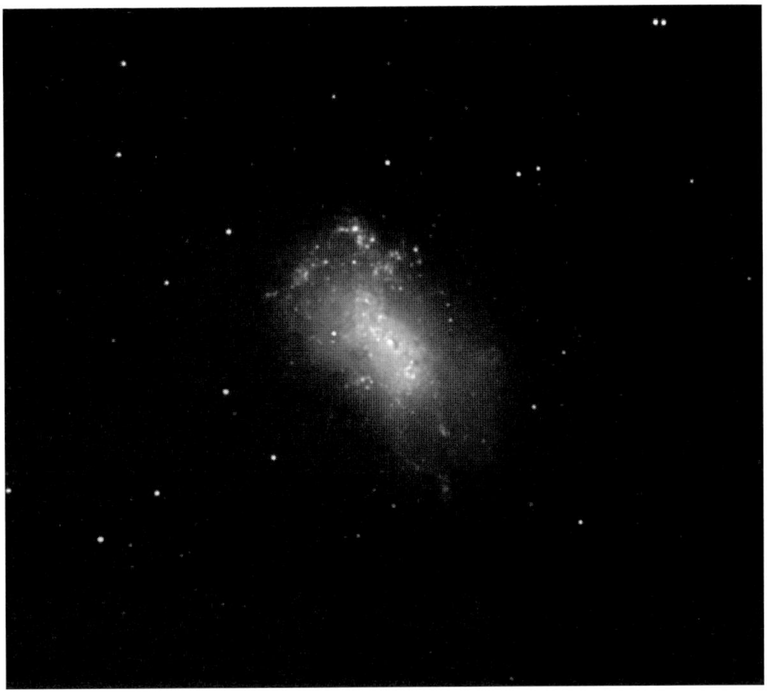

W. HERSCHEL: [Observed 27 April 1788] Very brilliant. Considerably large. Extended from south preceding to north following [from southwest to northeast]. With difficulty resolvable. Has 3 or 4 bright compressed spots. (H I-213)

GC: Very bright, considerably large, moderately extended toward position angle 15°, well resolved, star of magnitude 9 following [to the east] 5′.

NGC: Very bright, considerably large, moderately extended, double or [bifurcated?], well resolved, clearly consisting of stars, star of magnitude 9 following [eastward] 5′.

AS ONE CAN GLEAN FROM THE HERSCHEL, *GC*, and *NGC* descriptions tabulated above, NGC 4449 is one object that deceived 19th-century astronomers into believing they had resolved a "nebula" into stars. At the time galaxies as island universes were unknown to astronomers. Many of these systems belonged to what they considered the realm of "spiral nebulae," which seemed to fit a cosmological model introduced by the French mathematician Pierre Simon Laplace. Laplace's model postulated that our solar system condensed out of a nebulous vortex, and the spirals, many presumed, were solar systems being born. Indeed, not until 1923, when the outer regions of M31 were first resolved with the 100-inch reflector on California's Mount Wilson, was the true nature of the mysterious spirals revealed. In fact, until that time M31 was commonly referred to as the Great Spiral *Nebula* in Andromeda.

Unfortunately, no 19th-century telescope was powerful enough to reveal individual stars in a distant galaxy like NGC 4449. Looking at a photograph of this irregular system, it's easy to understand how the astronomers of that era were deceived; the galaxy looks like a clouded hourglass wrapped in knotty stellar filigree. One can only imagine the excitement these astronomers must have felt when they imagined the galaxy resolved into stars. Of course, the knowledge of the 19th-century observer was, in the words of French author Georges Simenon, "a little island in a great ocean of nonknowledge."

Today we know that NGC 4449 is an active star-forming dwarf galaxy seen at a 43° angle from face on. It has a bar running through it, much like the one we see in the Large Magellanic Cloud. The galaxy is surrounded by an enigmatic halo of diffuse hydrogen gas some 10 times larger than the galaxy's visible disk. At its estimated distance of 10 million light-years NGC 4449 has a modest linear diameter of 16,000 light-years (not including the invisible hydrogen halo) and an equally modest luminosity of 1.8 billion Suns. The galaxy is receding from us at a speed of 200 km per second. NGC 4449 has many H II regions concentrated primarily at its northern end; in photographs they appear as knots. Seeing these features helps us understand why William Herschel believed that, "with difficulty" he resolved this object, seeing specifically "3 or 4 bright compressed spots."

The Hubble Space Telescope has resolved hundreds of extended objects centered on the galaxy; they may be populous star clusters similar to those in the Large Magellanic Cloud. One bright central object may be a super star cluster only 6 to 10 million years old. Hubble also resolved intense regions of starburst activity. The galaxy's high star-formation rate is believed to have been caused by a merger or interaction with a companion galaxy. Radio observations have shown NGC 4449's hydrogen halo to have a disklike feature and a lopsided arm structure. In a series of computer simulations, Christian Theis and Sven Kohle (University of Kiel, Germany) have demonstrated that these features were created by a gravitational interaction between NGC 4449 and another dwarf galaxy called DDO 125. The

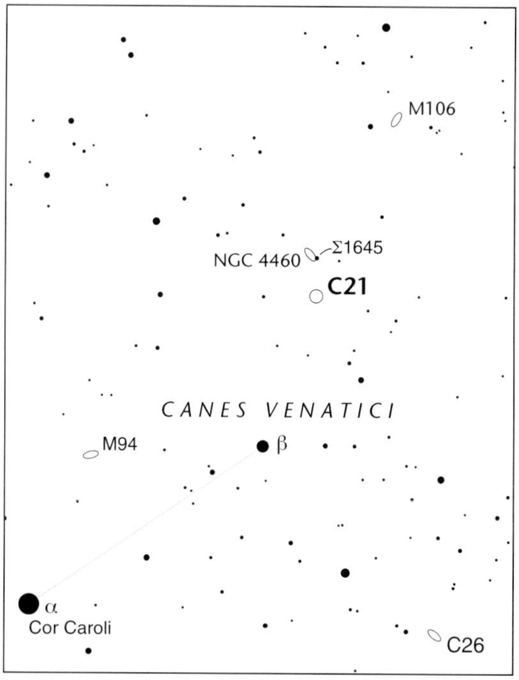

two galaxies have followed nearly parabolic orbits around one another, write the scientists, and were nearest to one another about 4 to 6 billion years ago, when they may have gotten as close as 80,000 light-years — nearly the diameter of our Milky Way. The origin of the hydrogen halo remains a mystery; it may have been drawn out from NGC 4449 by an earlier interaction.

NGC 4449 can be found just 3° north-northwest of 4th-magnitude Beta (β) Canum Venaticorum, midway between two 8th-magnitude stars (one roughly 1° to the east and the other a similar distance to the west). The galaxy shines with the light of an unfocused 9.6-magnitude star and has a high surface brightness, making it an easy find. In fact, NGC 4449 is visible in 7×35 binoculars as a tiny diffuse glow, though dark skies, patience, and averted vision are required to see it. The view through the 4-inch at 23× is quite remarkable. The galaxy appears as a bright 5′-long oval adorned with a brightly dappled center. A dif-

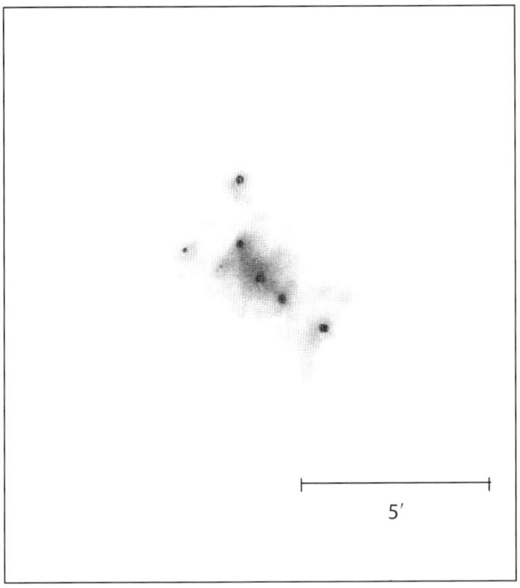

fuse but obvious glow of patchy light surrounds the galaxy's central region. At times this outer envelope seems to be bristling with whisker-like structures, though I believe this is an illusion caused by the alignment of various patches. The dappled texture is most strongly enhanced along the galaxy's major axis.

With a modest increase in power (72×), clumps of nebulous matter pop into view, branding the entire galaxy with a distinct X-shaped pattern. NGC 4449 is at its best at 189×. Several starlike spots line the galaxy's major axis, and a detached segment seems to waver in and out of view to the north. The galaxy's southern end has a roughly triangular shape whose apex points toward the rather robust core. After looking at the galaxy through her 13.1-inch reflector, Barbara Wilson called NGC 4449 the Box Galaxy because of its unusual squarish shape. She, too, noted its mottled interior.

Without question, I would rate NGC 4449 as a superb object for small-telescope users who would enjoy seeing details in a distant galaxy. Thus I was surprised to read that Christian Luginbuhl and Brian Skiff required a 10-inch telescope to bring out its central spots. Similarly, Roger Clark describes a view he had through an 8-inch Cassegrain reflector — under the dark skies of Hawaii, no less — as nothing more than an elongated glow with a faint extension to the north. Can a small telescope reveal the knots? I believe the brightest might show up in a high-quality 60-mm refractor.

While in the area, be sure to challenge yourself by looking for NGC 4460, an 11th-magnitude galaxy about 40′ north of NGC 4449. You may have to use moderate power to seek it out because the galaxy's dim elliptical form all but hides behind the glow of the nearby double star Struve 1645 (Σ1645), or ADS 8561, whose components shine at magnitudes 7.4 and 8.0, respectively, and are separated by 10″. Once you locate NGC 4460 using moderate power, try to detect it with low magnification. Doing so takes practice.

22

22

Light Blue Snowball
NGC 7662
Type: Planetary Nebula
Con: Andromeda

RA: $23^h\ 25^m\ 53.9^s$
Dec: +42° 32′ 06″
Mag: 8.3
Dim: 32″ × 28″
Dist: 3,200 light-years
Disc: William Herschel, 1784

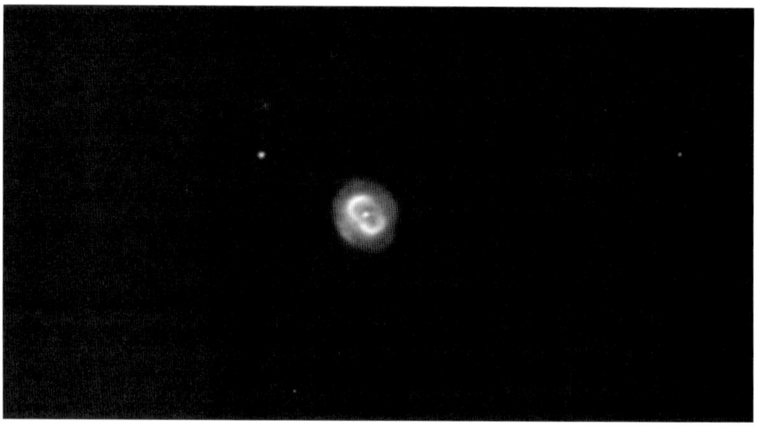

W. HERSCHEL: [Observed 6 October 1784] Bright, round, a planetary [with a] pretty well defined disk. 15″ diameter with a 7 feet reflector. (H IV-18)

GC: Planetary, very bright, pretty small in angular size, round, blue.

NGC: A magnificent planetary or annular nebula, very bright, pretty small in angular size, round, blue, variable nucleus.

NGC 7662 IS A BRIGHT DOUBLE-SHELLED planetary nebula with a mysterious central star. The nebula's core is so condensed that city observers with 7×35 binoculars have spied it as a "star" shining just 25′ southwest of 5th-magnitude 13 Andromedae, which lies about 15° north of the Great Square of Pegasus. A magnitude-7.9 star lies just to the planetary's east, and the two form an obvious pair. Through the Genesis NGC 7662 remains stellar at 23×, but with the magnification increased to 72× a pale aquamarine nebula suddenly swells to prominence, displaying a sharp core, a dense inner ring, and a diffuse outer envelope. At 189× the sharp core reveals itself not as a central star but as two bright arcs of nebulosity, one to the northeast, the other to the southwest. These surround a small cavity of dim light marking the center of the annulus. A diffuse aquamarine shell surrounds this bright inner ring and is elongated in the same direction. "This planetary was observed 16 times by William and John Herschel," notes Barbara Wilson. "They must have loved looking at it!" In the *General Catalogue* there is the following note: "According to Mr. [William] Lassell this superb 'planetary nebula' is bi-annular, consisting of a nucleus and *two oval rings*." Note, too, that John Herschel was the first of many observers to notice a blue tint to NGC 7662.

NGC 7662 has long been known as the Blue Snowball Nebula. Veteran *Sky & Telescope* editor Dennis di Cicco has credited Leland Copeland with originating this nickname. "Looking like a light blue snowball" is how Copeland described NGC 7662 in the magazine's February 1960 issue. "The key here is that the words 'blue snowball' don't have quotation marks and the first letter in each word is not capitalized," di Cicco stresses; "Copeland was just describing the object in his own words." Interestingly, Copeland's important qualifier — the word "light" — has subsequently been dropped in the decades that have passed since he drew attention to the nebula's curious color. Without the qualifier, one expects to see a remarkably blue color that really isn't there. In my notes I describe NGC 7662 as having a "*slightly* bluish glow"; Christian Luginbuhl and Brian Skiff say it has a "*faint* bluish light"; and, in his *Astronomical Objects for Southern Telescopes*, Ernst J. Hartung calls it a "vivid *pale* blue disk" (all emphases mine). But how much of this is suggestion?

A planetary nebula shines when its diffuse gases fluoresce after being energized by strong ultraviolet radiation from its hot central star. That star's extremely high surface temperature can be anywhere from 30,000 to several hundred thousand degrees Kelvin. (NGC 7662's central star has a surface temperature of 75,000° Kelvin.) The visible-light spectrum of a planetary nebula is dominated by two spectral lines of doubly ionized oxygen, whose wavelengths are 4959 and 5007 angstroms, respectively; these lines are in a portion of the visible spectrum that some refer to as "green" and others "blue-green." Thus we should not be surprised, perhaps, that astronomer David A. Allen used the words "rich green" to describe his view of NGC 7662 through the 60-inch reflector atop Mount Wilson — the numerous "blues" reported in the previous paragraph notwithstanding. Your eye, the nebula's altitude, the clarity of the air, and your telescope's aperture, magnification, and design — all these factors and more may affect the color you see when looking at a planetary nebula. Some would argue that no star, let alone a deep-sky object, should show any color, since the night-sensitive rods are colorblind. But it is conceivable that bright, condensed objects like NGC 7662 do emit enough light to stimulate the eye's color-sensitive cone cells.

NGC 7662 is one of the many planetaries imaged by the Hubble Space Telescope, whose repaired optics are expanding our envelope of knowledge about these enigmatic objects. Bruce Balick (University of Washington) and his colleagues recently combined HST images and ground-based spectra of NGC 7662 to study the nebula's internal motions. They found that the northeastern region of the planetary's bright interior ellipse is blueshifted while the southwestern part is redshifted, and they concluded that a good three-dimensional

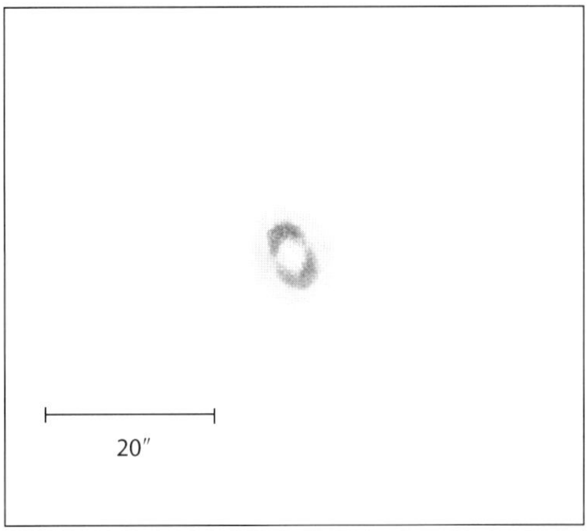

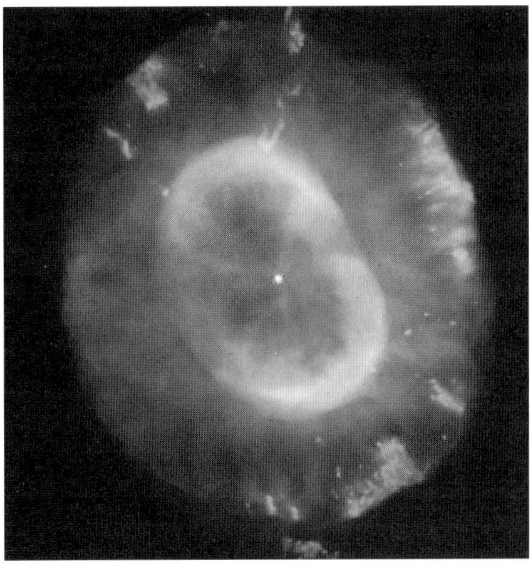

model for the ellipse is an expanding prolate ellipsoid (a football with rounded tips). The football is tilted by about 45° to our line of sight, with the northeastern tip approaching us and the southwestern end receding from us. The outermost regions of the surrounding halo mimic the shape of the football in the planetary's center.

One unsolved planetary-nebula mystery involves fast-moving knots of gas called FLIERs (*f*ast *l*ow-*i*onization *e*mission *r*egions). FLIERs can be identified in the outer regions

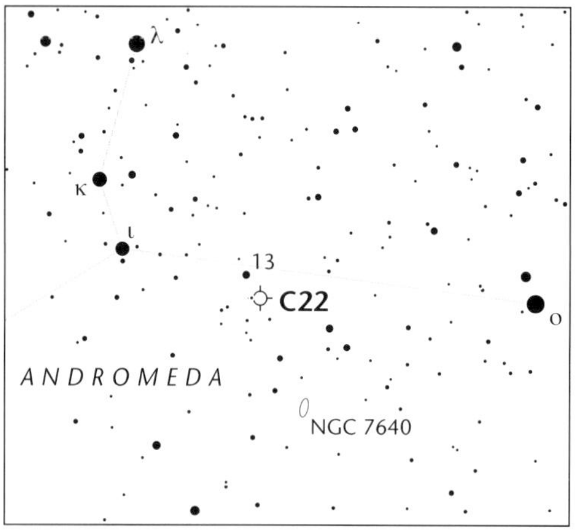

of several planetaries, where they appear as small knots, generally in pairs on opposite sides of its central star and often along the nebula's symmetry axis. In NGC 7662's case the FLIERs appear to be much younger than their parent nebula. On the basis of their anomalous chemical abundances Balick speculates that the FLIERs in planetary nebulae were probably dredged up from highly processed material deep inside the nebula's precursor star, then flung outward like small spitballs — an astonishing feat, he says, for a nearly solid star about to become a crystalline white dwarf. (See my Caldwell 15 essay for more on FLIERs.)

NGC 7662's central star is enigmatic in another way as well; it long has been suspected of being variable. On 80 different dates between 1897 and 1908, Edward Emerson Barnard found this star varying between 12th and 16th magnitude. The periods of greatest brightness, he says, were not long lasting, and they occurred at irregular intervals. This variability is controversial among astronomers, however, and Barnard himself harbored doubts. "Familiarity with the changing aspect of a star involved in nebulosity, due to moonlight, bad seeing, etc., has made me extremely cautious in the matter." Thus William Sheehan quotes the famous astronomer in his biography *The Immortal Fire Within: The Life and Work of Edward Emerson Barnard.* Photographically based magnitude estimates made since 1934 have shown the Light Blue Snowball's central star varying from magnitude 11.5 to 13.0, with 12.5 being the average. Some argue that the apparent variation is merely the result of atmospheric instabilities, a possibility Barnard himself hinted at. In his *Celestial Handbook* Robert Burnham Jr. suggested that amateurs with fairly large telescopes try to solve the controversy. But I am not aware of any attempts to do so. Therefore I would like to repeat Burnham's plea and remind amateurs, especially those with CCDs, that monitoring the brightness of the Light Blue Snowball's central star would make a great autumn or summer project, the results of which would be of value to the astronomical community.

Having said that, I do not know of any amateur who has actually seen NGC 7662's central star with his or her own eye at the telescope. Certainly a 13th-magnitude star should be discernible in the large instruments that adorn the fields at today's star parties. The problem, as I see it, seems to be that the hole at the center of NGC 7662's "doughnut" is glazed over with a sheen of light, especially at low magnifications. The key, then, must be power. We need to push the magnifications of our telescopes to the limits in order to penetrate this inner sanctum. How much power is needed? From Houston, Texas, Barbara Wilson viewed NGC 7662 at 356× through a 13.1-inch f/4.5 reflector yet failed to see the star. Roger Clark suggests an astoundingly high power of 720× for telescopes 2 to 8 inches in diameter. "This object provides a good example of how the upper limit to useful magnification depends on image brightness," Clark says of NGC 7662. "The lesson is that if you are observing an object and feel that higher magnification may bring out more detail, go ahead and ignore the 'accepted' magnification limit of 60× per inch of objective diameter."

Without question, the central star is there in photographs and CCD images of the Light Blue Snowball. Furthermore, while Hartung could not directly see the star with his 12-inch reflector, he did perceive its presence as a continuous streak when he inserted a prism into the telescope's light path, spreading the star's light (and the nebula's) into a spectrum. Whoever does see the star will be rewarded by a rare sight: that of a hot, blue dwarf that has driven its own atmosphere away. As seen today from Earth, the nebula's shells are expanding outward at rates measured in tens of km per second; the bright inner shell already spans three-tenths of a light-year and its faint outer envelope extends some 2 light-years. Had its precursor been our Sun, the outer envelope of NGC 7662 would stretch halfway to the nearest neighboring star.

Before you leave this fabulous object, I'd like you to search for an asterism I chanced upon while seeking NGC 7662. It is a nearly exact replica of M73, a Y-shaped grouping of four 10th-magnitude stars in Aquarius. The four close-knit stars in the Andromeda Y appear in the *Millennium Star Atlas*. The 3′-long group lies 2° south-southwest of 13 Andromedae and just 30′ east of the 11th-magnitude galaxy NGC 7640. The Y is oriented northeast to southwest.

23

23

Outer Limits Galaxy
NGC 891
Type: Spiral Galaxy (Sb)
Con: Andromeda

RA: 02^h 22.6^m
Dec: +42° 21′
Mag: 9.9
Dim: 12.2′ × 3.0′
SB: 13.7
Dist: 31 million light-years
Disc: Caroline Herschel, 1784

W. HERSCHEL: [Observed 6 October 1784] Considerably bright, much extended, above 15′ long. 3′ broad. A black division 3′ or 4′ long in the middle. (H V-19)

GC/NGC: Remarkable, bright, very large, very much extended toward position angle 22°.

IN LONG EXPOSURES MADE WITH GREAT TELEScopes, the edge-on spiral galaxy NGC 891 looks like a flying saucer pitching sharply through a city of suns. The galaxy's supernatural appearance was naturally befitting for the television series *The Outer Limits*, which fathered the vast science-fiction genre we enjoy today. At the close of the show's 1963 premiere episode, "The Galaxy Being," images of galaxies (including NGC 891, which impressed me most of all) ebbed and flowed with the pulsing rhythm of the show's haunting music. The episode's plot — about an inventor who uses radio equipment to turn "three dimensional static" into a pulsing being from a distant galaxy — captured the imaginations of impressionable youngsters. More generally, the show strove to give us an "altruistic look at the awesome and mysterious universe." To this day I wonder how many people of my generation were turned on to astronomy by that TV series. That's why I have nicknamed NGC 891 the Outer Limits Galaxy.

Its mystique notwithstanding, NGC 891 has been well studied since Caroline Herschel discovered it more than 200 years ago. It is tilted a mere 6° from edge on. R. Brent Tully (University of Hawaii) places the galaxy 31 million light-years from Earth in the Triangulum Spur of galaxies, whose 40 members include NGC

1003, NGC 1023, and NGC 1058 (a nice line of systems in Perseus, about 4° southeast of NGC 891). NGC 891 is receding from us at a speed of 529 km per second, which is about 100 km per second slower than NGC 1003 and NGC 1023 but about the same as NGC 1058.

If the distance of 31 million light-years is accepted, then NGC 891's true diameter is quite large, measuring 110,000 light-years, and its mass totals some 150 billion Suns. If we could somehow tilt the galaxy or look down onto its hub, we'd probably discover a structure similar to that of the textbook spiral M81 in Ursa Major (though recent infrared observations suggest that NGC 891 may be a barred spiral). In size and mass NGC 891 is comparable to our own Milky Way. Its rotational velocities — the orbital speeds of stars, gas, and dust around the galaxy's core — are also very similar to those in our galaxy. Radio observations from the early 1990s show a steep rise in rotation velocity as one goes outward radially from NGC 891's nucleus while remaining within the disk. A maximum speed of more than 250 km per second is then followed by a dip roughly 6,500 light-years from the hub. The rotation curve becomes almost flat at about 10,000 light-years and remains so out to the edge of the galaxy's visible disk. As with many spiral galaxies, including our own, the rotation speeds seem far greater than one would expect if NGC 891 contained no other material besides its visible stars, gas, and dust. The implication is that NGC 891, like those other spirals, is embedded within a halo of mysterious "dark matter" that exerts gravitational forces but emits no measurable electromagnetic radiation.

In 1961 Edwin Hubble determined that NGC 891's nuclear bulge is smaller than that in NGC 4565 (Caldwell 38), a similar and arguably more famous edge-on system in Coma Berenices. Hubble also noted that NGC 891's dust lane, like that of NGC 4565, has an intricate pattern running perpendicular to the galaxy's midplane. Thin fingers of darkness are seen in silhouette against the galaxy's brightest parts; they extend at least 30″ (4,500 light-years) above and below the galaxy's plane. Energy must be pumped into the dust to keep it this far from the midplane despite gravity's tendency to pull it back. One energizing mechanism might be supernova explosions. These violent ends to massive stars' lives can blow holes through a galaxy's disk and send streamers of dust and gas out into its halo. But it seems unlikely that supernovae alone could generate enough energy to explain NGC 891's many dark, dusty fingers. Other phenomena, such as gravitational and magnetic interactions between interstellar clouds, must also

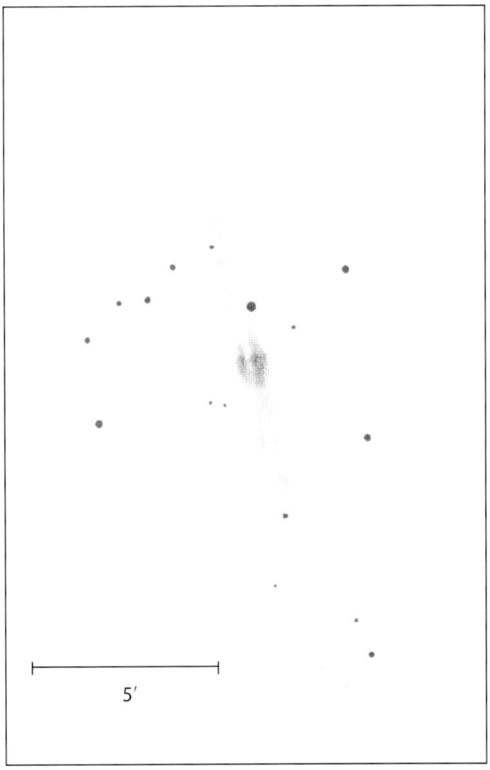

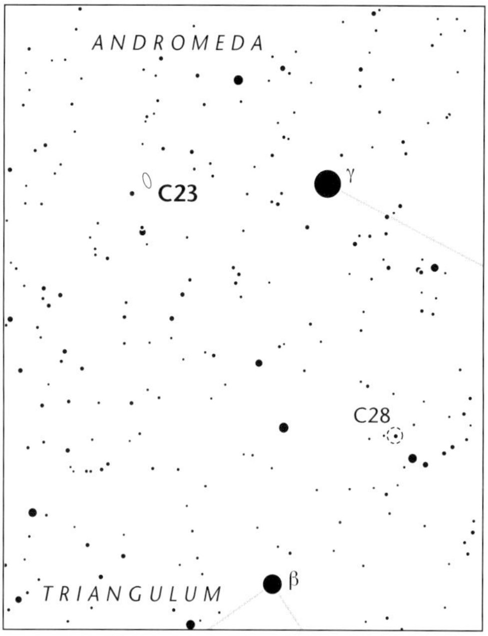

play a role. Eventually, though, gravity will in time pull the dust in NGC 891's "geysers" back into the galaxy's plane, just as volcanic ash falls back to earth after an explosive eruption.

As stated above, NGC 891 seems to contain copious quantities of mystery matter. Then again, maybe not. Using the European Space Agency's *Infrared Space Observatory*, two Dutch astronomers, Edwin A. Valentijn and Paul P. van der Werf, recently detected huge amounts of relatively warm molecular hydrogen gas in NGC 891. In the September 1, 1999, *Astrophysical Journal Letters* the scientists claim that the amount of gas they found — 5 to 15 times as much molecular hydrogen as atomic hydrogen — "matches well the mass required to resolve the problem of the missing matter." The molecular gas they detected resides in the galaxy's flat disk, not in a puffy halo, as most dark-matter theories postulate. But the scientists think this is a surmountable problem. If so, a prosaic substance — mere hydrogen — may be keeping the spinning spiral, and others like it, from flying apart.

NGC 891's visual magic is fully revealed only in photographs and CCD images. The magnitude-9.9 galaxy is a challenge for eyeball observers with small instruments because its dark equatorial clouds hide so much of its light. Interestingly, in 1940 Carl Seyfert demonstrated that NGC 891's dark lane is somewhat of an illusion. The lane's surface brightness is a modest 0.6 to 0.9 magnitude fainter than that of the galaxy's surrounding parts, so it only appears dark, just as hot sunspots appear dark against the even hotter face of the Sun. Regardless, NGC 891's 12′-long glow has such a middling overall surface brightness that dark skies are required to see it well. The field is easily located 3½° east of the fine double star Gamma^{1,2} ($\gamma^{1,2}$) Andromedae, the constellation's easternmost bright star. The galaxy is oriented north-northeast to south-southwest, its elongated form almost paralleling a chain of five stars (ranging from magnitude 6 to 8) about ½° to the southeast.

Christian Luginbuhl and Brian Skiff report seeing NGC 891 with difficulty in a 2½-inch telescope, saying it is "visible as a tiny streak of low surface brightness." With peripheral vision at 23× in the 4-inch, the galaxy is a ghostly sliver of light; it all but vanishes with a direct look. Once you locate NGC 891, try slowly rocking your telescope back and forth across the field and see if you don't catch glimpses of faint light extending from what appears to be a sharp central bulge. An increase in magnification (72× in my case) should then show that the sharp bulge is actually a magnitude-12.5 star just north of the true hub, itself a dim mottled patch of light. (The hub looks much brighter than that star in most photographs because the region is overexposed). I perceived the galaxy in its entirety more clearly at 72× than at 23×. Its dim, tapered body is peppered with several faint suns (stars belonging to our own Milky

Way) that pop in and out of view as the eye tries to trace the galaxy from end to end. But it's hard to separate the faint stars from the patches of light that mottle the galaxy's anemic extensions.

High power helps to separate the field stars from the galaxy. By dimming the galaxy to the point where it is almost invisible, you can enhance the stars superposed on it, which allows you to record their positions with confidence. Once that's done you can return to lower powers and record the position of the galaxy itself. The oval hub's dim glow has two condensations, one on either side of the dark lane; the western condensation seems the more conspicuous of the two, and the southwestern extension of the galaxy's disk is brighter than the one to the northeast. The galaxy's southwestern tip ends with a magnitude-13 star, beyond which a trail of faint suns streams off farther to the southwest. On the northeastern side a star of roughly magnitude 13.5 punctuates the eastern edge of the galaxy's midsection. Amazingly, not much detail is added with larger amateur instruments. Viewing through a 20-inch f/4 reflector at 222× from Columbus, Texas, Barbara Wilson could, however, clearly make out the galaxy's dark lane, the two bright condensations at its center, and five stars superposed on the galaxy.

In his *Celestial Handbook*, Robert Burnham Jr. notes that wide-angle photographs of the Milky Way strikingly resemble high-magnification ones of NGC 891. But it's almost impossible to accept the notion that this visually unassuming galaxy — this veritable wall of darkness centered on a dim and distant core of light — is remarkably similar in shape and size to our own home in the universe. How seemingly insignificant NGC 891 appears through our backyard telescopes. Consider, then, how remote and unimportant our Milky Way galaxy would appear through a telescope to an imaginary inhabitant of a world within NGC 891. In a sense our understanding of NGC 891's true nature should make us ponder afresh the possibility of extraterrestrial life. Like the "Galaxy Being" in *The Outer Limits*, NGC 891 seems to be sending us a silent message imploring us to "explore . . . reach out . . . [and] give thought to the mysteries of the universe."

24

Perseus A
NGC 1275
Type: Peculiar Galaxy
Con: Perseus

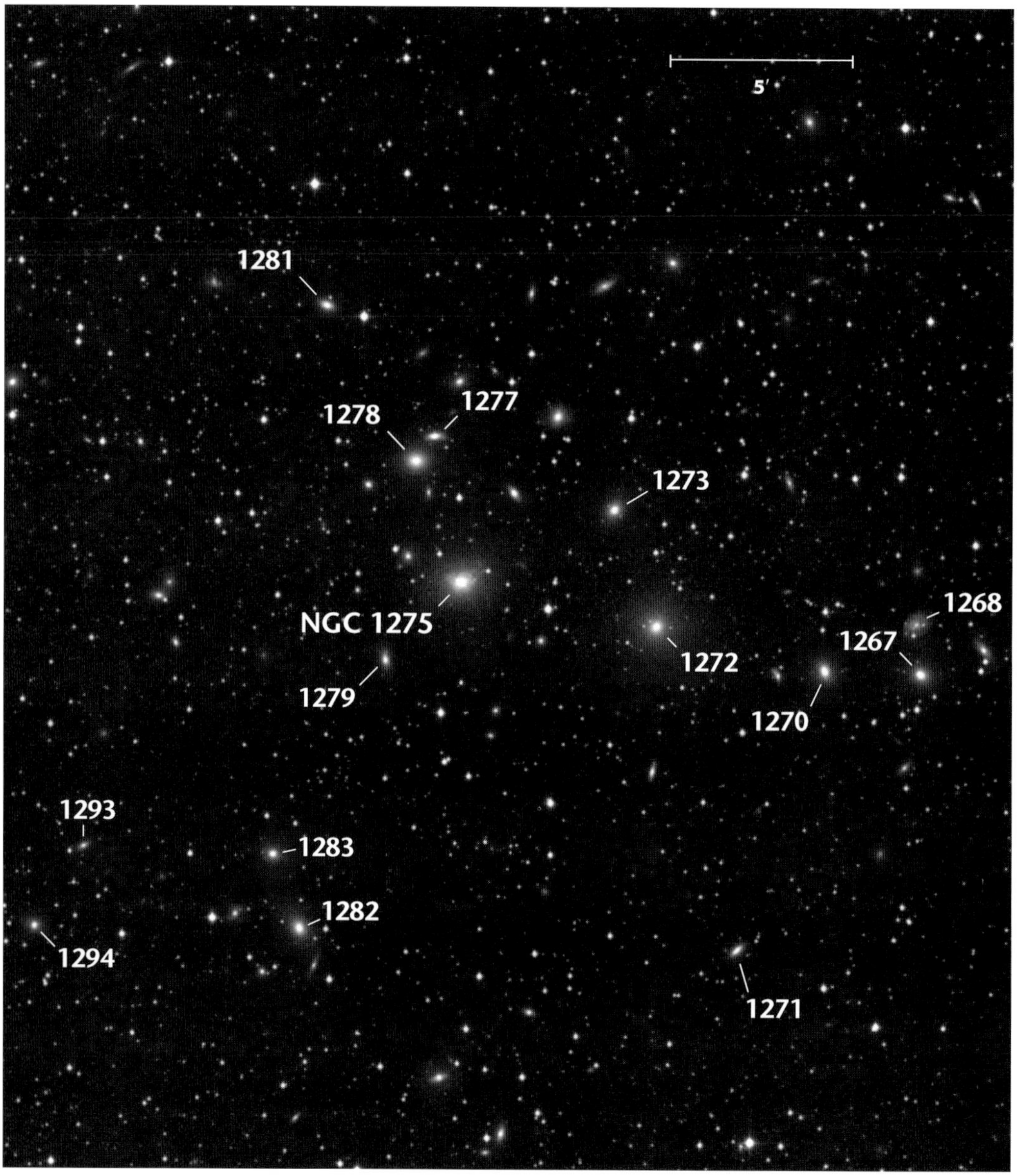

24

RA: $03^h\ 19^m\ 48^s$
Dec: +41° 30.7′
Mag: 11.9
Dim: 3.2′ × 2.3′
SB: 13.9
Dist: 230 million light-years

Disc: Heinrich Louis d'Arrest, mentioned in the 1867 *Siderum Nebulosorum Observationes Havnienses*

HERSCHEL: None.

GC: None.

NGC: Faint, small in angular size.

NGC 1275 IS THE BRIGHTEST AND MOST CENtral member of the Perseus Cluster of galaxies, a dense nest of some 12,000 systems roughly 230 million light-years distant — four times farther from us than the far better known and easier-to-observe Virgo Cluster. To see the Perseus Cluster, which also is known as Abell 426, we must peer through a thicket of stars and dust in the Milky Way's outer reaches. By penetrating this thicket with radio telescopes, astronomers have discovered that NGC 1275 and its family are cutting through a fog of intergalactic material; this is evidenced by long "wakes" of plasma trailing several of the cluster's galaxies. The intergalactic fog was probably ejected over the eons by the active nuclei in the cluster's feistier systems, or torn free from spirals passing near the cluster's dense core.

As with many spherical galaxy clusters rich in energetic ellipticals, the Perseus Cluster is also a strong emitter of X-rays. Four-fifths of the cluster's X-ray energy emanates from gases spread throughout the 3-million-light-year-wide cluster; the remainder comes from NGC 1275 itself.

NGC 1275 long has been known as a strong source of radio emission. It is listed in the *Third Cambridge Catalogue* of radio sources as 3C 84; early radio astronomers referred to it simply as Perseus A. In addition, Carl K. Seyfert included it in his 1943 list of galaxies that display exceptionally bright or starlike nuclei. In photographs taken with the powerful 200-inch telescope on Palomar Mountain, NGC 1275's nucleus appears very tight, almost stellar.

The galaxy's overall white-light appearance is that of a lenticular or elliptical system. A black "scratch" can be seen zigzagging across the galaxy's northern halo, and deep red-light images show a system of twisted filaments radiating in all directions. In a humorous way, these filaments suggest a child's drawing of an explosion in a spaghetti factory. The galaxy's filamentary "noodles" are fleeing into space with velocities on the order of 2,400 km per second. The enigmatic "scratch" has two parts: a bright streak that juts to the northwest before cutting sharply to the north, and a disjointed arc of matter that hugs the galaxy to the north, spirals to the east, then curves gently south. Bright H II regions and dark dust patches line both features. If we can free our imaginations and see this galaxy in three dimensions, the "scratch" looks like the flailing arms of an unseen spiral galaxy whose body has sailed obliquely into NGC 1275's "mouth." That scenario, in fact, is just what radio data suggest. As early as 1957 Walter Baade and Rudolph L. Minkowski proposed that NGC 1275 is an interacting pair of galaxies. The bright compact body appears to be that of an elliptical or lenticular galaxy, while most of the dust patches seem to be associated with the arms of a spiral system moving through NGC 1275 at a speed of 3,000 km per second. Tidal interactions have distorted the spiral's shape and may have ripped free the curved "tail" of stars to the east.

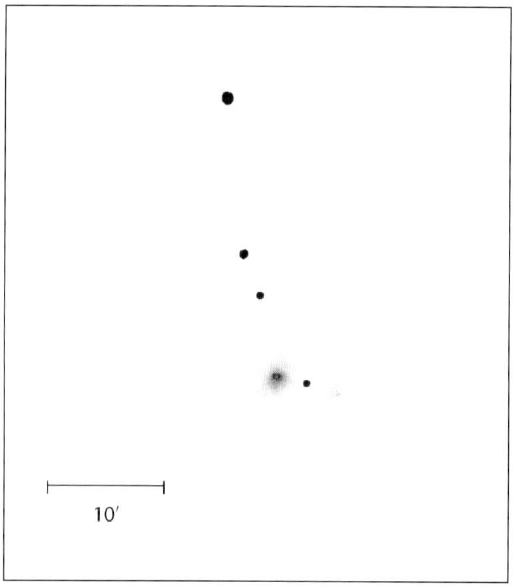

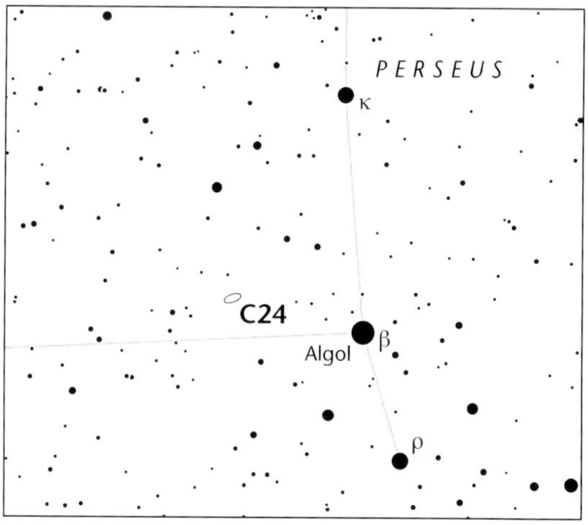

The proposed collision between NGC 1275 and its spiral companion may be fueling the galaxy's brilliant nucleus and its powerful radio and X-ray emissions. It appears to have played another role, one whose discovery had to await the launch of the Hubble Space Telescope. In 1991, Hubble's original Wide-Field/Planetary Camera resolved some 50 bright, blue objects around NGC 1275's core. These objects are believed to be massive, very young globular clusters perhaps only hundreds of millions of years old. At the time this finding seemed highly unusual because the globular clusters in our own Milky Way galaxy are among its oldest stars, having ages of roughly 10 to 15 billion years. Now, though, astronomers generally accept the notion that NGC 1275's young globular clusters formed during the collision or merging of the system's two constituent galaxies. HST has also revealed very young globular clusters in several other interacting systems, most notably NGC 4038 and NGC 4039 (Caldwell 60 and 61), a photogenic pair of colliding galaxies known as the Antennae.

Being a Seyfert galaxy, NGC 1275 is playing the role of a low-energy quasar whose nucleus episodically ejects matter at incredible velocities (6 percent of light speed and up). Although dramatic, these volcanic blasts are not unusual. It is estimated that 10 percent of all spiral galaxies are Seyferts and display similar behavior as seen today from Earth. The Seyfert stage, however, is believed to be transitory. If so, all spiral galaxies, including our Milky Way, might spend 10 percent of their lives in the Seyfert phase. Timothy Ferris provides a fine analogy in his popular book *Galaxies*:

> Imagine that you are looking out across a meadow full of fireflies on a summer evening. Charmed by the sight of the fireflies, you make a snapshot of the meadow using an exposure of, say, one second. The resulting photograph proves to be something of a disappointment. Each firefly lights up for only a fraction of the time that it spends in flight. The rest of the time the firefly is dark, storing up energy for another flash. A one-second exposure will capture the light of only a fraction of the fireflies, those that happened to be flashing at the moment when the photo was made.

Through the 4-inch, NGC 1275 is less a firefly and more a piece of starlit lint. The galaxy is the most obvious of the 19 cluster members plotted on the *Millennium Star Atlas*, and lies about 2° east-northeast of Algol. (A more detailed chart of the cluster appears on page 192 of Christian Luginbuhl and Brian Skiff's *Observing Handbook and Catalogue of Deep-Sky Objects*; it was used to label the cluster's galaxies by their NGC numbers on the photograph on page 100.) The galaxy is little more than a star with a faint coma, like a distant comet just starting to turn on. Luginbuhl and Skiff note that through a 6-inch telescope this coma measures 30″ and its size increases to 1′ in a 12-inch. They also note that the starlike nucleus appears to lie a little west of the coma's geometric center.

Identifying the cluster members surrounding NGC 1275 can be confusing and probably requires a large aperture; views of them through the Genesis were fleeting at best. Texas amateur Barbara Wilson had success with a 13.1-inch f/4.5 reflector and 115× from Twain Harte, California. She describes the cluster's members in the following way: "NGC 1272 is the 2nd largest, round but fainter than NGC 1275. NGC 1273 is smaller than NGC 1272 but has a higher surface brightness. NGC 1277 and 1278 at first appeared singular due to poor seeing, with NGC 1277 being very small and lying just northeast of NGC 1275. When the seeing settled down NGC 1278 was clearly separated from NGC 1277 — NGC 1278 is the brighter of the two, larger and elongated, and pretty faint. NGC 1281 is a very elongated, small galaxy near a magnitude-10 star. NGC 1270 is a small, slightly elongated elliptical southwest of NGC 1272."

The magic of the Perseus Cluster lies not in our visual impressions through modest backyard telescopes but in our understanding of the region's unseen turmoil. As with a snapshot of a busy city street, different scenes simultaneously unfold in the cluster before the mind's eye. By glancing into this tiny area of space we "see" one galaxy eating another, volcanic nuclei shooting debris into intergalactic space, and galaxies churning up wakes as they sail through an intergalactic sea of ionized gas. The snapshot shows a dramatic cosmic moment some 230 million years in the past — a relatively quiet time on Earth, when the earliest sharks ruled the seas and lizards stalked the fragile shores of Pangaea. It was a time when our ancestors cowered in the shadows of the night and could not contemplate the universe.

25

25

Intergalactic Tramp; Intergalactic Wanderer
NGC 2419
Type: Globular Cluster
Con: Lynx

RA: $07^h\ 38^m\ 08.5^s$
Dec: +38° 52′ 55″
Mag: 10.4
Diam: 4.6′
Dist: 275,000 light-years
Disc: William Herschel, 1788

W. HERSCHEL: [Observed 31 December 1788] Considerably bright. Round. Very gradually much brighter in the middle. About 3′ diameter. (H I-218)

GC: Pretty bright, pretty large, little extended toward position angle 90°, very gradually brighter in the middle, a star of magnitude 7 or 8 toward position angle 19°.

NGC: Pretty bright, pretty large, little extended toward position angle 90°, very gradually brighter in the middle, a star of magnitude 7 or 8 toward position angle 267°, 4′ distant.

OF THE 150 OR SO GLOBULAR CLUSTERS KNOWN or believed to be associated with our Milky Way galaxy, NGC 2419 is among the most remote. In 1918 Harvard College Observatory astronomer Harlow Shapley began determining the distances to these spherical swarms, each containing tens or hundreds of thousands of ancient stars. He found that globular clusters appeared to be distributed within a sphere 65,000 light-years in radius that was centered on the galactic center. This significant discovery demonstrated that our Sun was not at the center of the universe as many had previously believed.

Shapley knew NGC 2419 was a "distant" cluster and assigned it a distance of 99,000 light-years in his 1930 monograph, *Star Clusters*. But it was Walter Baade who first discovered just how extremely distant NGC 2419 really was. His 1935 observations from Mount Wilson placed the globular 182,000 light-years from the Sun — about as far as the Large Magellanic Cloud, the Milky Way's preeminent satellite galaxy. NGC 2419 indeed is a maverick, lying well beyond the edge of the Milky Way's starry disk. But even Baade's distance determination seems conservative with hindsight. More recent studies have pushed the cluster out even farther. The 1999 *Catalog of Parameters for Milky Way Globular Clusters* compiled by William E. Harris (McMaster University, Canada) places NGC 2419 some 275,000 light-years from the Sun. If true, this means the globular is fully 100,000 light-years more distant than the Large Magellanic Cloud. Even more remote are globular clusters Pal 3

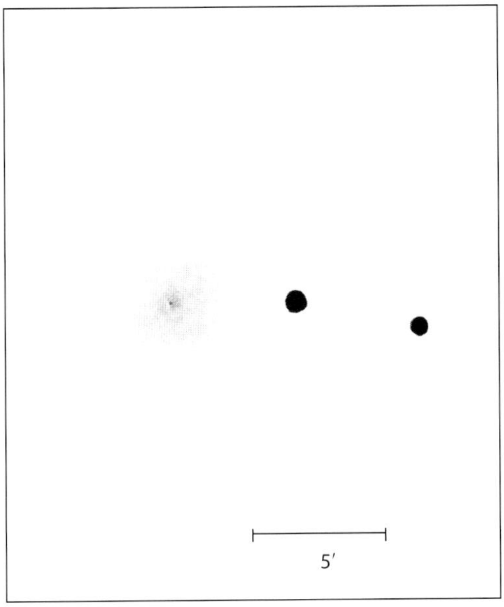

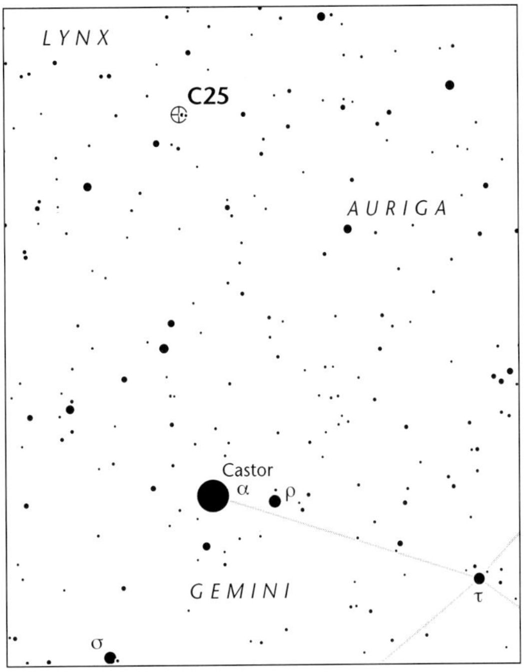

(300,000 light-years), Pal 4 (360,000 light-years), and AM 1 (400,000 light-years).

Some astronomers understandably refer to these extremely distant globulars as "intergalactic wanderers." And, though the common nickname for NGC 2419 is the Intergalactic Wanderer, Brent Archinal points out that we seem to have forgotten the cluster's original name: the Intergalactic Tramp, a name bestowed upon it by Shapley in 1944. "By 1958," Archinal writes, "Helen Sawyer Hogg, in her *Monograph* on clusters, indicated that [the Intergalactic Tramp] had become the accepted common name for the object (although she seems to attribute the name to Baade). I don't think it should go without mention that in many modern publications, Shapley's name has been changed to the Intergalactic Wanderer. I have not tried to trace the origin of this name but it would be interesting to do so. I suspect that this is just the 'politically correct' version of Shapley's name, because of differences in meaning that the word 'tramp' has now vs. the 1940s. Shapley was apparently trying to emphasize that NGC 2419 was so far from the center of the Milky Way, and was probably moving so slowly that it was almost not moving, or at least moving without any planned direction (like a railroad 'tramp,' or perhaps a 'tramp steamer' of his day). That is of course not quite the meaning of the word 'wanderer,' which implies something different."

Is NGC 2419 lost in space? No. "With steady improvements in the data," Harris writes, "and the growing understanding of things like dark matter and the mass profile of the Milky Way and other galaxies, [that idea] has rather faded away. All the known globular clusters in the Milky Way (including NGC 7006, 2419, and all the other remote ones) are now realized to be quite securely bound to the Milky Way." Harris does believe that free-floating intergalactic clusters could and probably do exist, but we haven't found any decent candidates yet.

Needless to say, NGC 2419 is not a knock-your-glasses-off spectacle. Considering its great distance, the fact that it shines as brightly as 10th magnitude suggests that the cluster has a high intrinsic luminosity. Indeed, images taken with large telescopes reveal a wealth of stars packed into a tiny disk only 4.6′ across. The brightest members shine individually with an apparent magnitude of 17.3 on average, while the far more numerous horizontal-branch stars (so called for their locations on a color-magnitude diagram) each shine at an apparent magnitude of 20.4 on average. So don't expect to resolve much in the way of stars once you locate this mysterious object. John Herschel didn't, it seems; on March 18, 1831, he called NGC 2419 simply "not very bright, large, pretty much elongated."

NGC 2419 lies about 7° north-northeast of the magnitude-1.6 star Castor, the second-brightest star in Gemini and a stunning multiple-star system for reasonably sized (or unusually sharp) telescopes. The cluster is nestled a mere 5′ east of two similarly bright stars, the westernmost of which is a fine 8th-magnitude double itself. (These stars are visible in the photograph on page 104). When I first searched the area using the 4-inch at 23× I made the mistake of looking for a small, concentrated glow and swept past the cluster several times. I did the same at 72×. Just when I was about to give up for the night, my eye caught sight of an extremely dim patch of light next to the easternmost of the two stars mentioned above. I immediately returned to 23× and had no problem seeing this diffuse mote of light because I knew exactly where to look.

At low power the cluster looks ethereal, like a dim and distant comet whose ices are just beginning to sublimate. At 72× the globular appears as a weak circular glow, largely uniform but with some suggestion of a central brightening and an irregular outer shell. The longer I looked at the cluster with 72× the more prominent and intriguing it became, though I discerned no further details. Higher power added little to the scene. Amateurs using telescopes in the 10-inch range might hope to see some patchiness toward the cluster's center, but little else. The object is a challenge for a 4-inch telescope under dark skies, so light pollution will probably render it invisible. That being said, don't let the fact that this cluster is so dim discourage you from trying to locate it. The object's beauty lies not in its appearance but in its distance. Without question NGC 2419 is the most distant globular cluster that can be seen with modest telescopes. AM 1 and the Palomar ("Pal") globulars are much too dim to be viewed with such instruments. However, these latter clusters are favorite challenges for amateurs using 24- to 40-inch reflectors at the annual Texas Star Party and other gatherings.

26

Silver Needle Galaxy
NGC 4244
Type: Spiral Galaxy (Scd)
Con: Canes Venatici

RA: $12^h\ 17.5^m$
Dec: +37° 48′
Mag: 10.4
Dim: 15.8′ × 1.7′
SB: 14.2
Dist: 10 million light-years
Disc: William Herschel, 1787

W. HERSCHEL: [Observed 17 March 1787] Very bright. Extended 60° from south preceding to north following [from southwest to northeast]. 20′ in length [and] 2′ broad. (H V-41)

GC/NGC: Pretty bright, very large, extremely extended toward position angle 43°, very gradually brighter in the middle.

NGC 4244 IS A VERY LATE-TYPE SPIRAL GALAXY seen exactly edge on. The galaxy displays no central bulge of any consequence, but it does have a tiny central condensation. In long-exposure photographs, such as one taken through the 200-inch reflector atop Palomar Mountain, the galaxy's entire surface is resolved into stars, with one obvious dark lane running through the galaxy's center along its major axis. The late astronomers Gérard de Vaucouleurs (University of Texas) and Erik B. Holmberg (Uppsala Astronomical Observatory, Sweden) determined that the galaxy must be nearby because its stars can be resolved throughout its surface. They estimated that the galaxy must be more distant than the farthest members of the Local Group but not so distant as the M81 and M101 groups. Modern studies support this hypothesis. With a modest linear diameter of 46,000 light-years and a total mass of 9.8 billion Suns, NGC 4244 is receding from us at 250 km per second while spinning about its core at 90 km per second.

NGC 4244 shines at magnitude 10.4 and can be found about 4½° southwest of 4th-magnitude Beta (β) Canum Venaticorum, just 30′ west of a pair of 7th-magnitude stars. The challenge is to see the galaxy with 7×35 binoculars, which I did with some difficulty under a dark sky, using averted vision. Even in binoculars, the galaxy appears as a wafer-thin wisp of light with an extremely low surface brightness. At 23× in the 4-inch NGC 4244 is, with just a glance, a compelling yet ghostly ellipse. A more concentrated visual effort reveals an uncanny sharpness to the galaxy's slender shape, which is why Arizona amateur Tom Polakis calls it the Silver Needle. The core appears obviously bright, while some beading is apparent along the length of the galaxy. Most noteworthy are two patches of concentrated light, one on either side of the core.

Through the 4-inch NGC 4244 looks best at 72×. The galaxy's northeastern tip points to an 11th-magnitude star, just southwest of which glows an irregular clump of unresolved starlight. The southwestern tip is punctuated by a pair of 12th-magnitude stars. Look carefully at these stars. Do you see a faint, diffuse patch of light about halfway back to the galaxy's core? This is yet another stellar concentration. How do we know it's not an enormous HII region? As early as 1961 Edwin Hubble remarked on this galaxy's lack of HII regions. He described the galaxy as being resolved "into stars throughout its surface. They are individual stars and not H II regions." Indeed, in a 1999 *Astrophysical Journal* paper, Charles Hoopes (New Mexico State University) and his colleagues note that "[m]ost of the HII regions are small and faint, except for two bright complexes on either end of the disk."

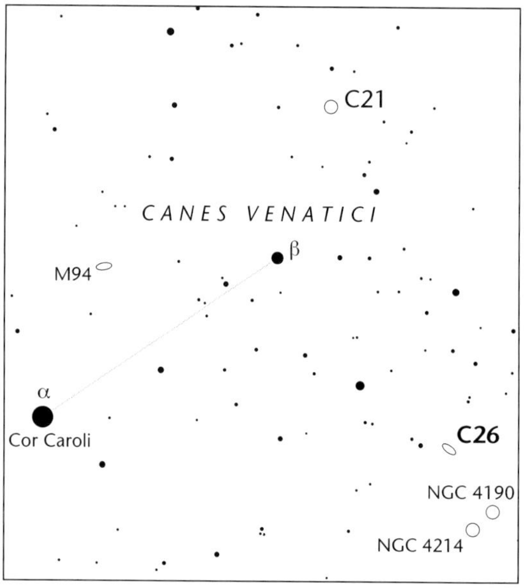

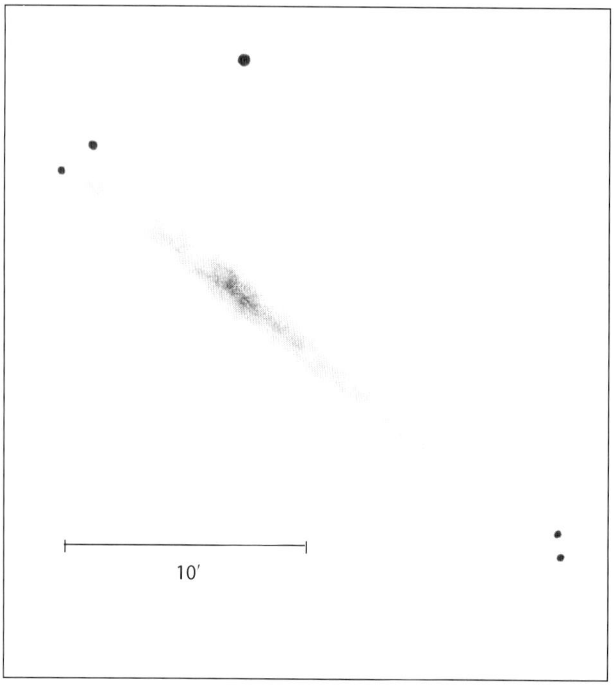

Before you leave this field, try to locate the magnitude-9.8 irregular galaxy NGC 4214 some 1½° to the south-southwest of NGC 4244. Seen together in a low-power field, the two galaxies look somewhat similar in shape and orientation to M81 and M82, but with diminished apparent size and brightness. Another galaxy, NGC 4190, lies only about 30′ northwest of NGC 4214, but it went unnoticed by me. This face-on galaxy shines at 13th magnitude and spans a tiny 1.6′. Christian Luginbuhl and Brian Skiff found NGC 4190 invisible in a 6-inch telescope and difficult to view in a 12-inch.

27

Crescent Nebula
NGC 6888
Type: Emission Nebula
Con: Cygnus

RA: $20^h\,12.0^m$
Dec: +38° 21′
Mag: 8.8 (nebula)
Mag: 7.7, var.? (central star)
Dim: 18′ × 13′
Dist: 4,700 light-years
Disc: William Herschel, 1792

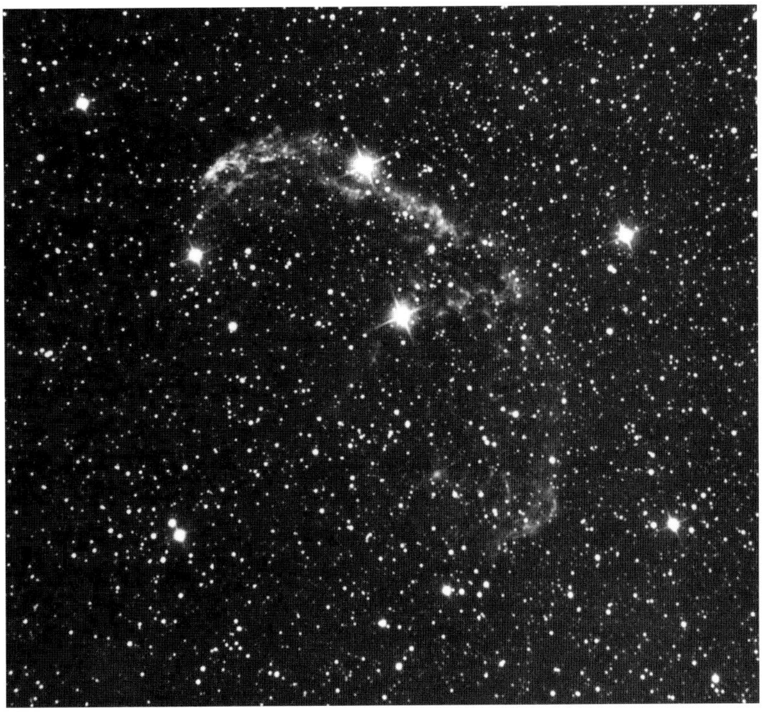

W. HERSCHEL: [Observed 15 September 1792] A double star of the 8th magnitude with a faint south preceding [south-westerly] milky ray joining to it. 8′ in length and 1½′ broad. (H IV-72)

GC / NGC: Faint, very large, very much extended, double star attached.

NGC 6888 IS COMMONLY KNOWN AS THE CRESCENT Nebula because, when seen through a backyard telescope, its brightest segments look like the dim spirit of a young Moon set against the glowing clouds of the Milky Way. Some publications simply list NGC 6888 as a magnitudeless "bright nebula." With its nature apparently so obscure, most amateurs probably shrug their shoulders at the thought of trying to hunt down the Crescent. Why look for some inconspicuous blip when modern star atlases show what seem to be much richer targets sprinkled across the Cygnus Milky Way? After all, NGC 6888 is but a tiny patch of light about 2¾° southwest of 2nd-magnitude Gamma (γ) Cygni, a bright star that lies at the heart of an enormously large and rich nebula complex spanning some 3°. The thought of tackling tiny NGC 6888 seems almost ludicrous in this context.

But the Crescent Nebula is one of the more intriguing objects in our galaxy. This annular nebula circumscribes a 7.5-magnitude Wolf-Rayet star known as HD 192163. Wolf-Rayets are extremely luminous and hot stars that shed mass at enormously high rates, in many cases after passing through a supergiant stage. This material travels at speeds as high as 3,000 km per second until it smashes into gases that previously constituted the star's outer atmosphere. The Crescent has long been believed to be the result of such a collision. Studies show it to be a prolate ellipsoid, 25 by 16 light-years across, that is tilted at a 45° angle to our line of sight. The Einstein and Rosat spacecraft revealed that X-rays emanate not from the nebula's cavity but from its visually brightest parts. The X-ray emissions originate in filamentary structures close to the nebula's inner border, suggesting that they are generated as material cools after being shocked by the fast wind from HD 192163. The nebula's outer edges, by contrast, show the first point of contact between the advancing Wolf-Rayet wind and the atmosphere of the former red supergiant.

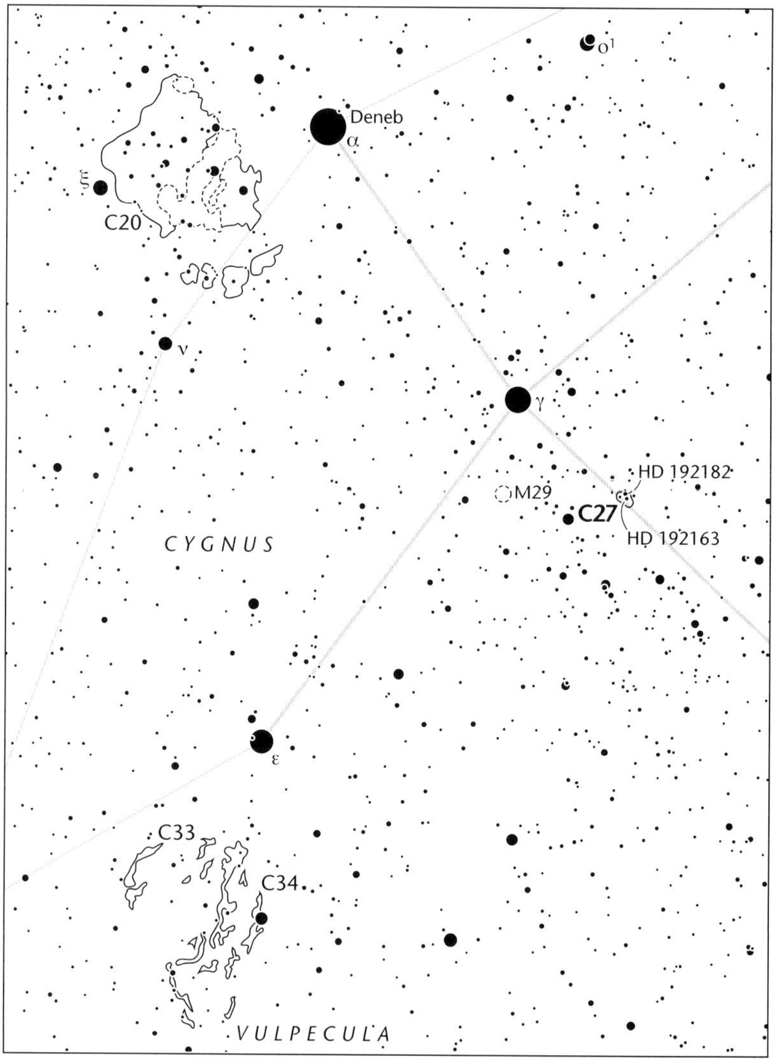

The revolutionary Infrared Astronomical Satellite has shown that a vastly larger (1.8° × 1.5°) shell surrounds the Crescent and is also centered on the Wolf-Rayet star. This extended structure may be the infrared signature of a 100,000-year-old supernova remnant. Since HD 192163 is believed to be a binary star, it's possible that the outer infrared shell was generated by the explosion of the primary star, while a very recent outflow from the secondary star formed the Crescent Nebula. Indeed, Wolf-Rayet stars are believed to be the progenitors of some supernova remnants, Cassiopeia A among them. But Argentinean astronomer Cristina E. Cappa (Argentine Institute of Radio Astronomy) and her colleagues have questioned the notion that the HD 192163 system has hosted a supernova explosion. In a 1996 *Astronomical Journal* paper they argue that HD 192163 could have ejected the extended outer shell during the star's life on the main sequence; material ejected by the star later in life then encountered the innermost wall of the outer shell and created NGC 6888. However, David A. Green (Mullard Radio Astronomy

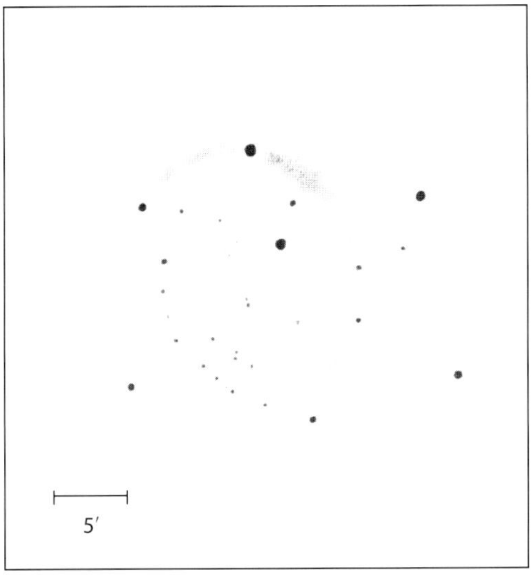

Observatory, England) still lists a possible supernova remnant in *A Catalogue of Galactic Supernova Remnants,* dated 1998. Although the jury is still out on the origin of the outer infrared nebula, it's safe to say that to see the Crescent Nebula is to see the results of several stages of energetic stellar activity.

Through 7×35 binoculars the nebula's environs look hazy, but since the remnant lies in a patchy, star-rich area of the Cygnus Milky Way and slices through a diamond of 7th- and 8th-magnitude stars, there is no way to confirm that this visual impression really constitutes a detection of the Crescent. On the clearest of nights the oval shape of NGC 6888 can be discerned — just barely — at 23× through the 4-inch refractor. The nebula's brightest segment is a sharp needle of light extending southwestward from the diamond's brightest star (HD 192182, a 7th-magnitude *G*-type star with an 11th-magnitude companion). So sharp is this needle when seen at low power that I initially mistook it for some weird diffraction spike — until I realized I was using a high-quality refractor that causes no such artifacts. In the 4-inch the star and the spike look like a distant quasar with a jet shooting outward.

The view looks much the same at 72×. But a concentrated effort (remember to breathe often and to look away from the eyepiece every now and then) reveals a patchy extension of very dim light continuing to the southwest before it gently curves southeast. Interestingly, I found this nebulous segment consistently easier to see than the curved remnant to the northeast, even though the latter looks brighter in photographs. I attribute this paradox to the fact that the nebula's northeastern section lies close to two of the diamond's stars and, at modest magnifications, the glare of these bright stars diminishes the view. By contrast, the southwestern extension, while fainter, lies in a region free from such troublemakers.

The ultimate visual challenge is to close the loop on the southeastern side of NGC 6888. I did this by using low power and averted vision. Take plenty of time to confirm what you see. I find that, with time, the thin vapors waver in and out of view. Also, if you compare the brightness of the shell's inner edge with that of the surrounding region, it should become clear that your eyes are picking up faint emissions from excited gases. This glow is more prominent on the southeastern side.

28

28

NGC 752
Type: Open Cluster
Con: Andromeda

RA: $01^h\ 57.6^m$
Dec: +37° 50′
Mag: 5.7
Diam: 75′
Dist: 1,200 light-years
Disc: William Herschel, 1786 (though probably preceded by Giovanni Batista Hodierna, before 1654)

W. Herschel: [Observed 21 September 1786] A very large coarsely scattered cluster of very [bright] stars. Irregularly round. Very rich. Takes up ½°, like a nebulous star to the naked eye. (H VII-32)

GC/NGC: Cluster, very large, rich in stars, stars large [bright] and scattered.

After looking at the data listed above you might well conclude that open cluster NGC 752 is one Caldwell object that Charles Messier should have noticed on his nightly comet hunts. However, the star-speckled object seemingly escaped his careful gaze, even though it is brighter than 6th magnitude and has an apparent diameter more than 1½ times that of the full Moon. The cluster is visible to the naked eye as a soft glow about 5° south-southwest of the golden 2nd-magnitude telescopic double star Gamma (γ) Andromedae. You can also use our galaxy's spiral companions in the Local Group as signposts; NGC 752 lies about 8° north-northeast of the famous Pinwheel Galaxy (M33 in Triangulum) and 14° east-southeast of the great Andromeda spiral (M31). With such competition vying for amateurs' time and attention, it's no wonder the relatively nondescript open cluster continues to go virtually unnoticed.

Nevertheless, NGC 752 is a noteworthy attraction. Certainly no binocular observer will be disappointed after turning his or her gaze to it under a dark sky. Because of its large apparent diameter and splash of bright 8th- to 12th-magnitude suns, telescopic views of NGC 752 will be best in a rich-field instrument at very low power. Charles Messier, who observed under the smoggy skies of 18th-century Paris, probably didn't have a chance of spotting this dim glow with the unaided eye. Furthermore, the telescopes he used to seek comets didn't afford the low power required to tighten up this loose aggregation of bright stars and make it conspicuous. Messier employed small, simple refractors and inefficient speculum-mirror reflectors, generally at magnifications ranging from 60× to 138×. Roger Clark states that the optimum magnification for detecting NGC 752 is 7× with a 2-inch telescope and 27× with an 8-inch. Clearly Messier was at a disadvantage. I can imagine him sweeping past NGC 752 and noticing a rich field of stars but moving on because the object did not look fuzzy. Of course, it's also possible that his telescope didn't provide a large enough field of view to make NGC 752 stand out as a discrete entity.

Brent Archinal has acquired an interesting naked-eye perspective on this cluster. While he was attending the Black Forest Star Party in north-central Pennsylvania, he set out to look for M33 with the naked eye — a time-honored test of sky darkness and transparency. "This is definitely one of the darkest sites I've seen in the eastern U.S.," he said, "second only to Spruce Knob, West Virginia. Something I noticed that might be worth pointing out is that if one tries to casually observe M33 with the unaided eye, it is easy to confuse it with NGC 752, which is in the vicinity and much easier to see."

Unlike most open clusters, which are tens or hundreds of *millions* of years old, NGC 752 is about 2 *billion* years old. This places NGC 752 among the oldest open clusters known, though it is much younger than 5-billion-year-old NGC 188 (Caldwell 1). Most of the 77 or so measured stars apparently populating NGC 752 are *F*-type subgiants that have moved off the main sequence and are evolving toward gianthood. An 18-year-long radial-velocity survey has confirmed that 15 red giants are members of the cluster. Rosat detected 49 X-ray sources in the cluster region, seven of which have been tentatively matched up with visible cluster members; three of these latter ob-jects are short-period binaries, one is a blue straggler, and one a so-called rapid rotator. These discoveries all paint a picture of a cluster whose stars have had considerable time to follow various evolutionary paths.

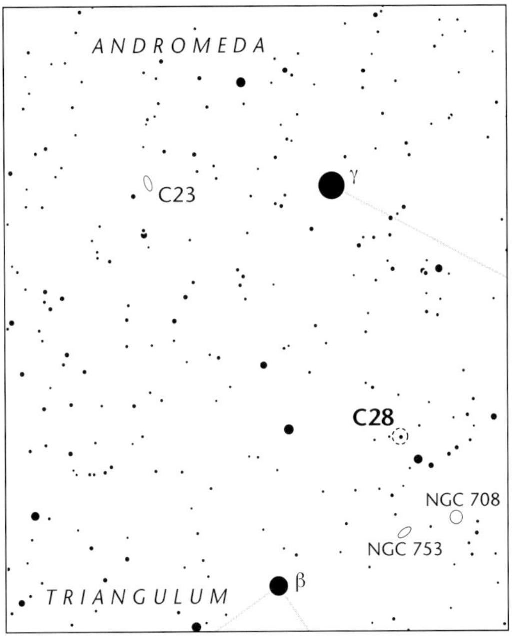

NGC 752's very open nature itself hints at the cluster's seniority. Unlike a globular cluster, whose hundreds of thousands of stars exert a mutual gravitation that keeps them moving in a tight pack, an open cluster tends to loosen as it ages. Stars lining the cluster's edge are lost to the gravitational pulls of passing stars, molecular clouds, and other clusters. These "lost" suns blend into the populated stellar streets of the Milky Way, where their identities eventually fade beyond recovery.

At 23× in the 4-inch NGC 752 seems to fill the field of view but, as alluded to above, it is hard to know where the cluster's boundaries end. The cluster has no central concentration; in fact, I see a curious double nature to it, each section seemingly a very wide open cluster of its own. To me these groupings look like celes-

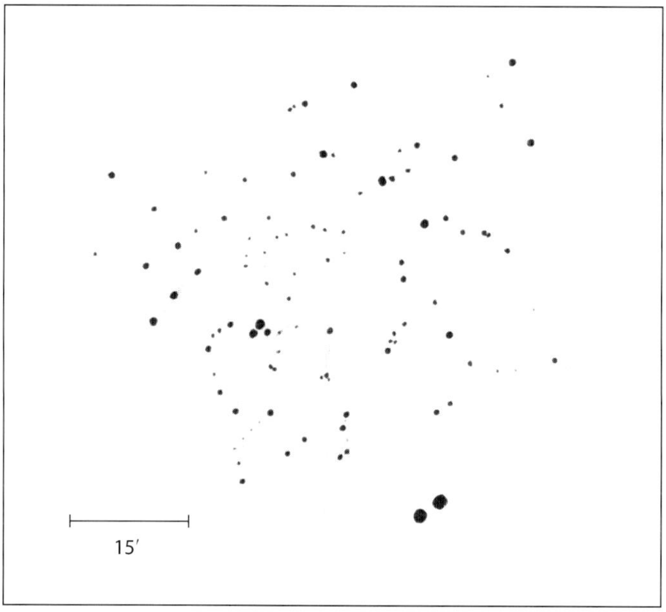

tial graffiti, and in a mirrored view (with east up and north to the left), the eastern group of stars spells out the word "YES" in a sprawling and erratic manner. Christian Luginbuhl and Brian Skiff note a fine pair of stars (yellow and red, respectively) to the south-southwest of the cluster center. These stars shine at magnitude 5.7 and 5.9, respectively, and are separated from one another by 3.6′. Several other colorful pairs adorn the cluster, and it would be worthwhile to spend an hour or so scouting them out. Bear in mind that the brightest star that appears to be in the cluster — a magnitude-7.1 *K*-type luminary in a tight stellar triangle — is not a true member of NGC 752.

Two small, faint galaxies lie about 1½° to the south-southwest of the cluster: NGC 753 (12th magnitude) and NGC 708 (13th). Each will challenge the user of a 6-inch or slightly larger telescope. NGC 708 is centered on a gaggle of eight other galaxies, all within ½° of its elliptical body.

29

NGC 5005
Type: Mixed Spiral Galaxy (SABbc)
Con: Canes Venatici

RA: $13^h 10.9^m$
Dec: +37° 03′
Mag: 9.8
Dim: 5.6′ × 2.9′
SB: 12.7
Dist: 69 million light-years
Disc: William Herschel, 1785

W. Herschel: [Observed 1 May 1785] Very bright. Considerably large, much extended from south preceding to north following [from southwest to northeast]. Suddenly much brighter in the middle. 6′ long. 1½′ broad. (H I-96)

GC: Very bright, very large, very much extended toward position angle 25°, very suddenly brighter in the middle to a nucleus.

NGC: Very bright, very large, very much extended toward position angle 66°, very suddenly brighter in the middle to a nucleus.

Most galaxies travel the depths of space gravitationally bound to other galaxies. NGC 5005 belongs to a group of 29 moderately distant galaxies that R. Brent Tully (University of Hawaii) and J. Richard Fisher (National Radio Astronomy Observatory) call the Canes Venatici Spur. It is a relatively large spiral with a diameter of 100,000 light-years (about that of our Milky Way) and a luminosity of 50 billion Suns. From our vantage point we see NGC 5005 inclined 36° from edge on. This offers us a good view of its intriguing spiral structure.

NGC 5005 is a late-type spiral whose visual appearance hints at the presence of a central bar. Photographs taken with large telescopes reveal an extremely sharp nucleus nestled in a bright lens-shaped core. Outside the core the spiral arms are moderately tight, with many knots and dark lanes. One prominent and peculiar dust lane in the galaxy's southwestern quadrant begins not in the nuclear region but on the periphery of the central lens. High-resolution images of the lens itself show dust lanes and luminous spiral filaments that can be traced as close as 7″ to the bright starlike nucleus.

NGC 5005 is a member of a special class of low-power Seyfert galaxies, which Timothy Heckman (University of Maryland) and Bruce Balick (University of Washington) have dubbed LINERs (*l*ow-*i*onization *n*uclear *e*mission-line *r*egions). Although not as energetic as other active galactic nuclei, LINERs share many of their characteristics; they emit X-rays and vary in brightness. As with all active galactic nuclei, many astronomers believe black holes power LINERs. LINERs are very common and may be

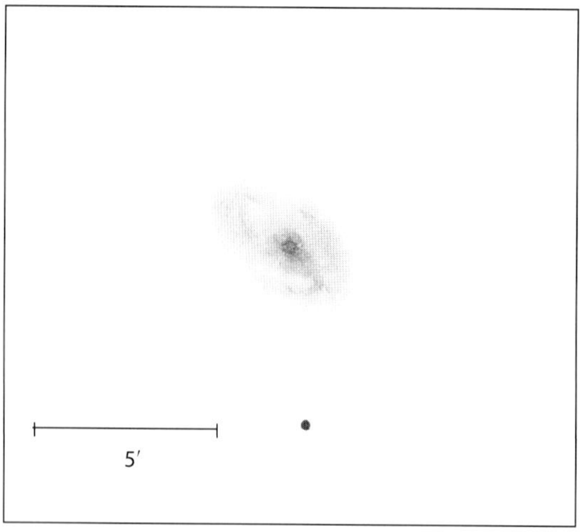

present in nearly half of all spiral galaxies.

Images of molecular gas in NGC 5005, made with the Owens Valley Radio Observatory millimeter-wave array, show three components in the galaxy's inner 20,000 light-years: a ring (or a pair of tightly wound arms) with a radius of about 10,000 light-years, a strong central condensation, and a bar that lies between the two to the northwest of the nucleus. Gas in the bar apparently is falling into the nucleus at a rate of about 50 solar masses per year and will eventually contribute 200 million solar masses to a disk surrounding the nucleus. This inflow of material may someday trigger starburst activity.

NGC 5005 is easy to locate, being only 3° east-southeast of the fine double star Alpha^{1,2} ($\alpha^{1,2}$) Canum Venaticorum (Cor Caroli) and 1½° due south of a sharp triangle of 6th- and 7th-magnitude stars (15, 16, and 17 Canum Venaticorum). Cor Caroli, the triangle, and the galaxy all fit in the same binocular field. The galaxy is visible in 7×35 binoculars, though only with averted vision. Take care when identifying NGC 5005, because just ½° southeast of it is its "twin," NGC 5033. NGC 5033 shines at roughly the same magnitude and has nearly the same dimensions as your Caldwell target; it, too, is visible in binoculars with averted vision.

Through the 4-inch at 23× NGC 5005 is a nice, slightly oval haze with a distinct brightening at the center. At times some fleeting detail appears in the galaxy's outer envelope. Medium power brings into view some ill-defined arcs of light, which, when drawn, should map out NGC 5005's spiral structure. Higher powers do little to enhance the view, and I find the galaxy looks best at 72×. After a prolonged study of the oval haze at 72× I was able to see an absence of light running from the northeast to the southwest on the northern side of the inner lens; it seems to separate the lens from a hazy arc of light. A little appendage or hook of light juts to the northeast from the inner lens, and it is here that the dark area mentioned above appears to emerge. The galaxy's southern extremity has a small sharp boundary, which could be a section of a spiral arm, perhaps glowing with light from H II regions. Interestingly, Texas amateur Barbara Wilson — who calls the galaxy "spectacular" though her 20-inch f/4 reflector at 288× — says the galaxy's outer extensions are long and faint and they fade into the background sky with no

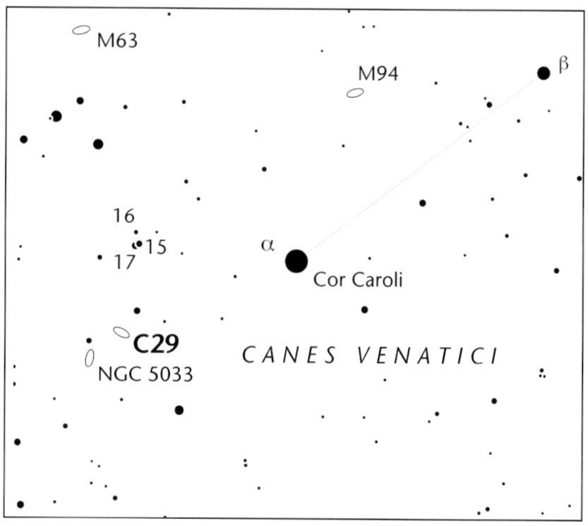

sharp edges. However, Christian Luginbuhl and Brian Skiff found that the halo faded more abruptly on the southeastern side when seen through a 12-inch telescope. The discrepancy appears to be a function of aperture and power.

Supernova hunters take note: on June 16, 1996, Claudio Bottari of Sava, Italy, discovered a magnitude-14.5 supernova (Supernova 1996ai) on a CCD image of NGC 5005. The new star was 24″ east and 4″ north of the galaxy's center, near an H II clump. Spectra taken shortly after the discovery suggested that the Type Ia supernova was heavily obscured by dust and gas. Nevertheless, it reached maximum light around 13th magnitude — well within visual range of amateur instruments.

30

30

NGC 7331
Type: Spiral Galaxy (Sbc)
Con: Pegasus

RA: 22^h 37.1^m
Dec: +34° 25′
Mag: 9.5
Dim: 9.7′ × 4.5′
SB: 13.3
Dist: 47 million light-years
Disc: William Herschel, 1784

W. HERSCHEL: [Observed 5 September 1784] Very bright, considerably large, much extended. Much brighter in the middle. Resolvable. (H I-53)

GC/NGC: Bright, pretty large, pretty much extended toward position angle 163°, suddenly much brighter in the middle.

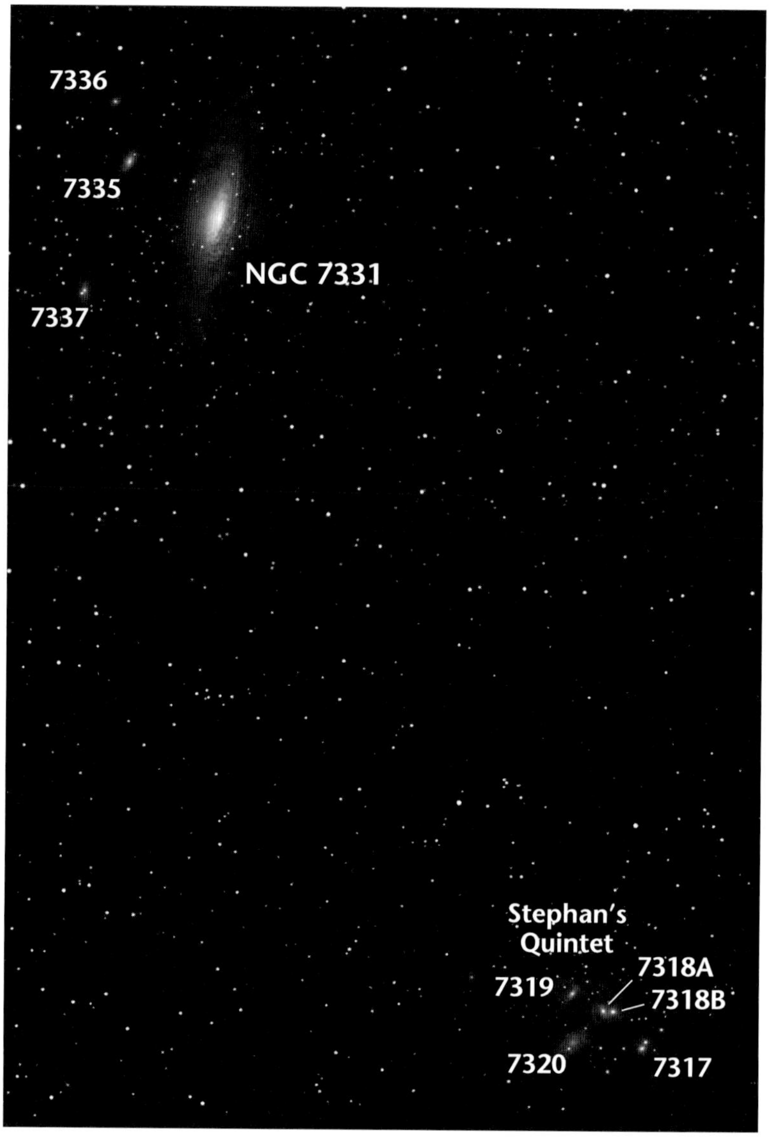

WHEN THE LAVENDER CURTAIN OF DUSK FALLS toward the western horizon in mid-December, the mythical winged horse Pegasus vaults high across the meridian. Many observers find this an opportune time to hunt down the great globular cluster M15 within sniffing distance of Enif, the star marking the Horse's nose, or the Andromeda Galaxy, M31, which immediately follows the Horse across the sky. But about 4½° north-northwest of Eta (η) Pegasi, "beneath" the flying Horse's front legs, lies another magnificent, though often overlooked, spiral galaxy: NGC 7331. From dark skies this galaxy can be glimpsed as a dim slash of light in 7×35 binoculars; challenge yourself and your friends to hunt it down this way at your next star party.

NGC 7331 belongs to the Pegasus Spur of galaxies, a small gathering of 35 systems that includes one member of the famous Stephan's Quintet (described below). NGC 7331's spiral disk is tilted 22° from edge on. It whisks away from the solar system at a generous 820 km per second. In detailed photographs the galaxy looks like an oblique aerial view of a hurricane at night; the glowing eye and spiral arms of the "storm" seem poised to snuff out the lesser stellar metropolises surrounding it.

The galaxy's modest apparent size and brightness belie its true enormity. In fact, NGC 7331 is among the largest galaxies known. If we accept a distance of 47 million light-years (a value that has been verified by the Hubble Space Telescope), NGC 7331's linear diameter becomes 130,000 light-years. NGC 7331 appears to have a total mass of 300 billion Suns. Thus it equals in size and mass the great Andromeda Galaxy and outranks our Milky Way.

As early as 1961, astronomers recognized that NGC 7331's structure is less like that of M31 and more like that of M63 in Canes Venatici or NGC 2841 in Ursa Major (though NGC 7331's spiral arms appear thicker and are more easily seen than those of the latter two galaxies). NGC 7331 is now classified as an Sbc galaxy, meaning it is a spiral with somewhat tightly coiled arms, a relatively small nucleus, and no bar. Spectroscopic data indicate that there has been large-scale star formation in NGC 7331's nucleus. Spectroscopy and photography also have helped astronomers determine which side of the giant spiral is closer to us. The data suggest that the northern end of the spindle is approaching us and that its western side — the one marked by the especially prominent dust lane — is closer; the spiral arms are trailing. The stars in the central 3″

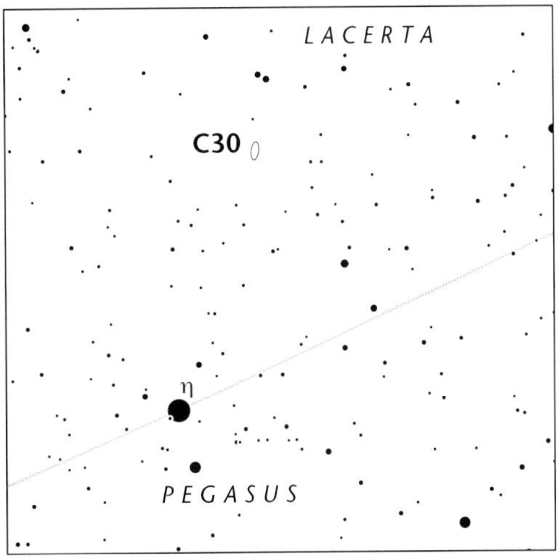

(650 light-years) of the galaxy's disk, while relatively young (about 2 billion years old), are very rich in "metals" (the chemical elements, heavier than helium, that stars produce via nuclear fusion). This region might represent a circumnuclear ring formed during a recent starburst episode. NGC 7331's starlike nucleus also probably experienced a secondary star-formation burst because, at an age of about 5 billion years, it is two to three times younger than the surrounding bulge. Rosat made the first deep X-ray observations of NGC 7331, and in the process discovered a nuclear X-ray source — a finding that bolsters the notion of a massive black hole at the galaxy's heart. In turn, that black hole may somehow be related to NGC 7331's modest *l*ow-*i*onization *n*uclear *e*mission-line *r*egion, or LINER.

It is curious that William Herschel believed that he had resolved NGC 7331. But as Larry Mitchell points out in Appendix C, Herschel suspected that *all* nebulae would be resolved into clusters of stars, if only they could be viewed with a sufficiently large telescope. Mitchell says this belief came as a result of Herschel's examination of the Messier objects, which he found either to be made of nothing

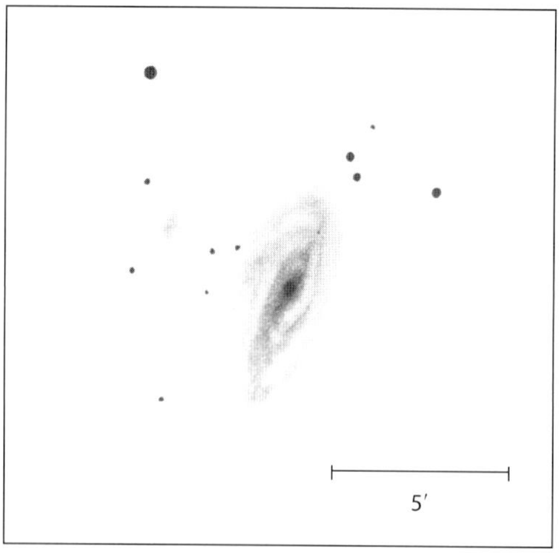

but stars or at least to contain stars and to give "every other indication of consisting of stars entirely." With their modest instruments, today's amateurs can understand why Herschel made the assumptions he did. For example, when viewing NGC 7331 through a 6-inch telescope, Christian Luginbuhl and Brian Skiff detected a stellar nucleus and grainy texture within the disk. Increasing the aperture to 10 inches, they report, brought out increased mottling, especially in the inner halo, and two 13.5-magnitude stars west of the northern tip of the disk. Herschel, with his much larger instruments and very high magnifications, must have detected these features, among others, and he may well have envisioned something akin to a distant cluster, perhaps one like M4 in Scorpius, whose center is a needle of starlight.

Through the Genesis at 23× NGC 7331 is wonderfully apparent as a small, elongated form nestled in a rich field of stars. With averted vision I perceive a tight lens-shaped core surrounded by a dim, elliptical outer halo or shell. At 72× and above, the galaxy becomes a complex mottled glow. Its nucleus appears especially sharp and is surrounded by fleeting shadows of dust. My drawing shows a great amount of detail, which becomes visible over several hours of continuous observing. With time, the galaxy's mottled disk transforms into a clearly discernible spiral pattern with intermittent dust lanes, the most prominent being the western furrow. Most impressive, however, is that some of the mottling within the bright arms corresponds to true patches of stars or H II regions. One prominent region appears just northwest of an approximately 12th-magnitude star lying east-southeast of the lens-shaped core.

NGC 7331 is haunted by several faint companions, four of which appear close to its western edge. I say haunted because, at 72× in the 4-inch, these companions seem to pop dimly into view with averted vision, only to vanish under a direct gaze. The brightest companion, NGC 7335, is listed in *The Deep Sky Field Guide* at magnitude 13.3, though I wonder if it is not brighter; I blindly (and correctly) positioned the tiny fuzz ball in my original sketch, which I made at the eyepiece, and wondered in the margin if it was really a galaxy. Luginbuhl and Skiff detected two more companions, NGC 7337 (magnitude 14.4) and NGC 7340 (magnitude 13.7), with a 10-inch telescope. The fourth, NGC 7336, apparently was too difficult for them. *The Deep Sky Field Guide* lists it at magnitude 15.8, but *Sky & Telescope*'s Roger W. Sinnott notes that this galaxy may not be so faint; the Lyon-Meudon Extragalactic Database, he says, attributes a blue magnitude of 15.4 to NGC 7336, and it probably is a little brighter visually.

About ½° south-southwest of NGC 7331, scrunched into a mere 3.5′ of sky, are the five challenging galaxies collectively known as Stephan's Quintet: NGC 7317, NGC 7318A, NGC 7318B, NGC 7319, and NGC 7320. Stephan's Quintet is of continual interest to astronomers. A 1995 study shows that radio emission from

NGC 7331 is extended much farther to the south than to the north, possibly indicating an interaction between NGC 7331 and at least one member of the Quintet. The Quintet's brightest member, NGC 7320 (magnitude 12.7), has a relatively small recession velocity of 786 km per second — a value rather similar to NGC 7331's; the other four galaxies (three of which display signs of strong tidal interaction) all have redshifts ranging from 5,700 to 6,800 km per second. Since recession velocity, or redshift, is proportional to distance in the Big Bang worldview, it appears that NGC 7320 is a foreground object that just happens to lie in the same direction as the other Quintet members. However, a small but vocal minority of researchers holds that redshift and distance may be unrelated in at least some cases, and they believe that NGC 7320 is physically associated with the rest of Stephan's Quintet.

Stephan's Quintet is also famous in another "heavenly" way. Woodruff T. Sullivan III explained this in the January 1997 issue of *Sky &Telescope*:

> In the well-known 1946 film *It's a Wonderful Life* the opening sequence involves a conversation between angels, who are represented as galaxies in the sky. Only on my latest (nth!) viewing of this film did I notice that the pictured galaxies greatly resembled Stephan's Quintet . . . . After some checking I discovered that the filmmakers indeed used a photograph of Stephan's Quintet, with NGC 7320 (first angel) talking to NGC 7318A and B (second angel). They were later joined by Angel Second Class Clarence, who is portrayed only as a foreground star, perhaps due to the difference in rank.

As Sullivan suggests, it is only appropriate that Clarence is not portrayed as a galaxy in this opening scene, for he had not yet earned his wings, which are obviously interpreted as the swooping, winglike spiral arms surrounding the "bodies" of the high-ranking angels. Luginbuhl and Skiff note that "with care, a 6-inch will show four of the [Quintet's] members, but it takes a 30-centimeter to show the fifth (NGC 7319; magnitude 13.1) even faintly." NGC 7319 has what astronomers call a Type 2 Seyfert nucleus (see Caldwell 24).

By the way, teachers, street-corner astronomers, and public-observing-session hosts might consider using Pegasus and its celestial delights to illustrate the night sky's utility as a time machine. Start with the four stars marking the corners of the Great Square of Pegasus: Markab (140 light-years distant), Alpheratz (97), Scheat (199), and Algenib (333). Next turn your telescope 15° north of the Square until you reach Caldwell 22, the bright planetary nebula NGC 7662 (3,200 light-years); this object could lead into a discussion of stellar evolution. The famous globular cluster M15 (31,000 light-years), 4° northwest of Enif, would then transport your audience out to the halo of our Milky Way galaxy. A tap of the tube could then take bystanders nearly 47 million light-years farther, to NGC 7331. NGC 7331 can also be fruitfully compared with M31, which equals NGC 7331 in size and mass but is 20 times closer to Earth. And finally, if your telescope is large enough to provide a decent view of Stephan's Quintet, you can then carry your guests back nearly 300 million years in time, discuss galaxy clusters, and, of course, talk about *It's a Wonderful Life.*

31

Flaming Star Nebula
IC 405
Type: Reflection and Emission Nebula
Con: Auriga

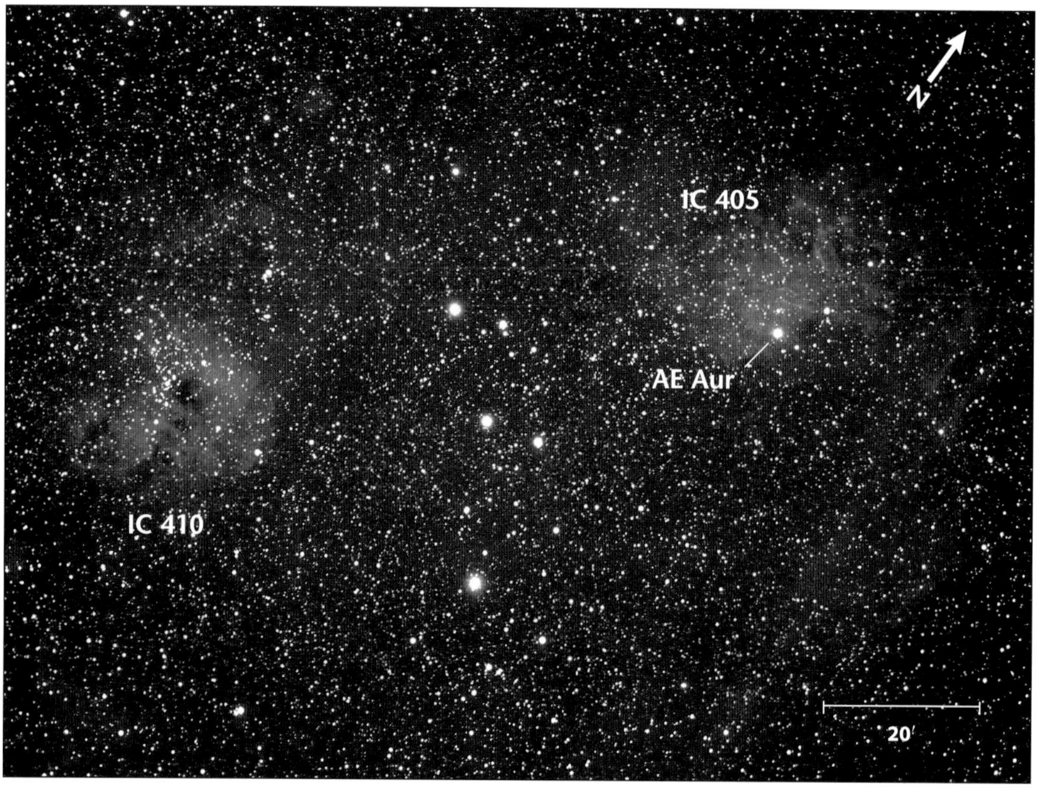

RA: 05^h 16.2^m
Dec: +34° 16′
Mag: ~6 (var.), star; 9.2, nebula (O'Meara)
Dim: 30′ × 20′
Dist: ~1,600 light-years
Disc: John M. Schaeberle, 1892

Herschel: None.

GC: None.

IC: Magnitude 6.7 star with pretty bright, very large nebulosity.

On some unknown date roughly 2.5 million years ago, when early members of the genus Homo roamed the East African savanna, an unusual drama was unfolding in the heavens. The action began in a dense nest of newborn suns now known as the Trapezium star cluster at the heart of the great Orion Nebula. There, a close encounter of two binary-star systems, each tightly bound, took place, and two of the four stars were flung away at high speeds while the remaining two settled down into a new orbit around one another. In time the two "runaway" stars escaped the confines of the Orion Nebula complex. Ronnie Hoogerwerf (Leiden Observatory, the Netherlands) and two colleagues arrived at this conclusion after perusing ultraprecise astrometry from the Hipparcos satellite to investigate the motions of two of Orion's best known "runaway" stars, Mu (μ) Columbae and AE Aurigae, and of Iota (ι) Orionis, a lower-velocity binary star whose two components are in a highly elliptical mutual orbit. This distinctive foursome, the astronomers say, is just what the binary-binary encounter scenario is most likely to produce.

Nowadays, when night falls and Auriga rises, we can see AE Aurigae — a variable star that fluctuates erratically between magnitudes 5.8 and 6.1 — burning among the gossamer filaments of IC 405 (Caldwell 31). Photographic and kinematic evidence confirms that AE Aurigae did not condense from the surrounding nebula (as most hot, young stars associated with nebulae have done) but is rushing through it like a woman running down a dark, cobwebbed hall with a candle. As the woman moves forward, any cobwebs before her will become illuminated by the approaching light, while those left clinging to the walls behind her will eventually dip back into the shadow of darkness. Likewise, AE Aurigae will continue to light up IC 405 for perhaps another 20,000 years, after which the light reflecting off IC 405 (or exciting its gases) will fade to near invisibility. Before it leaves the nebulous region, though, AE Aurigae will greatly alter the dusty clouds that surround it. Spectroscopic observations have shown the bright nebula to be incredibly tenuous, containing only 1,000 or so particles in every cubic centimeter. So, as the star "runs" through it, the pressure of starlight from AE Aurigae will push aside the surrounding gas just as our running woman would sweep away any cobwebs before her.

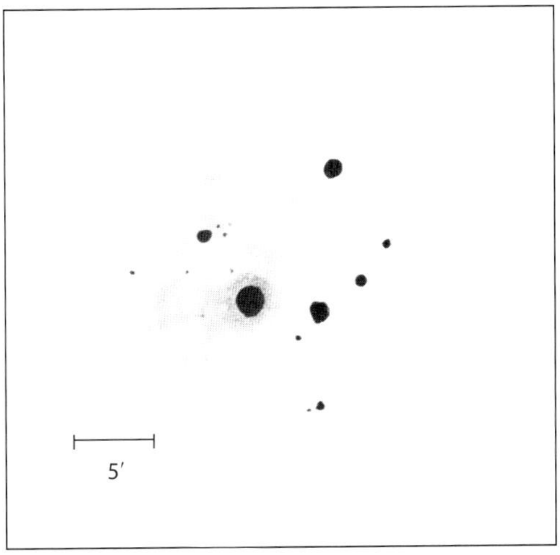

In a 1954 paper titled "The Radial Velocity of the Flaming-Star Nebula, IC 405," Nicholas U. Mayall (Lick Observatory) writes, "The galactic diffuse nebula IC 405 was discovered in 1892, first by J. M. Schaeberle, and then by M. Wolf and E. von Gothard, during photography of Nova Aurigae 1891." Wolf considered the inner structure of IC 405 on his photographs to be remarkable. Writing in a 1903 issue of *Monthly Notices* of the Royal Astronomical Society, he described the involved 6th-magnitude star as "a burning body from which several enormous curved flames seem to break out like gigantic prominences." He also thought it would be of interest for spectroscopists to observe this "*flaming star*." Thus did the nebula receive its name.

In photographs IC 405 is quite enormous. A 1°-wide pool of dust and gas, oriented southeast to northwest, surrounds AE Aurigae, though the star appears to rest on the middle of its southwestern edge. A faint, 1°-long extension of feebly glowing gas runs south of the brighter component's western side. Overall, IC 405, as recorded on the deepest photographs, looks like a bridal veil attached to a headpiece, or like Italy without the "toe" of the "boot."

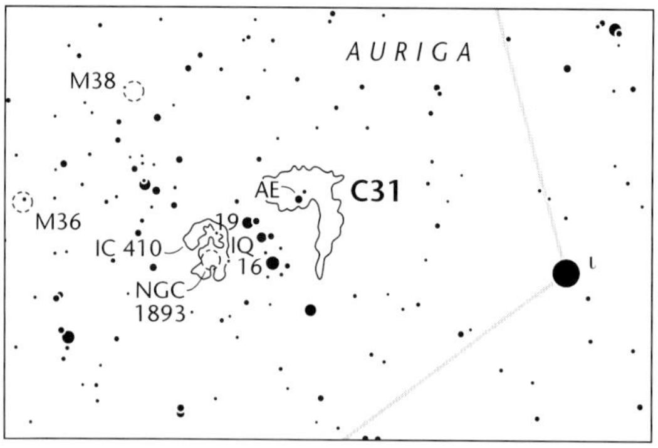

High-resolution blue-light images show the reflection nebulosity as a vast network of filaments. These features show a continuum spectrum believed to be caused primarily by starlight scattering off dust, and it is this reflection nebulosity that we can spy visually through our telescopes. Red-light images, however, show waves of ionized gas lined with splotches of dust that seem to be rippling away from the star in several directions. The long southern extension stands out in red light and is composed of faint parallel streamers, the easternmost of which has a relatively sharp outer edge. In the view of the late Robert Burnham Jr. "the appearance seems to suggest that this boundary is the edge of a zone which has been swept clear of nebulosity by the northward motion of the star." This motion might also explain the apparent "rippling" of gas ahead of the star in its present position. If the distance of 1,600 light-years is accepted for AE Aurigae, the cloud of reflective material extends at least 9 light-years from the star.

AE Aurigae is visible with the naked eye just northwest of an obvious tight, "hazy" grouping of suns, the brightest of which are 5th-magnitude 16, 19, and IQ Aurigae. AE Aurigae and its neighboring asterism, which *Sky & Telescope*'s Alan M. MacRobert likens to a "leaping minnow," lie outside a divine, well-defined brush stroke of Milky Way. When I train my 10×50 binoculars to that part of the sky, I'm surprised to find the magnitude-7.5 open cluster NGC 1893 (about 1° southeast of the "minnow") pulling my attention away from AE Aurigae and the naked-eye asterism. The reason for this is obvious. To the naked eye the asterism looks fuzzy, so curiosity causes me to raise my binoculars to it. Once I find the field in my binoculars, the asterism is no longer fuzzy, but nearby NGC 1893 is, and curiosity takes control once again. (The diffuse emission nebula interwoven with NGC 1893 in the photograph on page 122 is IC 410.)

Little has been published about IC 405's visual appearance. Burnham discusses only its physical and photographic qualities. The most descriptive discussion known to me appears in MacRobert's book *Star-Hopping for Backyard Astronomers*. There he writes, "In moderately large amateur telescopes the 'flames' appear as a large vague glow toward the star's east and north about 20′ across. To my surprise, [my] 6-inch seemed to show hints of it despite the light pollution. Quite a catch!" Roger Clark lists IC 405's apparent magnitude as 10.0 in his *Visual Astronomy of the Deep Sky*, and he suggests that relatively low powers (20× for a 6-inch telescope and 41× for a 12-inch) are best for viewing it. I found it to shine at magnitude 9.2 in the 4-inch at 23×; this magnitude applies to the reflection nebula, described by MacRobert, that lies immediately east and north of AE Aurigae.

Given its diffuse nature, IC 405 would seem to be a visual challenge, but it isn't. Most surprisingly, from Hawaii the object reveals itself in 10×50 binoculars as a faint puff of vapors southeast of AE Aurigae; averted vision is required, however. In the 4-inch at 23× the glow breaks down into finer details, but this requires intermittent viewing over a period of at least an hour. I find it helpful to gently sweep the telescope in a north-south direction while keeping my eyes relaxed and my mind alert. It also helps to take frequent notes and to make sketches to confirm any suspected details. With the wide field of view offered by the Genesis at 23×, AE Aurigae appears as the brightest star in a tight kite-shaped asterism oriented southeast to northwest; an 8th-magnitude star to the northwest of AE Aurigae marks the kite's tail. The nebula's most prominent portion is a wedge flaring southeast from AE. In blue-light photographs one can see dark nebulosity pinching into the southeastern side of this wedge, forming a distinct Y-shaped pattern, but I could not make out that detail in the 4-inch; perhaps those with larger telescopes can do so. A fainter nebulous glow surrounds the 10th-magnitude star just northeast of AE Aurigae, and a long, thin streamer runs along the kite's entire northeastern edge. Very faint patches of nebulosity hide inside the kite's southwestern edge.

Another of IC 405's surprises is that its brightest features take magnification well. The details are not apparent at first, but with patience the once-amorphous glow to the southeast becomes a sharp hook of nebulosity, while several patches appear to the northeast and north of AE Aurigae, almost suggesting a ripple pattern. The streamer along the kite's northeastern edge vanishes at high power, its surface brightness unable to withstand much magnification. The 8th-magnitude star marking the northwestern tip of the kite asterism is a nice double at high power. Its fainter companion is supposed to shine at about magnitude 10, though I found it closer to magnitude 12.5. Interestingly, the *Washington Double Star Catalog* lists 8th magnitude for both companions. Keep your eye on this star; it might be a variable.

32

NGC 4631
Type: Barred Spiral Galaxy (SBd)
Con: Canes Venatici

RA: 12h 42.1m
Dec: +32° 32′
Mag: 9.2
Dim: 14.7′ × 3.5′
SB: 13.3
Dist: 22 million light-years
Disc: William Herschel, 1787

W. HERSCHEL: [Observed 20 March 1787] Very bright. Much extended from south preceding to north following [from southwest to northeast] but near the parallel. Much brighter in the middle. 16′ long. (H V-42)

GC: Remarkable, very bright, very large, extremely extended toward approximate position angle 70°, bright in the middle to a nucleus, bright star nearby.

NGC: Remarkable, very bright, very large, extremely extended toward approximate position angle 70°, bright in the middle to a nucleus, magnitude 12 star attached at north.

TO THE NAKED EYE CANES VENATICI LOOKS anemic. Its brightest star, Cor Caroli, shines at 3rd magnitude, and its other stars are fainter than 4th magnitude. But the constellation hosts a dynasty of telescopic splendors: the "colorful" globular cluster M3; the dramatic Whirlpool Galaxy, M51; the delicate Sunflower Galaxy, M63; and the mysterious, black-hole-bearing Seyfert galaxy, M106. And there's more. In the obscure southwestern quadrant of the constellation lurks one of the most distinguished — yet often overlooked — deep-sky objects of its class: the edge-on galaxy NGC 4631. Shining at magnitude 9.2 and spanning nearly ¼°, NGC 4631 is, in fact, among the brightest and largest edge-on galaxies in the night sky, and a vivid reminder that Messier's catalog is not an exhaustive list of celestial greats.

At its distance of 22 million light-years, NGC 4631 is quite large, having a true linear diameter of 94,000 light-years and a total mass of some 50 billion Suns. We see the system inclined only 5° from edge on. But if we could tilt NGC 4631 over and view it face on, we would probably see a late-type barred spiral with very clumpy and loosely wound arms.

The late Gérard de Vaucouleurs was the first (in 1963) to suggest that NGC 4631 was a barred spiral. The radial-velocity data upon which his conclusion rested were considered highly suspect by some. However, in 1991 Jean-René Roy (Laval University, Canada) and his colleagues provided further evidence for a bar in the galaxy.

In exposures taken through large telescopes NGC 4631 looks like the rotting corpse of a galaxy. Warped and withered extensions seem to slump lifelessly on either side of its bulging core, while reams of dust seem to bore into the bundles of starlight that cover the galaxy's quilted surface. These features are all battle scars, for NGC 4631 is an outstanding example of an interacting galaxy undergoing intensive star formation. Its highly energetic core complex may have been triggered into activity by the tidal influences of its two nearest neighbors: NGC 4627, a dwarf elliptical or spheroidal galaxy merely 3′ to the north, and NGC 4656, a very irregular spiral about 30′ to the southeast. (Actually, two NGC numbers — 4656 and 4657 — describe different parts of this very peculiar object, which some amateurs know as the Hockey Stick. You can see NGC 4627 and the Hockey Stick in the photograph on page 126.) Radio telescopes have discovered five spectacular tidal spurs of hydrogen gas extending as far as 130,000 light-years from the disk of NGC 4631, while visible-light images have revealed a faint bridge of light connecting NGC 4631 to NGC 4627. Radio astronomers also have recently discovered two expanding supershells of gas in NGC 4631. The larger one, in the galaxy's eastern limb, spans 10,000 light-years in diameter, making it one of the largest known in a galactic disk. If supernova explosions created that expanding cavity of hydrogen gas, they must have numbered some 10,000 to 35,000. A more feasible explanation for the superbubble may be the impact of a fast-moving hydrogen cloud onto the galaxy's disk.

To find NGC 4631 look about halfway down, and a little southeast of, an imaginary line from Alpha[1,2] ($\alpha^{1,2}$) Canum Venaticorum (Cor Caroli, a superb double star) to Gamma (γ) Comae Berenices. The galaxy shines at magnitude 9.2, and with averted vision I have seen it (albeit with some difficulty) through 7×35 binoculars under dark skies. At 23× the galaxy is one of two equally bright elongated hazes ½° apart. (NGC 4631 is the more northerly one; the other is the Hockey Stick). Long and cigar-shaped, its pale light reminds me of a glowing meteor train. Its central region appears brightest, while its tapered tips gradually fade into darkness on either end. Increasing the magnification to 72× shows the galaxy in all its splendor. The disk is a beautifully detailed cloud of light and dark vapors entwined in a delicate embrace. Star clumps pepper the disk like snowballs tossed onto the side of a house. To fully appreciate this very complex sight, take the time to capture moments of good seeing, because a maze of visual treats awaits you. (The galaxy's central region itself warrants hours of attention.) Look for a triangle of bright patches just southwest of a 12th-magnitude star punctuating the northern rim of the galaxy; a dark but delicate grid of dust separates these patches from the other blossoms of light populating that region. Because of its dust complexes, Barbara Wilson considers NGC 4631 one of the finest galaxies in the sky. "This

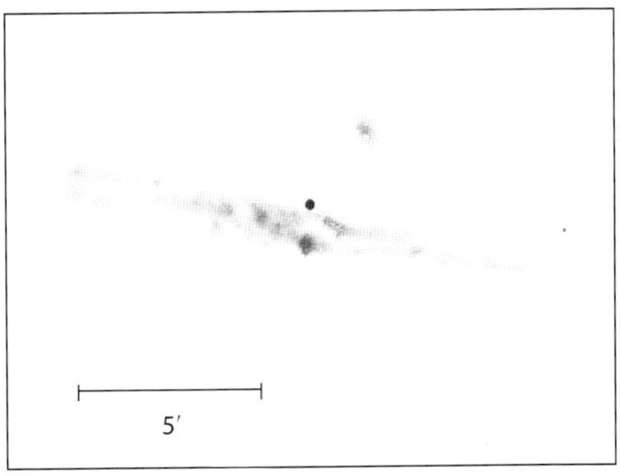

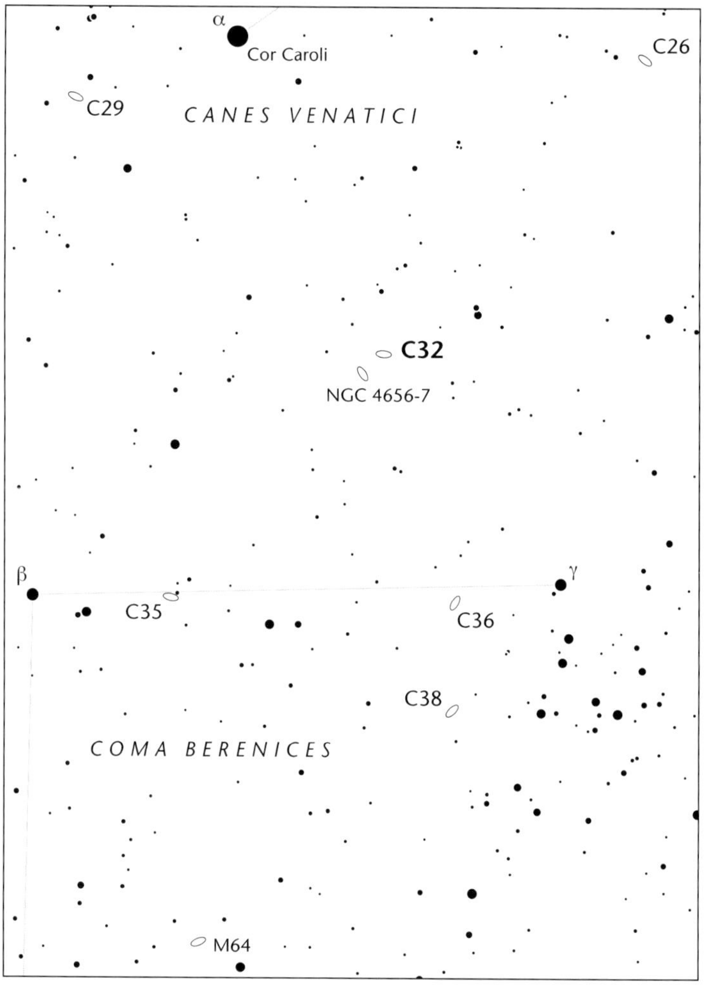

galaxy," she explains, "is the reason I bought a 20-inch telescope."

Christian Luginbuhl and Brian Skiff see the galaxy's western end tapering to a thin, long point near a 13.5-magnitude star. Interestingly, my eye tends to bend that tapered point into a gracefully curved tip that meets the star. Certainly this is an illusion, since photographs reveal the tapered end warping ever so slightly in the opposite direction, to the south. However, our mind's eyes like to "connect the dots." The galaxy's eastern end has a smoother, rounded shape, though larger telescopes show a fainter extension sharpening eastward to a point. Sinewy dust serrates the galaxy's interior portion, reminding me of a giant clam's open shell.

Surprisingly, the 4-inch shows 12.4-magnitude NGC 4627 quite well. During the 1998 Texas Star Party Houston amateur Al Kelly attached a Cook-book 245 CCD camera to his homemade 17½-inch f/4.5 Newtonian reflector to image the faint light bridge connecting NGC 4631 to NGC 4627. Skiff originally verified the existence of this feature photometrically, while Houston amateur Larry Mitchell is the first person known to have seen it with his own eyes. (He since has shown it to me through his 36-inch reflector.) This phantom feature resembles a strip of transparent tape adhering to both galaxies; it glows ever so slightly above the sky background, looking no more distinct than a patch of black ice.

33

Network Nebula; Bridal Veil Nebula (eastern segment)
NGC 6992 and NGC 6995
Type: Supernova Remnant
Con: Cygnus

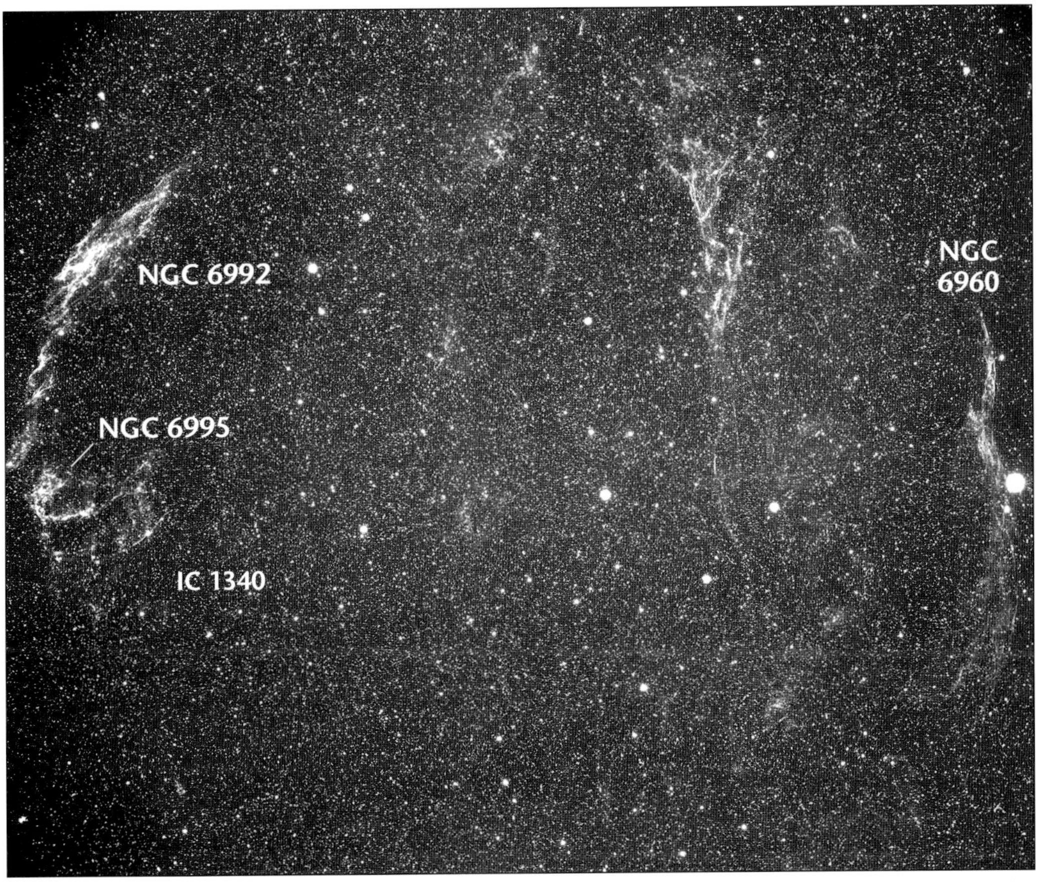

RA: 20^h 56.4^m (NGC 6992)
Dec: +31° 43′ (NGC 6992)
Mag: 7.5 (O'Meara)
Dim: 60′ × 8′
Dist: ~2,500 light-years
Disc: William Herschel, 1784

W. HERSCHEL (NGC 6992): [Observed 5 September 1784] Branching nebulosity, extending in R.A. near 1½° and in Polar Distance 52′. The following [eastern] part divides into several streams uniting again towards the south. (H V-14)

GC/NGC (NGC 6992): *Remarkable*, extremely faint, extremely large, extremely extended, extremely irregular figure, bifurcated.

NGC (6995): Faint, extremely large, nebula and stars in groups.

34

Filamentary Nebula; Bridal Veil Nebula (western segment)
NGC 6960
Type: Supernova Remnant
Con: Cygnus

RA: $20^h 45.7^m$
Dec: +30° 43′
Mag: 7.9 (O'Meara)
Dim: 70′ × 6′
Dist: ~2,500 light-years
Disc: William Herschel, 1784

W. HERSCHEL: [Observed 7 September 1784] Extended, passes through Kappa [52] Cygni. By the Newtonian view above 1° long. By the front-view near 2° long. See note: The front view is a method of using the reflecting telescope different from the Newtonian, Gregorian and Cassegrain forms. It consists in looking with the eye glass, placed a little out of the axis, directly in at the front, without the interposition of a small speculum; and has the capital advantage of giving us almost double the light of the former constructions. (H V-15)

GC/NGC: *Remarkable*, pretty bright, considerably large, extremely irregular figure, 52 Cygni involved.

ABOUT 15,000 YEARS AGO AN UNKNOWN STAR died spectacularly in Cygnus. The event, a Type II supernova explosion, would have briefly dominated the night sky on Earth. If prehistoric peoples were in the habit of gazing toward the heavens for meaning, they might have found it in the sudden appearance of this "new" star, which could have rivaled the full Moon in brilliance and remained visible for weeks in broad daylight. The creators of the oldest known human art, the Cro-Magnon, might have witnessed this stellar apparition, and conceivably they recorded the event in a cave painting somewhere — though if they did, the artwork remains undiscovered.

What *has* been discovered is the remnant of that powerful blast. Popularly known as the Veil Nebula, NGC 6960, NGC 6992, and NGC 6995 together constitute the corpse of a single supergiant star that perished in a supernova explosion. (More than 100 such corpses are known; doubtless many others await discovery.) Also known collectively as the Cygnus Loop, the entire wreathlike complex is an illusion of sorts. Its various pieces are parts of an expanding shell of dimly glowing gas measuring six Moon diameters in our sky, or 80 light-years in space; it appears looplike in part because we see more gas in a given square arcminute of sky when we look at the shell's edges than when we gaze through its middle. A photograph of the Veil seems to tell a simple yet haunting tale, one that has been echoed over and over again since time immemorial — that life, even for things cosmological, ends in death.

A Type II supernova announces the death of a supergiant star — one that starts life with at least 8 times the mass of the Sun. Betelgeuse, in Orion's shoulder, is a supergiant, as is Antares at the heart of Scorpius. Such stars age about 1,000 times faster than does our Sun. As it ages a supergiant star forges ever-heavier atomic nuclei, from carbon early in its life to iron near the end. Each nuclear-fusion stage releases energy and helps the star fight the ever-present pull of gravity. Once the star's core creates iron, however, nuclear fusion can proceed there no further. (Energy is consumed, rather than generated, by the fusion of elements heavier than iron.) Since energy is no longer being released, the outward pressure that supported the star stops. The star succumbs to the force of gravity. Within seconds the core of the once-mighty supergiant collapses into a sphere merely 10 km across. The rest of the star rushes inward, only to rebound off that dense core.

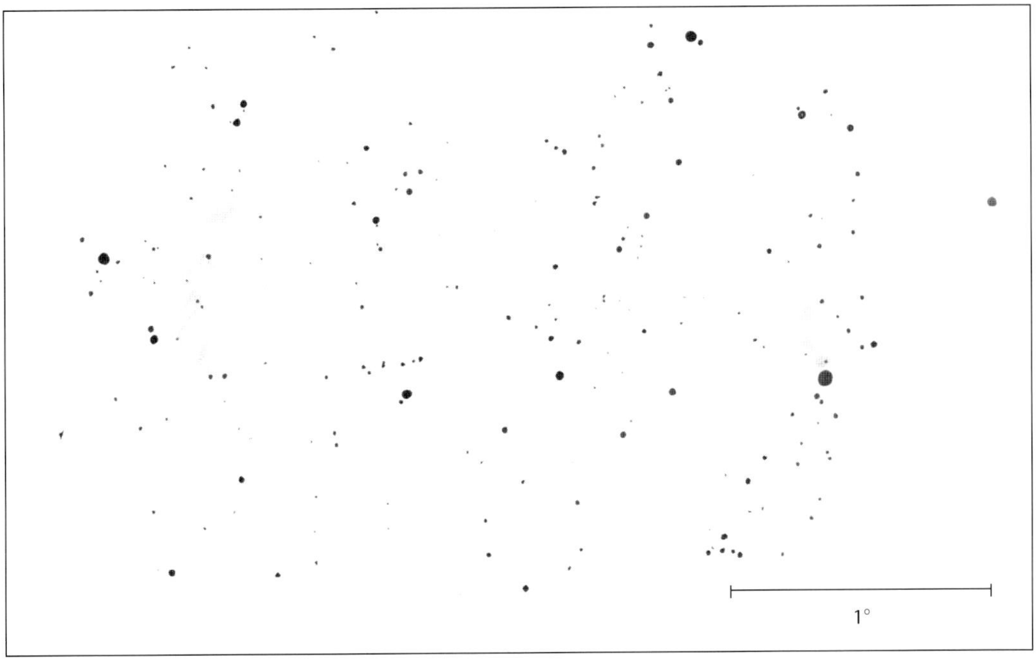

A quarter-second later, the star ends its life in a cataclysmic explosion that will shine with the light of a billion Suns — the luminosity of a small galaxy. The blast hurls several solar masses of material outward at speeds measured in thousands of km per second. Plowing through space, the ejecta collide with interstellar material, heating it up and causing it to glow. The ejecta will continue to expand into space until it fully dissipates and becomes part of the interstellar medium. Sprinkled among the fleeing debris are the newly forged atoms of heavy elements like gold, silver, and calcium. Many of the keepsakes we treasure — a wedding band; a great-grandmother's silver pendant; an Olympian's medal; a child's tooth — are made of atoms dispersed by supernova explosions before our solar system was born.

Of all the Type II supernovae to have occurred in our galaxy during recorded history, the most famous manifested itself on Earth in A.D. 1054. Chinese astrologers witnessed that event and recorded it as a "guest star" in the annals of the Sung dynasty. In the 18th century, the supernova's glowing remnant briefly fooled Charles Messier into believing he had chanced upon a new comet. The Crab Nebula, as the object is known today, became the first deep-sky object listed by Messier in his famous catalog of noncometary objects. Next to the Crab, the Veil Nebula in Cygnus is the supernova remnant most sought after from midnorthern latitudes, and it is a perennial favorite at summer star parties.

Pointing a telescope toward the Veil is easy; seeing it at the eyepiece may be difficult. Success depends on your observing site more than anything else. First locate magnitude-2.5 Epsilon (ε) Cygni, the star marking the eastern arm of the Northern Cross. About 3° south of Epsilon Cygni you'll find the magnitude-4.2 star 52 Cygni. The fainter, western segment of the Veil (NGC 6960) virtually "touches" 52 Cygni, running along a north-south line ever

33 & 34

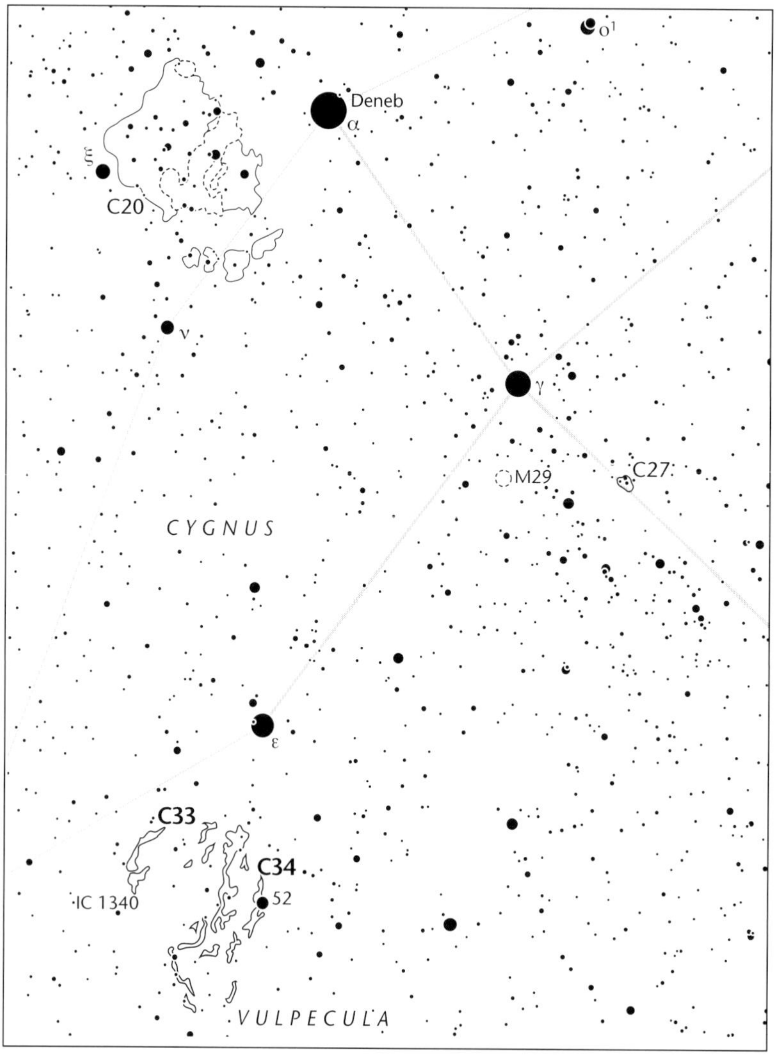

so slightly east of the star. The brighter, eastern segment (NGC 6992 and NGC 6995) lies 2.7° to the northeast of 52 Cygni. In its entirety the Cygnus Loop is enormous, measuring 3.5° × 2.7°. The two principal arcs mentioned here just fit inside the edge of the Genesis's low-power field of view, which spans nearly 3°. Under a dark sky NGC 6992 and NGC 6995 are easily visible in 7×35 binoculars, while NGC 6960 is more difficult but just within range. Together they look like fading X-rays of fractured chicken bones.

Some 20 years ago, I made several pointless attempts to see the Veil through the 9-inch refractor at Harvard College Observatory. (Light-pollution filters weren't available then.) Having failed with so mighty a telescope, I believed that seeing the Veil at the eyepiece was impossible — until the renowned nova discoverer Peter Collins finally showed it to me through 11×80 binoculars one year at Astro-Assembly, a gathering of amateur astronomers under the dark skies of Seagrave Observatory in Rhode Island. The experience opened my eyes to the corrosive power of light pollution, for what I could detect in binoculars under a dark sky I could not detect in a fairly large telescope under city skies. That same night, Richard Lynch, of the Skyscrapers Astronomy Club, showed me the Veil's feathery form through Seagrave's 9-inch refractor. I enjoyed seeing this stellar corpse from an observatory with the word "grave" in it. The experience has had an impact on me ever since. Here in Hawaii, for example, low-power views of the nebula still remind me of spirits fleeing an open grave.

Seen through the Genesis at 23×, the brightest portions of the Veil (NGC 6992 and NGC 6995) are remarkably obvious, looking like a cosmic ripple. Even a fast sweep of the region brings them clearly into view. In fact, during a comet hunt in December 1996 I chanced upon them, finding that they resembled a headless comet with a diabolically curved tail. This appearance made me wonder how Messier had missed such a splendor. But the answer seems obvious. Messier did not enjoy the wide fields of view provided by so many of today's telescopes. Through his narrow-field instruments the nebula, if detected at all, would have appeared like dim and stringy sections of the Milky Way and not anything like the condensed form of a comet. Remember, too, that the Veil resides on the eastern outskirts of the overwhelmingly complex Cygnus Milky Way. In his *Celestial Handbook* Robert Burnham Jr. argues that a wide-field eyepiece is essential for detecting this difficult object, which he likens to "a miniature Milky Way in itself."

Many works will state that small telescopes will not reveal much detail in the Veil's opposing segments. But this is not wholly true. What one sees depends in part on his or her location and experience. Through the Genesis under a dark sky, the Veil appears to me as intricately braided vapors encircling a hollow area of the Milky Way. The braided structure of the various filaments can be inferred from the numerous dark lanes that cross the glowing nebulosity. Especially apparent are the gaps between NGC 6992, NGC 6995, and IC 1340, the three segments that make up the Network Nebula. (IC 1340 was not included by Patrick Moore when he defined Caldwell 33, but clearly it is part of the Network Nebula.) I did view each independent section with medium and high power. The Network Nebula took 72× very well! I could not make out individual streamers, but I could infer their presence from the nebulosity's definite patchiness and its dark veins. *Sky & Telescope* editor Joshua Roth says he used to get "knockout views" of the Veil from California's Mount Pinos with a 4⅛-inch Astroscan and an ultra-high-contrast filter. Some of Mount Pinos's "old-timers," he adds, were surprised that this telescope could show the whole loop at 16×, and with some detail at that. He also finds the Veil pretty easy in 7×50 binoculars from a dark site.

The Filamentary Nebula (NGC 6960) is best seen by placing 52 Cygni in the center of the field. Look for a beautiful stream of vapor or a lock of silver hair flowing by the star, like a comet's dim ion tail. A very faint extension can be followed to the south, but doing so with a small telescope takes time and patience. I should note that I visually "climbed the ladder" of the Filamentary Nebula's twisting filaments at the 1994 Texas Star Party, where I climbed a ladder and peered through the eyepiece of a 36-inch reflector built (and then owned) by Florida amateurs Tom and Jeannie Clark. The view of the Veil through that telescope was as good as any photograph (better, in fact, because I experienced it firsthand) and the nebula possessed a rose-colored hue. It's rare to see such color in dim nebulosity, but aperture certainly helps, as does a narrowband filter favoring the nebula's fine-tuned emissions. Overall, the nebula's structure reminded me of coils of DNA.

Of course, to the peering eye of the Hubble Space Telescope even a tiny section of the Veil Nebula is inconceivably intricate, looking like a close-up photograph of colored smoke rising from an unseen cigarette. HST imaged an area where the supernova blast wave recently hit a cloud of denser-than-average interstellar gas. After comparing the HST image to a 1953 Palomar Observatory photograph, two Johns Hopkins University scientists recently concluded that the Veil is a good deal younger (about 5,000 years old) and closer to Earth (about 1,400 light-years) than widely believed. However, these revisions depend on a still-incomplete understanding of shock-wave propagation in the interstellar medium, and they have yet to gain universal acceptance.

Once you have contemplated the Veil Nebula's details to your eye's delight, return to low power and try to glimpse two fainter wisps inside the main loop. The westernmost wisp is the more prominent of the two. Look for a triangular "head" on the north end of this tadpolelike feature. Can you see the tadpole's "tail," at the southern end, curve northward? From a dark site, the entire inner arc is prominent enough to fool an unwary observer into believing that he or she has seen the Veil's easternmost segment. The perceived dimensions of these fainter, enclosed features are tentative because they tend to follow star streams, and it is hard to judge where the nebulosity ends and the star streams begin.

Like you, no doubt, I have wondered how the various names associated with this nebula came about. None of the names appear to be associated with pre-photographic (i.e., purely visual) descriptions. William Herschel wrote of "branching nebulosity" that "divides into several streams," while the Rev. T. W. Webb simply refers to NGC 6992 as a "large nebulosity in a curve" and to NGC 6960 as a "nebulous ray." The terms "Network Nebula" and "Filamentary Nebula," however, were in use at the beginning of the 20th century. They ostensibly describe the appearances of the eastern and western segments, respectively, on high-resolution photographic plates. Writing about the "Nebular Hypothesis" in his 1914 work *The Problem of Volcanism*, Joseph P. Iddings discusses several "[c]louds of luminous matter . . . separated by rifts of darkness whose character is unknown." Among them is "another great cloud-like nebula of this kind . . . the 'network' nebula in Cygnus." His use of quotation marks reveals that "Network Nebula" was already in use by the 1910s. Iddings goes on to describe the nebula as "a cirrus cloud of inconceivable extent." Iddings, then, might have been the first person to associate the nebula with Cirrus, another occasionally invoked moniker. In his popular 1938 work *Astronomy*, Forest Ray Moulton describes the nebula's appearance on a wide-field plate by Lowell Observatory astronomer Wilbur A. Cogshall as "Veil-like"; his may have been the earliest use of that term. Indeed, by 1957 the "Veil" was the common name used by Bart Bok for NGC 6960. "Cygnus Loop" was bestowed upon the nebula after Rudolph Minkowski photographed the entire supernova remnant with the 48-inch Schmidt camera on Palomar Mountain. The names, then, appear to follow progress in astrophotography.

By the way, supernovae occur in our galaxy roughly once every hundred years on average, which suggests that we are well overdue for another. So keep your eyes peeled to the night sky, or, as Jack Horkheimer, TV's Star Gazer, would say, "Keep looking up!"

35

NGC 4889
Type: Elliptical Galaxy (E4)
Con: Coma Berenices

RA: $13^h\ 00^m\ 08^s$
Dec: +27° 58.5′
Mag: 11.5

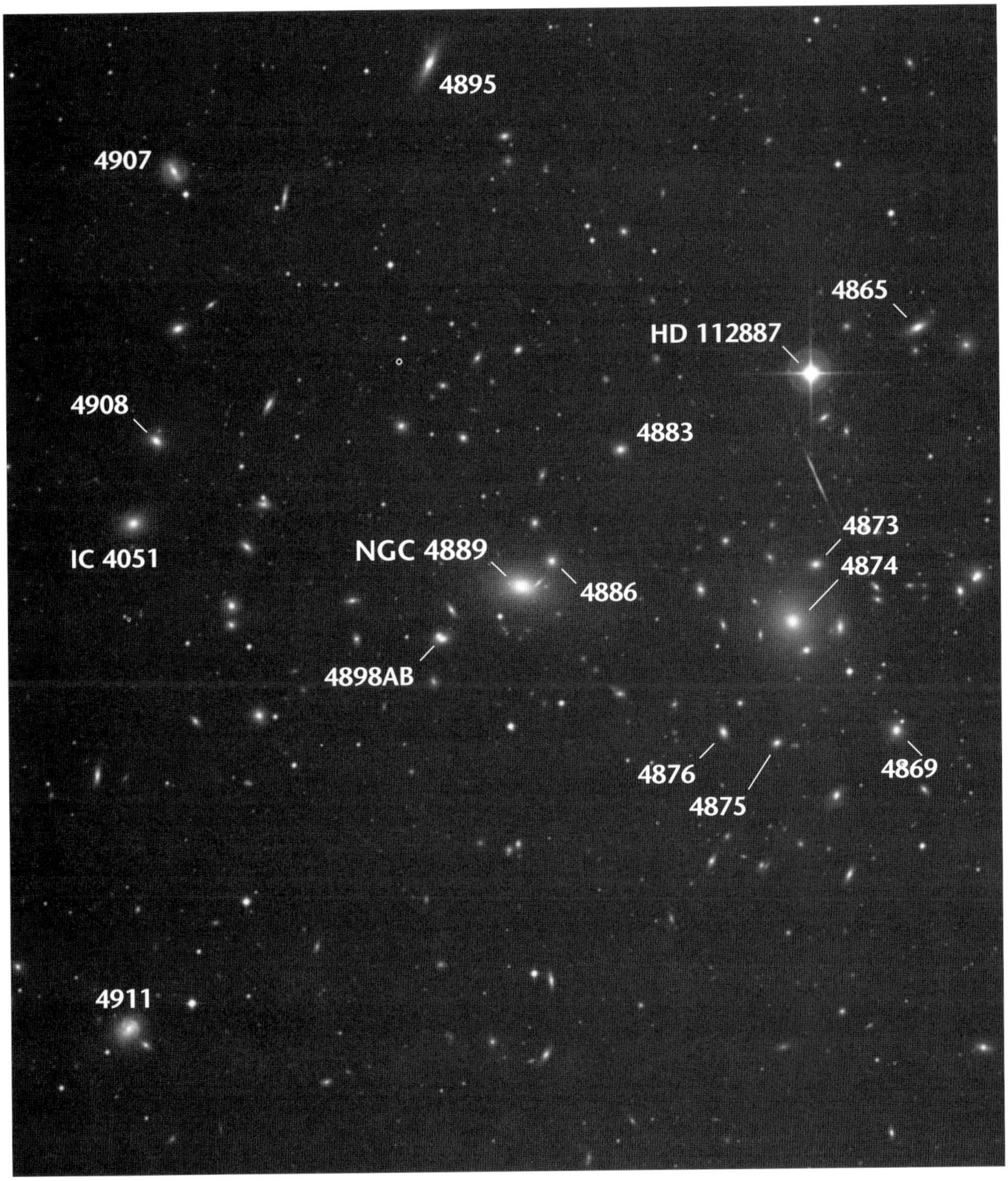

35

Dim: 2.8′ × 2.1′
SB: 13.3
Dist: 300 million light-years
Disc: William Herschel, 1785

W. HERSCHEL: [Observed 11 April 1785] Faint. (H II-391)

GC: Pretty bright, pretty much extended, bright in the middle, a star of magnitude 7 to the north. 4th of 5.

NGC: Pretty bright, pretty much extended, bright in the middle, a star of magnitude 7 to the north.

AMONG THE SILENT DESIRES OF MANY AMATEUR astronomers is the ability to peer ever deeper into the universe. We have, it seems, an innate longing to reach for the "beginning," or at least to better understand it. "Going up that river was like traveling back to the earliest beginnings of the world," said the fictitious explorer Marlow in Joseph Conrad's *Heart of Darkness*. Visual observers go "up that river" by seeking out fainter and fainter objects, objects that will let them look farther back in time. Theirs is a visual journey spurred on by that same, almost primitive, inner voice that must have inspired explorers of old to conquer uncharted territories. But unlike those Earthbound voyagers, today's amateur astronomers can travel phenomenal distances with the help of their telescopes.

Anyone who has seen all the Messier objects has witnessed some remarkably distant galaxies. The farthest Messier objects, those in the Virgo Cluster of galaxies, lie 55 million light-years away. Many of the objects in the Caldwell Catalog take us several steps beyond the Virgo Cluster, but none takes us farther than NGC 4889 (Caldwell 35) in Coma Berenices. By spying this tiny and dim object we thrust our reach 300 million light-years into intergalactic space; the feeble light we detect from the galaxy's pale hulk embarked on its silent journey when the first primitive seed plants sprouted from Earth's bleak and solemn soil and the earliest amphibians crawled out of the water to drag their bellies across our planet's virgin shores.

NGC 4889 belongs to the Coma Cluster of galaxies and is the easternmost of the two galaxies that dominate the cluster's center; its near-equal, NGC 4874, is a mere two-tenths of a magnitude fainter and, like NGC 4889, appears to be an intense source of radio emission. The Coma Cluster is a rich, spherical agglomeration of at least 1,000 bright elliptical and S0 galaxies and perhaps 10,000 dwarf galaxies. NGC 4889 itself is a giant elliptical or "cD" galaxy with a compact core, a smooth, lens-shaped body, and an extensive halo. Loosely speaking, the galaxy's structure mimics that of the Coma Cluster itself, for the cluster has a compact core centered on NGC 4889 and NGC 4874, a smooth halo of smaller elliptical and S0 galaxies perhaps 7 million light-years in radius, and an outer envelope, 40 million light-years across, that harbors many of the cluster's spiral galaxies. HST images of NGC 4889 taken in 1998 reveal two prominent companions superimposed on, or embedded within, the galaxy's extended envelope; a peppering of fainter objects also haunt the immediate environs. The HST data are now being used to determine what physical processes were occurring when galaxies were forming within the cluster's core.

In 1957 Caltech astronomer Fritz Zwicky identified 804 galaxies with photographic magnitudes brighter than 16.5 within 2⅔° of the Coma Cluster's center. The *Millennium Star Atlas* shows nearly three dozen of these galaxies within 1° of NGC 4889 and NGC 4874. In 4-inch and smaller telescopes, these near twins are the most obvious members of the Coma Cluster, along with NGC 4921, which shines at

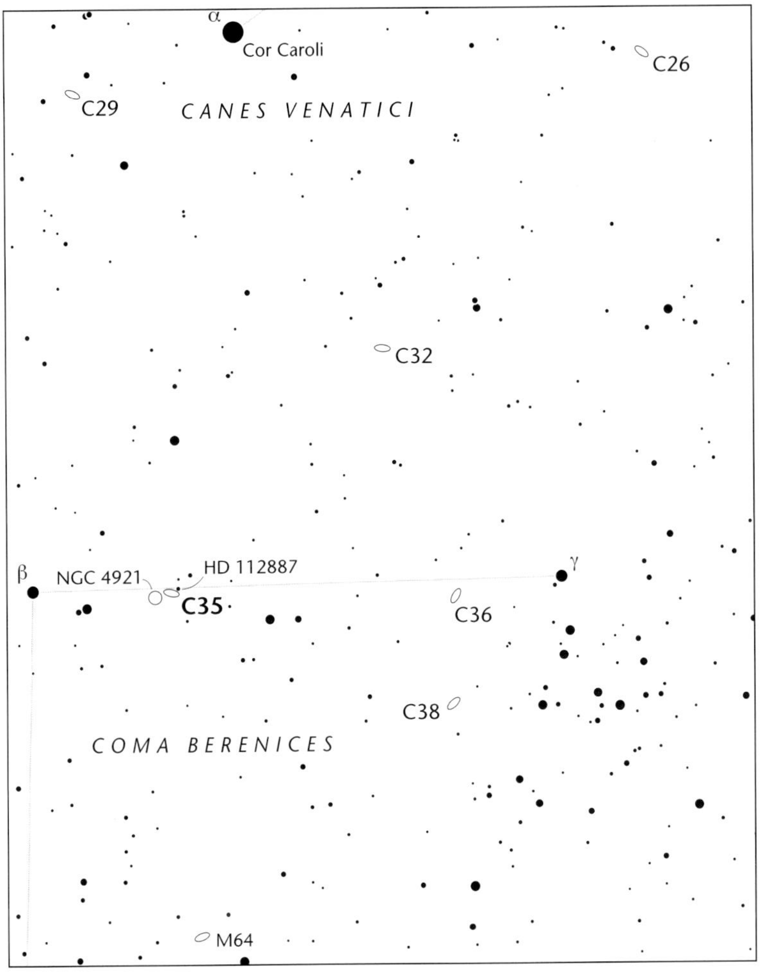

magnitude 12.2 some 18′ to the east-southeast; the rest of the galaxies in the 2°-wide field are all fainter than 13th magnitude.

NGC 4889 lies 2.6° essentially due west of 4th-magnitude Beta (β) Comae Berenices. In the Genesis at 23× NGC 4889 is a small, faint, slightly out-of-round glow with a smooth outer halo that gradually brightens to a sharp core. The galaxy is almost lost in the glare of HD 112887, a 7th-magnitude star 9′ to northwest. The harder-to-see NGC 4874 is a swollen spot of haze merely 6.5′ south of that same star. At 72×, and with time and much patience,

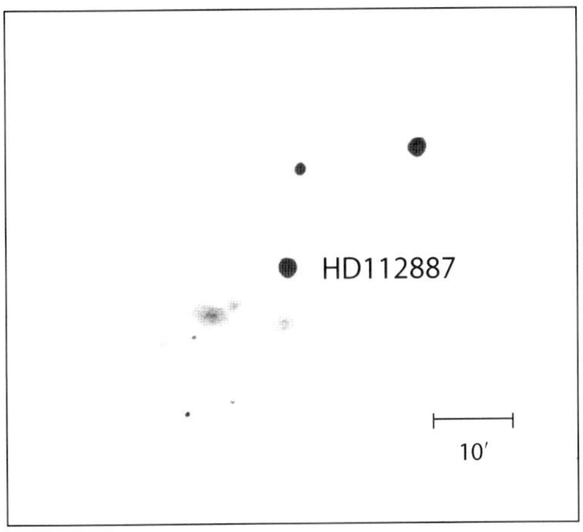

NGC 4889 appears elongated northwest to southeast. Under excellent seeing conditions its "extensions" break away from the main body and stand alone as two independent fuzzy spots: NGC 4886, a compact 13.9-magnitude galaxy to the northwest, and NGC 4898AB, a magnitude-13.6 pair of tightly interacting galaxies. I sketched these companions without knowing of their existence, believing at first that they were perhaps part of a linear bar belonging to NGC 4889. This observation demonstrates the human eye's ability to plumb the universe's depths with a small telescope. It also shows that compact 13th-magnitude galaxies can be seen through small telescopes under dark skies. The fact that *any* of the Coma Cluster's galaxies can be spotted from Earth in a 4-inch telescope implies that they are quite sizable. Indeed, NGC 4889 is immense. At its estimated distance of 300 million light-years its 2.8′-wide body measures 240,000 light-years — about 2½ times the size of our own Milky Way. By dint of the universe's expansion, NGC 4889 whisks away from us at an astounding 6,500 km per second.

Of course, viewing the Coma Cluster through larger instruments is mind-boggling. Texas amateur Barbara Wilson used a 13.1-inch f/4.5 reflector at 166× to see the entire field centered on NGC 4889 and NGC 4874 "full of faint nebulous round objects at the limit of vision." She saw dozens more when she moved the telescope one field-of-view diameter to the northeast. She ended her commentary with a single word: "Fascinating!" The Coma Cluster is indeed a vast jungle of galaxies, for some 30,000 of them can be spied within 6° of the cluster's core down to magnitude 19.0 — posing an imaging challenge to any amateur who owns a moderate-sized telescope and a CCD camera and who wants to explore this celestial heart of darkness.

36

NGC 4559
Type: Mixed Spiral Galaxy (SABcd)
Con: Coma Berenices

RA: $12^h\ 36.0^m$
Dec: +27° 58′
Mag: 10.0
Dim: 11.3′ × 5.0′
SB: 14.3
Dist: 32 million light-years
Disc: William Herschel, 1785

W. HERSCHEL: [Observed 11 April 1785] Very bright, very large, much extended from north preceding to south following [from northwest to southeast] 10′ or 12′ long. 4 stars in it. (H I-92)

GC/NGC: Very bright, very large, much extended toward position angle 150°, gradually brighter in the middle, 3 stars following [to the east of the object].

IN COMA BERENICES MORE THAN A DOZEN GALaxies shine at 10th magnitude or brighter. Among these are the famous Black Eye Galaxy, M64, and the constellation's flagship spiral, NGC 4565 (Caldwell 38), a textbook edge-on system. Just 2° due north of NGC 4565 (and perhaps overlooked because of its proximity to it) lies NGC 4559, a remarkable example of a spiral system in an intermediate morphological class.

The galaxy classification scheme developed by Edwin P. Hubble in 1925 is simple enough. First it divides galaxies into three major types: *ellipticals* (E); *unbarred spirals* (S); and *barred spirals* (SB). Hubble's famous "tuning fork" diagram, which illustrated his galaxy types, also included a then-hypothetical transitional form of galaxy, which he dubbed simply "S0." S0, or *lenticular*, galaxies (since discovered) are intermediate in form between ellipticals and spirals. They are lens-shaped spheroids embedded in flattened disks that lack spiral structure. The unbarred and barred spiral families each branch off of the lenticular S0 form on Hubble's diagram. Each branch is then subdivided largely according to the size of the galaxy's central bulge and the tightness of its spiral arms; both parameters are described by lowercase letters that range from "a" to "d". An "Sa" or "SBa" spiral has the largest nuclear region and most tightly wound arms, while an "Sd" or "SBd" spiral has the smallest nucleus and most loosely wound spiral arms.

So where does NGC 4559 fit in? According to the so-called Revised Morphological Galaxy Classification System, an elaboration of Hubble's original scheme, the galaxy is a mixed SAB(rs)cd.

This means that in shape NGC 4559 lies somewhere between a barred spiral and an unbarred one (SAB), whose inner arms form a ring that surrounds the needle-sharp nucleus (rs) while farther out the arms fly freely open (cd).

Recent advances in our knowledge of the high-redshift universe are providing us with a better understanding of the history of galaxy morphology. For instance, the Hubble Space Telescope has shown that the distinction between ellipticals and spirals grows less definite with increasing distance, becoming nearly imperceptible in images that show how the universe looked when it was only one-tenth its current age. Other studies suggest that ellipticals formed remarkably early in the universe while spiral galaxies took much longer to form. The origins and evolution of galaxy morphology remain active (and contentious) fields of research.

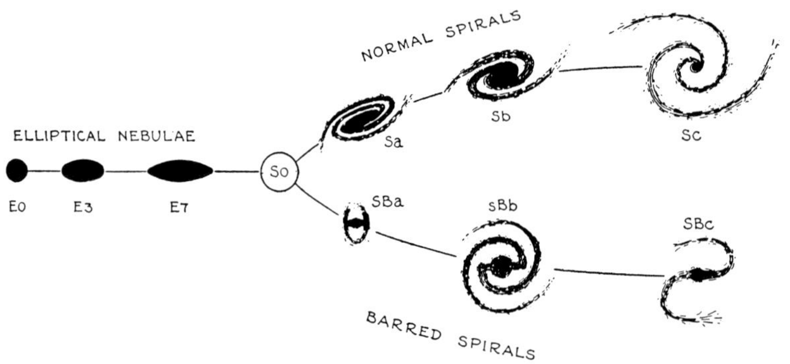

FIG. 1. *The Sequence of Nebular Types.*

NGC 4559 is a sizable galaxy, spanning 100,000 light-years, with a total mass of 40 billion Suns. Together with NGC 4565 it is a member of the Coma-Sculptor Cloud of galaxies, and it is receding from Earth at a speed of 820 km per second. The galaxy is inclined 69° from face on. In photographs NGC 4559 looks like a spinning saw blade whisking toward the observer at a skewed angle. Its untamed arms appear tormented, as if they have been stretched by three people in three different directions. The arms are dotted with bright H II regions that seem to cling to them like lint to a sweater.

In the mid-1990s the Rosat spacecraft revealed 17 discrete X-ray sources within NGC 4559. The most luminous one, sited in the galaxy's nucleus, is slightly extended, suggesting either an unresolved superposition of several compact sources or a diffuse source of emission. Another bright source was found in an outlying spiral arm, where it coincides with a conspicuous group of H II regions. Since little radio or infrared emission emerges from this location, the X-rays may have originated from a single, several-hundred-year-old supernova remnant.

Small-telescope users will have no trouble locating NGC 4559. The galaxy lies only 2° east-southeast of magnitude-4.3 Gamma (γ)

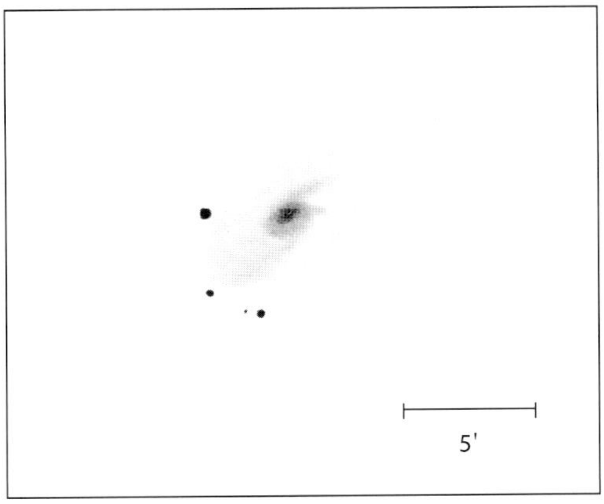

Comae Berenices, the obvious northernmost star in the constellation's famous inverted Y asterism. If your telescope is on an equatorial mount and has a 1°-wide field of view, you can locate the galaxy in two easy steps. First find Gamma Comae Berenices and place it at the northern end of your field of view. Next simply move two eyepiece fields to the east and NGC 4559 should almost be centered in your eyepiece. It is not difficult to see. In fact, I have spotted it (albeit with effort) in 7×35 binoculars as a dim but large glow. I have also chanced upon its polished 10th-magnitude form during a casual comet sweep with my 4-inch refractor at 23×.

Christian Luginbuhl and Brian Skiff call the galaxy "impressive" in a 6-inch telescope. At 23× in the 4-inch Genesis, the 11′-long object looks obviously oval-shaped, with three 12.5- to 13.5-magnitude stars forming an arc on the galaxy's southeastern edge. Another conspicuous star lies just beyond the galaxy's northeastern flank. Increasing the magnification to 72× will reveal NGC 4559's oval core, which gradually brightens to a starlike center. If the southern half of the galaxy appears slightly brighter and more "complete" than the northern half, this is the time to take some deep breaths and look for

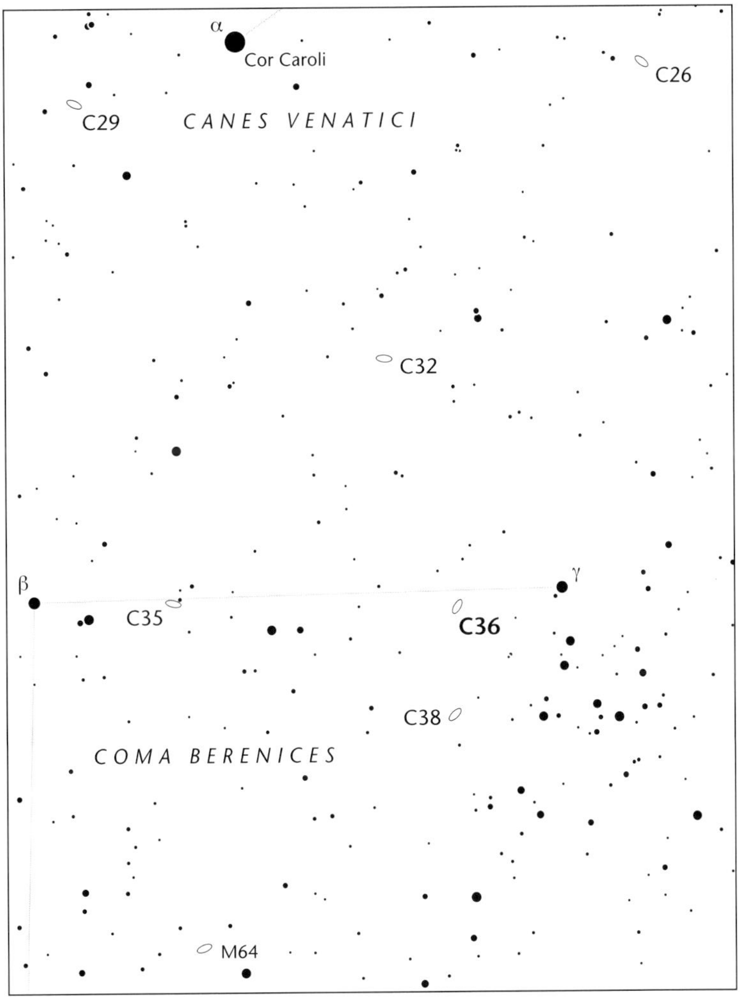

delicate variations of light. First look for a northwestward extension — a feeble spiral arm that appears to be pushing against an embryo of faint nebulosity. The strongest half of this arm is that nearest the core. More challenging is sighting the hyperfine dark lane separating the stubby southeastern arm from the oval core. Use the highest feasible magnification, lightly tap the telescope tube, and hyperventilate briefly. If this fails, stand up and walk around the telescope to regenerate your circulation, take a deep breath, return to the eyepiece, and try again. I suspected two spiral arms or two dust lanes after performing these rituals a few times, but I was not able to keep them in constant view. Maybe you will have better luck. Larger telescopes, of course, will reveal the galaxy's mottled texture. (There are seven *IC* objects, all condensations, in the galaxy.)

By the way, NGC 4559 is a good galaxy for supernova hunters to check in on every now and again. In 1941 a magnitude-13.2 supernova flared to prominence some 30″ west and 20″ north of the galaxy's nucleus. Such a luminous event could happen again, so be sure to place this galaxy on your supernova checklist.

37

NGC 6885
Type: Open Cluster
Con: Vulpecula

RA: 20^h 11.6^m (derived from Herschel's discovery position)
Dec: +26° 28′ (derived from Herschel's discovery position)
Mag: 8.1 (not including 20 Vulpeculae)
Diam: 20′
Dist: 1,950 l.y.
Disc: William Herschel, 1784

W. HERSCHEL: [Observed 9 September 1784] A cluster of coarsely scattered stars, not rich. (H VIII-20)

GC/NGC: Cluster, very bright, very large, rich in stars, little compressed, stars of magnitude 6 to 11.

CALDWELL 37, THE OPEN CLUSTER NGC 6885 in Vulpecula, poses a problem. So, too, does open cluster NGC 6882, also in Vulpecula. On modern star charts, NGC 6882 is superimposed upon NGC 6885 (or vice-versa). But this was not always the case. That brings us to the main problem: star charts and references are not consistent in the way they plot or portray the two clusters, so it's difficult to know which object we are viewing when we point our telescopes toward these clusters' part of the sky. There is a history behind the confusion, and I will recount some of it below. But first let's look at the issues confronting amateurs who want to locate and identify Caldwell 37.

Decades ago, observers using the reigning star charts of their day would have had no problem identifying NGC 6885. The 1962 edition of the Skalnate Pleso *Atlas of the Heavens* shows NGC 6885 as a sizable object (20′ wide) centered on the 6th-magnitude star 20 Vulpeculae. So, too, does the 1966 edition of *Norton's Star Atlas* and the first (1981) edition of Wil Tirion's *Sky Atlas 2000.0*. The 1978 edition of Robert W. Burnham Jr.'s *Celestial Handbook* lists NGC 6885 as an object 20′ across, with 35 stars of 6th to 11th magnitude, 20 Vulpeculae included. No one had to worry about confusing NGC 6885 with NGC 6882 because NGC 6882 is not plotted on any of these early atlases, and Burnham didn't mention or list NGC 6882 in his work. To locate NGC 6885 all one had to do was point a telescope toward 20 Vulpeculae, look for a loose gathering of moderately bright stars, and enjoy the view. Patrick Moore undoubtedly had this straightforward experience in mind when he selected NGC 6885 for his Caldwell Catalog.

Now, a word of warning: if you do not want to lose yourself in a convoluted tale of historical folly, just follow in Moore's footsteps, point your telescope toward 20 Vulpeculae, and enjoy the view. If, however, you decide to read on, do not be surprised if you wake up one morning banging your head against the wall of a padded cell.

Okay, I see that my warning didn't deter you. Let's look, then, at how NGC 6885 and NGC 6882 are treated in some modern atlases. The 1987 edition of *Uranometria 2000.0* shows NGC 6885 surprisingly diminished in size, appearing as a tiny (7′) cluster centered on 20 Vulpeculae; NGC 6882 is superimposed on NGC 6885 (though with its center offset to the northwest of 20 Vulpeculae) and measures 20′. The two clusters also appear this way in the second (1998) edition of Tirion's *Sky Atlas 2000.0.* In other words, NGC 6885 is now a tiny cluster of stars centered on 20 Vulpeculae, which lies in the southeastern quadrant of the larger cluster NGC 6882. The *Millennium Star Atlas* also portrays NGC 6885 as a 7′-wide cluster centered on 20 Vulpeculae, but larger NGC 6882 is offset to the west-southwest. The 18th edition of *Norton's Star Atlas* (dated 1989) did not alter the picture painted by earlier editions of that work; NGC 6885 is still centered on 20 Vulpeculae, and NGC 6882 is still not plotted. On the far end of the spectrum, *Observe: The Herschel Objects* (published by the Astronomical League) positions NGC 6882 just south of the star 19 Vulpeculae, which is 20′ north of 20 Vulpeculae.

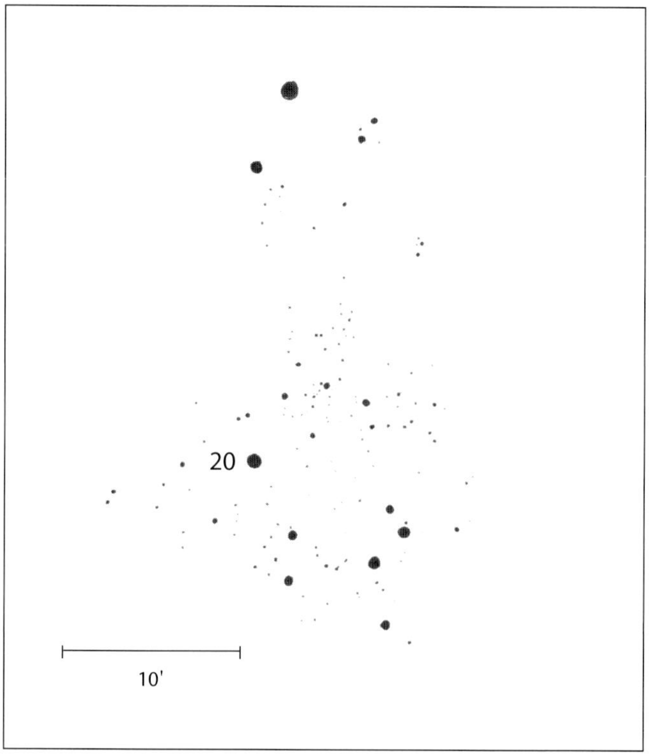

Clearly something is amiss, and the problem is only compounded by reading the various catalog descriptions of these objects. Texas deep-sky aficionado Barbara Wilson first brought this dilemma to my attention. "Notice," she said, "that the *NGC* description of 6885 does not match William Herschel's original description of this cluster." Of NGC 6885, William Herschel wrote: "A cluster of coarsely scattered stars, not rich." The *NGC* reads "Cluster, very bright, very large, rich in stars, little compressed, stars of magnitude 6 to 11." Why did the descriptions change, and does the discrepancy have anything to do with the incompatibility of the various atlas treatments?

The *NGC* is based on John Herschel's *General Catalogue of Nebulae* (*GC*), which was published in 1864, and not on William Herschel's original discovery data (at least not directly). In the *GC* John Herschel described NGC 6885 thus: "Cluster, very bright, very large, rich, little compressed," with stars ranging from 6th to 11th magnitude. Wilson suspects that John Herschel departed from his father's description of NGC 6885 to match what he saw in the sky. Johann Louis Emil Dreyer retained John Herschel's description in his 1888 *New General Catalogue of Nebulae and Clusters of Stars* (*NGC*). Neither William Herschel's description of NGC 6885 nor the *NGC*'s matches what modern atlases depict — a tiny cluster of faint stars around 20 Vulpeculae. Also note that William Herschel's description does not mention the presence of a 6th-magnitude star; that addition was made by John, who also noted the star's position. When the position for the 6th-magnitude star is precessed, we see that he was referring to 20 Vulpeculae. But a survey of John Herschel's original observing notes reveals that he did not say the cluster is "centered" on 20 Vulpeculae; he merely provides a position for the star.

Dreyer undoubtedly used the position of 20 Vulpeculae, however, to represent the "center" of NGC 6885. But 20 Vulpeculae is probably not the true heart of NGC 6885. For one thing, a quick look through a telescope reveals that no obvious 7′-wide cluster is centered on the star. (Of course, it's easy for anyone to visually force 20 Vulpeculae into the center of a bunch of stars in this rich Milky Way region, but it isn't immersed in a compact cluster.) For another, the Hipparcos satellite has pegged 20 Vulpeculae's distance at 1,140 light-years (plus or minus 20 percent), most likely making it a foreground star on the sightline to NGC 6885. In the September 1978 issue of *Sky & Telescope* Walter Scott Houston also placed 20 Vulpeculae "at the southeast edge" of NGC 6885, not at its core.

For its part, Volume 3 of the *Webb Society Deep-Sky Observer's Handbook* contains no listing for, or drawing of, NGC 6882, but under the listing for NGC 6885 contributor Edmund S. Barker did include this terse note: "Superimposed on 6882." A further note describes an observation of NGC 6885 with an 8-inch telescope at 170×: "the two clusters [lie] 6′ apart, the separation being somewhat arbitrary." The "two clusters" undoubtedly refer to NGC 6885 and NGC 6882, but, unfortunately, the book provides no clue as to where NGC 6882 lies.

Brent Archinal of the U.S. Naval Observatory has discovered one clue to the NGC 6882–6885 puzzle, but it raises as many questions as it answers. In 1930, Robert J. Trumpler cataloged the two clusters, as Archinal explains:

> In the region around 20 Vulpeculae two clusterings seem to be superposed: A loose clustering of 20-30 mostly faint stars, about 8′ in diameter, with center at 20h 7m 28s +26° 15.0′ (1900) [the very position (once precessed) given for NGC 6885 in *NGC 2000.0*] was identified with NGC 6882. The coarse group of a few bright stars clustered around 20 Vulpeculae, with a diameter of about 22′, was identified with NGC 6885. In each of these two clusters a physical relationship of the stars is indicated by the magnitude spectral class diagram, but the two clusters are evidently at different distances.

Trumpler went against the grain and made NGC 6882 a small cluster to the northwest of 20 Vulpeculae and NGC 6885 a large cluster around 20 Vulpeculae. The following year Per

Collinder reversed the two dimensions again in his catalog; his Collinder 416 (NGC 6882) measures 20′ and his Collinder 417 (NGC 6885) measures roughly 8′. The positions Collinder gives for the two objects are precisely the same positions that Trumpler gives for NGC 6882 and NGC 6885. In other words Collinder's position for NGC 6885 is the same as Trumpler's position for NGC 6885, but Collinder's diameter for NGC 6885 equals Trumpler's diameter for NGC 6882. And Collinder's position for NGC 6882 is the same as Trumpler's position for NGC 6882 but Collinder's diameter for NGC 6882 is roughly the same as Trumpler's diameter for NGC 6885. "My guess," Archinal says, "is that Collinder simply copied Trumpler's positions (since they match exactly) for NGC 6882 and NGC 6885, not realizing that his identifications were the reverse of Trumpler's." Wilson, who has also been trying to solve this puzzle for years, simply tossed her hands into the air after she searched numerous catalogs for answers, but found only more questions. "No one really agrees with each other on what NGC 6882 or 6885 is!" she says. "Every reference in [the] amateur literature gives conflicting information. It is obvious that the two clusters have been confused with each other since [the time of William] Herschel."

What a mess! But forget all the confusion. Let's wipe the slate clean and look at William Herschel's original discovery notes. (Such a step is fundamental in solving any identification problem; one must always refer to the original sources so far as is possible.) Just where did William Herschel position NGC 6885 and NGC 6882?

William Herschel made only one observation of each cluster on different (though subsequent) nights. First he discovered the object now known as NGC 6885. On September 9, 1784, he found NGC 6885 1′ east and 27′ south of 18 Vulpeculae. Interestingly, Herschel's discovery position of NGC 6885 is not centered on 20 Vulpeculae but on a point about 6.5′ west-southwest of that star. In fact, Herschel's discovery position for NGC 6885 matches the center of NGC 6882 as plotted on the *Millennium Star Atlas.* Furthermore, when I took an unbiased look through my telescope I saw a faint but dense patch of stars northwest of 20 Vulpeculae and a brighter but looser scattering of suns to the southwest of that star; Herschel's position for NGC 6885 is roughly midway between these two concentrations. Now at least one piece of the puzzle appears to make sense. Without question there is a cluster of stars (or the semblance of one) at Herschel's position for NGC 6885, and it is my recommendation that star charts place a 20′-wide NGC 6885 there. Seen another way, NGC 6882 in the *Millennium Star Atlas* should read NGC 6885.

Now let's look at NGC 6882.

On September 10, 1784, William Herschel discovered NGC 6882 just 1^m 12^s east and 12′ south of 18 Vulpeculae. Of it he penned "A cluster of coarsely scattered stars" — which is uncannily similar to his description of NGC 6885. Indeed, some astronomers have argued that Herschel's observation of NGC 6882 is simply a repeat viewing of NGC 6885. Supporting this allegation is the position of NGC 6882 when plotted on a star chart. William Herschel positioned NGC 6882 just south of the star 19 Vulpeculae, where no cluster is evident; this may explain why many of the early star charts do not show NGC 6882. Archinal found a small oval asterism at Herschel's position for NGC 6882 on the *National Geographic Society–Palomar Observatory Sky Survey* and once suggested that this was the missing NGC 6882. In September 1998, Archinal and I used the 31-inch reflector at the Warren Rupp Observatory, stopped down to give us the effective aperture of Herschel's telescope. We then employed an eyepiece that gave us a 15′ field of view (again, to match Herschel's) and swept the region. Both of us saw and studied this asterism, but it seemed unlikely that Herschel would describe such a distinct asterism as a "cluster of coarsely scattered stars." Archinal's thinking on the subject has evolved since, however. John Herschel never observed NGC 6882, so his records are of no help.

NGC/IC sleuth Harold G. Corwin (Infrared Processing and Analysis Center) has buttressed the notion of redundancy in the case of these two clusters. He posits that William Herschel made a 15′ error (one telescopic field diameter) in his September 10, 1784, observation. "Aside from the declination difference," he says, "the observations are consistent with the idea that he saw the same cluster both nights. This would make NGC 6882 [equal to] NGC 6885." Corwin further notes that William Herschel found seven objects on September 10, 1784. When Corwin analyzed Herschel's offsets for all of them — using the Smithsonian Astrophysical Observatory's star catalog for star positions and, when necessary, proper motions — he found that Herschel made significant errors in his positions for four of these objects. "I was looking for a pattern in the discrepancies between William Herschel's positions and the modern positions," he writes, "but do not see one. What I do see, though, is a possibility that his data on that night were taken with somewhat less care than normal. Distractions? Sleepiness? Who knows? In any event, there are significant errors in more than half of his positions for deep sky objects that night."

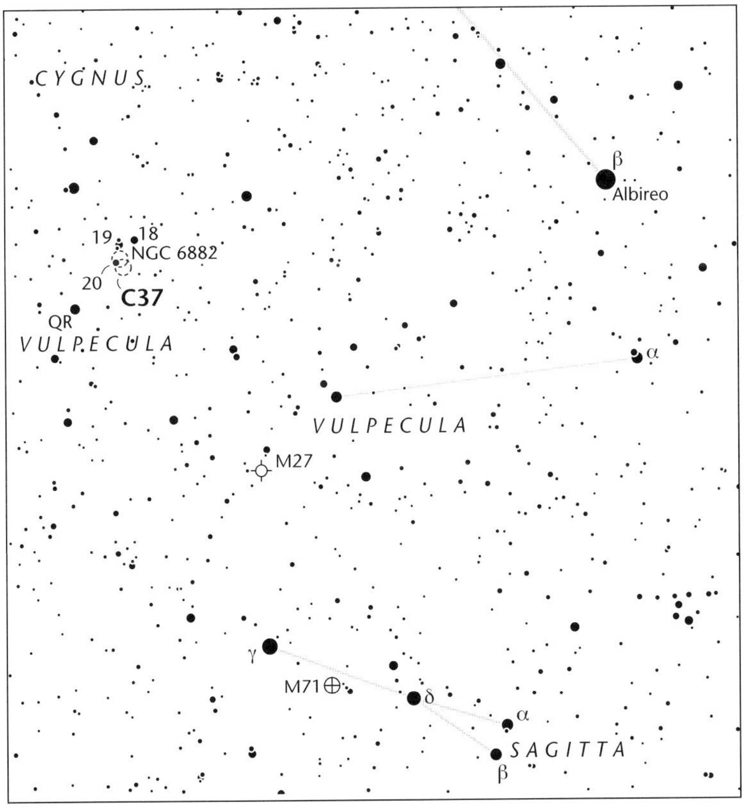

Corwin also notes that Dreyer documented a case in which William Herschel twice recorded the same nebula (NGC 4208 = NGC 4212) during the same sweep — probably within the same five-minute period — and entered the object twice in his lists. "So, it is possible for [Herschel] to have observed the same object twice within a short period of time without recognizing it as one he had seen before," Corwin says. "This reinforces my feeling that the 15 arc minute offset in the position of NGC 6882 is a simple mistake on [Herschel's] part.

But until we have the ability to go back in time . . . to peer over [Herschel's] shoulder as he was sweeping, I don't think we'll ever know for sure exactly what NGC 6882 is."

To find NGC 6885 first use your naked eye and look about 10° (one fist's width) southwest of Epsilon (ε) Cygni, the elbow of the celestial Swan's southern wing, or a similar distance (actually 9°) east-southeast of Beta (β) Cygni (Albireo), the famous telescopic double star marking the Swan's nose. There you will find the 6th-magnitude star 20 Vulpeculae huddled with two other similarly bright stars (18 Vulpeculae and 19 Vulpeculae) in a ½°-wide patch of sky; 20 Vulpeculae is the southernmost of the three. Two 7th-magnitude stars also accompany 19 Vulpeculae, and the entire naked-eye region looks like a speckled haze about the size of the full Moon. This haze is not caused by NGC 6885, which shines below the limit of naked-eye visibility at magnitude 8.1, but by an optical effect that spins closely knit stars into webs of imaginary wispiness. (Messier masters may prefer to star-hop to 20 Vulpeculae from M27, the Dumbbell Nebula, about 5° to the southwest.)

I should mention that some references include 20 Vulpeculae in the cluster's total brightness. The second edition of *Sky Catalogue 2000.0*, for instance, lists the cluster's photographic magnitude as 5.7. But this gives observers unrealistic expectations. Even under a dark sky one could easily miss NGC 6885 with a sweep of 7×35 binoculars because 20 Vulpeculae overwhelms its dim light, which looks like a burp of "gas" just southwest of that star. The cluster might well be lost entirely under city lights.

With the nearly 3° field of the Genesis at 23×, the region including 18, 19, and 20 Vulpeculae looks like a cluster being ripped apart by some unseen tidal stress; the densest population of stars near 20 Vulpeculae is a crumbling "comet" with a wide parabolic hood of starlight and a tail that flows northward toward 19 Vulpeculae. Look for a bar of fuzzy starlight 6′ northwest of 20 Vulpeculae, oriented east-northeast to west-southwest; this is the concentration of stars in NGC 6885 that Trumpler classified as NGC 6882. An obvious trapezoid of stars appears immediately to the southwest of 20 Vulpeculae.

At 72× a conspicuous pair of stars lies immediately to the northeast of 20 Vulpeculae, while a dozen or so fainter suns lie to the east. A rough line of equally spaced 11th- to 12th-magnitude stars slice through the fuzzy bar of light, which contains still fainter suns popping in and out near the limit of visibility. Doubling the magnification spreads out the stars farther, providing the best view of William Herschel's "cluster of coarsely scattered stars." In total some 35 stars populate the 20′-wide cluster when counted down to 12th magnitude; 60 members can be tallied down to 13th magnitude. Other suns make up the Milky Way backdrop. A nice 10th-magnitude topaz star lies just to the northeast of 19 Vulpeculae, itself a red-tinted type *K*3I-II giant.

Now return to low power. In a 2° field of view crisscrossing corridors of darkness can be seen in the region directly south of NGC 6885. The view is best if 18 Vulpeculae is placed at the northern end of the field of view. One corridor begins at 19 Vulpeculae and ends just east of the roughly 5th-magnitude variable (and tight double) star QR Vulpeculae, about 1.5° to the southeast. Another begins on the western end of NGC 6885 and ends to the west of QR Vulpeculae. A third corridor lies farther to the west and is paralleled by a sharp needle of dark nebulosity, also to the west. Two perpendicular dark alleys can also be seen in the southern end of this field. The challenge is to see the dark needle near the third corridor in 7×35 binoculars.

I thought I'd wrap up this discussion by going back to the beginning and leaving you with one final forehead-slapping bit of information. Despite everything that you've read, Archinal reminds us that it's still entirely possible that the clustering of stars to the northwest of 20 Vulpeculae is a separate physical group. A 1961 McDonald Observatory study of 15 stars in the area concluded that NGC 6885 might be "an association around NGC 6882," with both being at a distance of 1,950 light-years. And yet "the status of NGC 6885 is quite uncertain," says French cluster expert Noël Robichon, and Douglas Geisler (University of Concepción, Chile) sees "no strong indication" that either *NGC* object is really a star cluster. "So keep in mind," Archinal cautions, "that we just don't know" exactly how many clusters populate this unassuming but controversial corner of the northern Milky Way.

38

NGC 4565
Type: Spiral Galaxy (Sb)
Con: Coma Berenices

RA: $12^h 36.3^m$
Dec: +25° 59′
Mag: 9.6
Dim: 16.2′ × 2.3′
SB: 12.9
Dist: 32 million light-years
Disc: William Herschel, 1785

W. HERSCHEL: [Observed 6 April 1785] A lucid ray 20′ long or more. 3′ or 4′ broad from north preceding to south following [from northwest to southeast]. Very bright in the middle. A beautiful appearance. (H V-24)

GC/NGC: Bright, extremely large, extremely extended toward position angle 135°, very suddenly brighter in the middle to a nucleus equaling a star of magnitude 10 or 11.

MORE THAN ANY OTHER DEEP-SKY OBJECT, the sight of the edge-on spiral galaxy NGC 4565 jars my imagination, causing me to wonder about the unseen vistas hiding beyond its dusty confines. Compare the photograph of NGC 4565 above with that of the nearly face-on system IC 342 (Caldwell 5) on page 30. To me NGC 4565 looks more mysterious than IC 342. But why? Perhaps it's because order can be seen in long-exposure photographs of face-on spiral galaxies: multitudes of stars cling like nettles to well-structured arms; dust accumulates rather neatly in channeled lanes. Seen from above, a spiral galaxy is a recognizable visual vortex, a snapshot of celestial water pouring down a drain. Similar spiral structure abounds on Earth — in the unfolding of a flower, in swirling autumn leaves, or in the spinning arms of a hurricane.

Edge-on systems, in contrast, lack that immediacy of recognition. Most of their treasures are hidden behind turgid veils of dust. From our limited perspective it appears that we see more darkness than light, lending these galaxies a supernatural quality that attracts our curiosity. One might think that 19th-century astronomers would have seen these curious "nebulae" as their contemporaries saw the jungles of Africa from passing ships — "so hopeless and so dark," in the words of Joseph Conrad's Marlow, "so impenetrable to human thought." But in 1855 William Parsons, the third Earl of Rosse, described NGC 4565 as a "beautiful object" after seeing it through his 72-inch reflector. Using that powerful telescope five years earlier, Lord Rosse had drawn the equatorial dust lane in another edge-on system, NGC 891 (Caldwell 23). At that time he also discovered the spiral nature of some "nebulae" and likened them to nascent planetary systems. The clairvoyant John Herschel, however, had already hinted that extended nebulae were perhaps round nebulae seen edge on. NGC 891, in particular, gave him the impres-

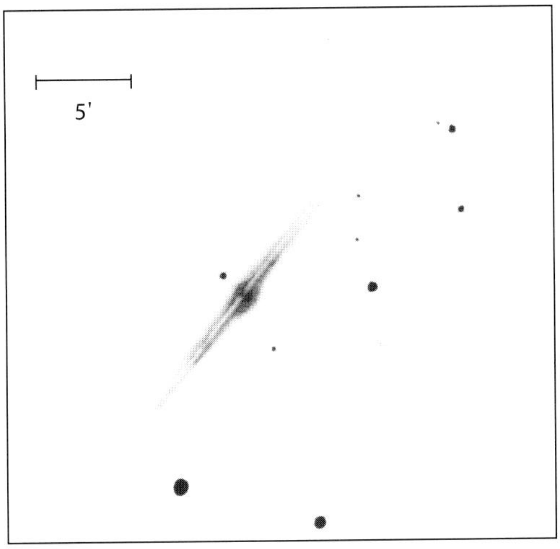

sion of a "thin flat ring of enormous dimensions, seen very obliquely." Since there was no reliable way at the time to determine the distances to these nebulae, their true sizes and natures remained a mystery. It wasn't until 1924 that astronomers finally learned that spiral nebulae weren't baby solar systems within our own galaxy but external Milky Ways of their own, scattered throughout a universe whose dimensions humankind was just beginning to grasp.

We now know that NGC 4565, like several other Caldwell objects, is a member of the cosmologically local Coma-Sculptor Cloud of galaxies. It is also the largest and most famous edge-on spiral galaxy in the night sky. If we accept its estimated distance of 32 million light-years, this moderately late-type spiral has a true linear diameter of 150,000 light-years and a total mass of 200 billion Suns. Interestingly, the Hubble Space Telescope has identified roughly 200 globular clusters around NGC 4565, a total virtually indistinguishable from that attending our own Milky Way. Recent radio observations have revealed that the galaxy's "rotation curve" — the orbital speed of its disk's constituents, plotted as a function of distance from the nucleus — mimics that of our galaxy; as with our Milky Way, NGC 4565's rotation curve indirectly argues for an extended halo of invisible "dark matter" enveloping the disk.

In high-resolution photographs the flattened disk of NGC 4565 displays bright and dark patches with much clumpiness and branching. As with NGC 891, photographs also show dark fingers of dust projecting above and below the equatorial plane of NGC 4565; this material may have been spewed out of the disk by supernova explosions, collisions of interstellar clouds, or magnetic forces. It would be quite a challenge, I imagine, to see these dusty projections at the eyepiece of even a large telescope.

Shining at a respectable magnitude 9.6, NGC 4565 is a ¼°-long needle of light about 2° due east of 17 Comae Berenices, a fine binocular double star in the diffuse, naked-eye Coma Berenices star cluster (also known as Melotte 111). It is a fabulous sight in a telescope of any size. The most dramatic view of NGC 4565 I can recall was through the 16-inch Boller & Chivens reflector at Harvard's Oak Ridge Station in Massachusetts. On one warm spring night in the mid-1970s I found myself galaxy-hopping with this telescope through the rich Coma-Virgo galaxy cloud. I did not know what to expect as I slowly slewed the telescope toward NGC 4565, for I had only seen photographs of the great spiral in books. Looking intently in the main eyepiece for a possible dim glow, I watched countless field stars sail across the eyepiece. Suddenly the sharp tip of a blade of light entered the field from the upper left. Deeper and deeper it cut into the field of view, until the galaxy's robust hub and girdle of darkness all but shattered the visual serenity that had preceded its appearance. I continued to slew the telescope, but the galaxy did not end — not until its leading edge began to exit the opposite edge of the field of view. This

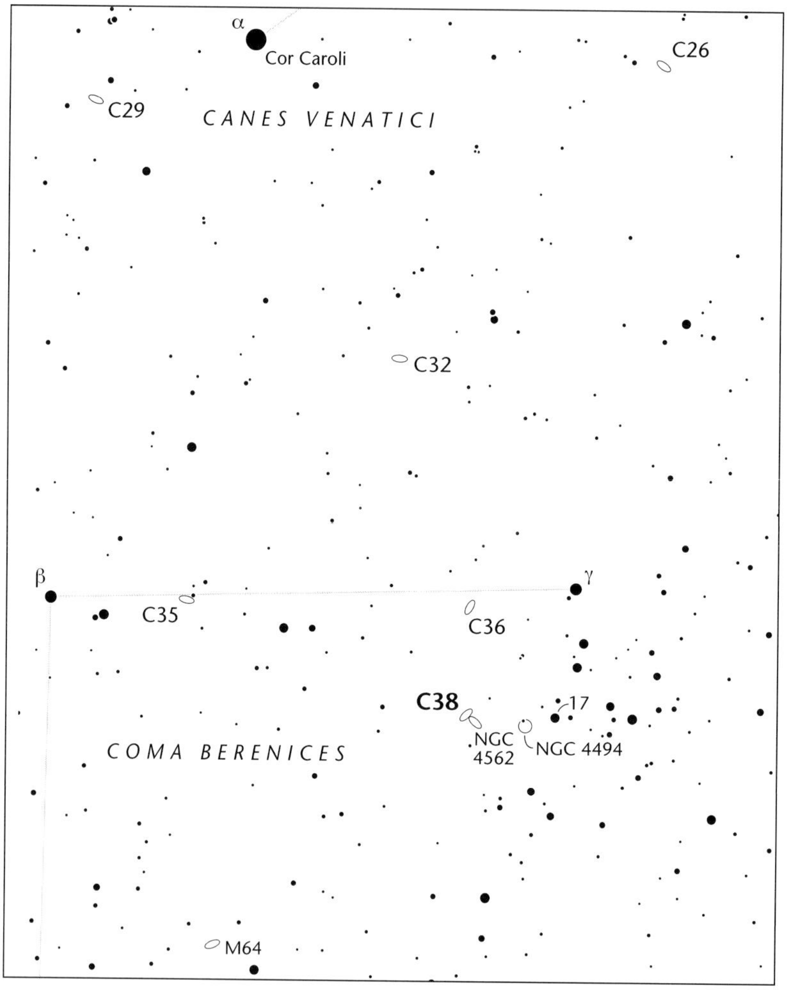

awe-inspiring experience has been branded into my mind, and to this day the sight of NGC 4565 reminds me why I pursue the stars.

Through small telescopes the galaxy is more pretty than awesome. Through the Genesis at 23× it is a long wafer of light oriented northwest to southeast. Two equally long threads of light extend from the galaxy's slightly swollen center, one on either side, like silk from a spider's abdomen. At 72× the galaxy's hub becomes more robust (a 13.5-magnitude star can be seen just northeast of it) and hints can be seen of the dark dust lane slicing through the center. High power and patience provide the best view; that's when the dark lane cleanly divides the bright hub into two distinct ovals. The southwestern oval is punctuated by a bright point of light, which is probably the "nucleus" mentioned in the *New General Catalogue*. With time, the dark lane can be followed across the entirety of the galaxy's equator, though it is more prominent on the southeastern limb, where the border between light and dark has a slight waviness to it. Be aware that NGC 4562, a magnitude-13.5 galaxy, lies 13′ southwest of NGC 4565. This tiny glow (which does not appear on *Sky Atlas 2000.0* but is plotted in the *Millennium Star Atlas* and *Uranometria 2000.0*) is commonly picked up through telescopes with moderately large apertures and mistaken for a comet. (NGC 4562 appears in the lower right-hand corner of the photograph on page 150.) Another galaxy that may be stumbled upon in the search for NGC 4565 is NGC 4494, a 10th-magnitude elliptical galaxy about ½° east-southeast of 17 Comae Berenices.

Perhaps the greatest visual challenge posed by NGC 4565, though, is to sight the edge-on spiral in 7×35 binoculars. Its thinness and its prominent dust lane both cause the galaxy to be close to the limit of detectability. I succeeded at an altitude of 4,200 feet under very dark Hawaiian skies. It was a very difficult observation, requiring a lot of time and patience for me to be confident I had seen the galaxy's slender body. *Sky & Telescope* editor Joshua Roth saw it in his 60-mm f/5.9 refractor from the Amateur Telescope Makers of Boston's dark-sky site in Westford, Massachusetts, and called it a "sweet little needle of light."

Since NGC 4565 and its rival, NGC 891 in Andromeda, lie at about the same altitude after sunset (at opposite ends of the sky) around the start of spring in the Northern Hemisphere, try observing them both on the same night. The two galaxies are essentially the same distance from us, and are roughly the same size, but NGC 4565 is much more conspicuous because its surface brightness is twice that of NGC 891. This difference in surface brightness can be attributed at least in part to extinction — the absorption and scattering of starlight by intervening dust, which lies in our Milky Way and within the edge-on spirals. NGC 891 lies on a sightline that skirts our home galaxy's midplane, and the passage through our galaxy's dusty disk dims NGC 891's light by 0.2 magnitude, or 17 percent. NGC 4565, by contrast, is a mere 3½° west of the north galactic pole — an essentially transparent window onto extragalactic space — and its light is reduced by only 0.05 magnitude (5 percent). Roth says he has yet to see NGC 891 in his 60-mm telescope.

39

39

Eskimo Nebula; Clown-Face Nebula; Lion Nebula
NGC 2392
Type: Planetary Nebula
Con: Gemini

RA: $07^h\ 29^m\ 10.8^s$
Dec: +20° 54′ 43″
Mag: 9.2
Dim: 47″ × 43″
Dist: ~4,000 light-years
Disc: William Herschel, 1787

W. HERSCHEL: [Observed 17 January 1787] A star of 9th magnitude with a pretty bright milky nebulosity all around. A very remarkable phenomenon. (H IV-45)

GC/NGC: Bright, small in angular size, round, a star of 9th magnitude, a star of 8th magnitude north following [to the northeast] 100″.

THROUGH SMALL TELESCOPES MOST PLANetary nebulae look like pale shells or symmetrical patches of light. Their enchanting central stars — whose ionizing ultraviolet rays energize the nebulae — usually lie just beyond the grasp of normal vision. But NGC 2392 in Gemini is an exception. Not only is it bright, being a mere 0.4 magnitude fainter than the famous Ring Nebula (M57) in Lyra, but its central star all but dominates the view. Actually it is somewhat misleading to say that NGC 2392, or any other planetary nebula for that matter, is "bright"; unlike every other type of deep-sky object, no planetary nebula is readily accessible to the unaided eye. NGC 2392 is unquestionably a telescopic wonder, however. Its singular appearance compelled the great William Herschel to call it "a very remarkable phenomenon." His son John examined the nebula in an 18¼-inch reflector and agreed; he saw it as an 8th-magnitude star "exactly in the center of an exactly round bright atmosphere." His use of the word "atmosphere" was appropriate for the times. In 1755 the Prussian philosopher and physicist Immanuel Kant hypothesized that the Sun and planets were born from a rotating cloud of gas and dust. By the 19th century, when John Herschel was turning his attention to his father's discoveries, geologists were using Kant's nebular hypothesis to explain the formation of the Earth. In his 1914 work *The Problem of Volcanism,* which devotes a chapter to nebular hypotheses, Joseph Iddings explains that geologists of the 19th century conceived a nebula as a "homogeneous gas revolving about a central nucleus and separating into annular zones that were supposed to condense into the planets and satellites of the solar system." Any views they might have had of NGC 2392, with its multiple shells and bright central core, certainly would have excited them. Of course, Kant was right in his thinking, as we know today, but the evidence that would ultimately verify his theory would not be found in planetary nebulae, which have nothing to do with planets or their formation.

NGC 2392's true grandeur surfaced with the advent of photography. Photographic plates — especially those taken with large telescopes in red or ultraviolet light — show the nebula's concentric shells primped with fine details. The central star lies at the core of a bizarre gyroscopic arrangement of bright gas, which in turn is surrounded by an annular fanfare of radial streaks. Such a sight compels our eyes to search for patterns, just as they would in the body and folds of a cloud in tumult. Once ignited, our imagination transforms the nebulous patches into something fanciful, like a face. Indeed, NGC 2392's most popular nickname is the Eskimo Nebula because of the way it resembles a face peering out from a fur-lined parka. Ultraviolet-light photographs tend to exaggerate the "nose" (the planetary's hot central star), thus the nebula's other popular nickname, the Clown-Face. Of course, we see what our own imaginations dictate. In his *Celestial Handbook* Robert Burnham Jr. says the "whole nebula irresistibly suggests the classic and unforgettable features" of the late comedian W. C. Fields. I have called NGC 2392 the Lion Nebula ever since childhood, for to me the pattern most resembles the face of a cartoon lion whose large and bristling mane surrounds its smiling countenance.

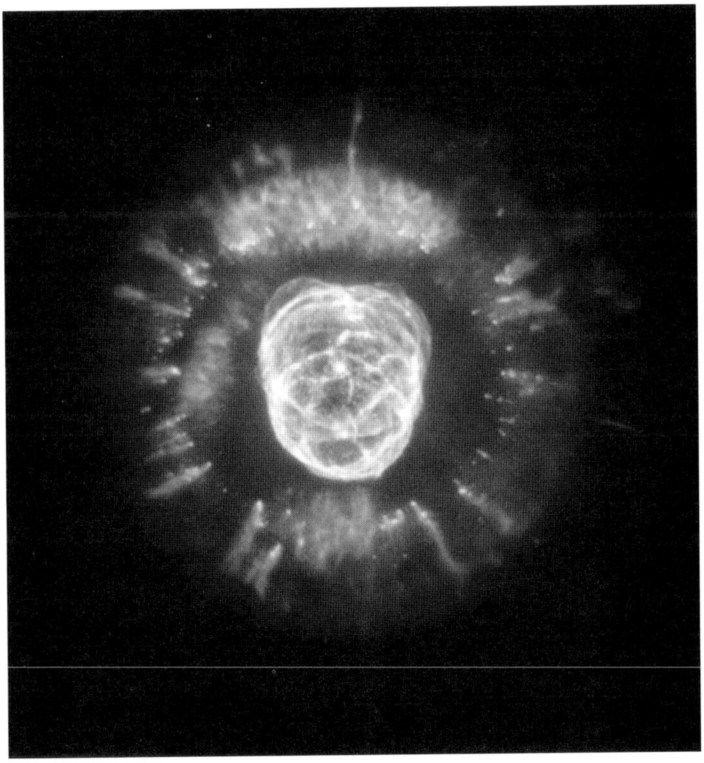

Imagination aside, what *is* NGC 2392, really? Planetary nebulae once were believed to be the abruptly ejected atmospheres of massive red-giant stars. However, this simple theory could not account for the wild shapes seen in these objects. There is a veritable zoo of heavenly creatures among the planetaries, including silkworms, butterflies, and eggs wrapped in kudzu. In the late 1970s, Sun Kwok (now at the University of Calgary, Canada) and his colleagues conceived of a new theory of planetary-nebula formation. They started with a red-giant star sloughing off its outer atmosphere and exposing a white-hot core (a white-dwarf star in the making). They then envisioned superfast winds streaming out from the hot central star and compressing and accelerating the slower circumstellar material that was ejected earlier; the interacting winds would create the shell shapes we see in planetaries today. Kwok's theory was vindicated when the

International Ultraviolet Explorer satellite found that many planetaries' central stars in fact have fast winds. He since has been able to invoke these winds in computer models that account for the detailed shapes of many planetaries. Of course, his models are a bit more complicated than described here, for they include asymmetric cocoons of matter, donuts of molecular gas and dust, and the effects of companion stars.

NGC 2392's singular appearance earned it the honor of being the first object imaged by the Hubble Space Telescope after the December 1999 repair mission, which replaced the orbiting observatory's six gyroscopes. Hubble resolved the nebula's inner portion into a weblike network of filamentary structures on the surfaces of two overlapping bubbles. The light-year-wide outer shell (the Eskimo's furlined parka) displays a halo of cometlike nodules whose tails radiate outward, away from the windy central star.

The appearance of a planetary nebula depends on its orientation with respect to our sightline. In NGC 2392's case we are peering more or less straight down the major axis of an inverted funnel of gas. The complex inner nebula (about 13″ in diameter) looks like two rings superimposed on one another, though slightly offset. And indeed that is what we are seeing. Planetary-nebula observer Jay McNeil of Houston, Texas, likens the inner nebula to "an old wooden barrel" (but a really big one, about two trillion km across) with both ends blown out. We see the translucent barrel's waist as well as the smaller rim at either end. The barrel has a major-axis to minor-axis ratio of about 2 to 1, and its long axis is inclined about 10° to our sightline. Support for this model comes from the expansion velocity of the Eskimo's equatorial region, which measures 60 km per second while wind escapes from the barrel's hollow ends at roughly twice that speed. Since the Eskimo is some 4,000 light-years distant, we are seeing it as it appeared when the nomadic Hyksos ruled Egypt in the Thirteenth Dynasty. If you were to somehow magically break the light-speed barrier and see the Eskimo as it really is today, it would appear measurably larger than 46″.

To find this exotic object, start with magnitude-3.5 Delta (δ) Geminorum, also known as Wasat. Meaning "the middle," this name refers to the star's position in Gemini, though it may also refer to the star's proximity to the ecliptic. This pale white *F*0 IV star lies 59 light-years distant and has a fine magnitude-8.2 *K*6 dwarf companion about 7″ to the northwest; the noted 19th-century observer Adm. William Henry Smyth saw this companion shining with a purple light. (Of course no two observers see exactly the same colors in stars, and their tints may be functions of atmospheric clarity, the telescope used, and, to some degree, the observer's imagination. That's one of the things that make our hobby so fascinating.) Delta Geminorum holds a special place in the hearts of those who knew the late Clyde Tombaugh. In 1930 Tombaugh discovered Pluto a mere ½° east of that star after comparing a pair of photographic plates he had taken during his yearlong search for Percival Lowell's Planet X. Delta Geminorum's "face" soon appeared alongside Pluto's in newspapers around the world. And Delta Geminorum, shining 44,000 times more brilliantly than dim 15th-magnitude Pluto, was every astronomer's guiding light to that distant world. Now, more than 70 years after its discovery, Pluto has left Delta Geminorum well behind. But the star's role in history will still shine in the memories of countless astronomy enthusiasts.

Today we can use Delta Geminorum to guide us to NGC 2392. It is the northernmost

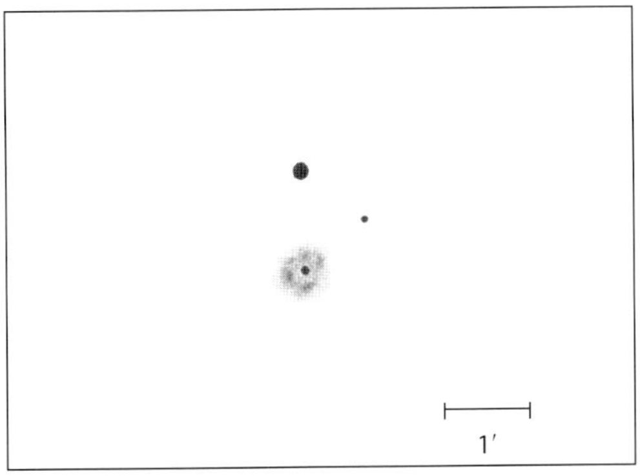

star in a large naked-eye triangle of similarly bright stars; the other two members are Lambda (λ) and Zeta (ζ) Geminorum. Delta Geminorum is also the northernmost star in a small kite-shaped asterism that also contains the 5th-magnitude stars 56, 61, and 63 Geminorum (itself a wide double). NGC 2392 is just 40′ southeast of 63 Geminorum, the easternmost star in the "kite." I generally draw a line between 61 and 63 Geminorum and point my telescope's sighting device just east of the halfway point (but a little closer to 63 Geminorum). At low power NGC 2392 will be in the field, looking like the southern member of a tight pair of 8th-magnitude stars separated by 1.6′. It's that simple. To be sure you're on target, use averted vision, and the planetary should appear to swell.

How did Charles Messier miss this spectacular object during his comet hunts? The answer is simple. NGC 2392 appears so nearly stellar at low power that it would not be difficult to sweep past it while looking for a diffuse object. In fact, I tried sweeping up the nebula at 23× with the 4-inch, as if I were Messier, and never once suspected its presence. And when I pointed the telescope directly at the Eskimo I had to use 72× to verify that I was seeing a nebula. Once I returned to 23× with that knowledge in mind, the nebula stood out as the "bloated star" noted by Smyth. Remember, too, that the planetary lies close to a star of similar magnitude. Any fuzziness Messier might have spotted here could also have been dismissed as an illusion. (Just such an illusion is posed by the double star M40 in Ursa Major; Johannes Hevelius mistook it for a nebula in the 17th century.)

The Rev. T. W. Webb penned something a little more entertaining about NGC 2392. "I was so much surprised at the result in my inferior telescope," he wrote, "that I cannot help supposing some temporary impediment to distinct vision at Bedford, for on coming accidentally across it in 1850 I found such a conspicuous nebulosity that I thought it was either damp on the eye lens or a telescopic comet." Webb's suspicions were understandable but unwarranted. With some effort it's not difficult for the users of even the smallest telescopes to see in NGC 2392 two obvious concentric shells of debris surrounding a tack-sharp central star — a type-*O7* white dwarf, which may have a hot unseen companion. The central star's brightness seems a tad uncertain, since apparent magnitudes have ranged from 9 to 10.5 in the modern literature.

If you plan to seek NGC 2392 with binoculars or a very low-power telescope, beware: you may think you've seen the nebula because it appears slightly swollen compared to other field stars. But this may be a "false positive" identification. The nebula lies 1.6′ south of a magnitude-8.5 star, and in binoculars you almost certainly will perceive the combined light of both objects. (I could just split the two with 10×50 binoculars, but their apparent separation was too small for me to reliably estimate the nebula's brightness with such a small aperture.) Note that our photo on page 154 shows the magnitude-8.5 star. In January 2001 I estimated NGC 2392's brightness with the 4-inch refractor at 23× and got magnitude 9.2, which

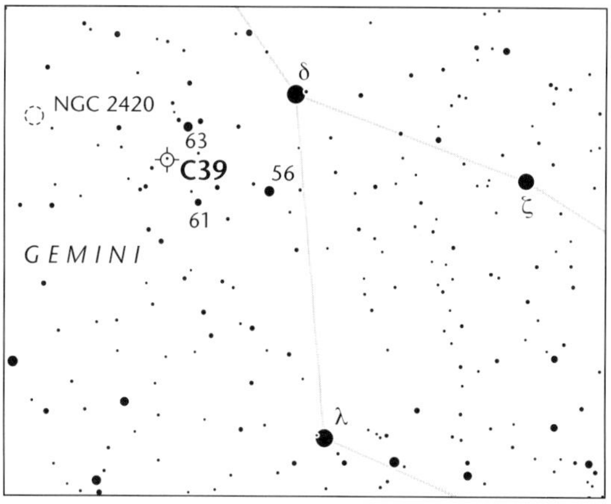

agrees with the value published by Christian Luginbuhl and Brian Skiff.

NGC 2392 takes magnification well, so use as much as your night-sky conditions will allow. Don't be afraid to crank up the power to 100× per inch of telescope aperture. (I did not do this with my 4-inch for the sake of consistency with the other observations reported in this book.) Through the 4-inch at 130× the nebula displays a crisp nucleus surrounded by a sharp inner shell and a fainter outer extension. Careful study of the inner envelope at this magnification shows two discrete regions of intensity to the northwest and north and a third knot or enhancement toward the southeast. These features tend to warp the shell's circular appearance into an ellipse. Pay extra attention to the northern enhancements, for I find they have a beaded or feathery texture. Through the 4-inch the outer halo is rather uniform, and none of the radial patterns seen in observatory photographs can be detected at 130×.

Owners of larger telescopes should be prepared to spend hours with this incredible planetary. At the 1998 Texas Star Party I studied NGC 2392 with McNeil and his 16-inch f/5.1 Dobsonian. We used astonishingly high powers, up to 1,200×, and were rewarded with some striking details. The central star appeared as a solitary sun floating in a bay of darkness fringed by a crisp inner ring. Three distinct beads and two arcs of light punctuated the inner ring, though there was no hint of duplicity. The outer halo appeared as a pale green sea that brightened suddenly into a mottled garland at the nebula's outer edge. Try as we might, we could not see the Eskimo's face, and I do not know of any visual observer who has.

Before moving on to the next Caldwell object, return to low power and move your telescope a little more than 2° northeast of the Eskimo, where you will encounter one of the sky's truly neglected deep-sky objects. NGC 2420 is a magnitude-8.3 open cluster whose southern edge sits right on the ecliptic. It is hardly ever mentioned in the amateur literature. In the 4-inch at 23× it's a lonely sight — perhaps two dozen 11th- to 12th-magnitude suns arranged like fine mesh around a dim 10′-wide cometlike glow. Increasing the magnification only enhances the effect. The cluster looks like a chipped and dusty diamond that has lost its luster.

40

NGC 3626
Type: Lenticular Galaxy (S0)
Con: Leo

RA: $11^h\ 20^m\ 04^s$
Dec: +18° 21.4′
Mag: 11.0
Dim: 3.0′ × 2.2′
SB: 12.5
Dist: 86 million light-years
Disc: William Herschel, 1784

W. HERSCHEL: [Observed 14 March 1784] Pretty bright, small, a little extended, brighter in the middle. (H II-52)

GC/NGC: Bright, small, very little extended, suddenly brighter in the middle.

NGC 3626 IS A CURIOUS OBJECT. AS IT IS ONLY marginally visible in small telescopes, one might wonder why Patrick Caldwell-Moore selected it over the wealth of other galaxies in Leo; in fact, NGC 3626 is the only Leo galaxy in his catalog. Of course, the Caldwell Catalog is not intended to be a list of 109 visually stunning objects but a selection of Moore's *favorite* deep-sky objects, whether they're telescopically dynamic, astrophysically intriguing, or visually challenging. The question, then, is how NGC 3626 fits into the picture.

NGC 3626 is a tiny 3′-wide object that shines with the light of an 11th-magnitude star. It is the second-brightest member of a dense gathering of 10 galaxies (ranging in brightness from 10th to about 13th magnitude) that populates Leo's triangular hindquarters. All these galaxies are plotted in the second edition of *Sky Atlas 2000.0*. All but two of them (NGC 3691 and NGC 3599) belong to the Leo Cloud of galaxies, an elongated archipelago of ten dozen island universes. NGC 3626 is a classic lenticular galaxy whisking away from us at 1,473 km per second. Less than 1° southwest of NGC 3626 lies another lenticular galaxy, NGC 3607, which shines a full magnitude brighter than our Caldwell target. (Why didn't Moore select that one instead, I wonder?) NGC 3607 also is part of a nice trio of galaxies, which is easy to locate. Anyway you can use this grouping as a steppingstone to NGC 3626.

To find NGC 3626 first locate the 3rd-magnitude stars Zosma, or Delta (δ) Leonis, and Chertan, or Theta (θ) Leonis, the westernmost stars in the Lion's hindquarters. The aforementioned trio is just east of the halfway point between the two stars. If you set your finderscope's cross hairs on the imaginary halfway point and use a wide-field eyepiece in your main telescope, the trio's brightest two galaxies (NGC 3607 and magnitude-10.8 NGC 3608) should be quite obvious. The third trio member, 12th-magnitude NGC 3605, will also be in the field, a few arcminutes southwest of NGC 3607. In the 4-inch at 23× all three galaxies pop into view, though NGC 3607 and 3608 are naturally more readily visible than NGC 3605.

Now, if you push your telescope 35′ due east of the center point between NGC 3607 and 3608, you will encounter a 10th-magnitude star. NGC 3626 sits alone just about 15′ northeast of that star. NGC 3626 is only 0.2 magnitude (20 percent) fainter than NGC 3608, so look for something similarly bright when you search for the Caldwell galaxy.

NGC 3626 is inclined 48° from face on. With the 4-inch at 23× and averted vision it looks like a slightly out-of-focus star south of a flattened keystone of dim suns. (With direct vision the galaxy just looks like another star.) Lenticular galaxies are known for being small and bright, packing a lot of light into their lens-shaped bodies. NGC 3626 is no exception, though its distance places it at a disadvantage. The galaxy is three and a half times farther from Earth than Leo's celebrated M65. If NGC 3626 were at M65's distance it would blaze at magnitude 8.3 (a full magnitude brighter than M65) and be a truly stunning sight, even outperforming the lenticular wonder NGC 5866 in Draco (see page 432). NGC 3626 is a large system, with a physical diameter of 70,000 light-years and a total mass of 120 billion Suns.

Alas, even at 130× the galaxy shows very little detail in the 4-inch. A dim outer halo surrounds a brighter lens-shaped oval that hugs the galaxy's needle-sharp nucleus. With a long, intense look I convinced myself that the inner lens's light distribution is not entirely uniform, though the slight intensity differences are fleeting at best. If yours is a small telescope, do not fret. Most lenticular galaxies are notoriously bland. Even with her 20-inch reflector Barbara Wilson saw no outstanding details in NGC 3626. Indeed, even in the plates taken with the 48-inch Schmidt telescope atop Palomar Mountain (one of which is reproduced on page 159), the galaxy looks like three gray ovals of varying intensity superimposed on one

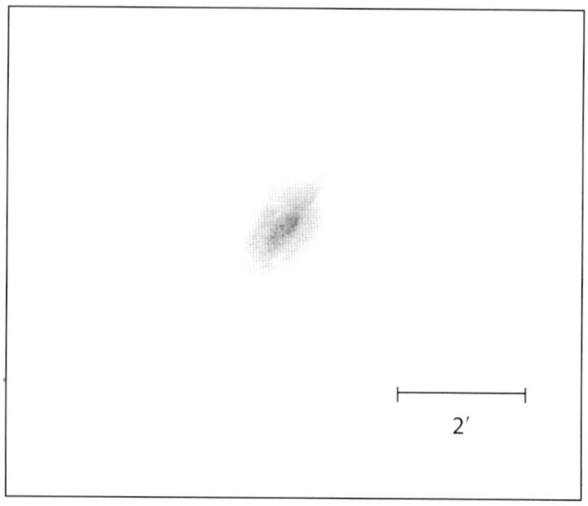

another; the brightest oval is skewed counterclockwise with respect to the smooth, symmetrical outer envelope. CCD images can penetrate the core that so often is burned out in photographs, and they reveal a pinpoint nucleus as well as a suggestive hint of obscuring dust in the eastern half of the inner lens.

In a 1998 *Astrophysical Journal* paper, Daniela Bettoni (Astronomical Observatory of Padova, Italy) and her colleagues reported a massive counter-rotating disk of molecular gas in NGC 3626. The gas is swimming against the tide, so to speak, of the stars in the galaxy's disk. This finding is not unprecedented. In 1992 Vera Rubin (Carnegie Institute of Washington) discovered that almost half of the stars in the disk of Virgo's NGC 4550 were orbiting in the direction opposite that of the galaxy's other stars. In other words, NGC 4550 appears to contain *two* disks, one spinning clockwise as seen from Earth, the other counterclockwise. (Anyone who has been to the Vatican will appreciate this discovery, for the Vatican has an intricate double-spiral staircase in which ascending visitors never meet descending ones, even though the spirals are concentric.) Other galaxies that show this cospatial counter-rotation are NGC 4826 in Coma Berenices, NGC 7217 and NGC 7332 in Pegasus,

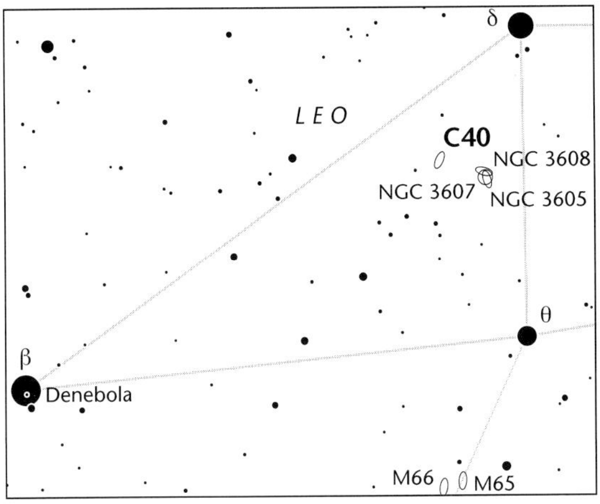

NGC 4546 in Virgo, NGC 4138 in Canes Venatici, and NGC 3593 in Leo.

How can a galaxy possess two disks with opposite angular momenta? NGC 3626's undisturbed morphology rules out a recent, disruptive merger between two spiral galaxies with opposite spins. But the counter-rotating gas might have been acquired gradually from the detritus of previous interactions. If the inward-falling material's orbit around the mass-gaining galaxy opposes that of the galaxy's preexisting disk, a counter-rotating structure is formed.

41

41

The Hyades;
Taurus Moving Cluster
Melotte 25
Type: Open Cluster
Con: Taurus

RA: 04^h 26.9^m
Dec: +15° 52′
Mag: 0.5
Diam: 5.5°
Dist: 151 light-years
Disc: known since antiquity

HERSCHEL: None.

GC/NGC: None.

NIGHTFALL HAS A MYSTICAL FEEL TO IT IN late December. The sky shines like black ice, trees look dark and skeletal, and the stars seem to shiver in the crisp winter air. Look halfway up the eastern sky and there is the haunting glow of the Pleiades, a cluster of hot, young stars shrouded in mist. Just above the horizon, mythical Orion throws a leg up over our fence of mountains (to paraphrase the great American poet Robert Frost) and slowly vaults into view with athletic grace. Sandwiched between these two stirring sights is the Hyades, a bright and alluring V-shaped arrangement of stars that skywatchers have pondered since the dawn of history.

Next to the Ursa Major moving group (the five innermost stars of the Big Dipper), the Hyades is the nearest open star cluster to our Sun, and it is the nearest moderately rich one. In classical mythology the V outlines the face of Taurus, the Bull, one of the sky's most ancient constellations. The stars of Taurus have been seen as a Bull since at least 4,000 B.C., when the Sun resided among them during the spring equinox. That the Sun united with a bull — a symbol of virility — at that pivotal time of year did not go unnoticed by the earliest civilizations, whose agricultural activities depended on the alternating seasons. The union heralded the start of spring in the

Northern Hemisphere, and people celebrated this special event with religious ceremonies and festivals. Although we no longer hold ceremonies to honor the union of Bull and Sun, one lasting symbol of the horned creature and its importance in human affairs can be seen on modern star charts: the Greek letter Alpha (α), which descended from the first letter in both the Hebrew and Phoenician alphabets, resembles a bull's head and horns.

Humans began associating the terrestrial Bull with the celestial Bull as early as the late Neolithic age. While touring Turkey after the August 1999 total solar eclipse, I saw some Neolithic cave paintings on display at the Museum of Anatolian Civilizations in Ankara. One shows a large red bull surrounded by several tiny black and red "stick figures" armed with bows and spears. I found myself mesmerized by the simple portrayal. Primitive hunters depended on the life cycles and migrations of large animals, and it's conceivable that this painting could have been used to invoke a spirit whose magical powers would help the artists with their hunt (at least that is what my guide inferred at the museum). Might this Bull have been Taurus? My guide did not know.

Anthropologists long have argued that the animals shown in cave paintings were created for religious rites and not to make artistic impressions. But the Turkish cave painting seemed to depict an actual (if fantastic) battle with a large beast of prey. It seemed no different than another one (dated circa 6,000 B.C.) that depicted an eruption of Hasan Dag, a volcano in central Turkey. The latter piece of artwork is now regarded as the earliest evidence of human interest in volcanic phenomena. Clearly art is subject to interpretation; that hasn't changed with the ages. In any case, while looking at the Neolithic battle scene I realized that our ancestors painted similar battle scenes in the night sky, and we benefit from their vivid imaginations today. Each winter night we can look up at the dark "cave" of the sky and admire the stick figure of Orion, the Hunter, poised to do battle with Taurus, the Bull. Wielding his club and thrusting his shield forward, the mighty Hunter successfully drives the Bull back, night after night, until he herds the horned animal into the pen of the western horizon.

In Greek mythology, seven Hyads were the daughters of Atlas (the Titan condemned to support the Earth and sky on his shoulders for eternity) and Aethra. The Hyades' paternity

makes them the half-sisters of the Pleiades, who were born to Atlas and Pleione; together the Hyades and Pleiades constitute the 14 daughters of Atlantis, or the Atlantides. Some works try to lessen Atlas's familial burden by dubbing the Hyades the Dodinides, or nurses of Bacchus. The sisters do have names, but mysteries remain. The Greek poet Hesiod (8th century B.C.) spoke of five: Eudora, Kleea, Koronis, Phaeo, and Phaesula. Seven are elsewhere attributed to Pherecydes, and their presumably corrupted names, according to Richard Hinckley Allen, are Aesula, Ambrosia, Dione, Eudora, Koronis, Polyxo, and Thyene (also known as Thyone). Other variations are invoked elsewhere. Alas, unlike the case with

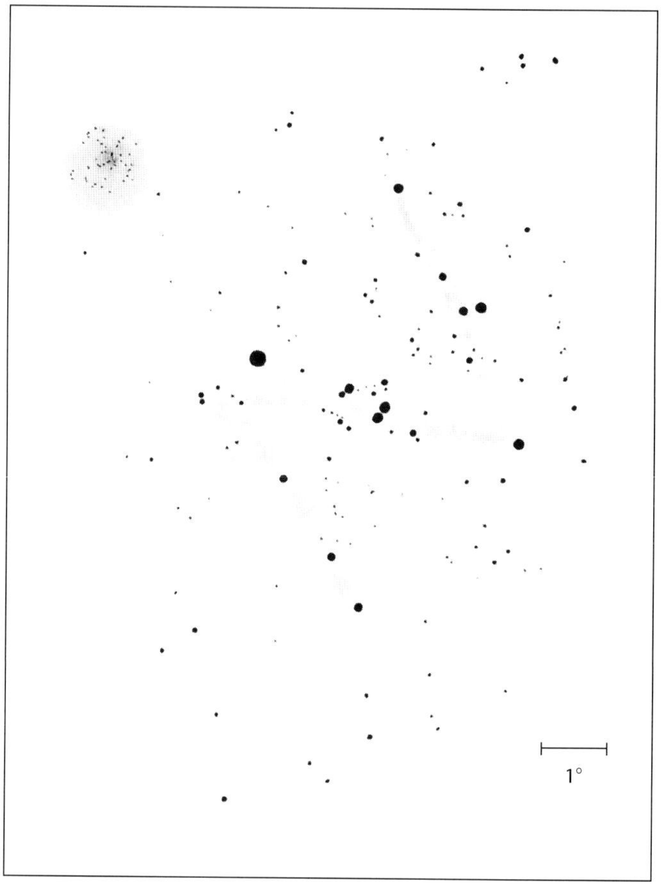

the Pleiades, no particular star in the Hyades has been assigned a particular classical name. This mystery cannot be solved by intuition, since the Hyades contains 11 stars brighter than magnitude 4.5 and some two dozen stars brighter than 6th magnitude. Nevertheless, unless a passage can be found in classical literature that assigns specific names to specific stars, we will have to rely on our imaginations. For instance, five "stars" (some are doubles or triples) in the celestial V have Greek letters: Alpha (α), Gamma (γ), Delta (δ), Epsilon (ε), and Theta (θ) Tauri. So we could (for fun) name four of the five Greek-lettered stars as follows: Alpha as Aesula or Ambrosia; Delta as Dione, Epsilon as Eudora, and Theta as Thyene. Ovid used Thyone when he referred to the whole group.

In Roman times the Hyades were also called Sidus Hyantis, the Rain-Bringing Stars. The phrase refers to the legendary grief the Hyades felt over the death of their terrestrial half-brother, Hyas; the sisters' tears were so profuse that they caused heavy rainfall on Earth. Here we can see how the Hyades became a bad omen for farmers and sailors. That myth no doubt refers to the ancient rising of the Hyades during the Mediterranean region's rainy season, when storms raged on land and sea.

Today we associate the rising of the Hyades with brisk winter nights. The warm orange light of the giant star Aldebaran punctuates the tip of the V's southeastern branch. Aldebaran — from the Arabic word *al dabaran*, "the follower" [of the Pleiades] — is a gentle giant; its diameter is only some 40 times that of our Sun. By contrast, the *M*-type supergiant star Antares in Scorpius is 880 times larger than our Sun, and Orion's Betelgeuse spans 770 solar diameters. (Note, though, that the diameter of a supergiant star is a fuzzy concept, since it varies strongly with wavelength; the supergiants were measured by laser inventor Charles Townes and his colleagues at a midinfrared wavelength of 11 microns.) Like other giants of its class, Aldebaran varies slightly in brightness, from about magnitude 0.75 to 0.95. The orange giant travels through space with a 13th-magnitude red-dwarf companion 31″ to the east-southeast. (Be careful to not confuse Aldebaran's true companion with an 11th-magnitude field star 2′ to the north-northeast.) The best way to detect Aldebaran's true companion star is during a lunar occultation of Aldebaran itself. Aldebaran is the brightest star that the Moon can occult (other than the Sun!). These occultations occur in bunches, and magazines like *Sky & Telescope* announce the events in their pages. The last series to be seen from the United States ended in 1999. Observers in the contiguous United States will have to wait for the year 2015 to see the next series begin.

Despite Aldebaran's prominent position in the Bull's face, it is not a true member of the Hyades cluster. That much has been confirmed by the Hipparcos satellite. Launched in August 1989, the High Precision Parallax Collecting Satellite was one of modern astronomy's true technological wonders. Equipped with two telescope systems aimed 58° apart, the rotating spacecraft was designed to measure star positions, brightnesses, and motions. Most fundamentally, by utilizing the Earth's orbit as a baseline for triangulation, Hipparcos pro-vided astronomers with parallaxes, and thereby distances, for tens of thousands of stars. Of the 118,218 stars in the Hipparcos catalog, the distances of 22,396 are now known to an accuracy of better than 10 percent. (Before Hipparcos's data were released in 1997 the distances to fewer than 1,000 stars were known so precisely.) Among those 22,396 precisely triangulated stars are many in the Hyades. The satellite has ended a century-long debate over the cluster's distance — a critical quantity for calibrating the distance scale of the entire visible universe. Hipparcos has pinpointed the Hyades' center of mass at 151 light-years from Earth, plus or minus a mere 0.9 light-year.

Taking advantage of the Hipparcos data, *Sky & Telescope* editor and *Millennium Star Atlas* coauthor Roger W. Sinnott created the 3-D stereogram of the Hyades shown below. To see the cluster in three dimensions, hold the page against your face so each box is in front of one eye. Relax and gaze into the distance "through" the paper while slowly moving it away to reading distance or a little beyond. The two views should merge in your mind's eye. Note that Aldebaran is omitted; at its distance of 65 light-years (also measured by Hipparcos), Aldebaran is less than half as far away as the Hyades.

Of course, few will argue that Aldebaran's rosy hue does not attract our gaze toward the Bull's illustrious V. If you disagree, go outside, close one eye, and block the light of Aldebaran with your thumb while gazing outward with your open eye. (You can also do this with a star chart by simply covering Aldebaran with a finger.) Without Aldebaran, the Bull's face is lost, and the V suddenly transforms into the letter N or Z.

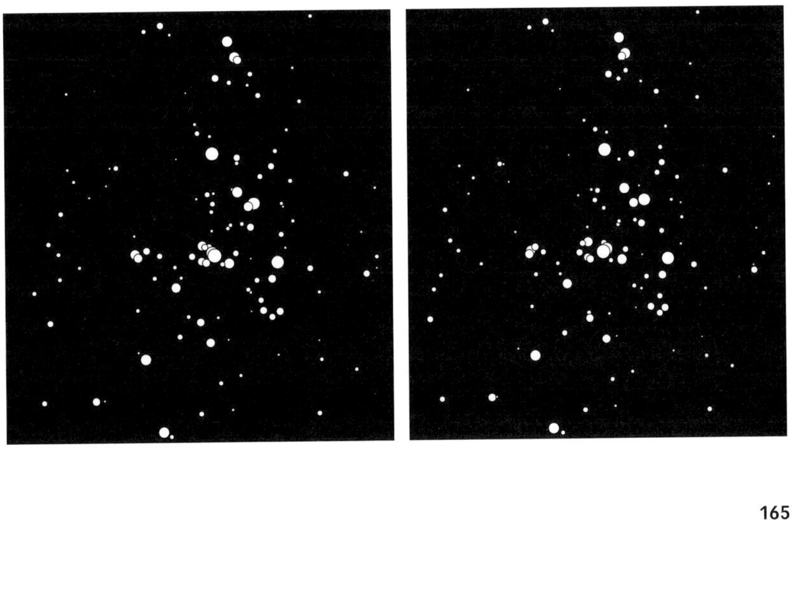

Owners of spectroscopes may want to test the instruments on Taurus's most prominent star. As Ernst Hartung notes in his *Astronomical Objects for Southern Telescopes,* Aldebaran displays a "fine absorption spectrum crossed by many dark bands and lines," some of which are visible with a 3-inch telescope. Eighteenth-century spectroscopic studies of Aldebaran revealed the presence of hydrogen, sodium, magnesium, calcium, iron, bismuth, tellurium, and mercury in the star's atmosphere. Of these findings Garrett P. Serviss wrote somewhat prophetically in his 1888 *Astronomy with an Opera-Glass*: "And so modern discoveries, while they have pushed back the stars to distances of which the ancients could not conceive, have, at the same time, and equally, widened the recognized boundaries of the physical universe and abolished forever the ancient distinction between the heavens and the earth. It is a plain road from the earth to the stars, though mortal feet can not tread it."

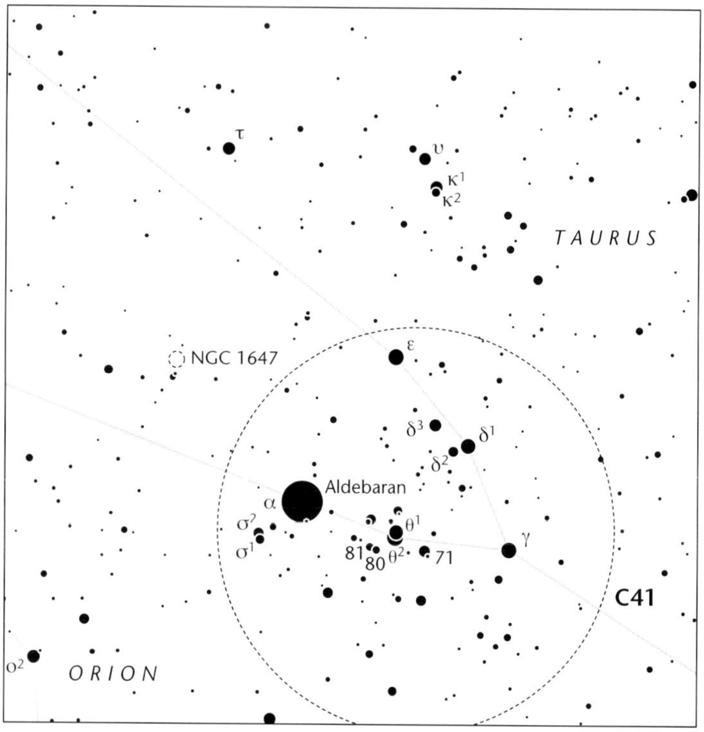

We skywatchers don't need feet to traverse the corridors of space; all we need are knowledge and imagination. So pick up a pair of binoculars and use your mind to probe the wonders of the Hyades. Even the smallest of binoculars and a little resourcefulness can transform the celestial V into a three-dimensional cityscape of bright lights. To mentally cross the cluster's heart you will need to travel about 10 light-years (95 trillion km); the outlying Hyads are spread over a region at least twice as wide. Now drop the binoculars and look at the Hyades with your naked eyes. Also known as the Taurus Moving Cluster, the Hyades are drifting through space at 46 km per second toward a point a few degrees east of Betelgeuse. According to European Space Agency scientist Jos de Bruijne, the cluster was closest to our solar system some 1,150,000 years ago, and in 50 million years it will be but a dim, ½°-wide telescopic target.

Now compare the Hyades' appearance with that of the Pleiades, using both the naked eye and binoculars. Although the Pleiades cluster is a mere 10° from the Hyades, it is 256 light-years more distant. And if, with their profusion of yellowish stars, the Hyades look older than the Pleiades to you, give yourself a pat on the back! The age of the Hyades is 625 million years old, about 10 times that of the Pleiades. (I wonder what Atlas would have to say about this state of affairs.) The Hyades may well have more in common with the 400-million-year-old Praesepe star cluster, M44 in Cancer, which also shares their motion through space.

Naked-eye stargazers and binocular users will find several attractive pairs of stars within the Hyades. Most obvious are the Theta Tauri and Delta Tauri groupings in the middle of each branch of the V. Theta^1 and Theta^2 ($\theta^{1,2}$) Tauri lie halfway between Aldebaran and Gamma Tauri inside a squashed ellipse of 5th- and 6th-magnitude stars. Theta^1 and Theta^2 Tauri shine at magnitude 3.8 and 3.4, respectively, and are 5.6′ apart on a roughly north-south line. Ptolemy may have seen these stars as one, but today they are not considered a particularly exacting test of visual acuity. Type-*A*7 Theta^2 Tauri is the brightest true member of the Hyades and is about 50 times more luminous than our Sun. Fainter Theta^1 Tauri is 9 light-years farther away. It is a type-*G*7 yellow giant, so be sure to compare it with that non-Hyad, Aldebaran. Through his 6-inch refractor Adm. William Henry Smyth described the colors of the Theta^{1,2} pairing as pearly white and yellow. The Delta Tauri stars ($\delta^{1,2}$ Tauri) are a much more open pair (18′) in the western branch of the V. Here we have a magnitude-3.8, type-*G*8 yellow giant and a hot *A*7 main-sequence star shining at magnitude 4.8.

Since naked-eye doubles abound in the Hyades, visual athletes can try for the following bronze-, silver-, and gold-medal pairs. You deserve a bronze medal if you can split Sigma^{1,2} ($\sigma^{1,2}$) Tauri, about 1° southeast of Aldebaran. The stars shine at magnitude 5.1 and 4.7, respectively, and are separated by 7.2′ along a south-southwest to north-northeast line. That's more than twice the angular distance between the components of the famous naked-eye double star Epsilon^{1,2} ($\varepsilon^{1,2}$) Lyrae (magnitudes 5.0 and 4.7, respectively). Christian Luginbuhl and Brian Skiff found it "relatively difficult" to split the Sigma Tauri stars, but I found the task relatively easy. In binoculars the Sigmas shine with a comparable pearly white luster. Astronomers once believed that Sigma^1 Tauri did not belong to the Hyades, but the Hipparcos satellite revealed that both components are true cluster members. The silver medal goes to anyone who can split Kappa^{1,2} ($\kappa^{1,2}$) Tauri in the Bull's ear. These two type-*A*7 stars (magnitude 4.2 and 5.3, respectively) are a bit tighter than the Sigmas, being separated by only 5.6′. The appreciable magnitude difference between the two components makes the task all the more difficult. In the 1923 edition of *Astronomy with an Opera-Glass*, Serviss thought it "improbable" that anyone could separate the Kappas with the naked eye. Of course, I wouldn't have included the Kappas if I didn't separate the stars myself and believe you could do the same. Try looking at the Sigmas and the Kappas between nautical and astronomical twilight (when the center of the Sun's disk is between 12° and 18° below the horizon, respectively). During this time of subdued illumination the contrast between the stars and the background sky is reduced, enhancing the eye's ability to resolve close pairs.

You get the gold medal if you can resolve 80 and 81 Tauri. These magnitude-5.6 and 5.5 stars are separated by about 8′; their faintness makes them more challenging than the Sigmas or the Kappas. (By the way, 81 Tauri is a telescopic double with a 9th-magnitude companion 2.7′ to the northwest.) Of course, I couldn't stop there, so I added another prize category. The Superman or Wonder Woman Medal goes to anyone who can separate the unnumbered pair of stars just 30′ northeast of Theta^{1,2} Tauri. These stars shine at magnitude 4.8 and 6.5, respectively, and are separated by a mere 4′ —

a tad wider than the 3.5′ separating Epsilon[1] and Epsilon[2] Lyrae. The Superman/Wonder Woman pair in Taurus challenges observers with a 1.7-magnitude (4.8×) brightness difference; furthermore, seeing the fainter sun requires a very dark sky. Good luck!

Finally, no binocular user should fail to look 3½° northeast of Aldebaran, where the ghostly form of open cluster NGC 1647 dominates the view. This fine 6th-magnitude cluster is centered on a pair of 9th-magnitude stars and can be seen from a dark sky as a naked-eye glow. The cluster measures 40′ across and lacks any central condensation. Certainly it is one of the winter sky's unsung showpieces, and I'm sure it would get a lot more attention were it not so thoroughly outranked by the more magnificent Hyades and Pleiades nearby.

42

NGC 7006
Type: Globular Cluster
Con: Delphinus

RA: $21^h\ 01^m\ 29.5^s$
Dec: +16° 11′ 15″
Mag: 10.6
Diam: 3.6′
Dist: 135,000 light-years
Disc: William Herschel, 1784

W. HERSCHEL: [21 August 1784]: Very bright, small, round, gradually much brighter in the middle. Resolvable. (H I-52)

GC/NGC: Bright, pretty large, round, gradually brighter in the middle.

GLOBULAR CLUSTERS, SCINTILLATING BOUQUETS of hundreds of thousands of suns, adorn the most distant reaches of our Milky Way. While bright stars and open clusters populate our galaxy's flattened disk, most globular clusters lurk in the galaxy's spherical halo, a cold and bleak domain of old suns and mystery matter that spans a quarter million light-years of space. But NGC 7006 (Caldwell 42) in Delphinus is exceptional even among globulars. It lies on the halo's very fringe, 135,000 light-years from Earth, and may be a part of the galaxy's outer halo, or corona. From our station on Earth, NGC 7006 lies nearly twice as far as M54, the most distant Messier globular, though it is twice as close as the most distant Caldwell globular: NGC 2419 (Caldwell 25), the famous Intergalactic Wanderer in Lynx. NGC 7006 lies 127,000 light-years from the center of our galaxy. Of the 150 or so globular clusters known within our Milky Way, NGC 7006 is one of the most remote; as of 1997 it was considered the eighth most far-flung globular from both the Earth and the galactic center.

If we consider the sheer immensity of the galactic halo and the paltry number of globulars that populate it, we can begin to understand the true loneliness of space. The distribution of the globulars, however, is not uniform. About one-third of our galaxy's globulars are concentrated toward the galactic center in the direction of Sagittarius, while the others are scattered far above and below the galactic plane. The famous Harvard College Observatory astronomer Harlow Shapley first suspected that the distribution of globulars was intimately related to the structure of the Milky Way galaxy, and in the 1910s and '20s he used the great 60- and 100-inch reflectors atop Mount Wilson in California to survey them. His most important contribution was his systematic study of variable stars within the clusters. Earlier work by Solon Bailey (also at Harvard) had shown that the majority of the variable stars in globular clusters brightened and dimmed over periods of hours. Now known as RR Lyrae stars, these "cluster variables" were of fundamental importance because they all had roughly the same intrinsic brightnesses. If an astronomer compared how bright (on average) a cluster variable appeared through a telescope to how bright it *really* was, he or she could determine the distance to the variable, and hence to the galaxy or cluster it is in.

Shapley's observations of 11 RR Lyrae stars in NGC 7006 showed the cluster to be about five times as remote as M3 or M5, the most distant globulars known at the time. After discovering and tracking both RR Lyraes and Cepheid variable stars in other globulars across the heavens, he placed his cosmic yardstick between the Earth and the clusters and mapped the three-dimensional structure of our host galaxy. With globulars lying tens or hundreds of thousands of light-years from our Earth, our galaxy swelled to magnificent proportions in Shapley's eyes.

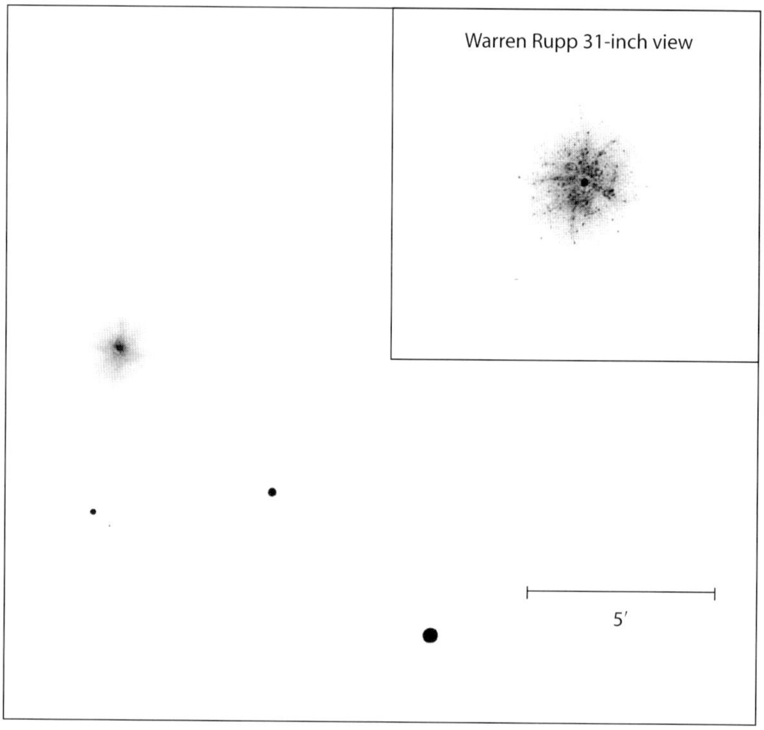

These were pioneering years, however, and Shapley did not know then, as we do now, just how dramatically our galaxy's interstellar medium dims the light of globular clusters at low galactic latitudes. Shapley's underestimation of this effect led him to overestimate the distances of globulars, and with them the size of our galaxy. For instance, Shapley had positioned NGC 7006 about 180,000 light-years from Earth, roughly 33 percent farther than we know it is today. But his data and analyses gave the first reliable indication that the Milky Way was a much larger system than previously had been believed. His results also offered astronomers their first real glimpse of the Milky Way's fantastic three-dimensional structure. Armed with the distances to globulars, Shapley had the power to dethrone the Sun from its central position in our galaxy, just as Copernicus had done to the Earth in our solar system four centuries earlier.

Small-telescope users will have to keep in mind the role this dim and distant globular cluster played in helping Shapley model our galaxy, since its visual appearance is understandably underwhelming. The cluster resides at a relatively low galactic latitude (–19°), so its brightness is dimmed by intervening dust. We see its 3.6′-wide disk shining at a modest apparent visual magnitude of 10.6. NGC 7006's surface brightness is also modest; on average, each square arcminute of its disk shines with the light of a magnitude-13.2 star. Some patience is required to find NGC 7006 because the cluster resides in a region devoid of bright naked-eye landmarks. The closest such landmark is magnitude-4.5 Gamma (γ) Delphini, the nose of the celestial Dolphin, 3.5° due west

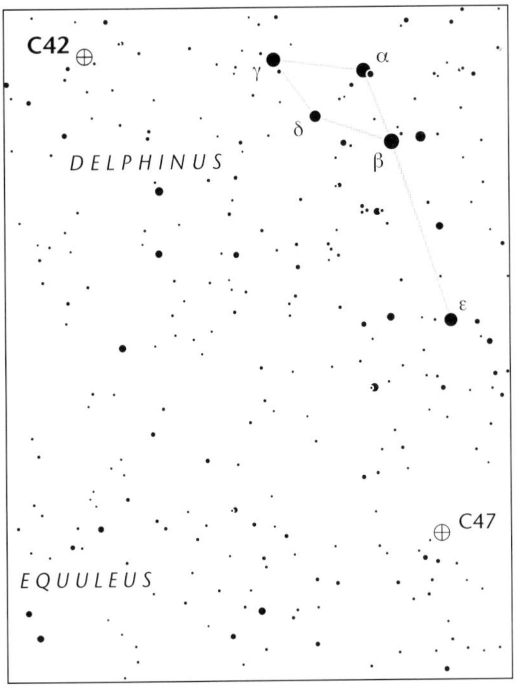

of the cluster. Owners of equatorially mounted telescopes will have no problem finding the cluster, since they just have to sweep their telescopes eastward with a low-power eyepiece. (In a more relaxed vein, Walter Scott Houston suggested setting the telescope on Gamma Delphini and waiting 15 minutes for the globular to drift into view.) Those using altazimuth mounts should refer to the finder chart here. Note that NGC 7006 is part of an upside-down and tilted T of 6th- to 7th-magnitude stars. First use binoculars to locate this asterism; then, using the widest field your telescope provides, sweep from Gamma Delphini to the T. If you have a Telrad finder, place the right edge of the widest illuminated ring on Gamma Delphini; the left edge of the outer ring will slice through the T. Don't expect to see the cluster in binoculars. I tried, with hand-held 7×50s and 10×50s, but failed. Houston tried with 16×40s mounted on a tripod, and, though he knew exactly where to look, he could not see the cluster through them.

Although NGC 7006 has been seen in telescopes as small as a 3-inch, doing so is a challenge. At 23× in the 4-inch Genesis under dark skies, the cluster appeared as a faint glow whose body had only a little more substance than that of NGC 2419. To pull NGC 7006's glow out from the background, Scotty suggested using more than 50×; with his 4-inch refractor 100× seemed best to him. In the Genesis at 72× the cluster shows a definite roundness, with hints of a core and a breathlike extension or outer halo. In time the core becomes obvious, as does the outer envelope, which causes the cluster to swell slightly with averted vision, just like a "blinking planetary." At 130× there is no hint of resolution. The cluster keeps its form but takes on a pasty glow. The core begins to show stubby crosslike arms or appendages. NGC 7006's amorphous nature should not surprise us. After all, the globular is one of the most difficult to resolve, and its brightest star shines at a feeble magnitude 15.6. Christian Luginbuhl and Brian Skiff were unable to resolve the cluster with a 10-inch telescope even in good seeing, though they did resolve a few faint stars over a granular background with a 12-inch at 475×. Using a 13.1-inch f/4.5 reflector, Barbara Wilson found the cluster to be a very dense fuzzball with no perceptible individual stars. Photographs taken through large telescopes reveal the cluster to be a tight snowball of starlight with an almost hexagonal core. A prominent knot of stars lies to the south of the core. The cluster is situated in a rich Milky Way field, so it is hard to discern which stars belong to the cluster and which to our galaxy's populous disk.

Of course, with a *really* big telescope NGC 7006 reveals itself in ways that can only be imagined by the owners of small instruments. In September 1998, Brent Archinal and I observed NGC 7006 with the 31-inch reflector at the Warren Rupp Observatory in Mansfield, Ohio. The cluster was a scintillating mass of dim starlight. At high power we saw an unresolved, starlike inner core surrounded by a dark ring that defined the inner edge of a finely resolved, though jagged, outer core. With averted vision thin arms radiated out from the stellar "hole" like the limbs of a starfish wrapped around a rock. The outer halo appeared loose and fractured, as if its shell were so old and brittle that pieces of it were flaking off into space.

Large-telescope users might be interested in this challenge: a tiny cluster of galaxies lies just 5′ to the southwest of NGC 7006. None of these galaxies is plotted on the *Millennium Star Atlas*, but a few of them can be seen in the Digitized Sky Survey image on page 169. I believe they will certainly fall to large Dobsonians.

Before leaving the area be sure to return to Gamma Delphini, which is a beautiful double star. The magnitude-5.2 secondary lies 9″ to the west of the yellowish primary; the stars were 12″ apart in 1823. Adm. William Henry Smyth saw the secondary gleam with a "light emerald" hue. The colors Smyth saw recall those of the famous double star Albireo in the nose of Cygnus, the Swan. Interestingly, though, the spectral types of Gamma Delphini's components are *K*1IV (primary) and *F*7V (secondary), respectively — basically right next to each other on the *OBAFGKM* sequence of stellar spectral types. So any color contrast perceived by Smyth was undoubtedly due to psychology, or intensity threshold effects in the eye. I see two golden stars, with the fainter one having a cooler (bluer) hue. But the only important question is: what do *you* see?

43

Electric Arc
NGC 7814
Type: Spiral Galaxy (Sab)
Con: Pegasus

RA: 00^h 03.2^m
Dec: +16° 09′
Mag: 10.6
Dim: 5.5′ × 2.3′
SB: 13.4
Dist: 49 million light-years
Disc: William Herschel, 1784

W. HERSCHEL: [Observed 8 October 1784] Pretty faint, pretty large, irregularly round, easily resolvable. (H II-240)

GC/NGC: Considerably bright, considerably large, elongated, very gradually brighter in the middle.

44

NGC 7479
Type: Barred Spiral Galaxy (SBc)
Con: Pegasus

RA: 23^h 05.0^m
Dec: +12° 19′
Mag: 10.8
Dim: 3.9′ × 3.0′
SB: 13.4
Dist: 106 million light-years
Disc: William Herschel, 1784

W. HERSCHEL: [Observed 19 October 1784] Considerably bright, much extended, gradually brighter in the middle, 4′ long and 2′ broad. (H I-55)

GC/NGC: Pretty bright, considerably large, much extended toward position angle 12°, between two stars.

43 & 44

NGC 7814 (Caldwell 43) and NGC 7479 (Caldwell 44) are two of the many reasonably bright galaxies residing in or near the Great Square of Pegasus. Grand and conspicuous, the Great Square is autumn's hallmark asterism. It portrays the mythical winged horse born from a vile mixture of sea foam and the blood that dripped from Medusa's freshly severed head. The Square's four corner stars all shine brightly at 2nd or 3rd magnitude. Its vast, 15°-wide interior, by contrast, is bleak.* Naked-eye stars (mostly dim) can be seen within the Square, but the density of starlight there seems inconsiderable when compared to one of summer's hallmark asterisms, such as the Teapot of Sagittarius. That's because when we look toward Pegasus we look high above our galaxy's dusty, star-studded plane, where a thin veil of starlight hardly obstructs the heavens beyond. Galaxy hunters actually prefer this kind of bland celestial backdrop. To them, the Great Square is like a clean window through which they can gaze upon the universe's most distant reaches.

Deep-sky observers of all levels have learned to use the Great Square to their advantage. In a sense, the Square is autumn's Grand Celestial Station. Riding their telescopes like trains, observers can depart from any of the Square's corners for different extragalactic destinations. For instance, beginners can use the northeastern corner star, Alpha (α) Andromedae, or Alpheratz, as their departure point for M31, the Great Andromeda Galaxy; they just travel two stops eastbound — Delta (δ), then Beta (β) Andromedae — before heading northbound past Mu (μ) and Nu (ν) Andromedae to M31. Observers traveling to NGC 7331 (Caldwell 30) usually depart from Beta (β) Pegasi, or Scheat, the Square's northwestern corner, and head one stop northwest to Eta (η) Pegasi before taking the northbound express to their destination. And so it is with the two southern stars in the Great Square.

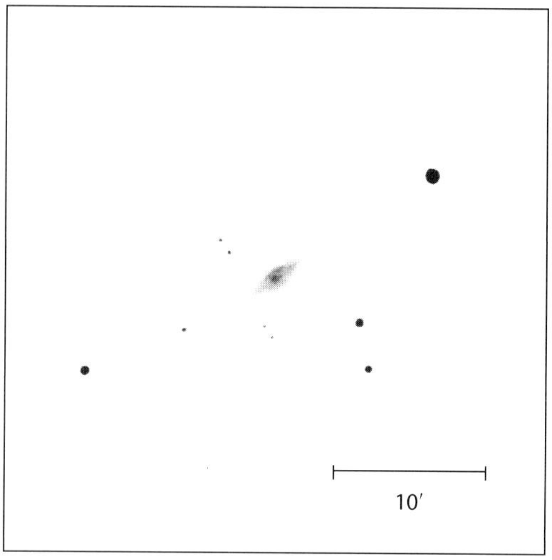

They lead us to our next two Caldwell destinations: the spiral galaxies NGC 7814 (Caldwell 43) and NGC 7479 (Caldwell 44).

NGC 7814 lies only 2½° west-northwest of Gamma (γ) Pegasi, the southeastern corner star. If you own a large telescope, don't mistake the 12th-magnitude galaxy NGC 14, which lies halfway between Gamma Pegasi and NGC 7814, for your target; both galaxies have bright central regions.

To find NGC 7814 start by surveying the Gamma Pegasi area with binoculars. Gamma Pegasi is the brightest star in a hooklike asterism of equally spaced binocular stars. I use this

*Some skywatchers judge the clarity of autumn nights by counting naked-eye stars within the Great Square. If they see a dozen or so stars, the naked-eye limit is about magnitude 6.0. Triple that number and the limit is down to roughly magnitude 6.5. Since the Square contains about 70 stars brighter than magnitude 7.0, I don't encourage observers under very dark skies to perform this exercise. They should rate the sky's transparency by scouting out a few particular stars of magnitude 6.5 and fainter.

hook as my "train route." Start at Gamma Pegasi and head 1¾° due north, where you'll encounter a magnitude-6.6 star. (Large-telescope users might want to stop here a moment and take a ¾° detour to the northeast, where they'll find the 11.6-magnitude elliptical galaxy NGC 57.) The next stop in the hook is the magnitude-5.5 yellow-giant star 87 Pegasi, followed by another magnitude-6.6 star just 1½° to its west-southwest. The final stop is less than 1½° to the south, at a magnitude-7.2 star; NGC 7814 lies 12′ southeast of

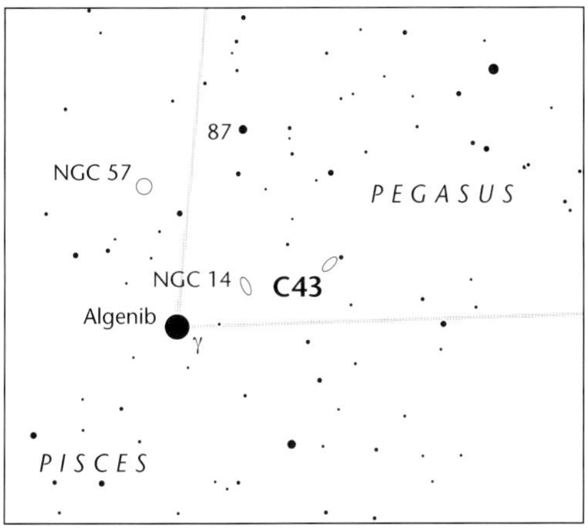

that star and 8′ northeast of a wide pair of 9.5-magnitude stars. If your telescope offers a 2° field of view, the star-hopping described here will be a cinch. A 3° field of view can take in a nice triangle made up of Gamma Pegasi, the magnitude-6.6 star north of it, and the galaxy.

NGC 7814 is an almost exactly edge-on spiral that belongs to the Pegasus Spur of galaxies, which includes NGC 7331 (Caldwell 30). Its recession velocity is 1,047 km per second. If its estimated distance of 49 million light-years is correct, NGC 7814's true diameter is 79,000 light-years. Most astronomers classify NGC 7814 as an early-type spiral, though some categorize it as a lenticular system. In photographs NGC 7814 displays a flying-saucer-shaped body that tapers in either direction along its major axis, which is delineated by a lane of dust. Because of the galaxy's extreme orientation, its dust lane obscures much of the light we would otherwise see from its disk. The galaxy's central bulge dominates the view. It has the mass and luminosity of a modest elliptical galaxy and appears to be flattened by rotation. By comparison, the disk contains only a few percent of the galaxy's total mass. Near-infrared observations, however, have revealed that the disk is substantially more significant than visible-light images suggest; large amounts of dust and gas exist at the disk's outskirts, and tightly wound spiral arms may be present.

A photograph of NGC 7814, taken in 1899 by James Keeler at Lick Observatory, inspired the famous astronomer-artist Lucien Rudaux (1874–1947) to write, "It bears a curious resemblance to an electric arc between two pointed electrodes." To me it looks as if several thousand excited fireflies have taken up temporary quarters on either side of the galaxy's wire-thin dust lane. Despite its conflagrant photographic appearance, the galaxy is not a blazing sight in small telescopes. Through my 4-inch at 23× I jotted down this immediate impression: "A terribly ghostly glow. Light pollution will kill it!" With a prolonged stare the galaxy displayed a tiny core with a diffuse halo. At 72× the galaxy reveals little more definition. A sharp, starlike core lurks inside the galaxy's distinctive oval glow, though no hint of dust can be seen near it. The oval's northeastern periphery looks brighter than its southwestern counterpart; this may be due to an unresolved arc of foreground stars close to the galaxy's northeastern border rather than some physical asymmetry within the bulge. The arc of light looks somewhat like a fuzzy white eyebrow that intensifies to a knot on the oval's southeastern end, then tapers off to form a short and extremely dim needle of light. No feature like this can be seen on the northwestern end.

Higher power does not enhance the view. In fact, with it the galaxy is much more difficult to observe because its dim light is being spread over a greater area. But with patience the view supports all that was seen at the other magnifications. The neighboring 7.2-magnitude star (which appears in the photograph on page 173) looks distinctly yellowish, but this might have been caused by smog from neighboring Kilauea, since the star has a blue-white spectral type of *A*0. Vog (volcanic smog) tends to redden starlight, and the effect is especially noticeable at high magnifications.

Out of idle curiosity, on one night in February 1997 I decided to try for NGC 7814 from inside the Dillingham Ranch house, which has large glass windows. Earlier experiments had shown that the windows did not visibly affect star images at 23×. But I wondered how the glass would affect dim nebulous objects like NGC 7814. (Apparently the notion of the Great Square as a "window" spurred me to try this.) Alas, I could not detect the galaxy through the windows, though I did see it once I moved my telescope outdoors. On one sparkling, clear night in November 1999 NGC 7814 was nearly overwhelmed by light from a four-day-old Moon. This suggests that light pollution will certainly compromise its visibility in small telescopes.

Our second Caldwell galaxy, NGC 7479, lies about 3° south of Alpha (α) Pegasi, or Markab, the southwestern corner of the Great Square; it's also 4½° east of magnitude-4.2 Xi (ξ) Pegasi in the Horse's mane. Observers with telescopes on equatorial mounts should take these "express routes" to the galaxy. Here are some other fun ways to find it. If the owner of a Telrad sighting device places Markab on the northern side of her finder's third and outermost ring, NGC 7479 will be just within the southern side of the second, or middle, ring. A Qwik Point owner should place his LED spot near the imaginary southern apex of a right triangle made with Markab and Xi Pegasi. You can also star-hop from 55 Pegasi, the westernmost of four 5th-magnitude stars forming a tight 1°-long trapezoid 5½° south of Markab. From 55 Pegasi, follow a chain of three 7th- to 8th-magnitude stars (each separated by less than 1°) to the north-northwest; NGC 7479 will be ½° northeast of the last star in the chain, at the end of another chain of 9th- to 11th-magnitude stars. (The former chain, but not the latter, is shown on our finder chart below.)

NGC 7479 is a highly asymmetrical barred spiral that belongs to the Pegasus Cloud of galaxies, a generous gathering of 51 island universes some 100 million light-years distant. The galaxy is slightly more than twice as far as NGC 7814, yet it shines with nearly the same brightness and appears nearly as large in angular terms. That's because the system is quite large intrinsically, having a total mass of

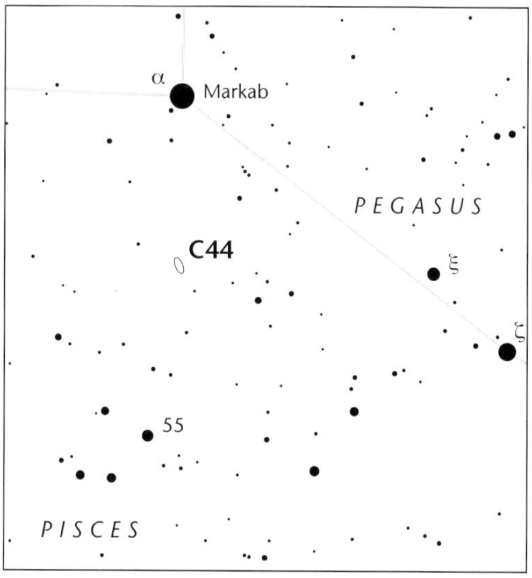

230 billion Suns and a physical diameter of nearly 120,000 light-years. NGC 7479 is 50 percent larger than NGC 7814 and about 20 percent larger than our Milky Way.

Part of the beauty of space is that we can look out upon the galaxies and see them at various angles to our line of sight. It's like looking at a snapshot of a room full of people waltzing in a circle; in such a photograph we can see all kinds of bodies at all kinds of angles. Our "snapshot" of NGC 7814, for example, shows a nonbarred spiral almost exactly edge on, while that of NGC 7479 shows a barred spiral halfway between edge on and face on. When we look at long-exposure photographs of NGC 7479 our gaze is immediately drawn to the galaxy's marked asymmetry. The dominant feature of its backward S form is a long, bright bar, which runs roughly north and south from the galaxy's central pip. The galaxy's two brightest arms are feebler extensions of the bar that look like stretched strands of taffy. The westward-curving spiral arm appears sharply defined, while the other arm looks flocculent, like frayed rope. One recent analysis of NGC 7479's gas dynamics implies that the dominant spiral arm is inclined to the galaxy's main plane. This suggests the spiral was tidally disturbed when a small satellite system merged with it. Such a merger might also account for the galaxy's long bar and its peculiar dust lanes, which rip through the bar like dissection marks made by a nervous student in an anatomy class. Feathery extensions of dust and gas also trickle off concentric regions of star formation on either side of the bar, and they fill in the otherwise dusty void with a mist of starlight.

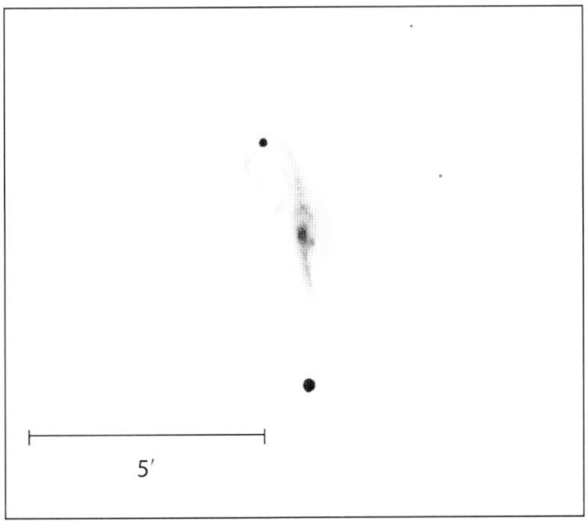

The galaxy is a fine sight in small telescopes but requires much patience. At 23× in the 4-inch NGC 7479 first appears as a very faint lens-shaped object, like a poorly erased pencil mark. But with a prolonged gaze, the galaxy's bar gradually materializes out of the haze like a white-robed figure approaching through a fog. In moments of fine seeing the galaxy seems to become transformed magically from a spindle of light to a pale cat's eye with a bright, vertical slit. I later discovered that my experiences with the 4-inch Genesis mimicked what Adm. William Henry Smyth perceived in 1830 through his 5.9-inch Tully refractor. In his *Cycle of Celestial Objects* Smyth describes NGC 7479 as being "very faint; but after long gazing under clock-work motion, it comes up, trending very nearly north and south, having a telescopic star at each extreme." John Herschel's 20-foot reflector revealed the galaxy as little more than a streak tapered at each end. With his view and Herschel's in mind, Smyth made the following deduction: "Now, the sphere being the general figure assumed in consequence of the particles mutually attracting each other, we can only suppose that the lenticular appearance before us, is a vast ring [of stars] lying obliquely to our line of vision."

Smyth was not totally off his rocker box. John Herschel's father, William, had discovered many nebulous objects that revealed themselves to be made of individual stars when scrutinized with his powerful reflectors. William Herschel concluded that his observations vindicated Immanuel Kant, who in 1775 had convinced himself that the small, elliptical, luminous patches then called "nebulous stars" were actually other "universes" beyond the Milky Way. William gained further conviction when he turned his telescope to numerous "nebulae" and resolved them into clusters of stars. He came to believe that all extended objects were distant stellar systems that could be resolved with large enough telescopes. Even his discovery, in 1785, that the Ring Nebula in Lyra was "perforated" didn't dissuade Herschel from his belief, for he still managed to spot stars within the luminous body of the Ring — or so he thought. Later discoveries would make Herschel reconsider his theory. But his ideas didn't go unnoticed, for, as we can see, Smyth's vivid interpretation of NGC 7479's structure was clearly an extension of Herschel's thinking.

We are lucky to have today's superior telescopes and lenses, for even at 72× the 4-inch Genesis reveals fine structure in NGC 7479. The galaxy's tiny core, for instance, becomes apparent, as does its bar, which has a needle-like quality. Spiral structure is near the limit of visibility at higher powers, as are some knots embedded in the galaxy's arms. Owners of monster Dobsonian telescopes can have a field day with this galaxy. In a 1997 *Astrophysical Journal Supplement Series* paper, Rosa M. Gonzalez Delgado (Space Telescope Science Institute) and her colleagues reported the discovery of 86 H II regions in the galaxy's two main spiral arms. The galaxy is literally bursting with star formation.

Supernova hunters should be on the lookout for new stars in this active galaxy. On July 27, 1990, the Berkeley Automated Supernova Search project discovered a roughly 16th-magnitude supernova (SN 1990U) lying on a possible H II region 22″ west and 54″ south of the galaxy's nucleus. The new star was a Type Ic supernova, one that fades slowly, dimming by about 3 magnitudes during the first 150 days after maximum light. Since the spectra of Type Ic supernovae show no obvious hydrogen or helium lines, they presumably occur when massive stars self-destruct after ejecting their outer envelopes of hydrogen and helium. Studying these events helps astronomers better understand how stars explode.

45

NGC 5248
Type: Mixed Spiral Galaxy (SABbc)
Con: Boötes

RA: $13^h\ 37.5^m$
Dec: +08° 53′
Mag: 10.3
Dim: 6.1′ × 4.6′
SB: 13.8
Dist: 74 million light-years
Disc: William Herschel, 1784

W. HERSCHEL: [Observed 15 April 1784] Very bright, considerably large, extended from north preceding to south following [from northwest to southeast]. Small bright nucleus. (H I-34)

GC/NGC: Bright, large, extended toward position angle 150°, pretty suddenly brighter in the middle to a resolvable nucleus.

BOÖTES, THE 13TH-LARGEST CONSTELLATION, hosts the 4th-brightest star in the night sky (topaz Arcturus) and a variety of variable and multiple stars. But the constellation is a virtual deep-sky wasteland for small-telescope users; sweeping its 907 square degrees with anything smaller than a 10-inch is visually unrewarding. Boötes lies too far from the plane of the Milky Way to be adorned with open clusters or nebulae. (One dim globular cluster, NGC 5466, is camped beside the southeastern edge of Canes Venatici.) And though dozens of galaxies haunt the otherwise hollow corridors of Boötes, only one is brighter than magnitude 11.0 — NGC 5248, a magnitude-10.3 spiral in the extreme southwestern corner of the constellation, just 10′ east of Virgo.

Robert Burnham Jr. calls NGC 5248 a "normal" spiral galaxy. But what is normal? As Timothy Ferris writes in his book *Galaxies*, "The study of galaxies by human beings has scarcely begun. Someone a century from now, reading what we thought about galaxies, would no doubt find much of it distorted, stunted or simply wrong." Yet, from surveys of the visible universe, astronomers have a pretty clear mental image of what constitutes a "normal" spiral: a symmetrical system, with a central nucleus in a spherical bulge rimmed by a broad, flat disk, itself a cosmic hurricane of star-studded arms adorned with clusters, star-forming regions, and nebulae both bright and dark. Like people, galaxies have several degrees of normalcy. It's normal for a spiral to have tightly wound arms or loosely wound ones. It's normal for them to have bars or to lack them. NGC 5248 is a classic example of a highly symmetrical system with a "mixed" disk structure that lies roughly midway between these extremes — which also is normal.

If NGC 5248's estimated distance of 74 million light-years is accepted, then the galaxy is indeed a physically large system with a diameter of 123,000 light-years and a total mass of 140 billion Suns. The spiral is receding from the solar system at a speed of 1,156 km per second. We see the galaxy inclined 45° from face on. Its extremely bright nucleus lies in a spherical lens laced with dust. Its two bright, dust-laden inner spiral arms look like water fleeing from a reciprocating lawn sprinkler, and they outline the central lens. In photographs taken with large telescopes, five very faint arms can be seen branching, like lacy filaments, from the central arms that amateurs can see. Two of the outer arms on the western side appear stubby, while two arms at the northwestern end wrap halfway around the galaxy; the remaining outer arm begins at the southeastern end of the galaxy's bright portion and spirals south through west before branching at its northwestern end. The brightest features in these otherwise very faint arms are composed of relatively young stars.

In 1998 Johan H. Knapen (University of Hertfordshire, England) and his colleagues imaged NGC 5248 with an adaptive-optics system and a near-infrared camera on the Canada-France-Hawaii Telescope, which sits on top of Hawaii's Mauna Kea volcano. The unprecedented, highly detailed images revealed a tightly wound and dusty spiral structure near the galaxy's quiescent nucleus. It ends near the inner edge of a starburst ring, which surrounds the nucleus at a radius of about 3″. Near-infrared observations made two years earlier by Debra M. Elmegreen (Vassar College) and her colleagues already had revealed hot spots more or less evenly spaced around this circumnuclear ring. Both groups' near-infrared images show details similar to those seen in Hubble Space Telescope ultraviolet images of the galaxy's nuclear region; the HST images show hot spots on small spiral arcs that collectively make up the starburst ring. The spacing of the hot spots implies that large-scale gravitational collapse is occurring along the periphery of the ring, where large populations of "super" star clusters have formed.

NGC 5248 is a bit of a challenge to locate, though it is not difficult to see. The galaxy is tucked into the extreme southwestern corner of Boötes, about 7° southwest of 4th-magnitude Upsilon (υ) Boötis in the Herdsman's

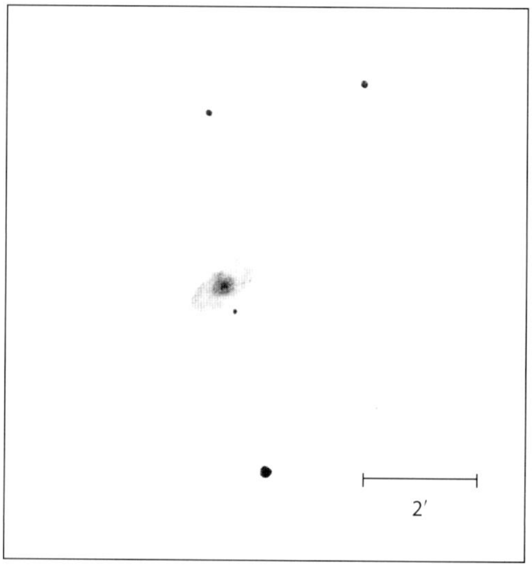

western knee. The spiral can also be found about 8½° east-southeast of 3rd-magnitude Epsilon (ε) Virginis, or Vindemiatrix. (In Roman days Vindemiatrix was Vindemiator; both names mean "grape gatherer.") To star-hop to the galaxy use your widest field of view and start from Vindemiatrix. First move 2° southeast to a solitary 6th-magnitude star. Another 2° hop in roughly the same direction will bring you to 5th-magnitude 59 Virginis. Place 59 Virginis in the northernmost edge of your widest field of view, then sweep 5° farther to the east. NGC 5248 should be near the center of the field. Those living under dark skies should try to spy the 10th-magnitude galaxy in binoculars. I had success with 7×35s from Hawaii, at an elevation of 4,200 feet, though it required much time and effort — which is the key to seeing anything dim.

As is the case with many deep-sky objects that appear so fine in photographs, the structure of NGC 5248 is deceptively challenging. At 23× in the 4-inch, NGC 5248 is a small, somewhat weak glow with a sharply condensed core surrounded by a pale oval disk oriented roughly east-west. As Christian Luginbuhl and Brian Skiff also note, the galaxy lies on the eastern side of an obvious 8′-long isosceles triangle of stars pointing south-southeast. The view intensifies at 72×, but not enough to bring out any structural details. A 13th-magnitude star borders the galaxy to the south-southwest. A possible knot of material pops in and out of view to the west. The galaxy's feeble surface brightness places small telescopes at a disadvantage. Still, the overall glow is mottled, making the galaxy look like a misty bog. Increased magnification only spreads out the galaxy's weak outer halo to the point of invisibility.

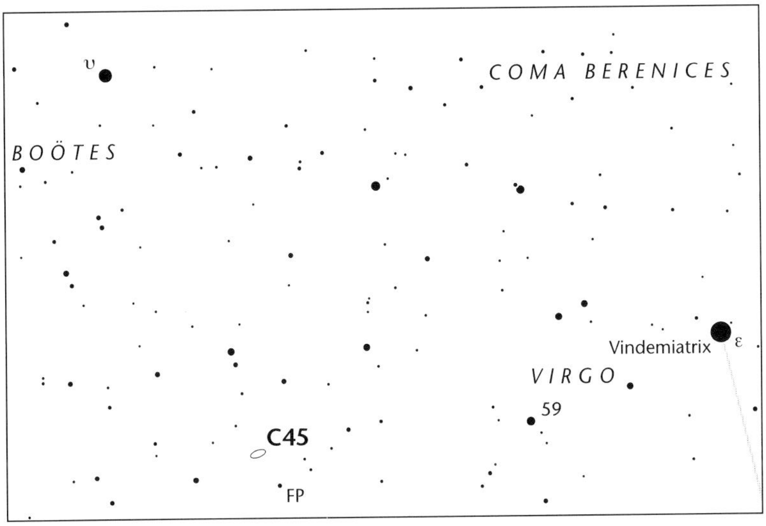

Amateurs with larger telescopes have had more success sighting various degrees of detail in NGC 5248. Using an 8-inch telescope at 390×, Jere Kahanpää of Jyväskylä, Finland, recorded a dark patch north of the nucleus. This observation is intriguing, since Luginbuhl and Skiff noticed a dark patch in a 12-inch, but it was *south* of the nucleus; the patch, they say, gives the galaxy a "black-eye" appearance. Alternatively, Ernst Hartung says that through a 12-inch the galaxy "shows only a conspicuous diffuse haze. . . . in a field with a few scattered stars." The nature of the galaxy is undeniable in a 20-inch. With such a telescope Barbara Wilson saw "beautiful" structure with the arms revealing many small knots.

The interesting variable star FP Virginis lies about 40′ southwest of NGC 5248. The star is one of those closely monitored by participants in the American Association of Variable Star Observers' photoelectric photometry program. Almost all of the stars on that program were observed by the Hipparcos satellite during its 3½-year mission. The Hipparcos team analyzed the satellite's extensive photometry on FP Virginis and found that the star's brightness variations are semiregular at best. The star varies between magnitude 6.6 and 7.1, possibly on two or more time scales.

46

46

Hubble's Variable Nebula; Herschel's Forgotten Fan
NGC 2261
Type: Reflection Nebula
Con: Monoceros

RA: 06^h 39.2^m
Dec: +08° 44′
Mag: ~10.0 (nebula; variable)
Dim: 3.5′ × 1.5′
Dist: 2,500 light-years
Disc: William Herschel, 1783

W. HERSCHEL: [26 December 1783] Considerably bright and fan shaped. About 2 minutes long from the center. (H IV-2)

GC/NGC: Bright, very much extended toward position angle 330°, nucleus cometic and equal to a star of 11th magnitude.

THERE IS A CURIOUS REFLECTION NEBULA IN Monoceros that, when photographed, looks like the head of a match just beginning to ignite. It's called NGC 2261, and it is one of the most unusual objects in the Caldwell Catalog. William Herschel discovered the fan-shaped glow in 1783. The nebulous object includes R Monocerotis, a 10th-magnitude "star" at the southern tip of the fan. (I've placed "star" in quotes, because, as we shall see, the star that ultimately illuminates the nebula cannot be seen at visual wavelengths.) While using the 6-inch refractor at Athens (Greece) Observatory in 1861, director and energetic variable-star observer J. F. Julius Schmidt discovered that "R Mon" fluctuates in brightness, though at irregular intervals. Historically the star has varied between magnitude 9.5 and 13.0. While comparing photographs of NGC 2261 taken between 1900 and 1916, American astronomer Edwin Hubble found that the nebula's shape and brightness also vary over time scales ranging from a few weeks to months. Ever since that discovery, NGC 2261 has been called Hubble's Variable Nebula, leading many to believe that Hubble actually discovered the nebula. Out of respect for Sir William, I like to call NGC 2261 "Herschel's Forgotten Fan."

Inspired by Hubble's discovery, Carl Otto Lampland began, in 1916, to monitor the nebula with the 42-inch reflector at Lowell Observatory. By 1951 he had produced a valuable series of 900 photographs. Like snapshots of a candle flame flickering in the wind, these images showed the nebula's shape and intensity changing over the years. (The photographs on page 183, taken more recently, only hint at the phenomenon.) Curiously, the variations in the nebula did not coincide with the irregular fluctuations of the "star" that supposedly illuminated it. Furthermore, individual details in the nebula could be "displaced" by as much as

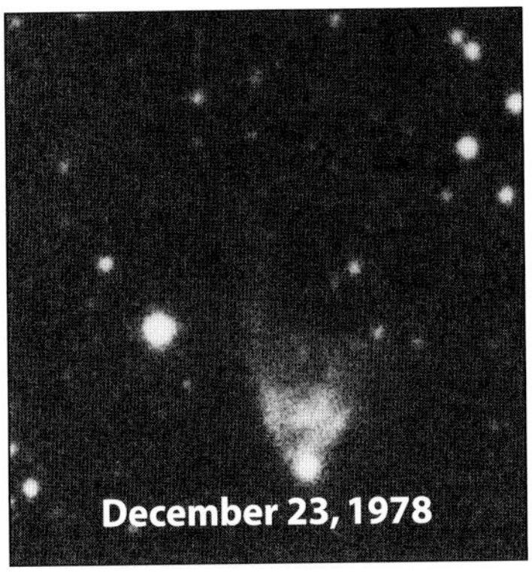

1″ in a mere four days. Clearly, Lampland contended, such changes could not be related to physical movements of material, since that would imply matter being propelled across space at supernatural velocities. Besides, any details that vanished in the cloud mysteriously reappeared, then disappeared, again and again, in the same locations. Lampland concluded that dark shadows — like those cast by the Moon onto the Earth during solar eclipses — were sweeping across the nebula to cause the observed effects. Such shadows, he believed, were cast by clouds of dust traveling very close to the nebula's light source.

Investigation continued. Using Lick Observatory's 120-inch reflector visually under excellent seeing conditions in 1959, George Herbig zoomed in on R Monocerotis and found it to be a tiny triangular nebula "about 5″ long, increasing in surface brightness toward the southern apex, fading out smoothly to the north, but with sharp, well defined edges on east and west. No star could be seen. In poor seeing or with low resolution, R Monocerotis would look simply like a diffuse 12th-magnitude star." Then in 1966 Frank Low and Bruce J. Smith announced that R Monocerotis was a "cocoon" nebula — a dusty envelope of gas (out of which planets might form) around a newborn star. R Monocerotis was the first nebula of this kind to be discovered; astronomers had predicted their existence just a year earlier. The team's evidence was the large amount of infrared radiation that R Monocerotis emitted relative to its output of visible light. The scientists hypothesized that the visible light ultimately came from a newborn star at the center of a dusty cocoon 200 astronomical units across. The 10th-magnitude "knot" we see in our telescopes as R Monocerotis, they argued, is that cocoon of dust being illuminated from within. In a sense this discovery was not new, for, as Edwin Hubble reported in 1916, the 19th-century observer William Lassell believed R Monocerotis was not a star — an opinion Edward Emerson Barnard shared after viewing the nebulous "star" with the 40-inch refractor at Yerkes

Observatory. Indeed, no matter what telescope or equipment was employed in the years to follow, the "star" always appeared to be associated with a distinct mass of dense nebulosity, at least in visible light. By 1984 astronomers were suggesting that a disk containing perhaps five Earth masses of dust surrounded the actual R Monocerotis star. NGC 2261 thus became a prime candidate for a site of planetary formation. The illuminating star was believed to be in an evolutionary state similar to that of our Sun some 4 billion years ago.

Astronomers continue to unravel the mysteries that have veiled this enigmatic object for two centuries. As explained in the November 1, 1997, *Astrophysical Journal*, Laird M. Close (University of Hawaii) and his colleagues have used the Hubble Space Telescope to confirm that R Monocerotis cannot be seen directly at visual wavelengths because dense clouds of dust and gas shield it from view. However, the researchers were able to penetrate the dust with an infrared camera and an adaptive-optics system on the 3.6-meter Canada-France-Hawaii Telescope (CFHT), and they found R Monocerotis to be a binary star about 2,500 light-years distant. The primary star, they concluded, is probably a young, hot *B*-type luminary 10 times more massive than our Sun, with matter accreting onto its surface. The secondary star lies just 0.7″ to the west of the primary and is 200 times fainter; it is probably an ordinary T Tauri star only 300,000 years old, with a mass of 1.5 Suns. Each star is buried in a dense doughnut of dust too small to be resolved by even the Hubble Space Telescope.

The HST and CFHT observations also suggest that the familiar fan-shaped nebula is really a large cone hollowed out by a jet of hot gas flowing out at right angles to the primary's disk, which is inclined some 20° from edge on. Deep within the cone, only 1 or 2 a.u. from the star, dark filaments spiral around the cone's walls like a double helix. Close believes these rotating filaments may follow twisted magnetic field lines, casting fleeting shadows on the walls of the surrounding cone. The patterns we see in NGC 2261 are complicated by the irregular shapes of the nebula's walls and by additional shadows cast by other very sluggish filaments within the shell. Close likens this

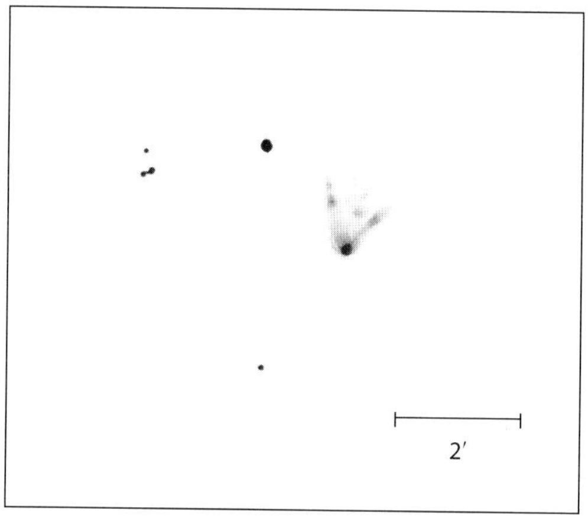

interplay of light and shadow to that of a rotating lighthouse beam and cloud banks at different distances. There is probably a symmetrical counterpart to the fan-shaped nebula on the southern side of the star, but it is heavily obscured from view by dust.

Of course, the dynamics of this fascinating object remain subjects for the imagination rather than the eye. Nevertheless, the nebula is a pleasure to view in telescopes of all sizes. As the Herschel and *NGC* descriptions imply, NGC 2261 looks like a comet with a fuzzy 10th-magnitude head and a fan-shaped dust tail. So it's not hard to imagine how the sight of NGC 2261 through backyard telescopes has caused many hearts to flutter; the nebula has probably been mistaken for a comet more than any other deep-sky object, and continues to fool (albeit tem-

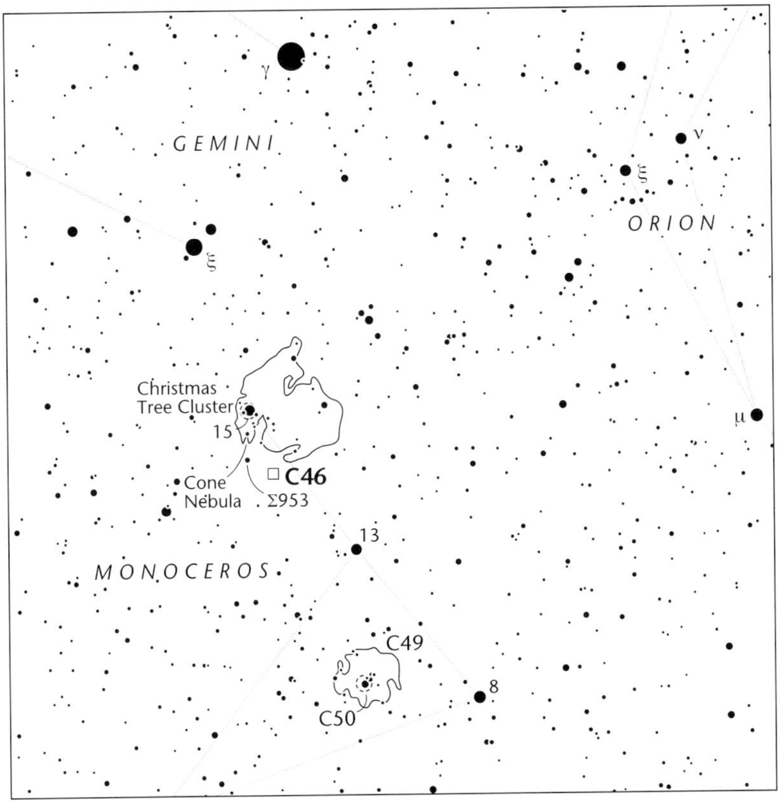

porarily) even the most practiced of observers.

NGC 2261 is easy and fun to find. First, note the feet of Gemini, which are marked by a roughly 10°-long line of four reasonably bright stars oriented northwest to southeast — 3rd-magnitude Mu (μ) Geminorum, 4th-magnitude Nu (ν) Geminorum, 2nd-magnitude Gamma (γ) Geminorum, and 3rd-magnitude Xi (ξ) Geminorum. Xi Geminorum also marks the northeastern end of a fainter 10°-long line of four stars that flows to the southwest; the other three stars are 5th-magnitude 15 Monocerotis, 4.5-magnitude 13 Monocerotis, and 4th-magnitude 8 Monocerotis. Raise your binoculars to 15 Monocerotis. Here you'll see one of the winter Milky Way's most celebrated clusters — NGC 2264, Leland Copeland's famous Christmas Tree, whose base, 15 Monocerotis, is a famous semiregular variable star that also goes by the name S Monocerotis. In his book *Star-Hopping for Backyard Astronomers*, Alan M. MacRobert says this luminous young star, whose spectral type is O7, is "about as blue a star as you'll ever see; they don't get bluer than this pale shade no matter how high the temperature goes." Of course, the color one sees depends on the sensitivity of one's eyes. For instance, Adm. William Henry Smyth saw the star as "greenish." (Bear in mind, though, that this man's vision would often ignite behind the eyepiece, allowing him to see the most vivid colors in the most mundane stars). In the 4-inch at 23× I could detect a thin stream of nebulosity flowing a few arcminutes southwest from S Monocerotis toward a small grouping of three 8th- to 9th-magnitude suns. In photographs this entire region is adorned with a symphony of nebulosity whose bright and dark folds seem to ripple across space like visible sound waves. The scene includes the majestic, though dark, Cone Nebula, a dense stalagmite of obscuring matter. (Had the ancients known of such a wonder, wrote Robert Burnham Jr. of the Cone, they would not have called it by "any lesser name than 'The Throne of God.'")

Hubble's Variable Nebula is about 1¼° to the south-southwest of the Christmas Tree cluster and ½° southwest of the 7th-magnitude double star Struve 953, which itself is roughly 1° south of 15 Monocerotis. (By the way, did you happen to notice that Herschel discovered NGC 2261 the night after Christmas?) Next to the symphony of dust and gas described above, Hubble's Variable Nebula looks like a single note of nebulosity, a puff of breath, an ignited match next to a raging forest fire. Or so it seems in deep, wide-field photographs. Visually, however, NGC 2261 is the most prominent glow among all this nebulous wonder.

So condensed is the light of the nebula around R Monocerotis that I have seen it in 7×35 binoculars, though with effort under a very dark sky. The difficulty is not so much in seeing a 10th-magnitude "star" with 7×35 binoculars from a dark site but in separating it from a neighboring 11th-magnitude star only about 2′ to the northeast. At 23× in the 4-inch I almost mistook the pair for a double star in one of my sweeps. Once you've located this "double" with your telescope, use averted vision and wait a moment; the nebula fanning off of R Monocerotis should swell prominently into view. Again, both the fan-shaped nebula and "star" at its apex are variable. Walter Scott Houston has noted that at times the nebulosity associated with R Monocerotis has been detected with a 3-inch telescope, while at other times a 10-inch was required. When I was viewing it in April 1997, the nebula was definitely fan-shaped at 72×. The fan was directed to the north-northwest of an obvious knot of nebulosity (namely R Monocerotis) at its southern apex. The fan's edges were best defined. I saw little detail inside the fan at this magnification, so the object looked like a comet with two sharp ion tails. At high power, however, the view was more complex — and most fascinating. The fan's eastern side was thin, like the shaft of a feather, while the western edge appeared less sharp. About halfway "up" the fan a faint arc of material curved "over" R Monocerotis like a single auroral flame. A dim knot of material could be seen near the center of this "flame"; from it an even fainter wisp of nebulosity seemed to flare before fading to the northwest. Do not fear using extremely high powers on NGC 2261, for it takes magnification well.

Those who own CCDs may want to follow in Lampland's footsteps and consider undertaking a long-term study of NGC 2261. Such a study could yield some interesting and valuable results. Various amateurs have already used their CCD images to create "movies" that dramatically illustrate shadows sweeping across the nebula. By the way, in 1949 Hubble's Variable Nebula became the first object to be officially photographed by the famous 200-inch reflector on Palomar Mountain.

47

NGC 6934
Type: Globular Cluster
Con: Delphinus

RA: 20^h 34^m 11.6^s
Dec: +07° 24′ 15″
Mag: 8.8
Diam: 7.1′
Dist: 57,000 light-years
Disc: William Herschel, 1785

W. HERSCHEL: [Observed 24 September 1785] Very bright, large, gradually much brighter in the middle. Easily resolvable. Beautiful object. (H I-103)

GC: Globular, bright, large, round, well resolved, clearly consisting of stars, stars of the 16th magnitude downwards, star of the 9th magnitude preceding [to the west].

NGC: Globular, bright, large, round, well resolved, clearly consisting of stars, stars of 16th magnitude and fainter, star of 9th magnitude preceding [to the west].

GLOBULAR CLUSTERS ARE THE MILKY WAY'S senior citizens. When our galaxy was an adolescent, some 10 to 16 billion years ago, thousands of globulars are thought to have formed from its original allotment of gas. The clusters may even have begun life as the cores of dwarf galaxies. In time, though, most of these primordial clusters perished in repeated, tragic encounters with each other or with the galactic center. Today, only about 150 globulars — the remnants of the largest dwarf galaxies? — are known to have survived these ravaging encounters. Delphinus, the small but illustrious celestial Dolphin, harbors two of these ancient galactic relics, and they are nearly cosmic twins: NGC 7006 (Caldwell 42) and NGC 6934 (Caldwell 47).

These Caldwell clusters are separated by

about 11° — roughly the width of a fist held at arm's length. NGC 6934 is about as far south of Epsilon (ε) Delphini, the Dolphin's tail, as NGC 7006 is east of Gamma (γ) Delphini, the Dolphin's nose (roughly 3¾°). The clusters have similar integrated spectral types (*F*7 and *F*6, respectively) and both display highly concentrated cores. NGC 6934 measures 118 light-years in diameter, while NGC 7006 is some 17 percent larger. Yet NGC 6934 is the brighter of the two clusters by a factor of 5. The reason for this apparent paradox is simple; at a distance of 57,000 light-years, NGC 6934 is a little less than half as distant.

NGC 6934 remained anonymous for billions of years, until William Herschel turned his telescope to it in 1785 and became the first human (on Earth, anyway) to perceive its primitive light. Herschel called NGC 6934 a "beautiful object" and found that it could be easily resolved at the eyepiece. But here is the historical rub. Resolving an object into stars had a different meaning and significance in Herschel's day than it does now. The degree to which a system could be resolved was not as important to Herschel as the mere fact that it could. Herschel made no distinction between open and globular clusters. In fact, he placed NGC 6934 into his Class I (bright nebulae), which included objects such as NGC 7331 (Caldwell 30), a spiral galaxy in Pegasus, and NGC 4697 (Caldwell 52), an elliptical galaxy in Virgo. Of course, Herschel did not know the true nature of any of these objects. He believed that all nebulae were collections of individual stars, and that resolving those stars was simply a matter of telescope size and power. Herschel's son, John, promulgated this theory, as Adm. William Henry Smyth notes in his *Cycle of Celestial Objects*: "[NGC 6934] is a mass of very small stars, and is therefore installed among Sir John Herschel's test objects for trying the space-penetrating powers of telescopes."

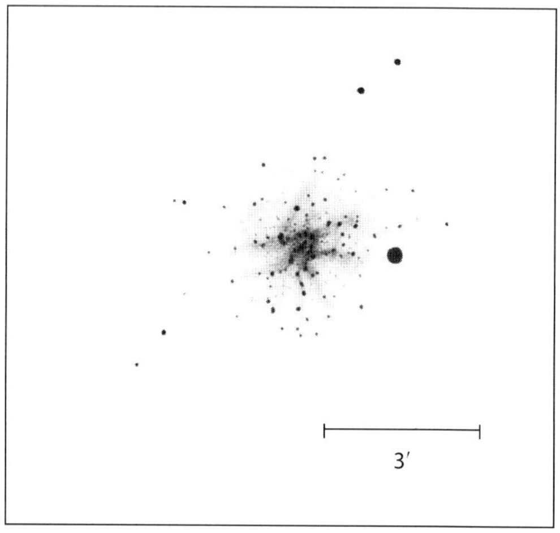

John Herschel was more prophetic than he could have imagined. Even today NGC 6934 remains a proving ground for the world's most powerful telescopes. The stars in NGC 6934's core are so tightly packed that for years even the thin atmosphere above Mauna Kea blurred them to the point of overlapping. Only when the 8.1-meter Gemini North telescope demonstrated its optical prowess in June 1999 did matters change. The latest of Mauna Kea's many superlative instruments, Gemini North "snapped" a few images of NGC 6934 with an adaptive-optics camera helping to compensate for atmospheric instability. The images were astoundingly sharp, showing individual stars at the cluster's very heart. As it turns out, Gemini North produced star images only 0.08″ across, and that's about as good as an 8-meter-class telescope ever can do at the near-infrared wavelengths that were used. The feat was comparable to a telescope in New York City resolving the headlights of an automobile in Mexico City, 2,000 miles distant!

Of course, the Hubble Space Telescope had previously flexed its visual muscles at NGC 6934. In 1999 HST revealed a population of blue stragglers in NGC 6934's core. Blue stragglers are stars that appear to be bluer and more luminous — and hence more massive — than the other stars on a globular cluster's hydrogen-burning main sequence. If they were really the same age as the cluster's other stars, blue stragglers should already have moved beyond the main-sequence stage to become swollen giants. But, at first glance, they are younger than their siblings. And they really are, in a sense. Astronomers believe that a blue straggler forms when two stars merge, as often happens in the crowded confines of a globular cluster's core; such a merger produces a single, young, massive star that hasn't evolved as much as its similarly sized siblings. Blue stragglers stand out on a globular cluster's Hertzsprung-Russell (or H-R) diagram, which plots the apparent magnitude of each star versus that star's color.

Because of their advanced ages, globular clusters generally lack bright, blue stars; such stars have already swelled toward red-gianthood (blue stragglers excepted). By measuring just how bright a globular's bluest main-sequence stars are, astronomers can estimate that globular's age. The HST data place NGC 6934's age somewhere between 13 and 16 billion years. Along with many other globular-cluster age estimates, this places a firm lower limit on the universe's age.

Finding NGC 6934 is easy. As noted above, it lies about 3¾° due south of Epsilon Delphini. If your telescope sports a Telrad sighting device, place the northern edge of the Telrad's outermost circle on Epsilon Delphini and the globular will be just inside the southern edge of the same circle. Through 7×35 binoculars the cluster seems impressively obvious, lying just north of a sideways hook of binocular stars. But NGC 6934's apparent brightness in binoculars is an illusion, since a similarly bright (magnitude-9.5) star lies only 2′ due west of the cluster; light from the star melds with that from the cluster to intensify the latter's appearance. (The star is visible just to the right of the cluster in the photograph on page 187.) At 23× in the 4-inch the two are barely separated; the cluster looks like the head of a tiny comet "kissing" the star to the west. A close pair of roughly 10th-magnitude stars lies about 7′ farther west of the magnitude-9.5 star. Sweeping the area with a 3° field of view reveals several long, faint, dark streaks running more or less north-south; the most prominent of these lanes lie to the east of the aforementioned "hook."

At 72× the cluster displays a dim core, which suddenly appears dappled with dense clumps of hazy starlight. The core is surrounded by a large, symmetrical glow. A prolonged look will enhance the core's apparent patchiness, especially with averted vision — and an occasional strong inhalation of air to replenish

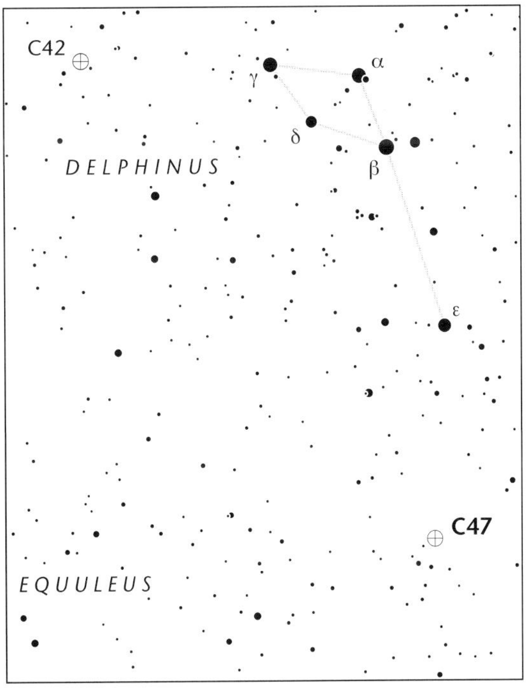

your eyes' oxygen supply. At high power the core looks like an airbag that has been punctured several times with an ice pick and is slowly deflating. Every now and then, at times of crisp seeing, the core bristles with a wedge-shaped mass of shimmering 14th-magnitude suns. Yet there is no central pip or obvious concentration within this triangular blotch of light. A dark lane north of the core helps create the triangular appearance. Starry arms hang limp from the cluster's body like lax tentacles. The most prominent arms curve to the north and to the south. The southern arm branches into two parallel streams or wisps of unresolved starlight. Another pair of arms diverges eastward, looking like a cockroach's antennae. On its west side the cluster has stumpy appendages that appear disrupted by the light of the 9th-magnitude foreground star. Overall, the cluster looks like a wet wad of cotton that has been tossed against the heavens again and again until it stuck.

48

NGC 2775
Type: Spiral Galaxy (Sab)
Con: Cancer

RA: 09^h 10.3^m
Dec: +07° 02′
Mag: 10.1
Dim: 4.6′ × 3.8′
SB: 13.1
Dist: 55 million light-years
Disc: William Herschel, 1783

W. HERSCHEL: [Observed 19 December 1783] Considerably bright, considerably large, very gradually brighter in the middle, having a nucleus or a bright compressed spot. Round. (H I-2)

GC / NGC: Considerably bright, considerably large, round, very gradually very suddenly much brighter in the middle, resolvable.

BEFORE THE INTERNATIONAL ASTRONOMICAL Union redefined the constellation boundaries in 1930, the spiral galaxy NGC 2775 (Caldwell 48) belonged to Hydra, not Cancer. It's quite entertaining to compare today's seemingly rigid and somewhat blocky constellation boundaries with the looser, more curvaceous ones used in the late 1800s. In fact, it's very much like comparing the ragged boundaries of the eastern United States with those boxy ones out west. It's as if curves had suddenly fallen out of favor with both celestial and terrestrial cartographers after the turn of the 20th century.

On equinox 2000.0 star charts, NGC 2775 can be found in the southeastern corner of Cancer, just ½° north of the Hydra border. NGC 2775 is the third most rewarding deep-sky object in Cancer, especially for observers with small telescopes; the other two are the famous Messier open star clusters M44 and M67. The best way to locate NGC 2775 is to look 3¾° east and slightly north of golden, 3rd-magnitude Zeta (ζ) Hydrae. In some works that star is part of Hydra's head; in others (this included), his neck. The 10th-magnitude galaxy also lies only 1¾° due east of an obvious binocular pair of 6th- and 7th-magnitude stars that makes a near-equilateral triangle with Zeta and another golden star, 5th-magnitude Omega (ω) Hydrae. Since NGC 2775 is rather small and moderately dim, you must star-hop slowly and carefully. The galaxy is just visible in 7×35 binoculars. Be sure you're relaxed when looking, and try to avoid any neck strain; otherwise, the galaxy will fade from view (if indeed you glimpse it at all).

In photographs NGC 2775 has a misty or curdled texture and almost appears featureless at a glance. But faint details do shine through. Most prominent are the galaxy's bright nucleus, its central lens, and its spiral arms, which look as thin and frail as old sewing thread. Tiny knots, like filigree, adorn the arms, which become frayed at the ends. Of NGC 2775's photographic appearance Edwin Hubble wrote in 1961: "Notice that the arms cannot be traced as individual structures but rather as separate segments . . . The arms, with their associated dust lanes, start abruptly at the edge of a completely amorphous central nucleus and lens. The boundary between the lens and the spiral structure is sharp." We see the galaxy inclined 39° from face on, so the arms appear to wrap around the central lens in an oblique, ringlike fashion. Despite its tender appearance, the galaxy is quite large. Accepting an estimated distance of 55 million light-years, the galaxy measures 73,000 light-years across and has a total luminosity of 17 billion Suns. It is receding from us at a speed of 1,135 km per second. The galaxy's spectrum is dominated by the light of old, cool stars.

NGC 2775 belongs to the Antlia-Hydra Cloud of galaxies, which has 106 members with known redshifts. It is also the largest member of the sparse NGC 2775 group, a few of whose members can be seen peppering the photograph on page 191. Just 11′ northeast of NGC 2775 is NGC 2777, an amorphous 14th-magnitude galaxy that is plotted on the *Millennium Star Atlas*. A 1981 Arecibo radio-telescope map of the group shows a cloud of neutral atomic hydrogen (H I) gas extending from NGC 2777 toward NGC 2775. There is no evidence for an interaction in visible-light photographs. Are the two galaxies tidally interacting? No, but a recent Very Large Array radio map of the complex reveals a neutral-hydrogen bridge between NGC 2777 and "U3," an anonymous companion galaxy that is about 3.5′ west-southwest from NGC 2777 and about 3 magnitudes fainter than NGC 2777. In a 1996 American Astronomical Society abstract, David E. Hogg and Morton S. Roberts (National Radio Astronomy Observatory) argue that radio maps of this neutral-hydrogen bridge clearly indicate that U3 and NGC 2777 are interacting.

At 23× in the 4-inch NGC 2775 displays a starlike core surrounded by a slightly oval haze that's oriented northwest to southeast.

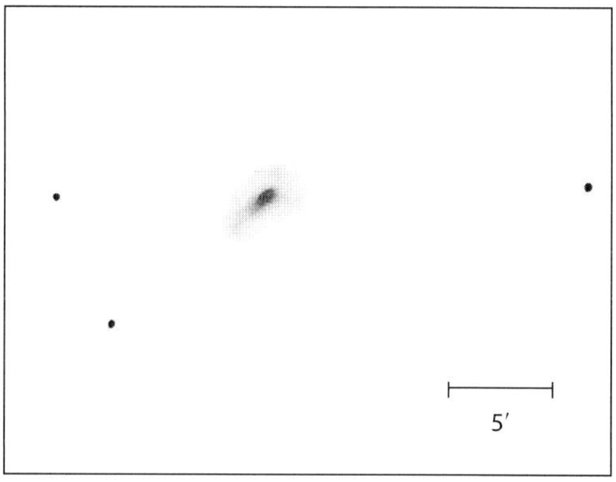

Many dim double stars populate the field, and some are tight enough to look fuzzy at low power — creating the illusion that the group's dim and distant galaxies can be seen in the 4-inch. At higher powers NGC 2775 remains largely elusive. The northeastern side may be brighter than the southwestern side, and hints of mottled clumps pop in and out of view in the galaxy's halo. All this detail is at the limit of vision with the 4-inch. But do concentrate on the galaxy's feeble light, because the southeastern extremity seems to have a faint arc suggesting spiral structure, though nothing remains visible long enough for me to be certain. See if the northern and western sectors don't also show some type of visual enhancements. Observers with telescopes as large as 20 inches have failed to see more structure.

On September 23, 1993, the Leuschner Observatory Supernova Search — which used an automated 30-inch telescope equipped with a CCD camera — discovered a supernova (Supernova 1993Z) just 15″ west and 42″ south of NGC 2775's nucleus. (The program ended in December 1994.) The "guest star" shined at about 14th magnitude in red light. Spectra obtained two days later with the Shane 120-inch reflector at Lick Observatory revealed that the supernova was a Type Ia about four weeks past maximum brightness. Type Ia supernovae are believed to be explosions, either of a single white-dwarf star that rapidly acquires mass from a companion star, causing a runaway thermonuclear reaction (the majority view), or of two white dwarfs that have spiraled into one another and merged (this view is held by a minority of astrophysicists).

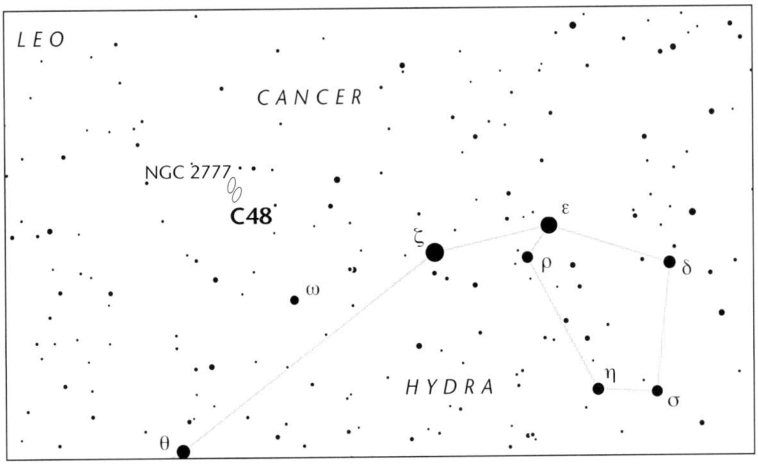

49 & 50

49

Rosette Nebula; The Great Nebula in Monoceros; Swift's Nebula
NGC 2237-8, 46
Type: Emission Nebula
Con: Monoceros

RA: 06^h 32.3^m
Dec: +05° 03′
Mag: ~5, northwestern segment (O'Meara)
Dim: 80′ × 60′
Dist: 4,900 light-years
Disc: (in pieces) Albert Marth, 1864; Lewis Swift, prior to 1883 (independently by Edward Emerson Barnard, 1883); Swift, 1886; and Barnard, 1890s

HERSCHEL: None.

GC: None.

NGC (2237): Pretty bright, very large, diffused, part of extremely large nebulous ring around [open cluster] NGC 2239 [= NGC 2244].

50

NGC 2244
Type: Open Cluster
Con: Monoceros

RA: 06^h 32.3^m
Dec: +04° 51′
Mag: 4.8; 4.4 (O'Meara)
Diam: 30′
Dist: 4,900 light-years
Disc: William Herschel, 1784

W. HERSCHEL: [24 January 1784] A beautiful cluster of scattered stars, chiefly of two sorts. The first large, the second arranged in winding lines, contains the 12th Monoc [12 Monocerotis]. (H VII-2)

GC: None.

NGC: Cluster, beautiful, stars scattered (12 Mon).

ONE OF THE HEAVENS' MOST OBSCURE constellations is dim Monoceros, a sprawling array of inconspicuous suns east of Orion. Of its more than 60 naked-eye stars, none shines brighter than magnitude 3.9; the constellation is as elusive as the mythical unicorn it represents. But halfway between the unicorn's nose and eye (8 and 13 Monocerotis, respectively), and just a jog to the southeast, we find the Great Nebula in Monoceros (Caldwell 49), one of the largest and most massive nebulae in the night sky. Better known as the Rosette Nebula, this wreathlike network of dust and gas is about as large as M42, the Orion Nebula, on the plane of the sky. But it is more than three times as far as M42, and therefore that much larger intrinsically; at its greatest extent the Rosette spans nearly 115 light-years, while the Orion Nebula spans only 40 light-years. In photographs the Rosette appears to ripple outward from a central hollow, as if a stone had been dropped into a pool of celestial vapors. A more elegant metaphor might be a snow-dusted rose seen by the frosty light of the Moon.

The Rosette Nebula surrounds the rectangular open cluster NGC 2244 (Caldwell 50), whose 100 members populate the nebula's ½°-wide central cavity. (Note that the position given in the table on page 194 is for 12 Monocerotis. That star appears to be on the southeastern end of the cluster's densest portion. But that is an illusion. Cluster members do extend farther to the southeast, though they are swaddled in nebulosity. Taking these "hidden" suns into consideration, Brian Skiff estimates that 12 Monocerotis marks the cluster's center as projected onto the plane of the sky.) Collectively these stars shine at magnitude 4.8, say sources, so they make a reasonable naked-eye target (and my own estimate makes them 0.4 magnitude brighter still). Ultraviolet radiation streaming from the cluster's young, hot stars excites the surrounding nebula, whose spectrum shows strong emission lines of hydrogen and singly ionized oxygen as well as the so-called "forbidden" blue-green lines of doubly ionized oxygen. But, unlike the cluster, the Rosette Nebula burns brightly only in photographs. Visually the nebula is dim and challenging, and while some have seen it with the unaided eye (myself included), others have argued that seeing it without optical aid is as likely as seeing a unicorn in New York City. But more on that later.

Glance at almost any photograph of the Rosette and you'll see dark veins of obscuring matter peppering the nebula's northwestern quadrant; it's as if someone had spray-painted the nebula with celestial graffiti. Magnified views of these features reveal the presence of tiny dark blobs. These blobs are known as Bok globules after the late Harvard College Observatory astronomer Bart J. Bok, who turned astronomers' attention to them in the late 1940s. (Bok called them "globules" because they reminded him of the small globules of fat floating atop the cream that the H.P. Hood Company delivered to his Belmont, Massachusetts, home.) Bok proposed that these diminutive dark masses, which span less than 10,000 a.u., were once attached to umbilical filaments of neutral hydrogen gas. Radiation from nearby stars separated, then compressed, them, and the orphaned dark clouds continued to contract under their own gravity until they formed the tiny globules we see today. These small black orbs may be crucibles of creation, for they are the primeval dust clouds within which new suns and new worlds may be forming. By studying Bok globules, we stand at the threshold of understanding the origins of new solar systems. Indeed, life itself may have been conceived within the confines of dusty cocoons like these.

The visual gem of the nebulous Rosette complex, however, is NGC 2244, the dazzling cluster at its heart. The cluster's physical diameter is 43 light-years. It is very young on the cosmic time scale, having formed within the Rosette Nebula less than 1 million years ago (about when the Trapezium stars formed in the Orion Nebula). More than 3 million years before these stars burned through their cocoons, the first hominids already had appeared on Earth. And by the time these stars bathed the Earth with their light, *Homo erectus* had migrated out of his native Africa and ventured onto Indonesian soil. We see the cluster quite clearly because the dust and gas at the Rosette's center have been forced away by strong winds from the cluster's hot, young stars. Blowing at supersonic speeds of 10 km per second, these winds have hollowed out the large cavity we see at the center of the nebula, creating an interstellar bubble. The powerhouse behind this activity appears to be a star 50 times as massive as the Sun and 1 million times as luminous. The cavity will continue to expand until the stars are free of nebulosity. Data from the now-defunct Infrared Astronomical Satellite, or IRAS, have yielded direct evidence of ongoing massive star formation. Furthermore, at the northeastern border of the Rosette Nebula, a large supernova remnant, the Monoceros Loop, has been found. Thus all stages of star formation, from the earliest to the latest, are present in this region.

For centuries the Rosette and its attendant cluster have been a historical enigma of sorts. Experienced observers may note that the nebula's NGC designation in the table on page 194 — NGC 2237-8, 46 — differs from the more popular usage of NGC 2237-9, 46. There is a reason for that, as we shall see. But first let's look at this remarkable object's history.

Some sources credit England's first Astronomer Royal, John Flamsteed (1646–1719), with discovering the star cluster now called NGC 2244 in 1690 or thereabouts. However, this appears to be a simple case of "close but no cigar." A look at Flamsteed's 1729 *Atlas Coelestis* reveals a single 6th-magnitude star, later designated 12 Monocerotis, near the position of NGC 2244. But 12 Monocerotis does not belong to NGC 2244; it is a foreground star superposed on the southeastern limb of the cluster's densest portion. The brightest true cluster member shines at 7th magnitude and hence would have been too faint for Flamsteed's charts. Although the Rosette cluster has a total integrated magnitude of 4.8, its light is spread across an area of sky equal to that of the full Moon. For Flamsteed to receive credit for discovering the cluster, he would have had to make note of stars other than 12 Monocerotis, or of a general fuzziness near that star. To my knowledge Flamsteed did neither. The irrefutable discovery of the Rosette's cluster, then, belongs to William Herschel, who found it in 1784 during one of his now-famous sweeps of the heavens. Some works credit John Herschel with the cluster's discovery, but John merely rediscovered it 30 years after his father. Contrary to popular belief, neither William nor John Herschel saw any nebulosity around the cluster. Why they didn't remains a mystery.

An even greater mystery is that the renowned 18th-century French comet hunter Charles Messier obviously missed (or ignored) the Rosette and its cluster during his sweeps; neither object is listed in his catalog. Certainly he hunted this region for comets, because the 50th object in his catalog, M50, is an open cluster in Monoceros, and it is 4° farther south, and a magnitude fainter, than NGC 2244. Messier chanced upon M50 (which had already been noted by at least one other astronomer) while observing the comet of 1772, so, at least for him, it was an accidental find. As for the Rosette cluster, Messier certainly would not have mistaken it for a telescopic comet. But what about with the naked eye? Neither Messier nor the Herschels noted this apparent fuzzy naked-eye patch in the sky.

At times like this I have to wonder if perhaps Messier and the Herschels were great telescopic observers but not particularly keen naked-eye skywatchers. I raise this point because most observers today claim to have little trouble seeing the Rosette cluster as a fuzzy patch of light with the unaided eye. Even the 19th-century observer, Rev. T. W. Webb said NGC 2244 is "[b]eautiful; visible to the naked eye," though he too missed the nebulosity around it, as did Adm. William Henry Smyth. Did Messier and the Herschels largely ignore the naked-eye sky? Perhaps this would explain why Messier missed the 1758 return of Halley's Comet; as Messier searched telescopically from Paris, a German farmer and amateur as-tronomer, Johann Georg Palitzsch, picked it up a month before Messier — and he did so with the unaided eye. Or perhaps the answer lies in the biology of the eye itself. I have often wondered if our eye-brain system has evolved over the last few centuries to become more sensitive to faint cosmic light. This would certainly explain some other deep-sky mysteries, such as Galileo's missing out on seeing the Orion Nebula and Huygens's having found that the Orion Nebula "cannot be well observed except with large telescopes."

The Rosette Nebula has an equally intriguing history, for it was discovered piecemeal. The first person to notice any of it was Albert Marth, an assistant to William Lassell on the Mediterranean island of Malta. Using Lassell's 48-inch speculum-metal reflector in 1864, Marth noticed a faint star embedded in a small patch of bright nebulosity (NGC 2238) in the northwestern quadrant of what we now call the Rosette. Nineteen years later American comet hunter Edward Emerson Barnard picked up a larger part of the nebula in the northwestern quadrant (NGC 2237).

But unknown to Barnard the same piece had been discovered years earlier by another American comet hunter, Lewis Swift of New York. In his excellent Barnard biography, *The Immortal Fire Within: The Life and Work of Edward Emerson Barnard,* William Sheehan documents these independent discoveries:

> [On] January 23, 1883, Barnard encountered a large and fairly bright nebulous object near the star 12 Monocerotis, and entered in his observing book: "Found a large nebulous object, [near] a scattering cluster of bright stars; it is elongated southwest and northeast. Larger than the field of view." Finding no such object in any of the catalogs, Barnard assumed, naturally enough, that he had found a comet, and he kept it in his sights much of the night looking for the telltale signs of its motion. But it remained stationary. Several more nights of observation put its nebular identity beyond doubt, and on writing to Swift, Barnard learned that his mentor had himself seen the nebula many years before but had never published an account of it. As Swift explained: "I did not know for a certainty that it was never recorded but took it for granted from its immense size that it was although I never could find a description that fitted it." He [Swift] finally published a note about it in the *Sidereal Messenger* in 1884. He there described it as "a nebula of the dumb-bell type, and . . . very large, one of the largest visible from this latitude," and added that "it is unaccountable to me that so conspicuous a nebula should have so long been overlooked. About eighteen months ago Mr. Barnard, now connected with the Vanderbilt Observatory . . . in his sweeps for comets, picked it up and called my attention to it, thinking it might possibly be a comet." Barnard's discovery, then, was entirely independent [of Swift's].

Scanning the area again in 1886, Swift found a bright patch of nebulosity (NGC 2246) in the eastern quadrant. But it was not until Barnard was at Lick Observatory in the early 1890s that the full extent of the Rosette Nebula became known. First his visual observations with the 12-inch refractor confirmed that what he and Swift had seen "was simply a brightish

49 & 50

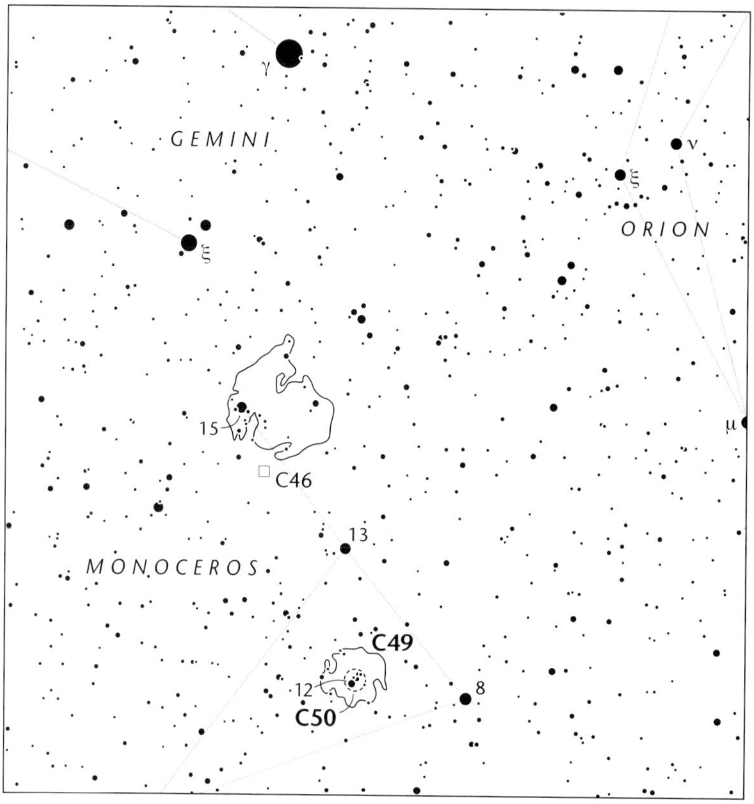

of the RNGC. NGC 2239 is in fact the same cluster as NGC 2244, thanks to a positional error by John Herschel. Archinal credits Canadian amateur Alister Ling for being the first to note the discrepancy, which Ling did in 1992 after considering John Herschel's 1833 description of this object: "The place of an 8th-magnitude star in the most compact part of a large, poor, but brilliant cluster." Ling astutely noted that there is no 8th-magnitude star at those coordinates, but if we add 1 minute to NGC 2239's putative right ascension, we find such a star in NGC 2244. John Herschel himself apparently caused the confusion, as Archinal explains: "He first explicitly states that this object, h 392 (later NGC 2239) is the same object as [HVII-2] (later NGC 2244) in his 1833 observations. He also describes no other such cluster in this region, so it is quite clear that he is talking about the one and only . . . large cluster here. But in the [General Catalogue] . . . he apparently noticed the discrepancy in right ascension. Instead of recognizing the 1-minute-of-time error, he assumed h 392 was a new object and listed these two objects separately, now as GC 1420 (h 392) and GC 1424 ([HVII-2]). Later references of course continued to incorrectly list this new object." In summary, if simplicity is to rule, the Rosette Nebula's designation should be NGC 2237-8, 46; the Rosette's central cluster should be

knot in a vast nebulous ring that entirely surrounded the cluster." Then his photographs brought out the entire wreathlike wonder and earned it the name "Rosette."

Thus, all the *NGC* pieces of the Rosette Nebula — 2237, 2238, and 2246 — are but parts of a single nebular complex. Brent Archinal adds a strong argument as to why the designation "NGC 2237-9, 46" is incorrect for the Rosette Nebula. First, NGC 2239 is not, and historically has never been, part of the nebula. (So references such as the second edition of *Sky Atlas 2000.0* need to be fixed to reflect that.) *Uranometria 2000.0* indicates that NGC 2239 is a small cluster of stars just west-northwest of NGC 2244. Wrong again, says Archinal, who describes the problem in a Webb Society monograph, *The "Non-Existent" Star Clusters* known as NGC 2244; and NGC 2239 should be struck from all catalogs and atlases.

To find the Rosette, use the same stellar guideposts that lead to Hubble's Variable Nebula (Caldwell 46). Follow the line of naked-eye stars that flows southwest of 3rd-magnitude Xi (ξ) Geminorum: 15, 13, and 8 Monocerotis. The Rosette is a fuzzy patch of light 2° east-northeast of 8 Monocerotis. I can see the cluster easily with the naked eye from Volcano, Hawaii, and I can see the nebulosity that way too, with averted vision; confirm your own naked-eye sightings with 7×35 binoculars. My naked-eye magnitude estimate of 4.4 for NGC 2244 includes the star 12 Monocerotis, while Archinal's estimate (magnitude 4.8) does not. Since there is no way to visually subtract 12 Monocerotis from the naked-eye view, my estimate better reflects what is actually seen.

I find it fascinating that no one noticed or suspected nebulosity around this naked-eye cluster until the late 1800s. That I can see the glow in 7×35 binoculars is even more astounding, since the nebulosity was first picked up with a 48-inch reflector, and only a tiny piece of it at that. In the 4-inch at 23× there is no mistaking the Rosette's existence, though the nebula looks like a pale specter of glowing gas riddled with dark lanes. One could argue that knowing the nebula exists helps one see it. But consider this: when Walter Scott Houston persuaded Canadian amateurs Fred Lossing and Rolf Meier to look for the Rosette with a 16-inch reflector, the team reported that they could only barely see the brighter parts. By the late 1970s, however, observers were seeing it consistently in 12-inch and larger telescopes, and Houston later saw it with a 5-inch "apogee scope." I can only conclude that the nebula is so close to the threshold of visibility that even the slightest light pollution or atmospheric contamination will obliterate it. I once thought that the sea-level atmosphere was dense enough to render the nebula invisible, but this is not true. While observing the 1999 Leonid meteor shower from the southeastern coast of Hawaii Volcanoes National Park, I took a break to look at the Rosette Nebula with my naked eye and had absolutely no trouble seeing it. Light pollution, rather than altitude, seems to be the biggest obstacle to seeing the nebula.

Of course, any arguments about the nebula's visibility were squelched during the nebula-filter revolution of the 1980s. An ultra-high-contrast (UHC) filter, for example, is designed to transmit the emission lines of doubly ionized oxygen at wavelengths near 5000 angstroms. Such a filter pulls the nebula out from the sky background and makes it visible to the otherwise naked eye — even from some light-polluted regions. Under truly dark skies like Hawaii's, however, there is absolutely no need for such a filter. The trick is to use averted vision or to block the cluster with a foreground object, like the roof of a house, and look for the nebulous glow; this latter trick is particularly effective at revealing the nebula's northwestern segment, which I estimated to shine at around 5th magnitude. Alternating between averted and direct vision will cause the nebula to swell and contract, respectively.

Altitude does help, of course, and I am amazed at how much detail the 4-inch reveals at 4,200 feet. At 23×, with a nearly 3° field of view, the nebula's northwestern section is most prominent and looks smoky gray. If I concentrate on that region alone, black knots of obscuring dust suddenly pop into view, and the longer I look, the more prominent they become. It's an amazing, almost mesmerizing, effect — seeing dark material slowly appear within a smooth glow, then overtake it, as if it had eaten its way out. I have to pull away from the eyepiece to break the spell, because the nebula starts to "morph" into luminous rafts populated by dark, slithering snakes.

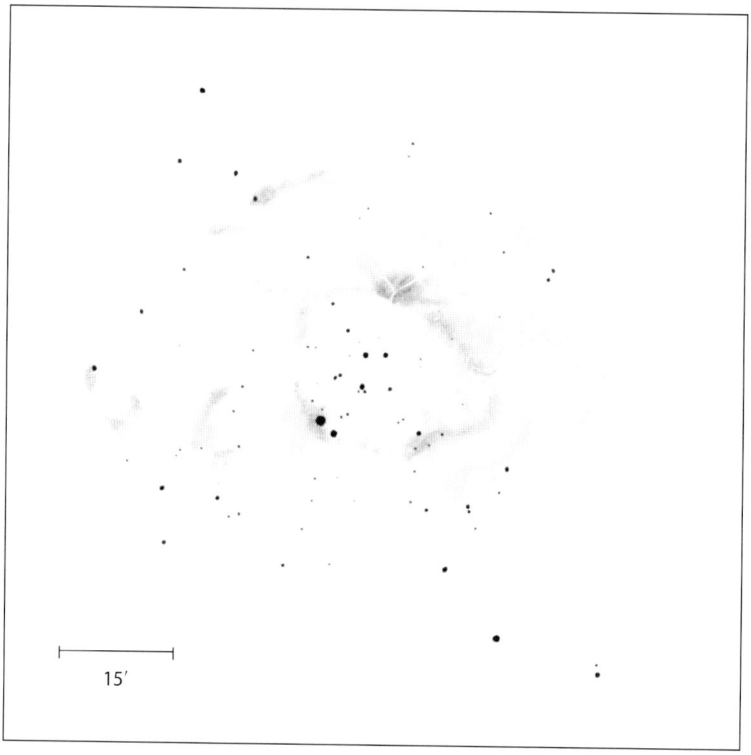

There are really two ways to observe the Rosette Nebula at low power. One is to concentrate on the dark squiggles and trace their patterns. The other is to take quick glances at the bright patches and record their appearances. Photographs tend to exaggerate the bright nebulosity, while eyepiece views tend to emphasize the dark. The latter nebulosity forms a complex web covering an equally complex matrix of glowing patches and stringy vapors. The western portion of the Rosette's inner edge is sharply angled, in marked contrast with its diffuse eastern side. The sky appears darkest beyond the nebula's northeastern perimeter. With a quick glance four prominent patches of bright nebulosity stand out to the northwest, northeast, southwest, and southeast. (Again, my estimate of 5th magnitude for the Rosette Nebula describes the brightness of only the northwestern segment.) However, the entire nebula seems skewed to the west, as if a gentle wind were pushing its vapors in that direction. The section on the southeastern end of the central cluster cavity is remarkably apparent and resembles a bank of fog pressing against a curved range of mountains. And an eerily cometlike glow can be seen on the northwestern side of the cavity; the "comet" has a braided "tail" that extends to the southwest. Just beyond its bright inner edge, the nebula's western portion contains a conspicuous peanut-shaped bay of darkness, out of which thin coils of emission nebulosity seem to extend westward. Amazingly, increasing the magnification of the Genesis beyond 23× does not help pull out any other details at all.

Unlike the view in photographs, the cluster itself seems free of nebulosity at the eyepiece of the Genesis. I could detect fuzziness only near the southeastern end, and this only with averted vision; I saw *no* fuzziness with a direct gaze. Also, the roundness of the central hollow in photographs is not well reflected in the eyepiece view. The cavity is largely oval and bows out to the west. Certainly this is an optical illusion, however. The cluster's six brightest stars are neatly organized into two rows of three stars each, with the yellowish nonmember, 12 Monocerotis, at the southeastern side.

At a glance the two stars at the cluster's heart are easy doubles, and the western double has a fainter companion to the southeast, making it a triple. Between these southern and central star pairs is another nice pair of equally bright suns oriented west-northwest and east-south-east. Barbara Wilson notes that with a 13.1-inch f/4.5 telescope at an altitude of 7,200 feet she could detect numerous faint suns encased in the nebulosity at 47×. The challenge I see, however, is for someone to detect the faint star that Albert Marth detected within a knot of nebulosity. Although he used a 48-inch telescope, I suspect the star is visible in much smaller instruments. Yet I know of no one repeating this remarkable feat.

51

51

The Scarecrow
IC 1613
Type: Irregular Barred Dwarf Galaxy (IBm)
Con: Cetus

RA: $01^h 04.8^m$
Dec: +02° 07′
Mag: 9.2
Dim: 18.8′ × 17.3′
SB: 15.5
Dist: 2.3 million light-years
Disc: Max Wolf, 1906

HERSCHEL: None.

GC: None.

IC: Faint, most extremely large.

THE IRREGULAR DWARF GALAXY IC 1613 (Caldwell 51) is the most challenging object to see in the Caldwell Catalog. It hides in the northwestern quadrant of Cetus, just ½° south of the Pisces border, where it escaped detection until Max Wolf discovered it photographically in 1906 with the Bruce 16-inch refractor at the Astrophysical Observatory in Heidelberg, Germany. Dating this important discovery was somewhat of a chore. The *Index Catalogues* credit Wolf with the discovery but do not provide a date or any references. Having exhausted some other resources, I turned the problem over to Brent Archinal, who ran up against the same wall. Archinal recruited the help of Harold Corwin, who also came up empty handed. But Corwin did suspect we'd find the discovery in the journal *Astronomische Nachrichten*. Archinal followed Corwin's advice and checked that journal's indexes covering dates from 1895 to 1908 (Sidney van den Bergh had noted in *Astronomy & Astrophysics Review* that IC 1613 was known in 1907). One of the last papers Archinal looked at appeared in the October 17, 1907, *Astronomische Nachrichten*. It was a paper by Wolf entitled "Ein Nebelfleckhaufen und Nebelreichtum in Sagittarius." It described what Wolf thought were some unusual patches of nebulosity in Sagittarius. "It turns out," Archinal says, "that these patches of nebulosity are obviously Barnard's Galaxy (NGC 6822) [Caldwell 57]. More importantly, this paper compares these patches to a nebula near 26 Ceti discovered earlier by Wolf, and reported in the November 9, 1906, *Monthly Notices* of the Royal Astronomical Society." Sure enough, in the *Monthly Notices* letter — "Extended Nebula near 26 Ceti" — Wolf announces his discovery of what we now call IC 1613:

> On four plates taken with the Bruce 16-inch telescope in September I find an extended nebula, which attracts attention both by its situation and its appearance.
>
> Practically all extended nebulosities are found in or near the Milky-way, but this object is about 70° distant from the plane of the Milky-way. Its densest part is situated at
>
> [RA] = $0^h\ 57^m.4$ [Dec] = + 1° 20′   (1855.0)
>
> Round this is spread faint nebulous matter of varying intensity. At several points small clouds of greater density shine forth. Further out, the intensity is so feeble that it is impossible to exactly define its limits. It seems that these will be materially altered by lengthening the exposure. My exposures extend to four hours. With these the nebulous cloud extends about 40′ in declination and about 30′ in R.A. [Three] B.D. stars
>
> [whose positions are given here]
>
> are all involved in the cloud.
>
> Southwards it reaches nearly to the B.D. star
>
> + 0°.207   0h 56m.9   + 1° 0′
>
> but it cannot be certainly detected up to the star. Especially towards the west the nebulous matter promises to go much further.
>
> Under the microscope the brighter parts of the mass are filled with numerous very small spots and short trails, so that the appearance is very different from that presented by the Milky-way nebulae. It seems possible that this object is a multitude of very small planetary nebulae collected in a cluster, which a more powerful instrument than mine may perhaps resolve.
>
> *Astrophysical Observatory, Heidelberg:* 1906 October.

Archinal updated to equinox 2000.0 the positional information supplied by Wolf and discovered that Wolf's position for the galaxy "was off." He also noticed that the size listed in table on page 202 (18.8′ × 17.3′) is noticeably smaller than Wolf's estimate of 40′ × 20′. "Did Wolf's exposure go so deep that it showed outer regions not indicated by the size in modern catalogues?" Archinal asks. "Is that why his position is off? It is also interesting that [Wolf] describes what are (now) obviously lots of H II regions within the galaxy."

That Messier and the Herschels failed to spot this loose assemblage of dim suns is not surprising. Although IC 1613 shines at 9th magnitude, its integrated brightness is deceiving. The galaxy's light is spread across an area of sky about the size of a quarter Moon, so each square arcminute of its surface shines with an apparent magnitude of 15.5 — more than six times fainter than Pluto's maximum brightness. This breath of extragalactic light was far beyond the visual grasp of Messier and his contemporaries, and it was too dim and large for the powerful but narrow-field Herschel telescopes. Many amateur astronomy guides pass over IC 1613, claiming it is too dim and difficult for amateur telescopes, but this is not necessarily the case. Before delving into that controversy, let's first look at the wonder of this tiny system, which astronomers often refer to as the dimmest member of our Local Group of galaxies.

IC 1613 lies 2.3 million light-years distant, so it is as far as M31, the Great Andromeda galaxy. It measures 12,400 light-years across, which is comparable to the sizes of NGC 147 and 185 (Caldwell 17 and 18), two of the dwarfs attending M31. But the Andromeda dwarfs are much easier than IC 1613 to see, because they have higher surface brightnesses. IC 1613 has the intrinsic luminosity of a mere 80 million Suns, making it one of the least luminous galaxies known. Like the Andromeda dwarfs, IC

1613 is one of the few galaxies known to be approaching our Milky Way. And though its approach speed of 234 km per second is comparable to that of NGC 147, it is not part of the Andromeda Galaxy subsystem. To understand why, we have to travel in our minds above the plane of our Local Group. Imagine a circle whose radius equals 3 million light-years. If we place the Milky Way at the center of this circle, we see the Local Group dominated by two groupings of galaxies. One huddles around the Milky Way at the center; the other lies at the 9:30 position, a little more than two-thirds of the way out from the center. (The latter group comprises the Andromeda Galaxy, its four brightest dwarf attendants, and M33, the Great Spiral in Triangulum.) IC 1613 would be in the same concentric circle as the Andromeda Galaxy grouping, but 40° away from it, in the 7:30 position; the spatial separation between the Andromeda Galaxy and IC 1613 — nearly 2 million light-years — is far too great for the objects to be considered closely associated. In fact, IC 1613 is but one of dozens of dwarf galaxies that populate the Local Group. Barnard's Galaxy (Caldwell 57) is another, and astronomers believe that many more of these diminutive galaxies have yet to be discovered. One of the problems, of course, is that if a loosely structured dwarf galaxy is seen against the plane of the Milky Way, its stars are hard to differentiate between those of our own galaxy.

Fortunately, IC 1613 lies well away from the stellar madness of the Milky Way. We see it against a relatively star-poor backdrop in Cetus. In photographs taken with large telescopes, IC 1613 looks, ironically, like a close-up shot of the galactic center taken with a 100-mm lens, but with one major difference: while the system has an oval appearance overall, its interior looks stretched, fragmented, shorn, and tattered. In other words, IC 1613 looks like extragalactic road kill. This analogy is not far from the truth. Throughout their histories irregular galaxies have taken severe gravitational beatings from passing galaxies or larger neighboring ones. Left alone, the irregulars would rearrange themselves into more symmetrical systems, which is what IC 1613 is trying to do. That's why I call IC 1613 the Scarecrow. In a remote way it reminds me of the scarecrow in the "Wizard of Oz"; after the witch's winged monkeys tore the straw man apart, he started to piece himself back together.

Indeed, IC 1613 may be a highly flattened spiral system seen tilted 27° from face on. But its spiral structure, if any, is certainly different than that of classical spirals like the Pinwheel Galaxy (M33) or the Whirlpool Galaxy (M51). IC 1613's features are more like those of the Large Magellanic Cloud, for both systems display a prominent central mass, slight spiral structure, and a well-resolved axial bar. Actually, astronomers find that IC 1613 most closely resembles the transition-type dwarf galaxy DDO 216 in Pegasus as well as Barnard's Galaxy. So irregular galaxies may not be so disorganized as first impressions imply.

IC 1613's brightest stars are concentrated in the galaxy's northeastern quadrant, which is separated from its core by an oval swath of dust that is rimmed by what appears to be a spiral arm. This quadrant also contains a huge association of giant blue stars and may be a cradle of star formation. Indeed, the galaxy is riddled with star-forming regions. Spectroscopic observations conducted with the Canada-France-Hawaii Telescope atop Mauna Kea have revealed 20 stellar associations in the galaxy's core and its northeastern and northwestern quadrants. Studies of the northeastern quadrant have shown that its youngest stars are between 5 and 20 million years old, while its oldest stars have ages up to 10 billion

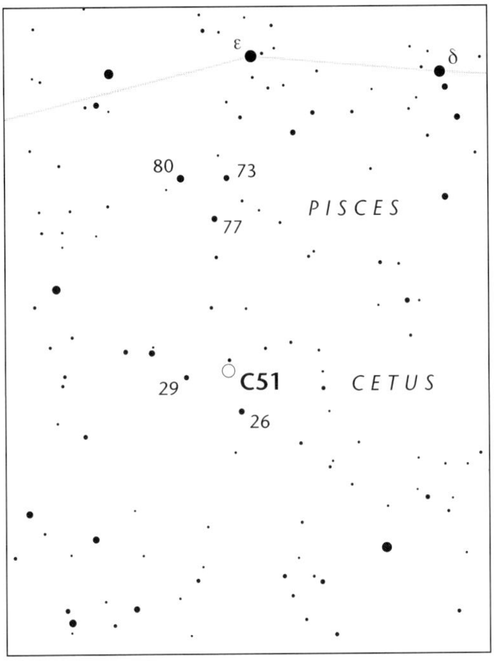

years. A total of 43 star-cluster candidates have also been located, of which 18 are probable open star clusters. (IC 1613 contains no bright globular clusters.) Hubble Space Telescope observations of roughly 700 square light-years in the galaxy's core support the notion that IC 1613 is still birthing stars. For the past 250 to 350 million years, the galaxy has had a roughly constant star-formation rate of about 35,000 suns per year, which is about 50 percent lower than it was 400 to 900 million years ago.

The galaxy is also rich in supergiant stars, and spectroscopic observations with the William Herschel Telescope on La Palma in the Canary Islands have revealed six interlocking interstellar shells in IC 1613. The shells are gigantic, with diameters ranging from 220 to 1,000 light-years, and they are expanding at speeds up to 50 km per second. Stellar winds associated with supernova explosions seem to drive this expansion. The region could be the first stage in the formation of a supergiant shell with a diameter greater than 3,000 light-years. Others have found that expanding "superbubbles" cover the whole of the galaxy. Each superbubble contains at least one stellar association, and the more massive stars in these associations seem to be quite important in shaping the interstellar medium of the galaxy.

HST also imaged Cepheid variable stars in IC 1613, and those stars confirmed R. Brent Tully's distance determination of 2.3 million light-years, which was derived by more indirect means.

Finding the field containing IC 1613 is not hard, but seeing the galaxy is another thing. Start by finding the 4th-magnitude stars Epsilon (ε) and Delta (δ) Piscium in the southernmost Fish's cord. The center of a tight equilateral triangle of roughly 6th-magnitude stars — 73, 77, and 80 Piscium — lies just 2½° south-southeast of Epsilon Piscium. Confirm the presence of these stars with binoculars. A curved line of three stars (magnitudes 6.3, 8.0, and 7.2, respectively) trickles 2½° south of 77 Piscium, bending ever so slightly east, then west, in the process. IC 1613 is immediately south of the third and southernmost (7th-magnitude) star, which places it inside of another 1°-wide triangle comprising the just-mentioned 7th-magnitude star and two 6th-magnitude stars (29 and 26 Ceti). Be certain you have the right field, because 29 and 26 Ceti are part of a northeast-southwest trending line of stars. For those who own an equatorial mount, IC 1613 is 3½° due south of 73 Piscium in the first triangle. All these stars are visible in the finder chart on page 206.

I tried unsuccessfully to see IC 1613 on several nights in 1997. How utterly absurd, I thought, that Patrick Caldwell-Moore would have selected such an object for his catalog. How did he expect anyone to see such a hideously dim galaxy? I knew of no guide that

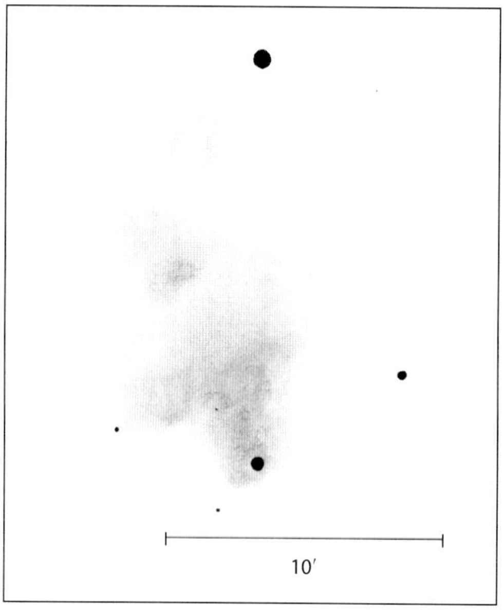

includes a visual description of the dwarf system, so I had no way of knowing what to expect. But I believe this "not knowing" was good, albeit frustrating, because I had no preconceived expectations. In the early morning hours on one extremely windy and clear August night that same year, when the region neared the meridian, I saw IC 1613 with the Genesis at 23×. In Hawaii, as on the mainland, cold fronts can bring incredible transparency, suddenly turning our night skies into dizzying starscapes of galactic wonder. When this happens I feel as if I am looking into the very soul of night. Peering into a telescope is difficult when the naked-eye sky is grabbing hold of your heart. Nevertheless, I immediately set my sights on IC 1613 and very easily located it. If there is a lesson to be learned from this celestial adventure of mine, it would be that the visibility of a celestial object (be it a faint star, a comet, a tight double star, a galaxy, a nebula, or whatnot) cannot be determined solely by effort alone. The atmosphere's clarity and the observer's physical condition also matter, as does his or her level of experience.

At 23× IC 1613 shimmers just above the sky background; it is definitely a "tube tapper." At first it appears as a pale circular glow sandwiched between the 7th-magnitude star and two 11th-magnitude stars 11′ to the southwest. With a concentrated effort and averted vision, the galaxy displays a mottled form. Its most noticeable concentration is about 5′ northeast of the southernmost 11th-magnitude star. Two knots, joined by a tendril of nebulosity, lie to the northwest and southeast of this "core" of stellar material. Another knot is displaced several arcminutes to the northeast. The galaxy is much more difficult to observe at 72×. I find it best to take numerous short glimpses over a long period of time, say 20 minutes, alternating photon collection with rest. When I do this, the galaxy appears more triangular in shape, and its bar becomes a scintillating strand of light with pale, mottled chunks. The knot to the northeast is clearly separated from the bar. Here is that gulf of darkness mentioned above, and a region of star birth. Imagine the processes going on in the celebrated Orion Nebula but across a much larger area and in another galaxy.

Owners of large Dobsonian telescopes should challenge themselves not only by trying to see this fascinating system, but also by hunting down the numerous anonymous galaxies we can see around *and* through it. Do not be intimidated by their minuscule photographic appearance or by the fact that they are not plotted on the *Millennium Star Atlas*. They should be readily visible at the eyepiece of a large reflector.

52

NGC 4697
Type: Elliptical Galaxy (E6) / Lenticular Galaxy (S0)
Con: Virgo

RA: 12^h 48.6^m
Dec: –05° 48′
Mag: 9.2; 9.0 (O'Meara)
Dim: 7.1′ × 5.4′
SB: 11.6
Dist: 76 million light-years
Disc: William Herschel, 1784

W. Herschel: [Observed 24 April 1784] Very bright, very large, suddenly much brighter in the middle. Resolvable nucleus. (H I-39)

GC / NGC: Very bright, large, little extended toward position angle ~45°, suddenly much brighter in the middle to a nucleus.

NGC 4697 is an enormous elliptical system in the southern extension of the Virgo Cluster and somewhat of a problem child. The problem is its morphology. In 1925 Edwin Hubble introduced a system for categorizing galaxy types, in which he recognized three main varieties: ellipticals, "normal" spirals, and barred spirals (see the discussion on page 139). Ellipticals, Hubble said, were amorphous objects "ranging from globular to discus shaped," and he organized them according to their flatness. The most spherical galaxies he called E0, and the most flattened ones he called E7. Hubble's scheme has been slightly revised today, but the framework is essentially the same. Until recently astronomers recognized NGC 4697 as a typical elliptical system, though they disagreed on its exact shape; various sources still list the galaxy as either an E4, an E5, or an E6 system. Recent observations have upped the ante, suggesting that NGC 4697 is an early-type disk galaxy with a lenticular (lens-shaped) body — an S0, in short (see Caldwell 53 for another example).

Both ellipticals and lenticulars can be lens-shaped, and both have very little gas and dust (in fact, some ellipticals may have none). Unlike lenticulars, however, ellipticals show no central bulges, nor structures like disks or spiral arms. A classical elliptical galaxy is simply an oval "shell" filled with ancient starlight and surrounded by a halo of senior globular clusters. In ellipticals, stars concentrate toward the center and form a nucleus of old, low-mass suns. William Herschel aptly described this nuclear concentration of light as "suddenly much brighter in the middle." Herschel also believed he could resolve the nucleus of NGC 4697, but his eyes fooled him; the galaxy is 76 million light-years away, more than 30 times the distance of the Andromeda Galaxy. Resolving its individual stars would have required supernatural vision. (In Herschel's defense, let me note that many ellipticals do appear mottled through the eyepiece. I have yet to encounter an explanation for this illusion, though I believe a sprinkling of field stars can contribute to it.) There is one other key difference between ellipticals and spirals. Ellipticals lack the merry-go-round simplicity of spiral disk structure and dynamics. Stellar orbits in an elliptical galaxy are haphazard; imagine an airport where planes are landing and taking off in all directions without the guidance of an air-traffic controller.

In photographs NGC 4697 looks very much like the textbook definition of an elliptical — completely smooth, its brightness fading evenly from center to edge. Or does it? New technology has opened new windows on the universe, giving us fresh looks at old wonders. The Infrared Astronomical Satellite, for instance, discovered a stellar disk (one contributing 2.1 percent of the galaxy's total luminosity) lying along NGC 4697's major axis. If we stick to the strict definitions of elliptical and lenticular galaxies, this seems to tilt NGC 4697 toward the lenticular category. Further evidence for NGC 4697's lenticular nature appeared in a 1996 *Astronomy and Astrophysics* paper by Herwig Dejonghe (Ghent University, Belgium) and three colleagues. Using the New Technology Telescope in Chile, they found a nuclear dust lane 3.4″ (1,300 light-years) from NGC 4697's center. Along with earlier evidence for a strong disk component to the galaxy, this bolstered the notion that NGC 4697's spheroid may simply be a lenticular galaxy's nuclear bulge. As astronomers scrutinize more elliptical galaxies, they may discover that stellar disks and dust lanes are commonplace in *all* late-stage ellipticals. If so, we'll need another way to discriminate between highly flattened ellipticals and true lenticular systems.

NGC 4697 is the first object in the Caldwell Catalog to lie south of the celestial equator. It is also, surprisingly, the catalog's only galaxy in the constellation of Virgo. But it is a beauty. In his *Astronomical Objects for Southern Telescopes*, Ernst Hartung describes NGC 4697 as being "an easy object even for 7.5 cm [3-inch]. The small elliptical['s] very bright centre is surrounded by a hazy envelope." The galaxy is indeed very bright, and its concentrated glow is visible in 7×35 binoculars. My binocular magnitude estimate (9.0) agrees reasonably well with the value listed in the table on page 207.

To find the galaxy start from Porrima, or Gamma (γ) Virginis, one of the finest double stars in the sky for small apertures. Just ½° east-southeast of Porrima, a 6th-magnitude star marks the northern end of a 5°-long string of five binocular stars. This string, which is oriented nearly north-south, ends with a magnitude-6.3 star. Our target is only about ½° northeast of that star. You can also start at the top of the "sail" of Corvus, marked by the roughly 3rd-magnitude stars Delta (δ) and Gamma (γ) Corvi. Now look for a 5th-magnitude star forming the apex of a northward-pointing, near-equilateral triangle. This star marks the southwestern end of an obvious 10°-long chain of six roughly 5th- to 6th-magnitude binocular stars, which includes 21 and Chi (χ) Virginis. NGC 4697 is only about ½°

northeast of the sixth star in the chain (the magnitude-6.3 star mentioned above) and about ½° south of the ecliptic. (Note that M104, the Sombrero Galaxy, is about 3½° due east of the second star in this chain.) Some star charts show several other galaxies near NGC 4697, but there is no mistaking any of them for Caldwell 52 because our target outshines even the brightest of these neighboring systems by nearly two magnitudes.

To see NGC 4697 with the Genesis at 23× I simply aim my Qwik Point sighting device at the aforementioned 6.3-magnitude star and *voilá*, the galaxy is immediately apparent. It sits inside the arms of a wide Y-shaped asterism of roughly 9th- to 11th-magnitude suns (with the open end of the Y oriented to the northeast), looking like a fuzzy olive suspended within a martini glass. Structure is visible at a glance in NGC 4697; a dense yet fuzzy core of light is surrounded by a fainter circular glow. A prolonged look brings out the galaxy's slightly oval shape, which is elongated along a northeast-southwest axis.

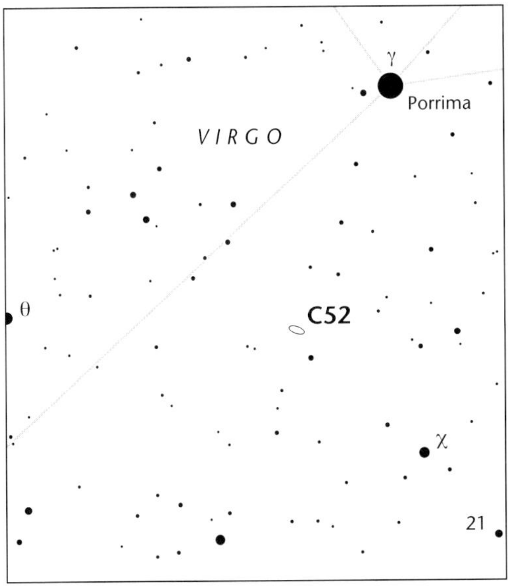

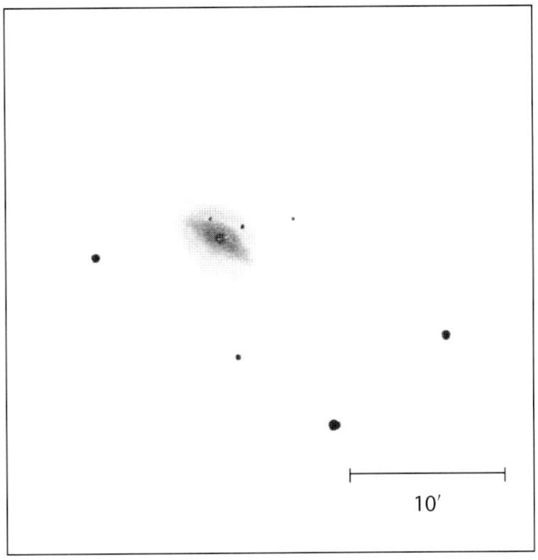

The view at 72× is much more complex. The galaxy's core stands out prominently. Here is that concentration of ancient starlight, surrounded by a swarm of unseen globular clusters. At this magnification we can start to appreciate the system's physical nature. This amorphous sphere of starlight is an impressive 150,000 light-years in diameter and shines with a total luminosity of 70 billion Suns. The galaxy is inclined 44° from face on and is whisking away from us at about 1,200 km per second.

Field stars of diminishing brightness bracket NGC 4697. A roughly 12th-magnitude star partners it to the east, while a much fainter sun, perhaps magnitude 14.5, glows like a dim knot equidistant to the southwest. A roughly 13th-magnitude star borders the galaxy to the north-northwest, with an equally dim sun farther out. Photographs show numerous fainter stars across the face of the galaxy — lending credence to Herschel's illusion of "resolution." Sometimes a "star" or mottled bit of light can be seen in the galaxy's northern sector. At times the galaxy's major axis appears more distinct than its minor axis. Is this an illusion, created, perhaps, by an alignment of foreground stars near the limit of vision? Or could this truly be the nominally elliptical galaxy's "hidden" disk? One problem with this theory is that I have seen this illusion with true ellipticals like M32. That being said, if NGC 4697's lenticular appearance is an illusion, it is an illusion shared with Texas deep-sky aficionado Barbara Wilson. Using a 20-inch f/4 reflector at 72×, Wilson saw this "absolutely beautiful fabulous brilliant oval" with a "long wispy brightness like a barred spiral . . . so it's a huge oval with a long bar-type brightness through the center." Perhaps amateurs with CCD cameras can look for this bar — a feature presumably blown out in long-exposure photographs — to determine if it is real or an illusion caused by faint unresolved foreground stars. NGC 4679 appears to be interacting with a faint unnamed galaxy 5.9′ to the west-northwest. This galaxy should be a challenge to observers with large telescopes.

53

The Spindle
NGC 3115
Type: Lenticular Galaxy (S0)
Con: Sextans

RA: 10^h 05.2^m
Dec: –07° 43′
Mag: 8.9
Dim: 6.9′ × 3.4′
SB: 12.1
Dist: 22 million light-years
Disc: William Herschel, 1787

W. HERSCHEL: [Observed 23 February 1787] Extremely bright, considerably large, much extended, 45° from south preceding to north following [from southwest to northeast]. Has a bright nucleus 2′ long, faint [branch ?] 5′ long. (H I-163)

GC/NGC: Very bright, large, very much extended toward position angle 46°, very gradually then suddenly much brighter in the middle to an extended nucleus.

ONE OF THE MOST BEAUTIFUL GALAXIES IN the heavens awaits small-telescope users in the obscure southern constellation Sextans. NGC 3115, the famous Spindle, lies about 20° south of Regulus, but few amateurs point their telescopes at it. Sextans is a boxy southern constellation with stars no brighter than magnitude 4.5, which isn't a problem if you live under a dark sky. But if you have to contend with city or suburban lights, navigating that starless tract is like rowing a boat blindfolded. This is unfortunate, since NGC 3115 is a splendid object. It is essentially no farther south than the great Orion Nebula, and at upper culmination it is fully one-third of the way up the sky from horizon to zenith at midnorthern latitudes.

At a glance it seems odd that NGC 3115 escaped Messier's careful gaze. After all, the galaxy is close to the ecliptic, placing it within the comet haystack scanned by most 18th-century comet hunters. In fact, Messier discovered objects farther than NGC 3115 from the ecliptic. The galaxy is certainly obvious enough in today's instruments. In fact, NGC 3115 is a grander sight than many galaxies in Messier's list. So how did he "miss" it? Let's first consider what NGC 3115 would have looked like through Messier's less-than-perfect telescopes. Messier used about a dozen different telescopes ranging from reflectors with inefficient speculum mirrors (the largest having an 8-inch aperture) to simple refractors with apertures up to 3½ inches. Certainly, the galaxy should have been easily discernible in a 6- or 8-inch telescope. But ponder these words by Adm. William Henry Smyth, who describes his view of NGC 3115 through a 6-inch refractor: "It is remarkable that this object was very clearly distinguished in my telescope; for [Herschel] says it was scarcely perceptible in his 20-foot when he gave it only six inches of aperture." Of course, the clear glass of Smyth's 6-inch refractor was obviously superior to Herschel's stopped-down speculum mirror. Had Messier used a 6-inch or smaller scope with a speculum mirror during his sweeps of Sextans, the galaxy would not have been an obvious sight. Walter Scott Houston also had some trouble seeing the galaxy in a small instrument: "Though I have observed NGC 3115 with a 4-inch telescope from Connecticut, averted vision was occasionally needed to find it." ("Scotty" adds, however, that the galaxy appeared "small but sharp" in a 5-inch "apogee" scope.) I believe a combination of factors were against Messier. When we go out to find NGC 3115, we usually look when the object is highest in the sky — that is, when it is near the meridian. But Messier did not have that luxury. He was a comet hunter sweeping the sky shortly after sunset or shortly before sunrise. So his sweeps through Sextans would have occurred when the constellation was either dipping toward the setting Sun or emerging from the dawn. The combination of his marginal optics and less-than-perfect observing circumstances plausibly would have placed the galaxy out of visual reach. Then again, perhaps Messier, like many of us, simply shied away from the Sextans void.

The origin of the galaxy's nickname, "The Spindle," is somewhat of a mystery. In his 1978 *Celestial Handbook*, Robert Burnham Jr. writes that NGC 3115 is "[p]opularly called the 'Spindle Nebula,'" but he fails to credit a source for the name. That Burnham uses the word "nebula" instead of "galaxy" could imply that the word "spindle" predated our knowledge of galaxies. Then again, in the early and mid-1900s, the two words were used interchangeably, even though astronomers knew the difference. For instance, M31, the Great *Galaxy* in Andromeda, was still being called the Great *Nebula* in Andromeda when I was young. Neither Smyth, the Rev. T.W. Webb, nor the Herschels describe NGC 3115 as a spindle.

In the 1962 *Larousse Encyclopedia of Astronomy*, Lucien Rudaux and Gérard de Vaucouleurs make no reference to the Spindle, but they do call NGC 3115 an "Elliptical Nebula." This led me to believe that the name was probably applied to the object between 1962 and 1978, when Burnham's *Handbook* was published. So I checked the 1968 edition of Ernst Hartung's *Astronomical Objects for Southern Telescopes* and found a possible solution. Of NGC 3115 Hartung writes, "This elliptical extra-galactic nebula looks like a bright pointed spindle." Burnham occasionally references Hartung's work, so it is possible that Burnham picked up the word "spindle" from Hartung.

Yet the "spindle" appears so prominent in photographs of NGC 3115 taken earlier this century, I found it hard to believe that some astronomer would not have referred to the galaxy by that name. I mentioned this suspicion to Brent Archinal, and he once again came to the rescue by finding a paper by Frank G. Pease entitled, "Photographs of nebulae with the 60-inch reflector 1911-1916," which was published in a 1917 issue of the *Astrophysical Journal*. Pease described a 100-minute plate of NGC 3115 thus:

> This is a bright spindle with an oblate center, which measures about 30″ × 35″. The disk lies in p=45°, is about 6″ wide, and strong and continuous for about 45″ either side of the nucleus; then come several interruptions and knots. The whole lies within an elliptical halo of nebulosity 3′ × 1′.

So Pease and Hartung may be independent sources for this popular nickname.

Hubble classified NGC 3115 as an E7 galaxy. He based his opinion on the way the galaxy appeared on plates taken with the 60-

and the 100-inch telescopes atop Mount Wilson. But the situation changed when the 200-inch reflector on Palomar Mountain began operation, for it soon revealed a thin equatorial plane, indicative of a disk, in NGC 3115. Based on the new evidence, NGC 3115's classification was changed, hovering somewhere between a transitional E7-S0 and a pure S0. As with NGC 4697 (Caldwell 52), changing technology changed a galaxy's classification. Of course, this is how science evolves.

Today there is no question that NGC 3115 is a pure lenticular (S0) galaxy with a very bright nucleus and modest disk inclined 24° from edge on. Redshift measurements show the galaxy receding at 700 km per second. If R. Brent Tully's distance of 22 million light-years is correct, NGC 3115 has a true linear diameter of nearly 40,000 light-years and a luminosity of about 7 billion Suns. The Hubble Space Telescope has confirmed the existence of a supermassive black hole in the core of NGC 3115 — one with the mass of 2 *billion* Suns. That's twice the mass estimated by the University of Michigan team that discovered the black hole in 1992 using ground-based telescopes. The confirmation was detailed in the March 10, 1996, issue of *Astrophysical Journal Letters*, and it marked the third time the Hubble Space Telescope had documented the existence of a supermassive one in a galaxy core.

Speaking of black holes, how can you find NGC 3115 in the dreaded Sextans void? Actually, the situation is not that bad. You just have to forget about Sextans and concentrate on the brighter stars of Hydra. Those whose telescopes have setting circles should start with 2nd-magnitude Alpha (α) Hydrae, or Alphard. Center Alphard in your field of view; move your telescope 9° due east, then 1° due north; and look in the eyepiece. You should see NGC 3115 about ½° due west of a pair of stars (with magnitudes of 6.6 and 7.5, respectively) oriented northeast-southwest. Do not confuse this pair of stars with another, brighter pair, 17 and 18 Sextantis, 1° to the southeast. A simpler, shorter route for star-hoppers starts from 3.6-magnitude Lambda (λ) Hydrae. The stars 17 and 18 Sextantis lie only 4° due north of

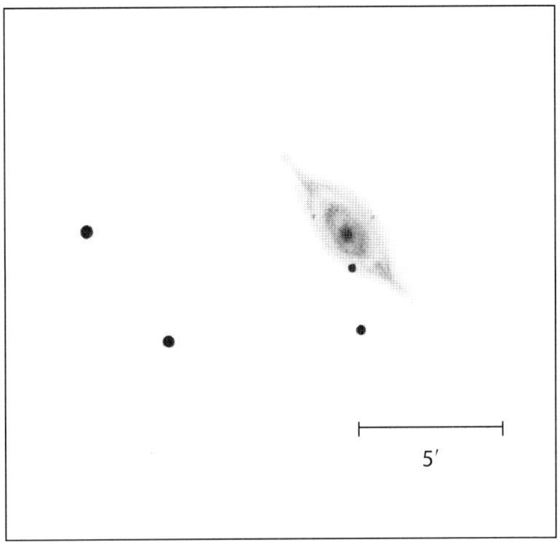

Lambda. So owners of a Telrad sighting device simply have to place Lambda Hydrae on the southern end of the device's luminous outer circle; 17 and 18 Sextantis will be on the northern end of the same circle. You can hop 1° northwest to the fainter pair of stars and then to NGC 3115. Houston also preferred this approach. "Not only is [Lambda Hydrae] an easy naked-eye star," he said, "but a distinct asterism formed by several surrounding stars will offer positive identification even in small finders."

Under clear, dark skies NGC 3115 is *easily* visible in 7×35 binoculars, and the view only gets better with a telescope. At 23× in the Genesis the 9th-magnitude galaxy immediately reveals its 7′-long spindle, which is oriented northeast to southwest. Look for a bright central condensation and a tapered halo. Webb called NGC 3115 "[v]ery distinct, with much

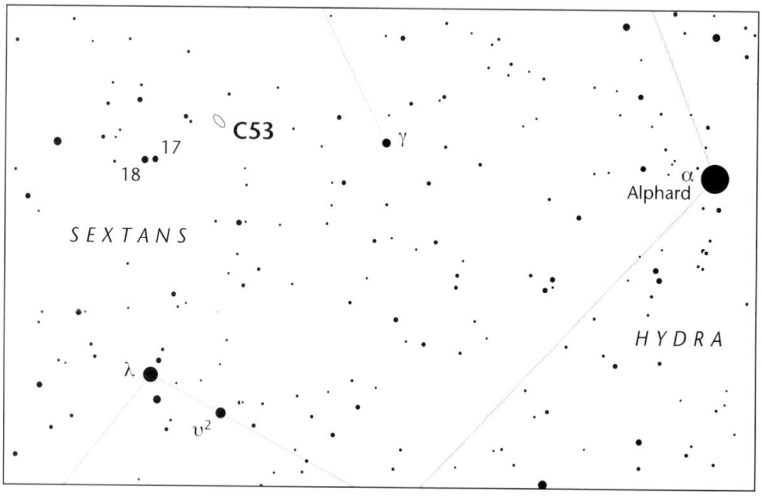

brighter centre, bearing magnifying unusually well." And indeed it does. At 72× the galaxy's nucleus looks layered, evincing a bright core surrounded by an oval haze. A sharply defined plane extends from either side of this inner halo, recalling Saturn with its rings edge on. Concentrate on the outer edges of this "ring plane." Do you see some fuzzy patches? High power brings out a fantastic sight. The galaxy's core is sharply defined, with a brilliant stellar bead embedded in a lens of light. A fainter, though obvious, oval halo surrounds the interior lens, then tapers alongside the razor-sharp linear extensions, which emanate from either side of the lens. The northeastern extension does not appear to connect to the lens, while the southwestern one does. Does this imply the presence of dust? Barbara Wilson suspected a dust lane on the galaxy's southern side, and I can understand why; the region just outside the inner lens is not as bright as the outer half of the lens. Dust may be responsible for this.

The details seen in NGC 3115 through small instruments certainly rival those obtained with larger telescopes, as Houston noted: "I think it fascinating to find an object that shows well in 5-inch binoculars, looks similar in a 12-inch instrument, and still perplexes professional astronomers working with the largest telescopes in the world."

A magnitude-11.5 star lies 3.3′ south of NGC 3115's core; it "balances" on the lip of the galaxy's outer halo. Note that a 14th-magnitude star is 1.3′ south of the galaxy's center so that you do not mistake it for a supernova. To date only one supernova has been discovered in NGC 3115 (Supernova 1935B), and it appeared about 36″ west and 60″ north of the galaxy's nucleus. But you always should keep an eye open for "surprise" stars.

54

NGC 2506
Type: Open Cluster
Con: Monoceros

RA: 08^h 00.0^m
Dec: –10° 46′
Mag: 7.6
Diam: 12.0′
Dist: 10,000 light-years
Disc: William Herschel, 1791

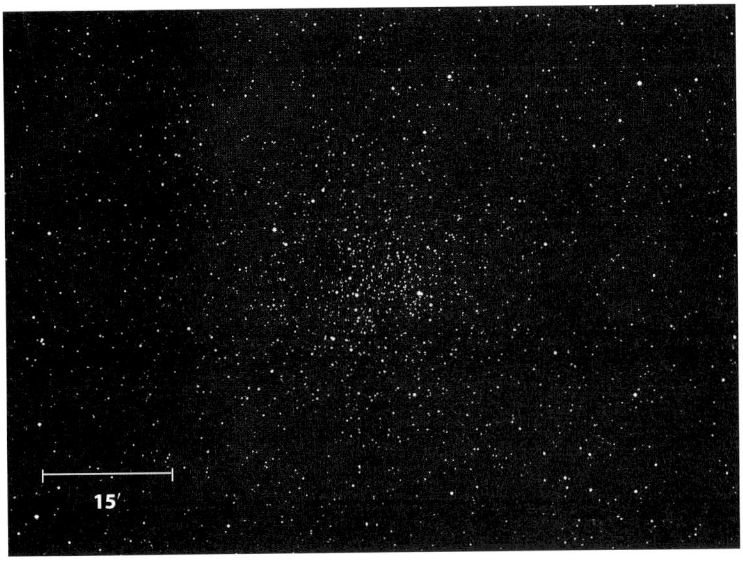

W. Herschel: [Observed 23 February 1791] A very compressed and very rich cluster of stars. The stars are of two sizes, some considerably large and the rest next to invisible. The compressed part 5′ or 6′ in diameter. (H VI-37)

GC/NGC: Cluster, pretty large, very rich, compressed, stars of 11th to 20th magnitude.

Go out on a clear winter's evening, when Sirius is sparkling like fire-fused crystal high in the south, and sweep your gaze upward along the winter Milky Way. Among its frosty ribbons of starlight are some of the heavens' most magnificent open star clusters, including M46 and M47 in Puppis and M48 in Hydra. And halfway between M46 and M48 lies one of the region's most understated galactic clusters, an old salt of the river Milky Way whose core of aged suns seems like a collection of faded memories. NGC 2506 is among the oldest open clusters known. In 1981, Robert McClure (Dominion Astrophysical Observatory, Canada) and his colleagues derived an age of 3.4 billion years for the cluster, making it a contemporary of M67 in Cancer. More recent studies, however, have dropped that value to about 2 billion years, making NGC 2506 closer in age to NGC 752 (Caldwell 28).

Old open clusters are noticeably deficient in "metals," the term astronomers use to describe all chemical elements heavier than hydrogen and helium. For instance, on average each star in M67 has only 82 percent as much iron as our Sun. Given their roughly comparable ages, one would expect that NGC 2506 might have a similar metal abundance. But it doesn't. NGC 2506 is much poorer in metals than M67, with its average star containing only one-third as much iron as our Sun. Current models of open-cluster formation show that the farther a cluster is from the galactic center, the poorer its metal content. It is the cluster's position in the galaxy, not its age, which determines its metal content. NGC 2506 currently lies near the direction of the galactic anticenter at a relatively high galactic latitude, and it is nearly 10,000 light-years from the galactic center as well as 1,600 light-years above the galactic plane. Together with its position, its low metal abundance places this cluster into a class of objects that includes NGC 2420 (the prototype), NGC 2158, NGC 2204, NGC 2243, and Melotte 66.

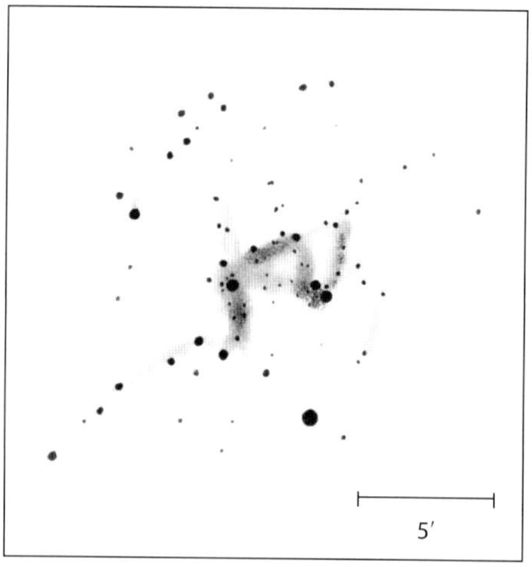

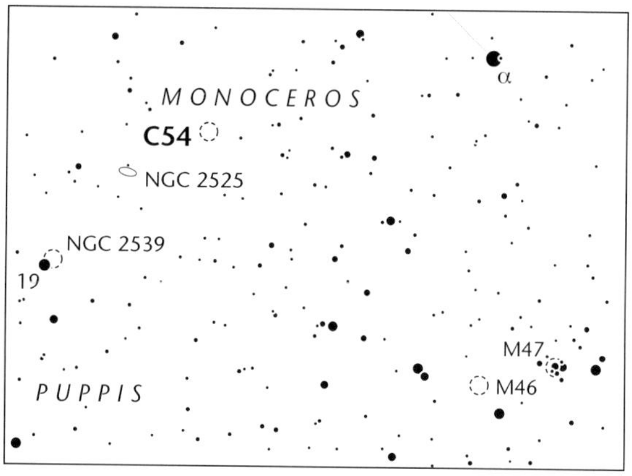

About half of the stars on NGC 2506's main sequence are binaries, as are more than 20 percent of the cluster's 800 or so total members. Some *blue stragglers* — stars that apparently linger on the main sequence — also have been confirmed as members of this cluster. Most astronomers believe that a blue straggler forms when two closely orbiting stars merge, producing an unusually massive single star that has yet to depart the hydrogen-fusing main-sequence stage of stellar evolution.

NGC 2506's apparent size of 12′ is deceivingly small. The cluster lies 10,000 light-years from Earth; that's twice as far as M46 and more than 5 times the distance of M48. NGC 2506 spans some 37 light-years, making it 16 percent larger intrinsically than M46 and nearly 50 percent larger than M48. If we could bring NGC 2506 as close as M48, it would shine at 4th magnitude and span a full degree of sky, making it a near rival to M44, the famous Beehive Cluster in Cancer. And if we could bring NGC 2506 as close as the Beehive Cluster (515 light-years), we would see a spectacular 1st-magnitude object 4° across. Poets would have been inspired to glorify this scintillating fury of grizzled suns, and we amateurs would have had the chance to ponder another cluster as magnificent as the Pleiades.

We can at least imagine such glory as we peer at NGC 2506 through our telescopes. To find it I follow a river of suns that meanders eastward from Sirius, past 4th-magnitude Gamma (γ) Canis Majoris and beyond the clusters M46 and M47, until I get to 5th-magnitude 19 Puppis — a journey of more than 20°. Confirming 19 Puppis's identity is simple, because the star is superimposed on the southeastern side of another dramatic open cluster, 6.5-magnitude NGC 2539. The star is also the southernmost in a 4°-long chain of three stars pointing to the north-northwest. That chain's other two stars shine at magnitude 6.3 and 6.2, respectively; the magnitude-6.2 star, at the chain's northern end, is also a visual double star. You will need to confirm that you've correctly located these stars in your finderscope or in binoculars before carrying on. NGC 2506 makes a near-equilateral triangle with chain's northernmost two stars, to the west.

In 7×35 binoculars NGC 2506 looks unimpressive, almost inconsequential — a celestial afterthought. But stick with it for a moment, because with averted vision a bright core should emerge, followed by a diffuse outer halo. The halo even appears speckled with dim starlight, but this is an illusion created by Milky Way field stars. Through the 4-inch at 23×, NGC 2506 is a small, tight cluster, a knot of stars protruding from the rich Milky Way background. With concentration the cluster looks like a dim and distant globular cluster. The west side of its dense, asymmetrical core is punctuated by what appears to be a single sun. The asymmetry favors the north-northeast, where there is an obvious arc of stars, many paired.

At 72× the cluster's brightest stars form loosely spiraling waves of light, with obvious clumps marking the wave crests; using my imagination, I see a frothy sea just starting to churn beneath a waterspout. The star that I noticed on the cluster's western side at low power is actually a pair of 11th-magnitude suns. They sit on the western bank of a V-shaped bay of relative darkness. Look about 3′ east of this star pair, at the cluster's very core. There you'll find another fainter and wider pair of stars anchoring a fuzzy bar of starlight, oriented north-south, whose northern end fades into a knifelike point. The cluster's northwestern section is populated by meandering chains of stellar clumps that lie against a background haze of unresolved suns. Some of the clumps make distinct cursive shapes, like the Greek letters delta (δ), upsilon (υ), and mu (μ). Some star chains flow like coronal streamers, especially to the north. At high power, many of these arms take on a delicate spiral nature. The southwestern section of the cluster is dominated by the dark bay, which seems to leak starlight to the southeast and southwest. Christian Luginbuhl and Brian Skiff confirm the absence of starlight in the southern section, writing, "In 25 cm [a 10-inch telescope] the cluster is irregularly round with a blank spot on the [southern] side." With a 12-inch telescope they estimated that the densest portion of the cluster spans about 8′ and shows about 100 stars, most of which are very faint. At higher powers the cluster's outer envelope starts to blend in with the Milky Way background.

Before leaving this region of sky, be sure to return to NGC 2539 next to 19 Puppis, for it is a much more pleasing object visually. Also, while sweeping en route to NGC 2506 I encountered the 11.6-magnitude galaxy NGC 2525 sandwiched between two faint stars about 2° northwest of 19 Puppis.

55

55

Saturn Nebula; Venus Nebula; Ghost of Saturn Nebula
NGC 7009
Type: Planetary Nebula
Con: Aquarius

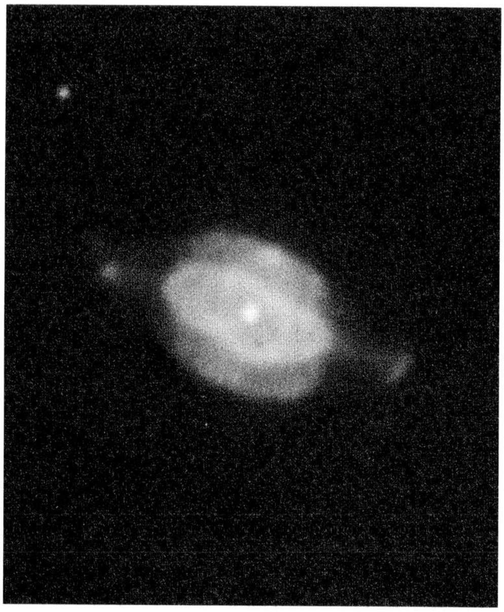

RA: $21^h\ 04^m\ 10.8^s$
Dec: –11° 21′ 49″
Mag: 8.0
Dim: 44″ × 23″
Dist: 1,400 light-years
Disc: William Herschel, 1782

W. HERSCHEL: [Observed 7 September 1782] Very bright, nearly round planetary disk not well defined. (H IV-1)

GC / NGC: A magnificent object, planetary, very bright, small, elliptic.

ON THE EVENING OF AUGUST 29, 1780, Pierre Méchain, Charles Messier's contemporary and rival comet hunter, discovered a faint glow in Aquarius 4° southwest of 4.5-magnitude Nu (ν) Aquarii. Messier searched for Méchain's find the following October (on the evenings of the 4th and 5th) and determined its position relative to Nu Aquarii. Now known to be a globular cluster, this object became the 72nd in Messier's growing list of non-comets. On those same two evenings, Messier discovered a "[c]luster of three or four faint stars, which, at first glance, resembles a nebula." Again Messier determined the new object's position relative to Nu Aquarii, and this tiny asterism became the 73rd object to enter his catalog. The irony of these discoveries and their follow-up observations is that about 1¼° due west of Nu Aquarii is one of the heavens' brightest planetary nebulae — NGC 7009. And it entirely escaped both Messier's gaze and Méchain's.

But this is not surprising. Although NGC 7009 is bright, it is also small (its brightest part spans only 23″), and it can be easily mistaken for a star at low power, especially when seen through a small telescope. The nebula did not fool William Herschel, though; he discovered it in 1782, finding it "very bright." He also classified the object now known as NGC 7009 as a *planetary* nebula because it was "nearly round," like a planet, and bore magnification well, like a planet. However, the nebula's hallmark ring-like extensions remained undetected until Lord Rosse (William Parsons, third Earl of Rosse) scrutinized the nebula with his 72-inch reflector. In his 1850 paper, "Observations on the Nebulae," Rosse described NGC 7009 as having a fairly uniform luminous disk with "ansae which probably indicate a surrounding nebulous ring seen edgeways." Indeed, as Robert Burnham Jr. notes in his *Celestial Handbook,* "Rosse's drawing portrays the nebula with a somewhat greater degree of symmetry and perfection than is actually the case; it is shown as a perfect nebulous miniature of Saturn." Not surprisingly, Rosse dubbed it the Saturn Nebula. (Arguably it should be called the Ghost of Saturn Nebula to parallel Caldwell 59, the Ghost of Jupiter Nebula in Hydra.)

Equally fascinating is Adm. William Henry Smyth's 1836 impression of NGC 7009 as seen through his 6-inch refractor: "This object is bright to the very disc, and but for its pale blue tint, would be a very miniature of Venus." (Thus my other suggested name, the Venus Nebula.) Herschel, and later Smyth, believed planetaries to be solar systems in formation, and one can sense the excitement engendered by this belief in the remainder of Smyth's description:

> If this object be only equally distant from us with the stars, its real dimensions must be such as would fill the whole orbit of Uranus. Now a globular body of the magnitude of the orbit of Uranus, would contain within its periphery more than 68,000 millions of globes as large as our Sun!

One can only imagine what Smyth would have penned had he seen a photograph of this extremely bizarre planetary taken with a large telescope. In such photographs the nebula is a ghostly array of nebulous rings and baubles, a pendant of celestial jewels strung together by diaphanous vapors. At the core of this gaseous locket is an aged sun's nearly naked heart. Unlike the central stars of many planetaries, NGC 7009's is fairly easy to see in small telescopes. It shines at magnitude 11.5 and resides in the hollow of a bright elliptical ring inclined 84° from our line of sight. That ring has a sharply defined exterior and a scalloped interior, which together suggest the "lips" of a smiling clamshell — especially since the northern and southern boundaries of the ring are brighter than its eastern and western extremities. A fainter, elliptical ring surrounds the bright inner ring, and it is adorned with a mysterious alignment of bright knots and condensations along its northern and southern edges. Most enigmatic, however, are the two evanescent rays protruding east-northeast and west-southwest from the nebula. Each ends at a bright knot or bead.

NGC 7009's progenitor was probably a main-sequence star of about 2 solar masses. It is now a hot, bluish dwarf with a surface temperature of about 55,000° Kelvin. Strong ultraviolet radiation from the central star causes the nebula to fluoresce. To most eyes the nebula glows with a blue or green tint, the visual signature of doubly ionized oxygen atoms. The hot central star's rotation may have caused the irregularities along the inner edge of the inner ellipse. Actually, the inner nebula contains two rings of material, probably related to different episodes of mass ejections. The rings are not aligned. This may be due to the central star precessing, its rotational axis wobbling about once every 30,000 years. An unseen companion object could be driving this precession; if so,

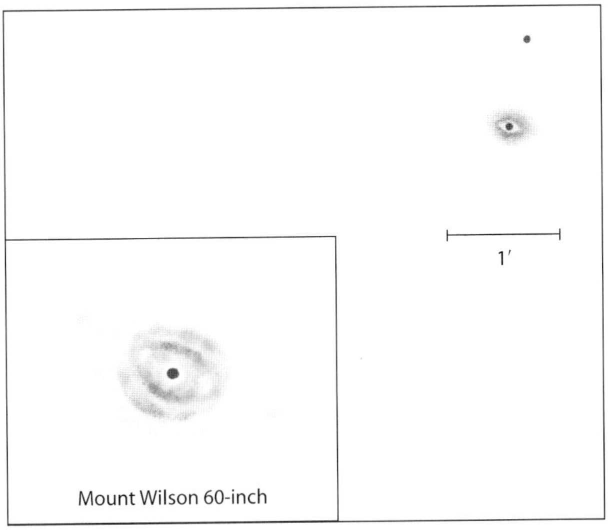

Mount Wilson 60-inch

the companion would have to be located about 4.5 solar radii from the central star to account for the misalignment of the inner ellipsoids.

A recent discovery also revealed a dim inner pair of condensations that resembles the well-known outer pair, with one knot on either side of the central star. The condensation pairs may have originated in separate shells of ejected matter that have since been pushed outward by the central star's ultraviolet radiation or particulate winds. The International Ultraviolet Explorer satellite found no evidence of light-scattering dust particles in NGC 7009.

In April 1996 the Hubble Space Telescope imaged NGC 7009, giving astronomers an unprecedented look at this complex planetary nebula. An enhanced color version of this image (a black-and-white version is shown below) shows a central star trapped in the cavity of a pale blue chrysalis of gas. Large hollow lobes of yellow gas extend from this chrysalis, making the inner nebula look like some sort of psychedelic sushi. The green outer envelope (the star's former outer layers) appears somewhat boxy and multidimensional, like a time-lapse exposure of an open beer barrel rolling in a circle. Thin green jets extend along the nebula's major axis and end in bright blood-red knots, as if the jets were recently cauterized veins. If the nebula is 1,400 light-years distant, these knots lie 0.15 light-year from the central star.

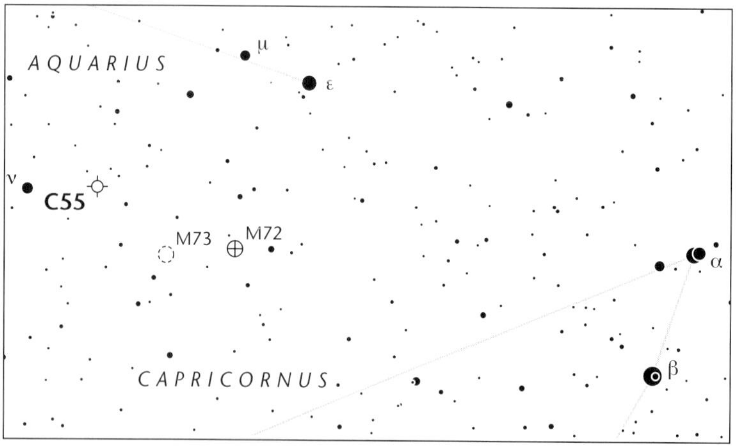

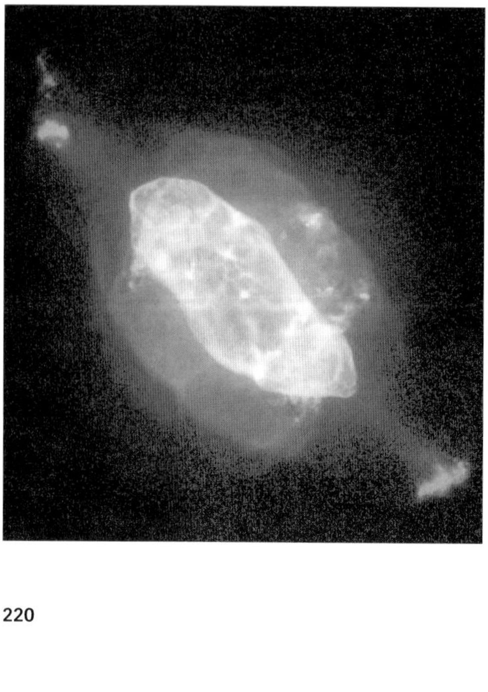

Despite its visual complexity, the HST image is giving astronomers a better understanding of the mysterious processes that transform a low-mass star like our Sun into a white dwarf. A computer model based on the HST data indicates that NGC 7009's progenitor star first expelled one solar mass of gas, which now constitutes the "barrel." This gas confines stellar winds flowing from the central star. The confined gases slide along the walls of the barrel's cavity until they converge at its tips. As Bruce Balick (University of Washington) explains, "Each ansa [terminal knot] is joined to the tips of the cavity by a long greenish jet of material, much like the hot gases that follow a bullet from the barrel of a rifle. . . . The tips act somewhat like a nozzle that form gas jets, like a stream of water from the nozzle of a garden hose. The dense tips of the jets might be material that the jets have plowed ahead of them as they push[ed] forward. However, the models also predict a high degree of gas turbulence, which is not observed. Future improvements in the observations and refinements in the models will help us to understand the physics of the outflows and the conditions of the jets."

The Saturn Nebula lies in the westernmost reaches of Aquarius, a little more than 3° north of the central Capricornus border. As mentioned above, it is about 1¼° due west of Nu Aquarii, which, in turn, is about 10° northwest of 3rd-magnitude Delta (δ) Capricorni. The bright nebula is easy to spot in 7×35 binoculars; it looks like an 8th-magnitude star. Through the Genesis at 23× the nebula is apparent (with averted vision) as a slightly swollen star. Concentrate long enough, and breathe rhythmically, and you should see the planetary's outer envelope materialize; look for a fuzzy, tightly wrapped collar around the nebula's bright center.

Overall, the nebula appears round, and it truly looks like a gaseous outer planet. Increasing the magnification to 72× enhances the tight inner ring and the pale green outer halo. (On the subject of color: like me, Christian Luginbuhl and Brian Skiff found the nebula pale green when viewed through a 6-inch. But the colors of planetaries are highly suggestive; most observers see blue in NGC 7009.) At high power, the central star appears as a bright point of light centered in the densely packed inner ring. The cavity is obvious, though it is not dark; the central star looks like a candle seen through a frost-covered window. Walter Scott Houston called the central star "a beacon," but that was after a cataract operation sensitized his observing eye to ultraviolet light.

A very faint elliptical shell surrounds the inner nebula, and it looks mottled. Perhaps this mottling is what Angelo Secchi observed in the 19th century when he "saw [NGC 7009] sparkle, and thought it a heap of stars." Unlike the case with so many other planetaries, increased magnification does not seem to greatly improve views of NGC 7009, though it does bring out the faint raylike extensions. Once spotted, these gossamer wings can be seen at the limit of resolution at 72× in the 4-inch. Burnham said that the jets can be seen with a 10-inch under good sea-level skies. Luginbuhl and Skiff clearly saw "spike-like extensions" with a 10-inch, "each seemingly as long as the disk." All these observations are amazing if you consider that it took a 72-inch telescope to discover the ansae. Describing the view through a 12-inch telescope, Luginbuhl and Skiff note that the core is unevenly bright with "bright curds on the [north] side and a darker area on the [east]." Here are the scalloped edges already being resolved in a modest amateur instrument. In a 17½-inch f/4.5 reflector, Barbara Wilson (who sees the nebula as green at 32×) resolved the knots at the end of each ansa at 154×. So here is a challenge for all observers: how small a telescope will show these knots? (Wilson also notes that the nebula's color fades to a pale greenish-white at 154×.)

In August 1986 I had the opportunity to use the 60-inch reflector atop Mount Wilson, California. Observing with that venerable instrument was like going back in time to the glory days when astronomers began to peel away the mysteries of the visible universe, like the skin of an onion, with the depth of their questions and the power of their telescopes. One of the first objects I turned the 60-inch to was NGC 7009. Photographs of this nebula, taken by the same telescope, inspired me to look and wonder. Of course, overexposed photographs from large telescopes tend to exaggerate brightness while obscuring fine details. To my trained eye, however, the same telescope made NGC 7009 appear ethereally lucid, a ghostly form that seemed trapped in a limbo of light, something half-solid and half-gaseous. At 400× the 11.5-magnitude central star blazed forth with hypnotic steadiness. It appeared to be caged by two sharply defined pale-green rings, each decorated with a string of irregular condensations. The object's hallmark extensions seemed to stream out from the green rings like spotlights. Each ended at a tight knot, as if two opposing beams from the same lighthouse were illuminating a distant wall of fog. No imagination was required to visualize Rosse's metaphor; here was Saturn with its rings seen edgewise, with bright moons adorning the tips of the hairline ring. And no effort was needed to look through Smyth's eyes and "convince the reflecting mind that it is a mark of devotion which we owe to our Maker, to study with earnestness the beautiful and harmonious works around us, however their immensity may at first bewilder us."

56

Pac-Man Nebula
NGC 246
Type: Planetary Nebula
Con: Cetus

RA: $00^h\ 47^m\ 03.3^s$
Dec: –11° 52′ 19″
Mag: 10.9
Dim: 4.6′ × 4.1′
Dist: 1,600 light-years
Disc: William Herschel, 1785

W. Herschel: [Observed 27 November 1785] Four or five pretty [bright] stars forming a trapezium of about 5′ in diameter. The enclosed space is filled up with faintly terminated milky nebulosity. The stars seem to have no [connection] with the nebulosity. (H V-25)

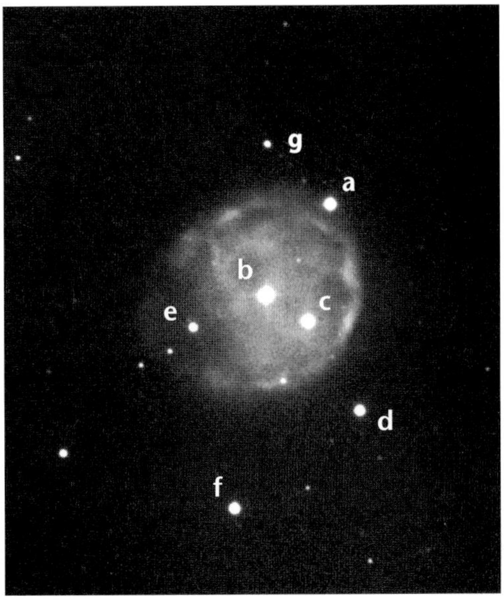

GC / NGC: Very faint, large, 4 stars in a diffused nebula.

NGC 246 is one of the Caldwell Catalog's most visually intriguing planetary nebulae. Look at the photograph above. Here we see a broken shell of gas centered on a 12th-magnitude star. Several other conspicuous stars are superimposed on the gas shell as well, and still others lie just beyond the nebula's confines. Now read William Herschel's discovery description. Does it cause you to raise an eyebrow? Where is the 5′-wide trapezium of stars? And isn't it strange that he would write "four *or* five pretty [bright] stars" [emphasis mine]? Note, too, that Herschel does not say that the nebulosity is centered on a 12th-magnitude star. Instead he says that the nebula fills the space within the trapezium. The *New General Catalogue* is more satisfying because it seems to better describe what is seen, at least in long-exposure photographs — namely four bright stars associated with a diffuse nebula (though these four stars do not form a trapezoid).

Here's another mystery, which could be related to the first. In the December 1984 issue of *Sky & Telescope*, Walter Scott Houston reported that, with an 8-inch f/5 reflector, skilled deep-sky observer Ronald Morales of Arizona easily saw NGC 246 "as a round, diffuse glow behind three stars of similar brightness. A fourth was glimpsed, but the central star was not seen." Really? Would a skilled observer like Morales miss seeing a 12th-magnitude star in an 8-inch telescope? You wouldn't think so. Yet Morales apparently is not alone. In August 1986 veteran observer Barbara Wilson didn't see the central star either, and she used a 13.1-inch f/4.5 reflector at 222×. A drawing in Volume 2 of the *Webb Society Deep-Sky Observers Handbook* does show the central star in an annulus whose periphery is dappled with the light of three obvious stars; the drawing is based on a view through an 8-inch telescope at 75×, and the stars in the drawing match those seen in photographs. Ernst Hartung also saw the central star. In his

Astronomical Objects for Southern Telescopes he said it was faint in a 12-inch. Now let's throw in the monkey wrench: when I observed NGC 246 for this book in December 1996 and December 1997, I had no problem seeing the central star in a 4-inch.

Before addressing these problems, we have yet one more puzzle to consider. Herschel wrote that the nebulosity is faint. "I question his description of it as faint," Houston argued, "since I can see it with the 4-inch Clark." But Hartung called the nebula "rather faint" in his 12-inch. Wilson calls it bright in the 13.1-inch, while Christian Luginbuhl and Brian Skiff state that "[t]his planetary nebula is faint, but can easily be found at low power in [a 6-inch]."

Let's tackle the last mystery first. Photographs tend to make NGC 246 appear deceivingly small, but it isn't. With a diameter of nearly 4.6′, NGC 246 is 60 percent larger than the famous Owl Nebula (M97), a planetary in Ursa Major. It also is a full magnitude fainter than the Owl. And anyone who has searched for the Owl knows how difficult the task is under anything but dark skies. The same is true for NGC 246, only more so. But the popular literature seems to have perpetuated a myth. Houston writes that "[m]odern estimates placed the . . . planetary at magnitude 8.5." Roger Clark makes this same claim in his *Visual Astronomy of the Deep-Sky*, and the Webb Society's *Handbook* lists NGC 246's magnitude as 8.0. Actually, NGC 246's *photographic* magnitude is 8.0, but its *visual* magnitude is 10.9. So the nebula appears much brighter (relative to comparison stars) in photographs than it does to the eye. Of course, one's telescope size, choice of magnification, and observing site also enter into this equation. But, truth be told, NGC 246's brightness is simply a demon to estimate, for the faint nebula has relatively bright stars superimposed on it.

The central-star puzzle is tougher to explain, though I suspect it is largely the result of an optical illusion. Of course, some planetaries' central stars do vary in brightness, and variations would certainly explain why Hartung thought NGC 246's central star faint in a 12½-inch reflector in the 1960s, and Morales and Wilson failed to see it in the mid-1980s, while I saw it so readily in the mid-1990s. Indeed, photovisual and photometric observations of the central star have revealed brightness variations over the decades. In 1930 the central star's photographic magnitude was as bright as 9, while it dipped to 11.2 in 1969. But this difference in magnitude is certainly not great enough to explain the above-cited discrepancies. After all, an 11th-magnitude star would practically blaze in a 12½-inch telescope.

I believe there is more to this mystery than meets the eye. Note that Morales saw three similarly bright stars, and a fainter, fourth star, superimposed on the nebula. His description of these stars matches photographs perfectly. Morales just did not realize that the second-brightest star of the four is the central star. The same goes for Wilson. You see, the eastern rim of the nebula is exceedingly faint and fully one-quarter of the shell is all but invisible, even in photographs. Visually, then, the central star looks off-center against the nebula. As for Herschel's observation, I believe he did not see the eastern part of the nebula. If so, his trapezium could be formed by the stars labeled **a**, **b**, **c**, and **g** on the photograph on page 223.

Astronomers have determined that NGC 246's 200,000° Kelvin central star has one of the most highly excited spectra known. Most prominent in this spectrum are lines of thrice-ionized carbon (C IV), five-times-ionized oxygen (O VI), and probably some singly ionized helium (He II). Nitrogen is absent, so this star is quite evolved and probably has undergone the triple-alpha process, wherein the star fuses helium to make carbon nuclei. The ultraviolet spectrum of NGC 246's central star displays remarkable P Cygni profiles. (P Cygni is a "novalike" variable that erratically changes brightness, and its pairs of slightly offset emission and absorption lines attest to an expanding outer atmosphere.) Spectroscopic observations of the nebula itself reveal that its outer shell is expanding more rapidly than its inner region; the acceleration is most probably caused by radiation pressure from the central star. NGC 246's central star may be near the point of exhausting its nuclear fuel. That means it soon will contract from its present radius (5 to 8 percent of our Sun's) to become a white dwarf barely larger than Earth.

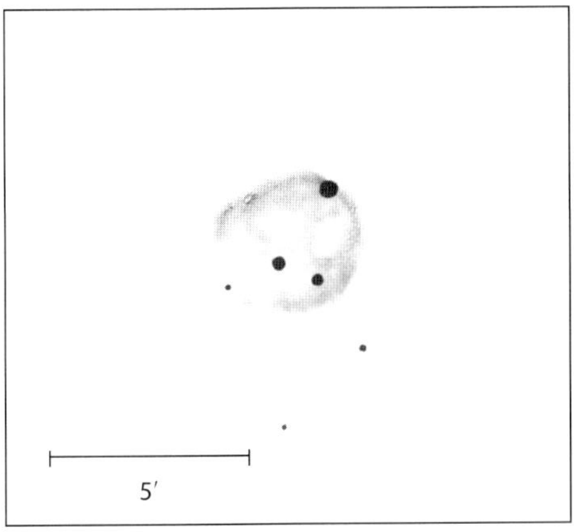

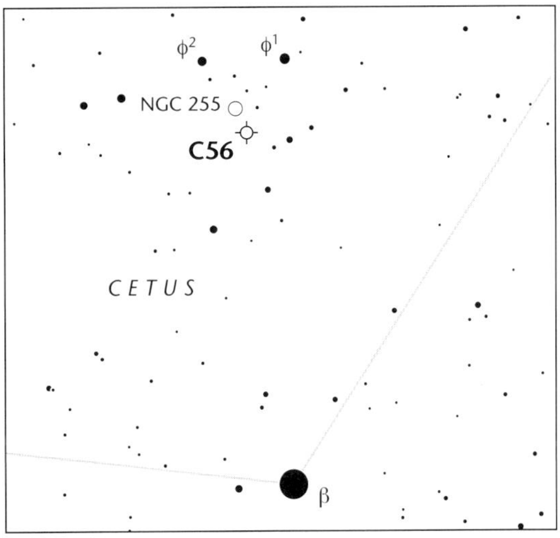

To locate NGC 246 first pick up 2nd-magnitude Beta (β) Ceti. About 7½° due north you'll find 4.8-magnitude Phi[1] (φ^1) Ceti and, 1½° farther east, 5.2-magnitude Phi2 (φ^2) Ceti. Confirm that you've located these stars with your binoculars, then point your telescope toward their midpoint. NGC 246, 1⅓° to the south of their midpoint, makes a nearly equilateral triangle with these two stars. Be careful, though, because the 12th-magnitude galaxy NGC 255 is just ½° north-northeast of your target. NGC 246 lies at a high southerly galactic latitude (–74°), so we see it through a virtually transparent part of the Milky Way.

At low power in the 4-inch, NGC 246 lies in a field of stars resembling a crooked Taurus. At first I passed over the nebula while sweeping for it, noticing only a small clustering of stars as I moved by it. Of course, the "cluster" is the tiny grouping of stars superimposed on the nebulosity. At 23× it's hard not to imagine that the nebulosity encompasses all the bright stars in the area, but a careful look reveals the planetary has two connected crescents of nebulosity that arc through an obvious triangle of bright stars. The southwestern rim of one arc curves between the stars labeled **c** and **d** on the photograph. This arc then tapers and fades to the east toward a void surrounding star **e**.

Now for yet another mystery. Looking back through my notes on NGC 246, I noticed that in my initial sketch of the region (dated December 3, 1996), the star labeled **f** on the photograph was by far the dimmest of the six I plotted. On the following evening, however, my sketch shows **f** just slightly fainter than star **d**. Curiously,

I did not notice **f** brightening, but instead recorded star **e** looking fainter. In other words, the apparent dimming of star **e** could have been caused by the brightening of star **d**. Admittedly, star **f** was rather outside my scope of interest (no pun intended!) because it is not associated with any nebulosity. So consider yourself warned about the reality of these fluctuations. But here's the clincher. While studying NGC 246 a little more than four years later (on January 15, 2001), I was taken by an obvious diamond of light: stars **a**, **b**, **c**, and **f**. When I compared my drawing to those made in 1996, I found that star **f** was clearly brighter in 2001 than star **d**, while star **e** was still its faint self. I encourage others to monitor these stars to see if such fluctuations are real. (By the way, star **b** *is* the nebula's central star.)

NGC 246 is at its finest at moderate power (72×). Most prominent is an arcing swath of nebulosity north of the central star; the two are separated by a dark void. This nebulous section ends abruptly on the northeastern side at a bright knot that looks like a fuzzy 13th-magnitude star. The eastern side of the planetary's shell cannot be seen in the 4-inch. I call NGC 246 the Pac-Man Nebula because, in photographs, its dark eastern side looks like the mouth of that voracious video-game character, which is about to devour the dim star **e**. The dark cavity north of the central star is the creature's eye.

The western border of the roughly circular nebula is well defined; it arcs gracefully through the brightest star (**a**) in the "cluster." A wedge of bright gas joins star **a** to the central star, and it forms an interstice between two voids: one (already described) to the north of the central star, and a dimmer one to the north of star **c**, the third-brightest star superimposed on the nebula. It's a challenge to see these features; averted vision and lots of time are required. I find it best to relax between efforts. That being said, I find NGC 246's "eyes" easier to detect than the Owl's. But don't be fooled into looking for something obvious. The nebula's dark eyes are difficult to define because of the bright intervening stars. They are mere suggestions. I recommend spending several nights with this object to average out any atmospheric vagaries that may affect your observations. Use high magnification to determine the brightnesses of the stars associated with the nebula. Houston said that objects like NGC 246 show up better in long-focus instruments since they tend to scatter less light into the field of view than do short-focus optics. He also suggested using a Barlow lens to obtain higher magnification: "A low-power eyepiece and Barlow seem to scatter less light and produce a higher-contrast view than a high-power eyepiece alone." See if you agree.

For those *big*-telescope users who like to compete with the Hubble Space Telescope for optical prowess, Howard Bond (Space Telescope Science Institute) notes that NGC 246's central star is also a resolvable binary. The 14th-magnitude companion is a type *G*8-*K*0 V star and lies 3.8″ from the primary. By fitting this star onto the main sequence, Bond derived a distance to the nebula (1,600 light-years) that, he says, "is one of the more secure distances known for a galactic planetary nebula, but it could be improved still further with a better estimate of the metallicity of the system." Assuming that distance is correct, NGC 246's physical diameter is 1.8 light-years — six times that of the Saturn Nebula (Caldwell 55).

57

Barnard's Galaxy
NGC 6822
Type: Irregular Barred Dwarf Galaxy (IBm)
Con: Sagittarius

RA: $19^h 44.9^m$
Dec: –14° 48′
Mag: 8.8; 7.9 (O'Meara)
Diam: 15.0′
SB: 14.8
Dist: 2.3 million light-years
Disc: Edward Emerson Barnard, 1884

HERSCHEL: None.

GC: None.

NGC (original): Very faint, large, extended, diffused.

NGC 2000.0: Very faint, very small, extended, diffused; the same as IC 4895.

NGC 6822 IN SAGITTARIUS IS ARGUABLY the most celebrated and important deep-sky discovery made by Edward Emerson Barnard — one of history's greatest visual astronomers. Barnard swept up this large and elusive glow in 1884 with the 5-inch Byrne refractor at Vanderbilt University Observatory. At the time Barnard had no idea of the nebula's extragalactic nature. Sadly, he would pass away in 1923 — just one year before Edwin Hubble would use the famous 100-inch reflector atop Mount Wilson to identify several Cepheid variables in it. When Hubble worked out the distances to these stars, he realized that NGC 6822, like the Andromeda "Nebula," is not part of our Milky Way but a self-contained stellar system far beyond. Barnard's Galaxy (as NGC 6822 would soon become known) had emerged from the stellar foreground of the Milky Way to stand out boldly as an independent galaxy — a small but important member of our Local Group.

The photographs Hubble took with the 100-inch resolved NGC 6822 to its core, allowing him to identify numerous variable stars, clusters, and nebulae. Ironically, Hubble's visual impression of the object through the same telescope was much different. Hubble wrote that NGC 6822 was "fairly conspicuous" at low power in a 4-inch finderscope but "barely discernible at the primary focus of the 100-inch." So a bigger aperture isn't necessarily always better. Indeed, observations of the object made with other large telescopes led to some confusion. For instance, compare the *NGC 2000.0* description with the original *NGC* description tabulated above. In the former the 15′ × 15′ galaxy is described as being "very small," not "large" as in the latter; it is also identified as IC 4895 there. In a 1925 *Astrophysical Journal* article, Hubble describes why:

Barnard's remarks obviously refer to the cluster as a whole and furnish the description in the [original] N.G.C. — ["very faint, large, extended, diffused."] During the next year, 1887, the region was examined with the 26-inch refractor at Leander McCormick Observatory. The relatively high magnification and restricted field conspired to render the cluster inconspicuous. It was missed entirely by the observer. The two brightest of the small involved nebulae, however, were seen and measured. Curiously enough, they were both attributed to Barnard and the discrepancy with the N.G.C. description was overlooked. When the *First Index Catalogue* was compiled, the brighter of these two objects was assumed to be N.G.C. 6822, although the description in the N.G.C. was not amended, and the fainter object was listed as a new nebula, "I.C. 1308, eF, eS, lE, gbM, 6822 p 12°" [excessively faint, excessively small, little extended, gradually brighter in the middle, NGC 6822 preceding toward position angle 12°]. H. A. Howe, with a 20-inch refractor, found only the brighter of these two small nebulae and remarked that N.G.C. 6822 was "very small" instead of "large," as described by Barnard."

The first photographs were made in 1906 and 1907 by Wolf at Heidelberg with the 16-inch Bruce camera and the 28.5-inch Walz reflector. He identified the two small nebulae as N.G.C. 6822 and I.C. 1308, and announced the cluster, Barnard's original discovery, as a new object which was duly listed in the *Second Index Catalogue* as "I.C. 4895, group of neb., 25′ diam." Wolf described the cluster as an exceptionally dense *Nebelfleckhaufen*, or cluster of small nebulae, similar to those more open clusters to whose study he has contributed so largely. This remarkable interpretation existed for fifteen years unchallenged and apparently unnoticed.

Note that Hubble described the galaxy as a "cluster."

In modern-day photographs NGC 6822 is reminiscent of the Large Magellanic Cloud, but its metallicity is more similar to that of the Small Magellanic Cloud. The galaxy seems to lack symmetry, and it has been classified as an irregular dwarf system. A prominent bar runs along its major axis. At the northern end a few clouds of glowing gas can be seen, while bright, bluish stars scatter out into a straggling arm on the other end. The distribution of the galaxy's neutral (cold atomic) hydrogen is rather complex and clumpy; the gas, "seen" by radio telescopes, appears to form a rotating disk. There is also evidence for some expanding shells. Indeed, Barnard's Galaxy contains many Wolf-Rayet stars — very hot, luminous suns with expanding atmospheres and intense stellar winds. Small groups of such stars have been found at the centers of the larger nebulous bubbles in NGC 6822. These larger nebulous structures are most likely the result of Wolf-Rayet winds colliding with the galaxy's more or less stationary interstellar medium. At each collisional interface the interstellar medium glows, producing a thin, luminous bubble.

Recent Hubble Space Telescope observations of individual stars in NGC 6822 reveal that star formation probably began there about 12 to 15 billion years ago. The star-formation rate has been close to constant (or declining) in the last few billion years, though episodes of enhanced star formation have occurred across the galaxy's body in the last 100 to 200 million years. The strength of this enhancement has been somewhat different from one region of the galaxy to another, being higher in the bar region than in the galaxy's outskirts, and higher still along the bar's northern and southern edges than in its center. In a 1999 paper in the *Publications* of the Astronomical Society of the Pacific, John B. Hutchings (Dominion Astrophysical Observatory, Canada) used HST data to show how the galaxy's western region may be the remnant of a tidal event that triggered the most recent star-formation episode.

If you live under dark skies, seeing Barnard's Galaxy should not be challenging. I had no problem sweeping it up in a 4-inch at 23×, given my telescope's nearly 3° field of view. In fact, I just pushed the telescope tube to the dwarf galaxy's general vicinity, and I had the object in focus before I had finished two sweeps of the sky. But start your search first with binoculars. Begin at 4th-magnitude Rho^1 (ρ^1) Sagittarii — the tip of the "spoon's handle" northeast of the Sagittarius Teapot. Move 5° northeast to a pair of 5th-magnitude stars, 54 and 55 Sagittarii, which lie only 30′ apart from one another. Just ¾° to the north-northeast of 55 Sagittarii is a magnitude-5.5 star. The galaxy is only ¾° farther to the northeast. But don't put down the binoculars. I challenged myself to see the galaxy with 7×35 binoculars and, after relaxing my gaze, could definitely pick it out against the Milky Way as a faint ghost — a uniform splotch of dim light. With these binoculars I estimated the galaxy's brightness to be magnitude 7.9 — nearly a full magnitude brighter than the accepted value listed elsewhere.

At 23× in the Genesis my first impression was of a triangular wedge of light balanced on a flattened pyramid of moderately bright stars. This is how NGC 6822 appeared with direct vision. With averted vision the galaxy swelled, resembling a miniature version of M24 (a true star-cloud in Sagittarius). At 72× the galaxy is a phantom glow speckled with a multitude of foreground stars. But when averted vision is used, a strong needlelike bar can be seen running down its length. Accepting a distance of 2.3 million light-years for NGC 6822, the true linear extent of this bar is 11,400 light-years, and the galaxy's total luminosity is about 260 million Suns. While observing I convinced myself that, beyond the dappled starlight of foreground suns, the galaxy's haze is mottled, broken into numerous patches. I also believed that the faint galactic haze before me was just at the limit of resolvability. This may not be illusory. Robert Burnham Jr. reports that the brightest individual stars in NGC 6822 shine at about

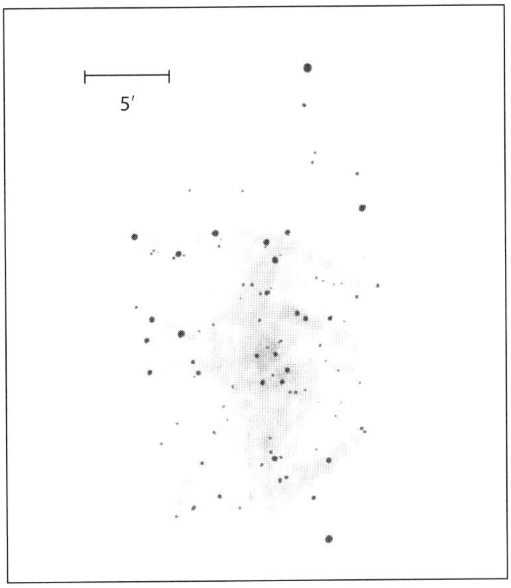

15th magnitude, and the estimates in his *Handbook* usually run on the faint side. NGC 6822 is one of the few galaxies that will reveal some of its individual stars to patient observers with small telescopes. IC 1308 is an emission nebulosity 6′ north-northeast of the galaxy's center.

The galaxy is all but lost at high power. Still, traces of its bar can be seen with keen averted vision. Note that the bar has some faint extensions of starlight that look like two Es standing back to back — a curious natural monogram for its discoverer. Oddly, but as the last-quarter Moon began to rise one night while I was looking at NGC 6822, the galaxy seemed to pop into view with sharper clarity at 23× and take on a pale green tinge. Is this

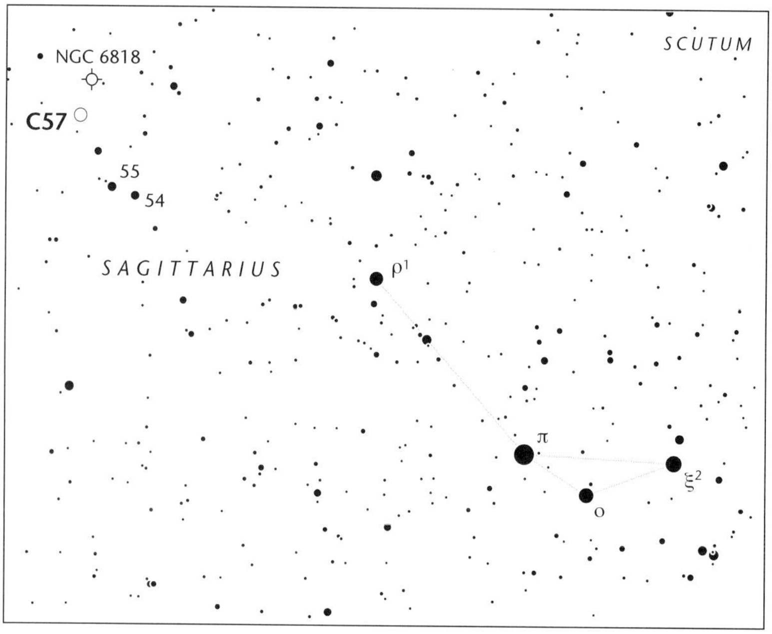

another optical illusion? Or did the background Milky Way fade away under the moonlight to reveal the galaxy?

The tiny (17″-wide) planetary nebula NGC 6818 lies about 35′ north-northwest of Barnard's Galaxy. It appears in the guise of a conspicuous "star." At 23× one could easily pass over the magnitude-9.3 glow because its light is so condensed. But the planetary swells slightly (if you know where to look) with averted vision. Moderate power brings out the nebula's nature, showing a bright, dense core surrounded by a more tenuous and slightly irregular halo. It's hard to tell if the stars I see close to the planetary at high power aren't really dense knots in its gaseous shell. I see no color and no sign of a central star in NGC 6818, even at high power.

58

NGC 2360
Type: Open Cluster
Con: Canis Major

RA: 07^h 17.7^m
Dec: –15° 38′
Mag: 7.2; 7.0 (O'Meara)
Diam: 14.0′
Dist: 3,700 light-years
Disc: Caroline Herschel, 1785

W. HERSCHEL: [Observed 4 February 1785] A beautiful cluster of pretty compressed stars near ½° diameter. Caroline Herschel. (H VII-12)

GC / NGC: Cluster, very large, rich, pretty compressed, stars from magnitude 9 to 12.

CAROLINE HERSCHEL WAS A BRIGHT LIGHT shining in her brother William's life. Standing 4 feet, 2 inches high and singing with a soprano's voice, she discovered eight comets; indexed, expanded, and corrected John Flamsteed's "British Catalogue" of 3,000 stars; and completed a catalog of 2,500 nebulae after her beloved brother's death. For her numerous contributions to the science of astronomy the United Kingdom's Royal Astronomical Society awarded her its gold medal in 1828. Her achievements grew out of her singular devotion for her brother and his work. "Almost inevitably Caroline's role changed from looking after William to helping him with his scientific activities," write her biographers John J. O'Connor and Edmund F. Robertson. As Caroline wrote:

> Every leisure moment was eagerly snatched at for resuming some work which was in progress, without taking time or changing dress, and many a lace ruffle . . . was torn or bespattered by molten pitch. . . . I was even obliged to feed [William Herschel] by putting the vitals by bits into his mouth; this was once the case when at the finishing of a 7 foot mirror he had not left his hands from it for 16 hours . . . "

When William turned his telescopes skyward in pursuit of new nebulae, Caroline would record his observations. The following day she toiled over the lengthy calculations necessary to reduce William's data.

In 1785, while assisting her brother in his great Deep Sky Survey, Caroline Herschel discovered a bright star cluster in Canis Major. At least that is what William's catalog entry for the object implies: "A beautiful cluster of pretty compressed stars near ½° diameter. Caroline Herschel." William assigned the number H VII-12 to the cluster. But his catalog's endnotes credit Caroline with discovering H VII-13 (NGC 2204), not H VII-12 (NGC 2360). Which object did Caroline discover? Most historians agree that Herschel simply erred in his endnotes and duplicated his reference to H VII-12. If so, H VII-13 is probably H VII-12. This conclusion was reached by Adm. William Henry Smyth, who, in his 1844 *Cycle of Celestial Objects*, wrote of H VII-12: "A tolerably compressed but extensive cluster, on the boundary between the Unicorn, and the Greater Dog. It was discovered by the indefatigable Miss Herschel, in 1785."

Once you locate Sirius, look for Caroline's cluster 3½° due east of 4th-magnitude Gamma (γ) Canis Majoris, the easternmost star in the Great Dog's head. Though small, the magnitude-7.2 cluster is easily spotted in 7×35 binoculars as a puff of celestial smoke a little less than ½° east of a magnitude-5.5 star. Under dark skies even small binoculars, held with a steady hand, should start to resolve sections of the cluster. The magnitude-5.5 star marks the eastern tip of a very thin, diamond-shaped asterism (3° × 1°) with Gamma Canis Majoris at the western end; the diamond's wide southern corner is a fine double star (magnitudes 6.0 and 6.6) oriented east-west and separated by some 5′. The diamond points right to the cluster. If you fail to find NGC 2360 that way, try Smyth's method. He wrote that NGC 2360 "can be fished up, under a moderately magnifying eye-piece, at 7°½ [east-northeast] of Sirius; where an imaginary line from Aldebaran passed over Bellatrix, will intersect it." (I've put Smyth's directions from Sirius in brackets, because his original text erroneously places the cluster "west-northwest" of the star.) Sharp-eyed observers under dark skies should challenge themselves to see NGC 2360 with the unaided eye. I've tried on some spectacularly clear nights at an altitude of 4,200 feet and can say only that the cluster seemed to bob above and below the very threshold of visibility. My magnitude estimate (7.0) was made with 10×50 binoculars.

The view changes dramatically with a telescope. NGC 2360 is in a fabulously rich star field and through the Genesis at 23× it appears much as Smyth describes it, namely "a singular group of very lucid specks, formed of stars nearly all of 10th magnitude." Most of these "lucid specks" huddle in a rather elliptical core that has a three-dimensional quality to it; the

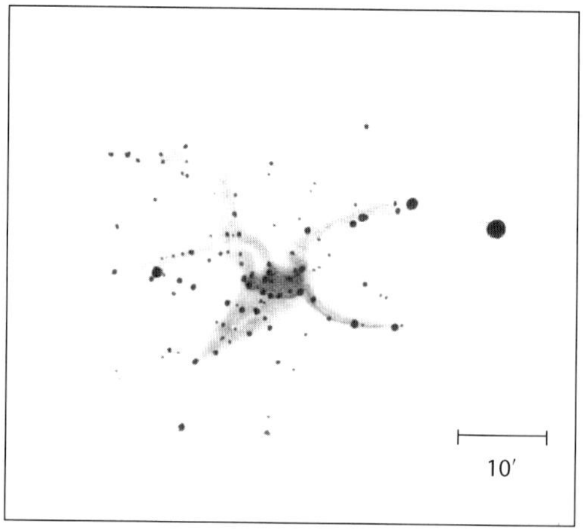

stars appear to be piled or stacked on top of one another in an orderly fashion. With a prolonged gaze the core starts to look like a butterfly's beaded chrysalis suspended by silken threads; it's as if we're about to witness some strange and wonderful transformation of starlight into star life. Star-studded arms project from the core to form a roughly sideways X. A 9th-magnitude star lies near the tip of the northeastern arm of the X, between a finely resolved cone of stardust extending to the southeast and a jagged trail of starlight that leads to the northeast. These arms have faint extensions that loop back in to the core, so the cluster really does look like a butterfly in flight. A singular thread of stars reaches out and nearly touches the 5.5-magnitude star to the west like a finger about to tap a shoulder. Ernst Hartung also saw NGC 2360's concentrated core and limber arms, which, he wrote, merged "into the rich field with long irregular rays." He called the entire spectacle "rich and delicate" as seen through his 4-inch telescope.

At 72× the cluster's central bar appears sharply uniform, with three or four horizontal rows of stars separated by prominent dark lanes, like those of M25 in Sagittarius. In fact,

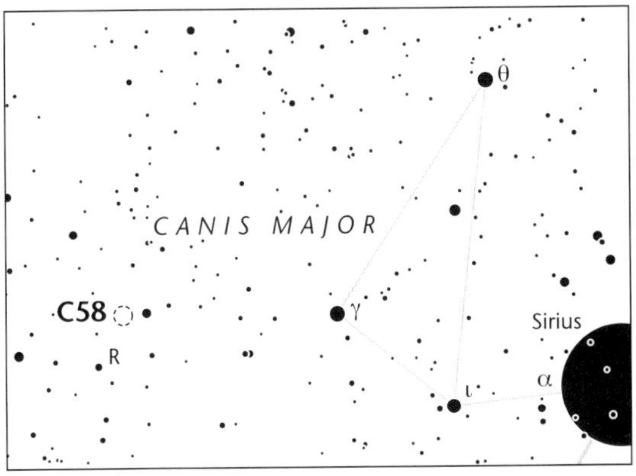

on really crisp, dark nights the entire cluster appears disrupted by intense black fissures lined with rays of sunlight. All kinds of geometric patterns can be imagined in the seemingly regimental array of stars — a triangle to the east, a straight line to the south, an ellipse to the west. Using my imagination, the bright core looks like the body of a rat, complete with head, feet, and a tail stretching to the southwest. And the stellar spectacle increases with aperture. Through her 17½-inch at 166×, Barbara Wilson finds NGC 2360 one of the nicest clusters in Canis Major, being "round and rich with bright straight star chains, pentagon shaped." Brent Archinal tallies 91 stars in the cluster, which spans 15 light-years of space. Although this seems rather sparse, it is similar to the star count in, say, M26 in Scutum, which measures 12 light-years across and has a total population of about 120 suns. NGC 2360's brightest star shines at magnitude 10.4, while its faintest member is a dim 17th magnitude.

As with many clusters in star-rich regions of the Milky Way, it is hard to define NGC 2360's boundaries; the cluster lies but 1½° west of the galactic equator, in a region wealthy with 10th- to 12th-magnitude suns. Indeed, in his *Celestial Objects for Common Telescopes* Rev. T. W. Webb describes NGC 2360 as a "[b]eautiful [cluster] . . . melting into a very rich neighbourhood, as though the Galaxy were approaching us." Webb's description may also allude to the cluster's three-dimensional quality.

Despite its visual splendor the cluster is aged. In the October (I) 1990 issue of *Astronomy and Astrophysics*, Swiss astronomers Jean-Claude Mermilliod and Michel Mayor estimate the cluster's age at about 2.2 billion years — about halfway between the ages of the Hyades (Caldwell 41) and M67 in Cancer. At this advanced age NGC 2360 has formed a sizable population of red-giant stars, which stand out on the cluster's color-magnitude diagram.

NGC 2360 is also about 2½° west of the Puppis border and some 5° west-southwest of M46 and M47, the famous open-cluster couple in Puppis. Less than 1° southeast of NGC 2360 you'll find another bright stellar treasure, R Canis Majoris, an Algol-type eclipsing binary that dips from magnitude 5.7 to magnitude 6.3 once every 1.14 days.

59

59

Ghost of Jupiter; CBS Eye Nebula; Rival of Uranus
NGC 3242
Type: Planetary Nebula
Con: Hydra

RA: $10^h 24^m 46.1^s$
Dec: –18° 38′ 33″
Mag: 7.8; 7.3 (O'Meara)
Dim: 45″ × 36″
Dist: 1,400 light-years
Disc: William Herschel, 1785

W. HERSCHEL: [Observed 7 February 1785] Beautiful, brilliant, planetary disk ill defined, but uniformly bright, the light of the colour of Jupiter. 40″ diameter. Second observation: near 1′ diameter by estimation. (H IV-27)

GC: Remarkable planetary, very bright, little extended toward position angle 135°, about 32″, blue.

NGC: Remarkable planetary, very bright, little extended toward position angle 147°, 45″ diameter, blue.

MANY DEEP-SKY OBJECTS LOOK LIKE GHOSTS in small telescopes, but only one boasts of that fact in its popular name: NGC 3242, the Ghost of Jupiter — an object amateurs love to "scare up" at star parties. In a sense this planetary nebula really is a ghost. It is a set of translucent gas shells blown off of a dying sun. The nebula's shells are expanding outward at speeds of roughly 30 km per second, and they will continue to do so until they dissolve into the interstellar medium like spirits fading from view. You can see this specter haunting the sky 1.8° south-southwest of magnitude-3.8 Mu (μ) Hydrae. Note that my magnitude estimate for the Ghost of Jupiter (7.3), made with 7×50 binoculars, is a half magnitude brighter than the generally accepted value. So here's your first Ghost of Jupiter challenge: try to detect the nebula with the unaided eye. The object is so bright and starlike that it can be seen with the eye alone from very dark-sky sites. This would be a great exercise for those attending the annual Winter Star Party in the Florida Keys, not only because Hydra is well placed in the sky on winter evenings, but also because the star party's latitude places the nebula higher in the sky than is the case at midnorthern locales.

Of course, one could argue that *all* planetary nebulae are ghostly glows and round like planets. But NGC 3242 appears about the same size as Jupiter, so the nickname makes perfect sense. That being said, there is a misconception and an enigma associated with the name. In April 1971 the late Walter Scott Houston wrote in *Sky & Telescope* that William Herschel discovered the nebula in 1785, which is true. He also wrote that "[a]ccording to Admiral [William Henry] Smyth, 'From its size, equable light, and colour, this fine object

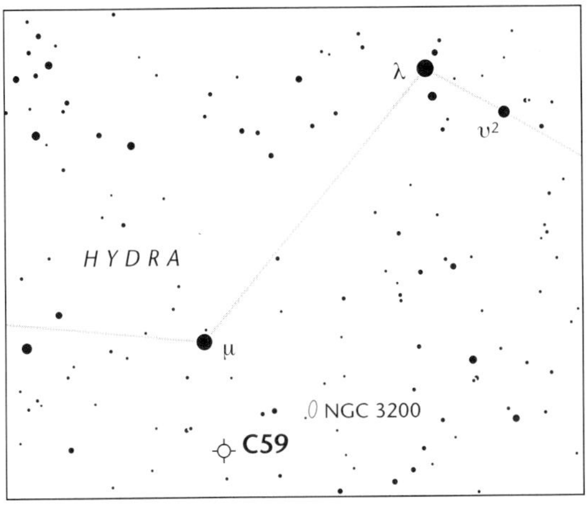

resembles Jupiter . . . '" Smyth's full account, penned in 1837, reads as follows: "A planetary nebula, pale greyish-white, nearly 2° south of m, about 20° south-west by west of Regulus, and in the middle of Hydra's body. From its size, equable light, and colour, this fine object resembles Jupiter; and whatever be its nature, must be of awfully enormous magnitude." Indeed, today, it's hard not to find a reference to the Ghost of Jupiter that does not credit Smyth with at least triggering the *idea* for the name — even though he never said "Ghost of Jupiter."

But the time has come to dispel that popular myth. Smyth was not the first person to draw attention to NGC 3242's resemblance to Jupiter. William Herschel, the object's discoverer, did that. On February 7, 1785, he wrote of the object: "Beautiful, brilliant, planetary disk ill defined, but uniformly bright, the light of the colour of Jupiter." Herschel's observation obviously inspired Smyth to take the analogy one step further by taking the object's size into consideration. So it is William Herschel, the creator of the term "planetary nebula," who deserves credit for marrying Jupiter and its "ghost" in our minds.

Now for the million-dollar question: Who first coined the moniker Ghost of Jupiter?

Houston didn't. In his May 1969 "Deep-Sky Wonders" column, he described NGC 3242 as a "bluish egg." And though Houston mentioned Smyth's observations in his April 1971 column, he did not then refer to the planetary as the Ghost of Jupiter. Indeed, though Houston had written about NGC 3242 numerous times since 1947, he did not introduce "Ghost of Jupiter" until his March 1984 column: "W. H. Smyth mentioned that NGC 3242 is similar in size and color to the planet Jupiter. Accordingly, it still is sometimes called the Ghost of Jupiter."

Ernst Hartung does not mention NGC 3242's popular nickname in his 1968 edition of *Astronomical Objects for Southern Telescopes*. James Mullaney and Wallace McCall include NGC 3242 in their 1978 *The Finest Deep-Sky Objects*, but they do not refer to it as the Ghost of Jupiter. And the moniker does not appear in Volume 2 of the Webb Society's *Deep-Sky Observer's Handbook* (1978). Clearly the name was not in popular use before 1978. Houston, like others before and after him, most likely got the name from Volume 2 of Robert Burnham Jr.'s 1978 *Celestial Handbook*. Of NGC 3242 Burnham writes, "In the small telescope it shows as a pale bluish softly glowing disc . . . appearing like a 'ghost of Jupiter.'" The mystery would be solved but for one nagging fact. Burnham, now deceased, placed "ghost of Jupiter" in quotes, which suggests (but does not necessarily prove) that someone else had coined the term. If so, who? For now, that remains a mystery.

Burnham also wrote that the planetary's bright inner ring resembles the outline of a human eye. This has led to the nebula's other popular nickname, the "Eye Nebula." Arizona amateur Steve Coe put a more modern spin on it, saying that through his 17½-inch f/4.5

reflector at 100× NGC 3242 resembles the eye-shaped logo used by the CBS television network. Indeed, the "CBS Eye" moniker has caught on fast and is now popularly employed as much as the "Ghost." The more streamlined name, the Eye Nebula, didn't fare so well because there is yet another Eye Nebula: the Cat's Eye (Caldwell 6), a bright planetary in Draco.

Smyth, who is infamous for his truly "colorful" descriptions of double stars, did not note NGC 3242's subtle blue or green hue, which so many amateurs enjoy today. I believe that since William Herschel did not attribute a particular color to NGC 3242, neither did Smyth. John Herschel first specified the nebula's color, seeing it as a "decided pale blue; at all events a good sky-blue." The Rev. T. W. Webb found it "of a steady pale bl[ue] light."

Heinrich Louis d'Arrest didn't see color in NGC 3242 but made out two "nuclei" near the limb. And Angelo Secchi, who studied NGC 3242 at 1,000×, saw "within a circular nebulosity two clusters, connected by two semicircular arches of stars, forming a sparkling ring, with one star on the hazy ground of the centre." William Huggins acquired a spectrum of the "Ghost," and this spectrum led him to conclude that Secchi's ring is made of gas, not solid matter.

Curiously, of all these great observers, only Secchi saw the planetary's central star. I say "curious" because in photographs the star blazes at magnitude 11.3. Granted, most visual estimates place it closer to 12th magnitude. Still, the central star should be obvious at the eyepiece of a fairly large telescope. Why, then, did John Herschel — who observed NGC 3242 from the Cape of Good Hope on four different occasions with an 18¼-inch reflector — not record it? After all, John, like his father, tried to resolve *all* nebulae into stars.

The central-star mystery lives on. I have seen very few accounts of amateurs seeing

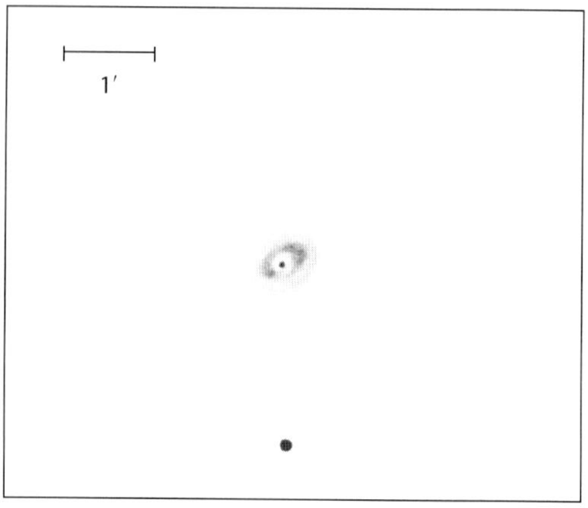

NGC 3242's central star through small telescopes. But consider the following sequence of notes from Houston's "Deep-Sky Wonders" column. In 1947 Houston said the central star was 11th magnitude. In 1971 he reported seeing the "11.5-magnitude" central star in a 6-inch at 150×. In 1978 he wrote that the central star is "much fainter" than photographic magnitude 11.4. And in 1984 he said that California amateur Tokuo Nakamoto estimated the star as magnitude 12.5. That same year "Scotty" was at the eyepiece, looking at NGC 3242 after having cataract surgery. This is how things looked through a 5-inch "apogee" scope at 39×: "The central star was easily seen with my eye that had its lens removed during cataract surgery. The star appeared almost as bright as the entire planetary in this eye, while it was hardly visible at all in my normal eye. This was surely due to a greater amount of ultraviolet (UV) light reaching the retina of the eye without its natural lens. Central stars in planetaries are generally strong emitters of UV." In their *Observing Handbook and Catalogue of Deep-Sky Objects,* Christian Luginbuhl and Brian Skiff record the central star's magnitude as 11.9, while *The Deep Sky Field Guide* (1993) lists it as 12.1. James Mullaney also has noticed the central star's "strange variations over the years" and suggests amateurs monitor it in 6-inch or larger telescopes.

When I observed the "Ghost" from Hawaii with my 4-inch in 1997, five things became apparent. (**1**) At 23× the nebula was so condensed that it appeared simply as a 7th-magnitude "star"; only with intense concentration could I convince myself that I was seeing a nebula and not a star. Therefore I believe it is possible to see it with the unaided eye. (**2**) At 72× the nebula's central star, its elliptical inner ring, and hints of its outer halo are visible if I stare directly at the central region without using averted vision. (**3**) The nebula, with its aquamarine hue, looks more like Uranus than Jupiter, so it could also be called the Rival of Uranus. (**4**) With a quick glance at high power, d'Arrest's two "nuclei" are readily apparent; they are the inner ring's enhanced extremities. And (**5**) with a prolonged glance at high power, Secchi's inner ring of "stars" can be seen to rotate as the eye follows those beads of light around.

Amateurs using low power might mistake the nebula's exceedingly bright inner ring for its central star. The planetary's high surface brightness takes magnification well, so be sure to crank your telescope up to at least 75× for each inch of its aperture. Direct vision reveals the central star because it suppresses the amount of light entering the eye from the nebula; with averted vision, the entire central ring and the star blend into one magnificent blur. Do use averted vision to trace out the full extent of the 45″-wide outer halo, though. As I mentioned, with a quick glance at high power the northwestern and southeastern tips of the elliptical inner ring shone like two stars. But the longer I looked, the more sausage-shaped the tips became, until they split into beads of light; the southeastern enhancement did so especially well. These are Secchi's shimmering pearls, the ones that start to "rotate" as the eye traces out the ring. Interestingly, Secchi made a similar observation of the beaded ring in M57, the Ring Nebula in Lyra; he saw it "glittering like star-dust." Equally interesting is that at 189×, NGC 3242's inner ring was essentially two sausages, and an increase to 216× showed a fine thread of light connecting the sausages at either end.

The Ghost of Jupiter is very similar in brightness, size, and distance to the Ring Nebula. In true physical terms the Ring spans 0.42 light-year (27,000 a.u.), while the Ghost measures 0.30 light-year (19,000 a.u.) across. Modern studies have shown that the Ghost's inner ring is an ellipse inclined some 70° to our line of sight. Most photographs show a bright and clumpy inner ring surrounded by a faint, smooth outer envelope. The Hubble Space Telescope has revealed a distinct asymmetry to the inner ring, with the southeastern limb being much brighter than the northeastern one. The inner ring's inner edge is scalloped in the Hubble view, and fine wisps of nebulosity cross the annulus like threads of finely woven silk. Beyond the nebula's outer envelope are the mysterious FLIERs (*f*ast *l*ow-*i*onization *e*mission-line *r*egions). These small knots of material — generally near a planetary nebula's symmetry axis, on opposite sides of the central star — have much higher outflow speeds than the nebula in which they are embedded. They also generally are much less ionized (that is, their constituent atoms have retained more of their electrons). The pale outer halo has a fine brushstroke quality to it, like the first layer of watercolor put to a piece of paper. Ground-based astronomers have also discovered what may be a dim halo of ionized helium surrounding the nebula's shell.

NGC 3242's central star may be a triple system consisting of a white-dwarf star, a very close brown-dwarf companion, and a more massive and distant companion. The 0.01-

solar-mass brown dwarf probably orbited within the distended atmosphere of the nebula's progenitor star, spinning up the progenitor's envelope and creating a wind that pushed material outward. The third member of this unusual stellar trio is believed to have about 1 solar mass and to take about 4,000 years to orbit the white dwarf. It could be responsible for the asymmetry seen in the inner ring.

If you're up for a challenge, sweep your scope a little more than 1½° northwest of NGC 3242 and see if you can pick out the barred spiral galaxy NGC 3200; it's a tiny (4.1′ × 1.2′) system that Luginbuhl and Skiff found "fairly faint" in a 10-inch. The galaxy's photographic blue magnitude is 12.3.

60

Ringtail Galaxy; Rattail Galaxy; The Antennae (northwestern component)
NGC 4038
Type: Irregular Barred Galaxy (IBPm)
Con: Corvus

RA: $12^h\ 01^m\ 53^s$
Dec: –18° 51.9′
Mag: 10.5; 10.5 (O'Meara)
Dim: 11.2′ × 5.9′
SB: 13.7
Dist: 83 million light-years
Disc: William Herschel, 1785

W. Herschel: [Observed 7 February 1785] Pretty bright, large opening with a branch, or two nebulae very faintly joined. The south is the smallest. (H IV-28.1)

GC/NGC: Pretty bright, considerably large, round, very gradually brighter in the middle.

61

Ringtail Galaxy; Rattail Galaxy; The Antennae (southeastern component)
NGC 4039
Type: Irregular Galaxy (IPm)
Con: Corvus

RA: $12^h\ 01^m\ 54^s$
Dec: –18° 53.1′
Mag: 10.3; 11.2 (O'Meara)
Dim: 9.8′ × 4.5′
SB: 13.0
Dist: 83 million light-years
Disc: William Herschel, 1785

W. Herschel: [Observed 7 February 1785] See Caldwell 60. (H IV-28.2)

GC/NGC: Pretty faint, pretty large.

60 & 61

WHILE SWEEPING WITH HIS POWERFUL REflector through Corvus, the Crow, on February 7, 1785, William Herschel chanced upon a nebula unlike any other he had seen. Of it he wrote: "Large opening with a branch, or two nebulae very faintly joined." What an utterly simple yet profound comment. It's as if Herschel were describing two butterflies dangling from a branch in some romantic embrace. How innocent he seems to us today, when we look at his records — we who live in a time of astrophysical sophistication. In his day Herschel had no way of fathoming what he had discovered. He could only describe what he saw and ponder its significance in the light of other faint and diffuse objects he had seen (no pun intended). Imagine his response had he suddenly learned the incredible truth: that he had stumbled upon two extragalactic cannibals, two savage galaxies ripping each other apart — each disemboweling billions of stars from the other and flinging them across several hundred million light-years of space, where they continue to shine like luminous entrails. Such is the case with NGC 4038 (Caldwell 60) and NGC 4039 (Caldwell 61), the finest and most popular pair of interacting galaxies in the heavens.

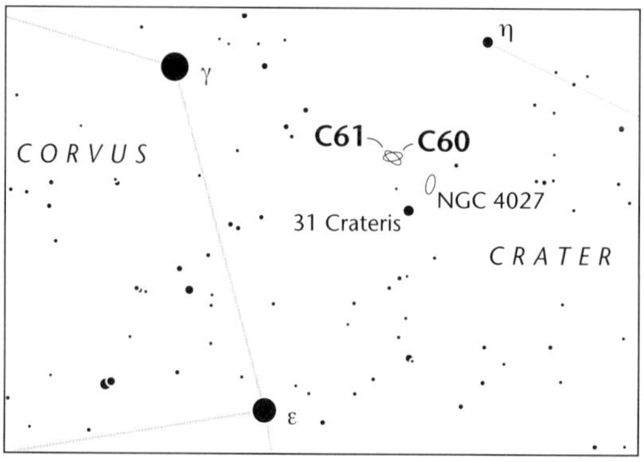

NGC 4038 and NGC 4039 are also the brightest of the five galaxies plotted in the second edition of *Sky Atlas 2000.0* near the star 31 Crateris (which lies within the modern borders for Corvus). To find the galaxy pair, look 3¾° southwest of magnitude-2.6 Gamma (γ) Corvi and about ¾° north and ever so slightly east of 31 Crateris. In photographs the galaxies look like two tadpoles bonded together at their eastern ends. Two long "tails" extend from the bridge connecting the tadpoles, one curving to the north, then east, the other to the southeast, then southwest. Computer simulations show that such filaments form during close encounters between spiral galaxies. When two spirals collide, their constituent stars do not smash together like cars; the distances between the stars in each galaxy are too great for that to happen. But a battle of tidal forces does ensue, and ultimately it disrupts each galactic system. Furthermore, while gravity yanks each galaxy apart, gas clouds collide, triggering star formation. Then, as the galaxies' respective nuclei move apart from one another, tides pull billions of stars from each spiral's disk. As a result, the galaxies lose mass and, with it, some of the gravitational pull that once held each of them together. The orphaned stars flee unimpeded into space, forming two taffylike streamers. Yet the battle has not ended. Like two warriors gathering strength, the galaxies separate, then orbit one another, until their mutual gravities force them back into battle once again. Today we see NGC 4038 and NGC 4039 in such a second round of engagement. Not until some time in the dim and distant future, perhaps a billion years from now, will the war end, with the two galaxies merging into a single chaotic stellar swarm.

A report in the *Astrophysical Journal* for April 1, 1997, confirms that massive star formation has been triggered by this extragalactic

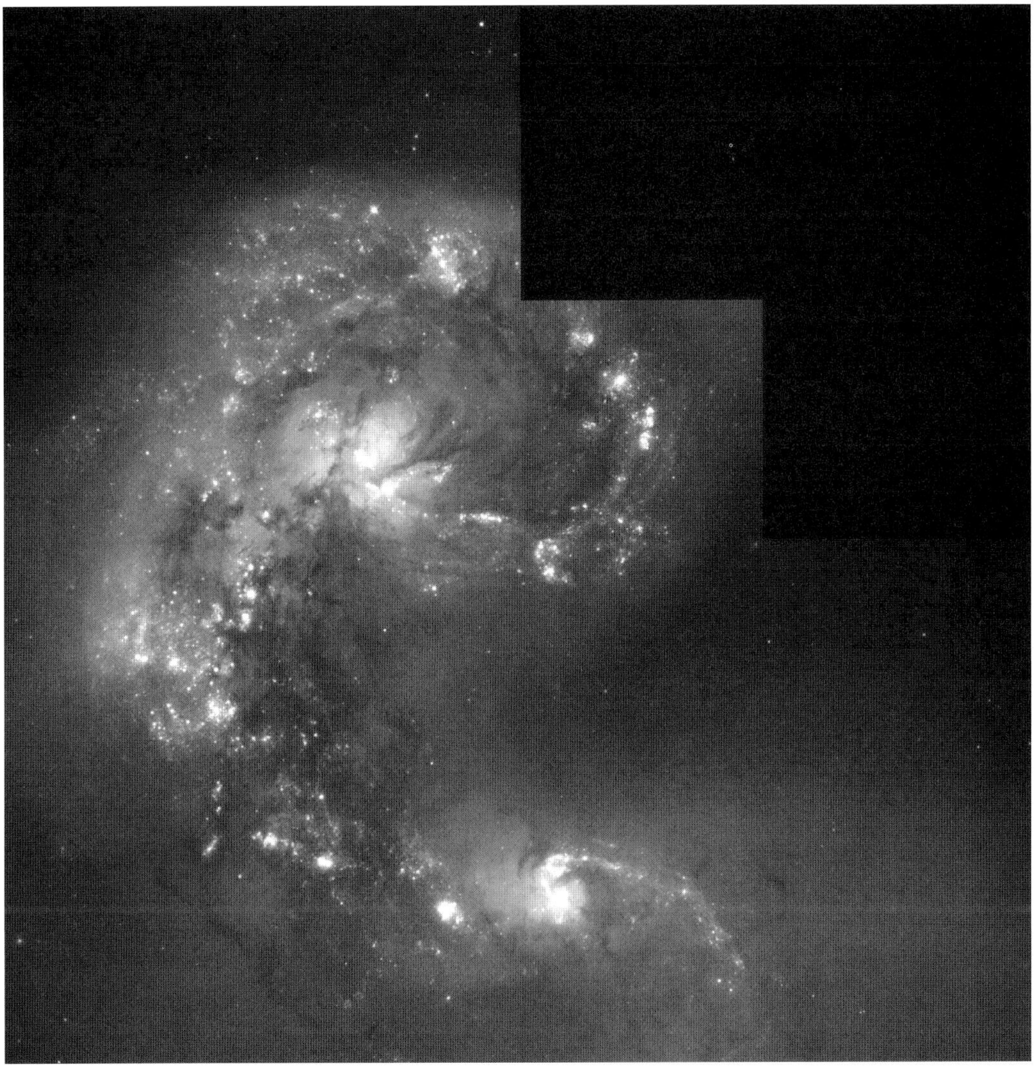

war. X-ray images taken by the Rosat satellite reveal that the most intense X-ray emission comes from each galaxy's nucleus and very knotty arms. The radiation may be coming from scores of X-ray binaries and supernova remnants. Indeed, the duo has been the site of recently recorded supernova activity, with two explosions witnessed within the last century — one in 1921, the other in 1974. (Supernova hunters should keep an eye on this active galaxy pair.) Images taken with the 2.1-meter telescope at Kitt Peak National Observatory have also helped astronomers identify blue knots of star formation in the distorted galaxy disks; some older stellar populations in their tattered bulges; and a large dust complex at the apparent interface between the two disks. Midinfrared spectroscopic observations along this interface verify that starburst activity has recently occurred there. More recently, and more spectacularly, the Hubble Space Telescope has revealed a brilliant fireworks show within the dynamic duo. Hubble uncovered more than 1,000 bright, young star clusters bursting to life as a result of the collision between NGC 4038 and NGC 4039. The brightest of these clusters contains roughly a million stars, all just a few million years old. One cluster, in NGC 4038's northern arm, looks like a miniature barred spiral galaxy itself. Indeed, some of the system's newborn clusters appear to be young globulars, while others may ultimately congeal to form globular clusters perhaps a billion years from now. The Hubble images also show unprecedented detail in the streams of reddish dust bleeding across the comma-shaped surface of NGC 4038 and funneling onto the core of smaller NGC 4039.

What of this drama, if anything, can one see through a small telescope? Certainly the details of a distant galaxy merger are far too difficult to see with anything but the largest of instruments — right? When Walter Scott Houston observed NGC 4038 and NGC 4039 through a 4-inch Clark refractor from the "indifferent skies" of Haddam, Connecticut, he saw "little more than an asymmetrical 11th-magnitude blur." One problem is that the galaxies have low surface brightnesses; to see them well, we need to be under dark skies and to have a lot of patience. Be aware that an 11th-magnitude galaxy, NGC 4027, lies about 45′ southwest of the Antennae — one of the duo's common nicknames — and is closer to 31 Crateris than our targets are. Houston suspected that many amateurs had sent him descriptions of that galaxy rather than of the Ringtail (yet another nickname for the galaxy pair). You've got the correct object in view if it is part of an arc of four roughly 9th-magnitude stars flowing from 31 Crateris to the north-northeast. (The first two stars form a roughly 5′-wide pair 20′ from 31 Crateris.) The Ringtail (or the Antennae, if you prefer) is just south of the fourth (and brightest) star in the arc. Those living under dark skies might want to challenge themselves by trying to see the pair in binoculars. This is an especially good challenge for those attending the Texas Star Party. I found them just barely at the threshold of visibility in 7×35 binoculars from Hawaii. I have not read of any other binocular sightings.

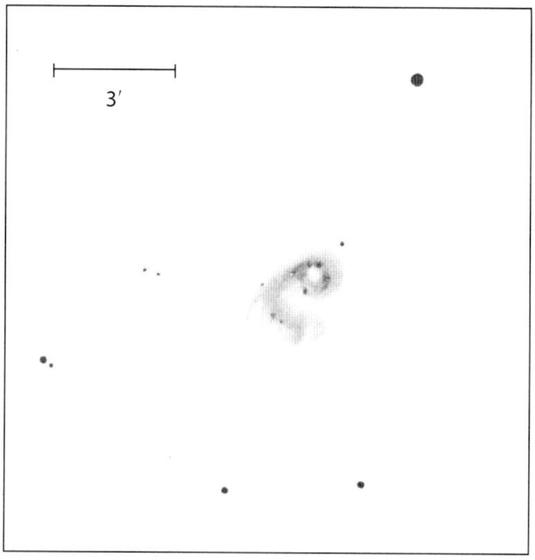

In the 4-inch at 23×, the Ringtail looks undeniably double. Or is it more accurate to say it is elongated, with a dark but tenuous wedge to the west? The northwestern component, NGC 4038, appears much brighter and larger than NGC 4039 to the southeast. (This supports the *NGC* descriptions on page 239.) I'm certain that the magnitudes given by the *Deep Sky Field Guide* (and used in the table above) for the Antennae are incorrect; they make NGC 4038 and NGC 4039 seem near-equals in brightness. Christian Luginbuhl and Brian Skiff's *Observing Handbook and Catalogue of Deep-Sky Objects* lists a magnitude of 10.7 for NGC 4038 but does not provide one for NGC 4039; Roger

W. Sinnott's *NGC 2000.0* has a visual magnitude of 10.7 for NGC 4038 and a photographic magnitude of 13 for the dim southern component. I estimated NGC 4039's visual magnitude to be in the ballpark of 11.2.

The overall view of the galaxy pair is best at 72×, because higher powers tend to overly spread out the light of this already diffuse object. This also explains why the Ringtail is such a poor sight from suburban locations; add light to the sky background and the galaxy pair simply fades from view. At 72× NGC 4038 looks like a clumpy wad of tissue, while NGC 4039 looks more like a tapered haze. The dark 1′-wide wedge of space separating the two components is exceptionally clear at this power. Try focusing your attention on the fainter member of the duo. Can you see that it, too, is not a uniform mass of light? Together the interacting spirals look like curdled milk stirred into coffee. Every now and again, I did suspect that I was seeing the southeastern section of the tidal tail. If the 4-inch at an altitude of 4,200 feet does indeed perform like a 10-inch at sea level, then I might not have been imagining it, for Houston says he could "easily see" the tail in a 10-inch reflector from Omaha, Nebraska. Using high power on one exceptionally clear and transparent night, I resolved the northwestern component's entire rim into a circle of knotty clumps; the most prominent one matches the location of the "barred spiral" cluster mentioned above. What a glorious moment, to be able to see traces of extragalactic star formation that has been in progress for perhaps only a million years.

If we accept the distance to this galaxy pair as 83 million light-years, then NGC 4038 has a linear diameter of 270,000 light-years (30,000 light-years broader than NGC 4039) and a luminosity of 40 billion Suns. We see NGC 4038 inclined 27° from edge-on, and the duo is receding from us at a speed of 1,640 km per second. The systems belong to the Corvus Cloud of galaxies, which has 63 members with known redshifts.

As alluded to above, the NGC 4038–39 duo has several common names, the most popular being the Ringtail (or Rattail) Galaxy and the Antennae. But one could think of others. Houston saw the two interacting galaxies as an apostrophe, or the "mother of all water fleas." Ron Morales envisioned them as shrimp. Luginbuhl and Skiff called the pair a cashew, while Tom Lorenzin saw it as a comma. I thought it looked like the Greek lowercase letter Nu (ν) on its side. And John Sanford says the galaxies look like a doughnut with a bite taken out of it. See if you can come up with your own imaginative moniker.

One final thought. If anyone questions that observing is a visual sport, or that you can better your visual threshold, consider the following account. When I first attended the Texas Star Party (TSP) in 1986, I was more of a skilled planetary observer than a deep-sky observer. One night Larry Mitchell showed me NGC 4676A and NGC 4676B (The Mice), an interacting pair of galaxies in Coma Berenices, through his 24-inch reflector. The Mice are similar to the Antennae, but they are fainter (13th magnitude each), and their tails are much harder to see — especially the southern one. "See the tails?" Mitchell asked me on that particular night. "Yup," I replied, "I see one bright one." "But do you see the *other* tail?" Try as I might, I couldn't. He laughed, and said, with an exaggerated Texan drawl. "Weeelll, Stephen, don't worry, we're going to make you a deep-sky observer yet." I continued to attend the TSP on and off over the next 13 years. Each year, I would observe, listen, and learn from Mitchell, Barbara Wilson, Brian Skiff, and many other great deep-sky observers. I began to view these objects with a planetary observer's patient eye. Then, at the 1999 TSP, Larry called me over to his telescope. It was early in the morning, I recall, and I climbed up the tall ladder to the eyepiece. "Whatch'a got?," I asked. "The Mice," he responded. As I casually looked at the little "critters," Mitchell asked, "See the tails?" "Yes," I replied, without giving the matter any thought. "*Both* of them?" he pressed. "Yes," I replied, again with no hesitation. "Well, I'll be darned, Stephen." Larry said. "Did you know you couldn't see that other tail 13 years ago!" A shiver ran up my spine when I realized he was right. "Welcome to the club, Stephen," Larry said. "I think you're *finally* — and *officially* — a deep-sky observer." That night I had grown in some spiritual way, and I had teachers like Mitchell, Wilson, and Skiff, to thank for it.

62

Milkweed Seed Galaxy
NGC 247
Type: Mixed Spiral Galaxy (SABd)
Con: Cetus

RA: $00^h\ 47.1^m$
Dec: –20° 46′
Mag: 9.2; 8.9 (O'Meara)
Dim: 22.2′ × 6.7′
SB: 14.1
Dist: 6.8 million light-years
Disc: William Herschel, 1784

W. Herschel: [Observed 20 October 1784] A streak of light, nearly in the meridian 26′ long [and] 3′ or 4′ broad. Pretty bright. (H V-20)

GC/NGC: Faint, extremely large, very much extended toward position angle 172°.

Observers looking for a "whale" of a time in the winter sky usually turn their telescopes toward the bright Seyfert galaxy M77 in Cetus, the celestial Whale (actually, Cetus is really a sea monster). But if you dip your telescope tube 3° south-southeast of Beta (β) Ceti, the 2nd-magnitude star marking the Whale's tail, you'll see a sight overlooked by many: the nearly edge-on spiral galaxy NGC 247. In nearly half a century of writing for *Sky & Telescope* magazine, Walter Scott Houston only mentioned this galaxy three times, and then only in passing. Robert Burnham Jr. discusses it in his *Celestial Handbook*, but his description of its visual appearance (a "faint haze") is uninspiring. Furthermore, one of the southern sky's most stunning galaxies, NGC 253 (Caldwell 65), lies only 4½° due south of NGC 247. So it's easy to understand why NGC 247 has been ignored. That is, until Patrick Caldwell-Moore made it the 62nd object in his Caldwell Catalog.

From Hawaii NGC 247 is visible in 7×35 binoculars. It looks like a thin thread extending from the pinprick of a magnitude-9.5 star that punctuates its southern tip. How can something so faint be visible in small binoculars? Well, that's easy. The galaxy isn't faint; it just has a low surface brightness, so its beauty is most apparent under clear, dark skies. I estimated the galaxy's apparent visual magnitude to be 8.9 — fully 30 percent brighter than the *Deep Sky Field Guide*'s value, which, in turn, is 1.5 magnitudes (4 times) brighter than Burnham's value. Christian Luginbuhl and Brian Skiff confirm my estimate to within 10 percent; they list magnitude 8.8. Together, NGC 247 and its attendant 9th-magnitude star look like a silken fiber attached to a milkweed seed adrift in the cool winter wind. With alternating direct and averted vision, the "fiber" glints, as if by the light of the Moon.

NGC 247 is among the closest galaxies to our Milky Way. It belongs to the Sculptor Group of galaxies, the closest aggregation of island universes to our Local Group. It is also a member of the South Galactic Pole Group of galaxies, which includes NGC 55, NGC 253, and NGC 300 (all Caldwell objects). NGC 247 lies 6½° from the South Galactic Pole, so its light is hardly absorbed at all by dust in our own Milky Way. In photographs, the highly inclined galaxy displays a bright central mass with a short bar (oriented northeast-southwest) and ill-defined spiral arms studded with knots. One long arm extends to the north before looping east, then dipping south, where it reconnects with the central mass. Whether it's illusion or reality, the sight leaves the impression that an enormous dark hollow lurks in the disk. A second arm shoots due south off the stubby bar and then curves just beyond the 9th-magnitude star. This sudden jog makes the arm look like a broken bone that hasn't properly healed. Accepting R. Brent Tully's distance estimate of 6.8 million light-years, the galaxy measures 34,600 light-years across. It is receding from us at 159 km per second.

The Rosat satellite picked up X-rays from hot gas within NGC 247's disk and from what is probably a cluster of young, low-mass stars

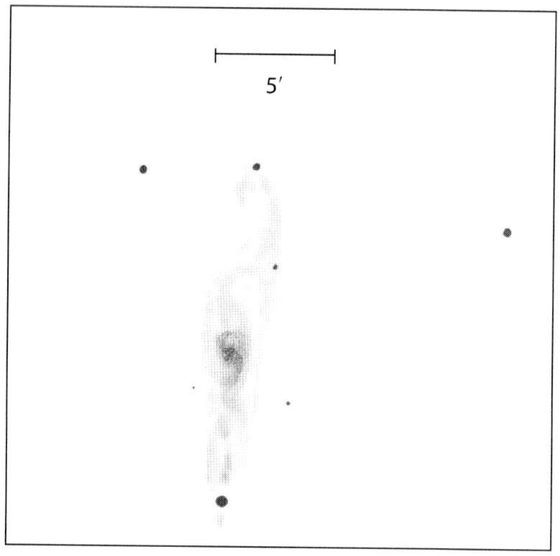

or brown dwarfs within the galaxy. Dwarf galaxies like NGC 247 are believed to be dominated by dark matter, a significant fraction of which is thought to exist in their halos as small, compact objects like brown dwarfs or giant planets — the so-called MACHOs (for *ma*ssive *c*ompact *h*alo *o*bjects). NGC 247 offers astronomers the possibility to observe a halo of MACHOs being formed. The material associated with the X-ray emission weighs in at about 1 billion solar masses, or 10 percent of the galaxy's total mass.

The view through the 4-inch Genesis does not disappoint. With a glance at 23× NGC 247 immediately displays a noticeable core, with faint extensions to the north and south. These extensions appear to swell and contract as I alternate between averted and direct vision, first appearing as a thin thread of light, then as a luminous tassel seemingly tacked to the sky by the 9th-magnitude star. The galaxy's patchy edges are difficult to trace. Its southern extension is more noticeable, and there's a suggestion of a break in the northern extension, which makes it easy to imagine that a dark void lurks there.

At 72× NGC 247 sits in a triangle of 9th- to 11th-magnitude stars. Little else seems to shine

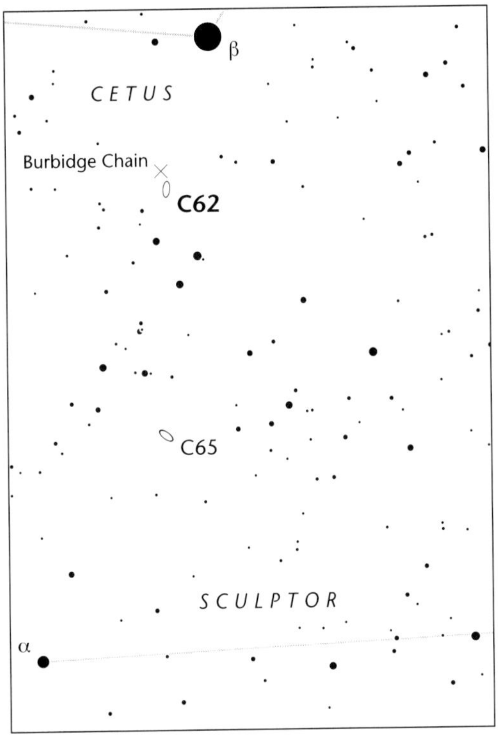

through the narrow haze. But give the galaxy time, because there is a wealth of fine detail to be seen. Clumps of starlight, twisted arms, and dark clouds gradually reveal themselves to a patient observer. Reserve high power for the galaxy's nuclear region, which is another treasure trove of tiny details. At high power the nucleus is a bright bead of scintillating wonder immersed in a tight ellipse of light. Just beyond the eastern edge of the central mass is a hyperfine "arm" or patchy structure that hugs the central mass before extending to the south. It looks like a continuation of the northern loop, but this may be an illusion. I cannot see the bar in full, but I can infer its presence from the way the details are angled in the central ellipse; they seem to match the orientation of the bar known to scientists. Also, one spiral arm to the north does have a strong barlike component, but I don't see exactly the same thing to the south. The southern arm has two distinct clumps separated by a fuzzy "string" of light that sweeps out beyond the magnitude-9.5 star. All this detail is remarkably fine considering how small an aperture I used to make the observations.

By the way, the trick to seeing the dark void in NGC 247's northern limb is to observe it at low power, which concentrates light and increases the contrast between luminous and dark areas. Low power is especially effective in bringing out the faint eastern "wall" that confines the dark "hole."

Long-exposure photographs made with large telescopes show NGC 247 superimposed on a group of fainter, far more distant galaxies. And just 5′ to the northeast of the galaxy's northernmost tip begins the Burbidge Chain — a string of five very dim (14th- to 15th-magnitude) and peculiar galaxies aligned north-south within an area only 10′ across. Seeing the Chain is a challenge for large-telescope users.

63

63

Helical Nebula; Helix Nebula; Sunflower Nebula
NGC 7293
Type: Planetary Nebula
Con: Aquarius

RA: $22^h 29^m 38.4^s$
Dec: –20° 50′ 13″
Mag: 7.3; 6.0 (O'Meara)
Dim: 12′ × 10′
Dist: 522 light-years
Disc: Georg Friedrich Julius Arthur von Auwers (published in 1862)

HERSCHEL: None.

GC/NGC: Remarkable, pretty faint, very large, elongated or bi-nuclear.

THE ECLECTIC GERMAN ASTRONOMER GEORG Friedrich Julius Arthur von Auwers (1838–1915) is most noted for his work on Procyon's proper motion; his observations of Venus's 1874 and 1882 transits across the face of the Sun; and his discovery of T Scorpii, a nova in the globular cluster M80. But he also published, in 1862, a catalog of William Herschel's nebulae and clusters — a work he began preparing for when he was only 16 years old. Printed in Volume 34 of the *Königsberg Observations*, the work also includes a list of 52 "very widely diffused nebulosities," including von Auwers's discovery of an immense object in Aquarius. That object is recognized today as the largest and closest planetary nebula known in the heavens — NGC 7293, the famous Helical or Helix Nebula. For his contributions to science, a lunar crater has been named in von Auwers's honor.

The Helix lies about 11° northwest of 1st-magnitude Alpha (α) Piscis Austrini (Fomalhaut) and 1¼° west of 5th-magnitude Nu (ν) Aquarii. Published visual magnitudes range from "6th" to 7.4. Given the object's brightness and its proximity to the globular cluster M30 in Capricornus (the planetary is about 12° east-northeast of the cluster), one has to wonder how the greatest observers, from Charles Messier to John Herschel, failed to detect it. But the answer is simple. First, the Helix is enormous, spanning an impressive 12′ × 10′ on the sky. Second, it has a low surface brightness. Third, its center is a gaping "hole." And finally, the object is most readily apparent with binoculars or a rich-field telescope. By condensing its light, such instruments increase the contrast between the nebula and the sky background, making it easier to see. By contrast, many of our astronomical ancestors used telescopes that had limited fields of view. The late Walter Scott Houston concurred: "The bulk of my correspondents indicated that the Helix was more readily seen in binoculars and finders than in telescopes. . . . [James] Meketa believes that the Helix might be seen with the naked eye under excellent conditions."

Today, large, low-contrast objects are the bane of urban and suburban skywatchers. Throughout my childhood, the Helix was one of my visual nemeses. It was right up there with M74, that large and diffuse galaxy in Pisces. I grew up in Cambridge, Massachusetts, and did most of my observing with either a 4½-inch long-focus reflector or the 9-inch Clark refractor at Harvard College Observatory. For 20 years I searched for the Helix on the best nights, when the object was high above the horizon, but I

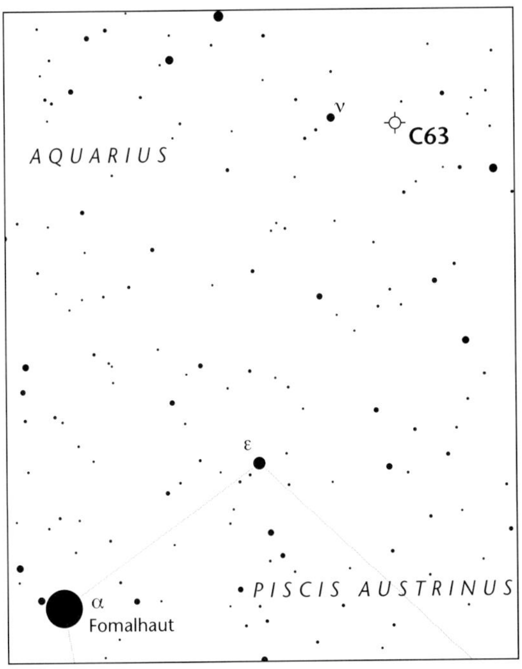

had two strikes against me: city lights and a limited field of view. I did not succeed in seeing the Helix until July 27, 1997, after I started the Caldwell project under the dark skies of Volcano, Hawaii. I did not know what to expect at first, because until then I had only viewed the Helix in photographs. So imagine my surprise (and disgust) when I raised a pair of 7×35 binoculars to the field and saw the Helix without a hitch! For the first time since childhood I felt a sense of primal awe and astonishment. "It's glorious," I penned. "It's enormous. And with averted vision I suspect a hole." Because of this shocking sighting, NGC 7293 became what I called the most "surprising" object in the Caldwell Catalog. It became even more so when I also "suspected [having seen] it with the unaided eye" that same night.

Given that provisional detection, I was astonished to come across no positive naked-eye sightings of the Helix in the literature. The closest account I found was a statement by Hans Vehrenberg in his *Atlas of Deep-Sky Splendors*. There he writes that the Helix "is of such low surface brightness that even on dark and extremely clear nights it is barely distinguishable as a very faint patch." Unfortunately, we do not know if the sentence pertains to the telescopic or naked-eye view. But once again I surprised myself by finding support for my tentative naked-eye detection in an unexpected place. While researching an article on the gegenschein for the May 2000 issue of *Sky & Telescope*, I came across the following entry, dated October 3, 1983, in my Hawaiian diaries (I was only 26 years old at the time): "The Helix Nebula was visible to the unaided eye as a very faint almost imperceptible haze." That night I was testing the naked-eye limit of the night sky from a 4,000-foot-high site in Waimea. I saw stars as faint as magnitude 8.0 and estimated the Helix's apparent magnitude to be 6.0. I did not even seem impressed that I did this (remember that I was foremost a planet observer then), nor did I bother looking at the Helix with binoculars. The object just happened to be in the field that I was watching. This "discovery" also alerted me to the fact that

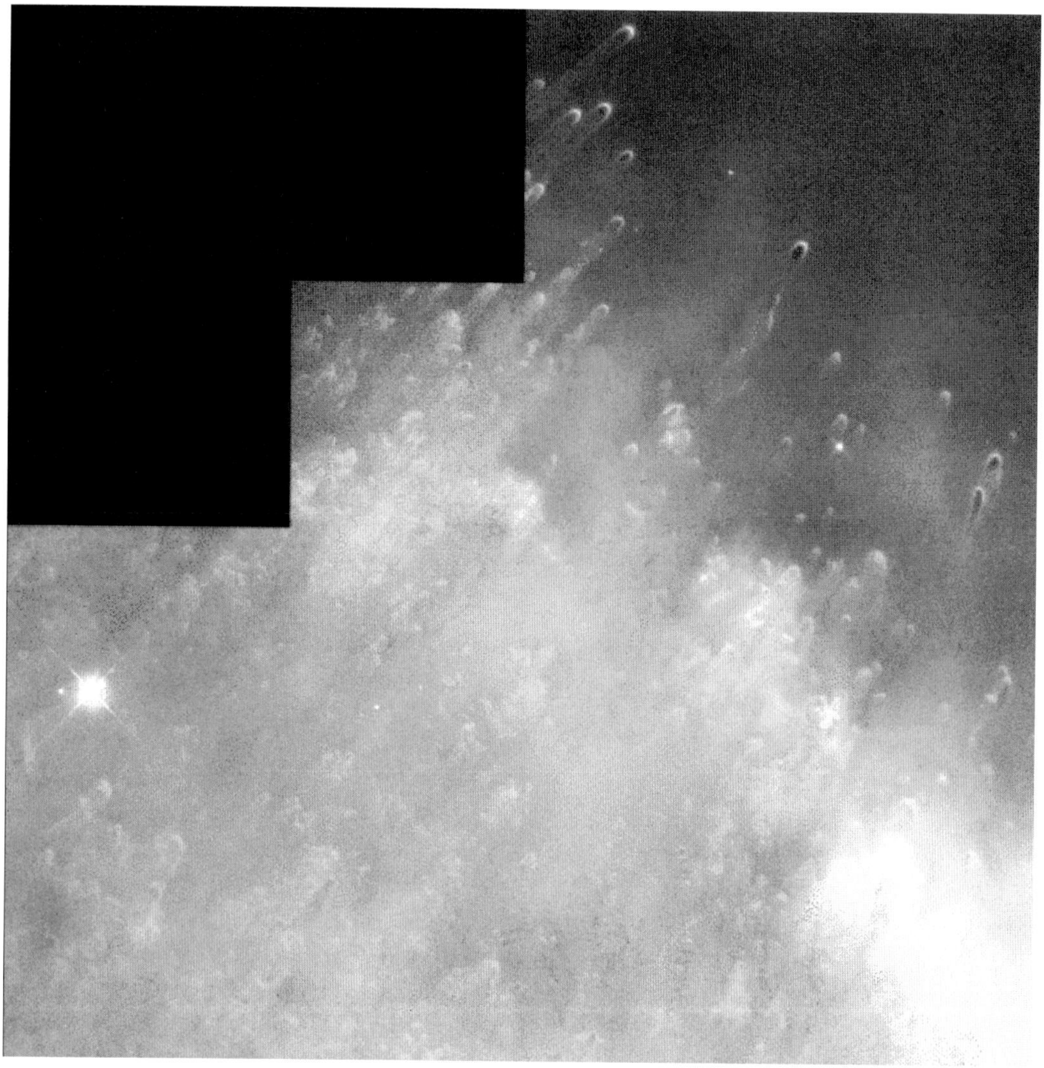

my eyes had aged; what I "suspected" to be visible in the year 2000 "was visible" seventeen years earlier. Of course, as Houston says, "[a]tmospheric clarity obviously played a major role for seekers of the Helix. Observers on mountains did much better than smog fighters." Here in Hawaii we don't have industrial or automotive smog, but we do have volcanic smog, or "vog." Personally, I believe that while our eyes deteriorate with age, we can balance that with the "experience factor." The more you look, the more you learn how to see, and thus the more you can see. Since I can't accept the fact that I'm aging, I've decided to blame the vog.

The Helix appears so large and diffuse because it is so close. Over the last 20 years, distance estimates have varied considerably, with the majority clustered around 420 light-years. John Meaburn (Victoria University, England) gave a value of 522 light-years in a 1998 paper in the *Monthly Notices* of the Royal Astronomical

Society. Accepting that distance, we find the nebula measures 1.8 × 1.5 light-years. The Helix's proximity also makes it unique among the planetaries, because it allows astronomers to study its small-scale structures with ground-based telescopes and extremely fine details with the orbiting Hubble Space Telescope.

In ground-based images the Helix looks like a typical annular planetary nebula. The vacuous region inside the ring shows evidence of oxygen atoms being excited by ultraviolet radiation flowing out from the extremely hot (120,000° Kelvin) central star. That star shines at magnitude 13.4 (though it may be variable), so it's a bit brighter than Pluto is when closest to the Sun. The ring itself glows with the light of excited nitrogen and hydrogen atoms. Everything looks normal until you look closely at the ring and see that it has a helical structure with two twists. But this is an optical illusion. In January 1999 a team of astronomers released a three-dimensional, globe-like map of the Helix based on data from a submillimeter-wave telescope atop Mauna Kea. Close inspection of the 3-D image reveals that the Helix Nebula behaves much like a double-headed garden sprinkler. For each blob of gas that was ejected on one side of the nebula, a similar structure was ejected on the opposite side. The researchers say this behavior likely has been caused by a companion star that affects how the Helix's progenitor ejects its outer layers.

The Helix is also the only planetary in which gaseous knots with cometlike tails have been observed. This does not necessarily make the Helix unique; more distant planetaries probably have similar but irresolvable features. A few hundred of the Helix's knots were first detected from the ground, but HST increased the count to 3,500. None of the knots is found close to the central star. Rather, they form a thick ring, with the most visible knots lying along the inner edge of the luminous annulus, trillions of km from the central star. At first, astronomers believed these features might be evidence of a gigantic cloud of comets like the Oort Cloud postulated to surround our Sun. But it turns out that these features are much too large to be comets. Each gaseous knot is at least twice the size of our solar system, and each tail stretches for more than 150 billion km. The gaseous knots display crescent-shaped heads that point to the central star. Meaburn says that each crescent is an ionized skin on the surface of a dusty molecular globule. And since the "comet tails" point away from the central star (which is ionizing the crescents), it's clear that each globule is being shaped by a wind from that star. In fact, we're seeing a sort of cosmic rear-end collision in which hot gas spewing from the surface of the central star rushes through a shell of colder gas that the star ejected when it was still a red giant, some 10,000 years earlier. When the hot, rarefied gas mixes with the cooler, denser gas, instability results, and the cooler cloud fragments into these gaseous knots. A close look at the HST image shows the cool ring writhing with these ghoulish "comets." The ring looks as if it's effervescing like an Alka-Seltzer tablet in water. Astronomers expect the gaseous knots, each several billion km across, to eventually dissipate into the cold blackness of interstellar space. "If this phenomenon is common among stars," says Rice University astronomer C. Robert O'Dell, who acquired the images, "then our galaxy could be littered with trillions of these objects." The Helix's knots are moving outward at about 15 km per second on average. This radiating array of pseudocometary pellets led Houston to coin the object's other popular nickname, the Sunflower Nebula.

So how does this celestial marvel hold up

to the scrutiny of amateur telescopes? My 4-inch seems to perform just fine on it at 23×. The first impression I got was that of a ghastly, pale ring of ashen light trapped within a triangle of 10.0- to 11.5-magnitude stars. The northwestern and southwestern stars each have a roughly 12.5-magnitude companion separated by some 3′. (The brighter member of the southwestern pair is a fine double star in its own right, though it is best seen at higher magnification.) The triangle's third star lies just to the east of the nebula and shines at about magnitude 11.5, though it looks much fainter in photographs. With averted vision, the Helix's edges appear sharply defined. The northwestern side looks thinner than the southeastern side. Look for two enhancements in the luminous ring; one, to the northeast, connects to another, to the southwest. Try averted vision and see if these enhancements don't look like a pair of ghostly lips blowing you a kiss. With time, they appear mottled. The central cavity is quite transparent at low power.

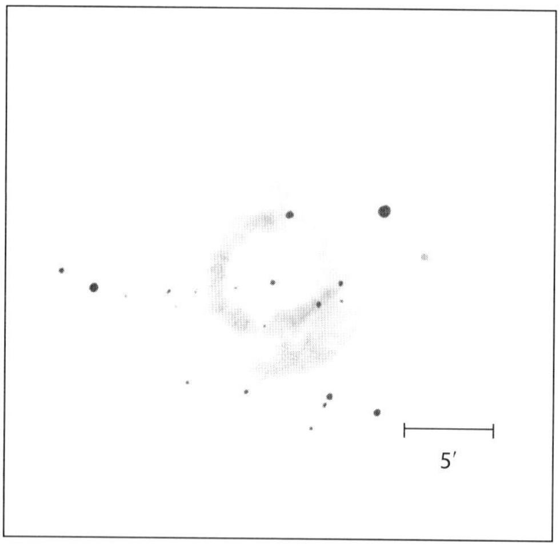

With a lot of patience you can see a fair amount of detail in the nebula at 72× (the object all but disappears with anything much higher). First, I suggest you draw the surrounding star field before you go out; use the photograph below as a guide. That way you can pencil in nebular details without having to worry about getting the stars right; you can later tweak the sizes of the stars to reflect their relative brightnesses in the eyepiece. Now, working your way from the inside out, first check the "hole" for the magnitude-13.4 central star. It takes time for me to see it with my 4-inch, but there's no mistaking it once I do. If you have a sturdy mount, try looking for the central star with a higher magnification, say 140×. You will not see the nebulosity, but that's not what you're after at this point.

Next look for the dense inner ring with 72×; it's the brightest part of the entire nebula. The ring contains several background stars and nebulous enhancements, especially to the northwest and southwest. To see the outermost nebulosity, try placing the Helix outside your field of view, then slowly bringing it back (the eye is more sensitive to light in motion). The nebula's broad eastern portion has a dim but sharp edge that is not difficult to spot in this way. The most challenging part of the nebula lies to the west, where a tail of gas appears to curve northward from the southern rim of the inner annulus. This tail meets the 10th-magnitude star bordering the nebula to the northwest, then curves back to the inner ring and connects with its northern edge. This last wisp of gas is the most difficult portion to see, but if you can trace it you will have mapped the Helix's full "double twist." If you fail to see this dim wisp, return to low power and look again. Low power will condense its light and make it more apparent. Do not expect to see all this detail at once with a 4-inch. My drawing at upper left is a composite of the Helix as seen at low, medium, and high power.

I have a special challenge for the owner of

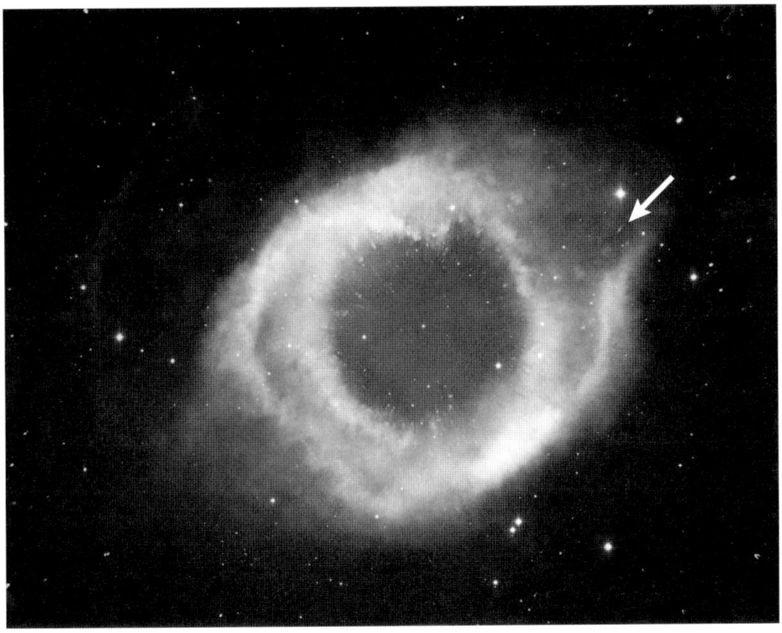

any large telescope that can provide a wide field of view. Almost all photographs of the Helix reveal traces of the cometlike structures; yet I have heard no visual reports of them. I believe that if these structures can be photographed, they can be seen. So there's your challenge. Meanwhile, Arizona amateur Steve Coe has another: "If you are looking for a real observing challenge for your next observing session, the Helix can supply a toughie! The picture of the Helix on page 195 in Robert Burnham Jr.'s *Celestial Handbook* shows a very faint edge-on galaxy at the edge of the Helix Nebula. It is located at the end of the Helix that opens up near a pretty bright star [the aforementioned 10th-magnitude star]. Using the 17.5-inch on a night I rated 9/10 at 8,000 feet in the Central Mountains of Arizona, I could just pick it out at 300×." (This tiny galaxy is arrowed in the photograph above.) Good luck!

64

64

Tau Canis Majoris Cluster;
"Northern" Jewel Box;
Mexican Jumping Star;
Liquid Pyramid
NGC 2362
Type: Open Cluster
Con: Canis Major

RA: 07^h 18.7^m
Dec: –24° 57′
Mag: 4.1
Diam: 6′
Dist: 5,100 light-years
Disc: Giovanni Batista Hodierna, before 1654

W. HERSCHEL: [Observed 6 March 1785] A very beautiful cluster of pretty [bright] stars, very rich, contain[s] the [star] 30 Canis [Majoris]. (H VII-17)

GC: Cluster, pretty large, rich.

NGC: Cluster, pretty large, rich (30 Canis Majoris).

IN THE LOWER BACKBONE OF CANIS MAJOR, Orion's greatest hunting dog, is one of the most inconspicuous yet visually stunning open clusters accessible to observers in both hemispheres. NGC 2362 is called the Tau (τ) Canis Majoris Cluster by most, but I call it the "Northern" Jewel Box. I place "Northern" in quotes because the cluster is north of the original Jewel Box cluster (Caldwell 94) but south of the celestial equator. The two Jewel Boxes have much in common. Both have Greek-letter designations; the Jewel Box is also known as Kappa (κ) Crucis. Both shine at 4th magnitude; both are roughly 5,000 light-years distant; and both are among the youngest star clusters known, with ages of a few million years. In his *Astronomical Objects for Southern Telescopes,* Ernst Hartung writes: "This most beautiful cluster contains about forty stars grouped around the very bright white 30 [Canis Majoris] which is set like a jewel in their midst."

Shining at magnitude 3.8, the Tau Canis Majoris Cluster is about 13° southeast of dazzling Sirius and nearly 3° northeast of 2nd-magnitude Delta (δ) Canis Majoris. The cluster is only 0.7 magnitude fainter than M44, the famous Beehive Cluster in Cancer. Yet look at the star called Tau Canis Majoris with the naked eye and you'd never know a cluster existed around it. That's because the stars in NGC 2362 are packed into an area only 6′ across (1/12 the Beehive's angular diameter).

Contrary to popular belief, William Herschel did not discover the Tau Canis Majoris Cluster; he rediscovered it. The honor goes to Giovanni Batista Hodierna (1597–1660), whose observations appeared in his *De Admirandi Coeli Caracteribus* — a catalog, published in 1654, of some 40 deep-sky objects. The catalog somehow escaped Herschel's notice, and indeed the modern world's, until Giorgia Foderà Serio, Luigi Indorato, and Pietro Nastasi brought it to our attention in the February 1985 issue of the *Journal for the History of Astronomy*. Hodierna, an astronomer at the court of the Duke of Montechiaro, used a simple 20× Galilean refractor for his observations. Most interesting, Hodeirna also predated Herschel with a "plurality of worlds" concept, for he believed that all nebulous objects were clusters of stars that astronomers had resolved to various degrees. (Actually, this belief has roots that stretch back to Galileo, who had resolved the Milky Way, and "nebulous patches" within the Milky Way, into a number of stars.)

Both William and John Herschel singled out NGC 2362, calling it a "beautiful cluster"

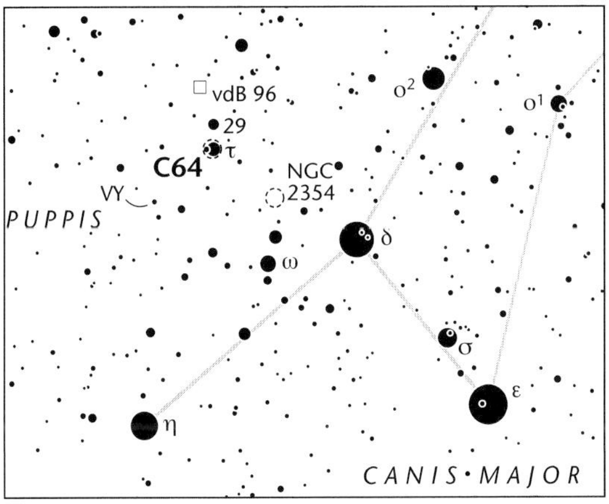

and a "fine cluster," respectively. In his 1844 *Cycle of Celestial Objects*, Adm. William Henry Smyth describes it under the heading of 30 Canis Majoris, writing "a star with a companion [29 Canis Majoris] in cluster . . . on the Greater Dog's back. . . . The whole has a beautiful appearance, the bright white star [30 Canis Majoris] being surrounded by a rich gathering of minute companions, in a slightly elongated form, and nearly vertical position."

Several sources list NGC 2362 as a star cluster and a nebula. *Sky Catalogue 2000.0* lists NGC 2362 in its section on open clusters and again in its section on bright nebulae. But look at the *NGC* description in the table above. Note that there is no mention of nebulosity. *Sky Catalogue 2000.0* also includes a Trumpler code for the open cluster and that code includes the letter *n* for nebula. This, too, is a mystery, because in his original 1930 work (Lick Observatory *Bulletin* Vol. 14, No. 420) Robert J. Trumpler did not use an *n*. In other words, Trumpler did not see any nebulosity in NGC 2362. In fact, color-magnitude diagrams of the cluster reveal little if any reddening or absorption.

In 1949 the late Harold Johnson of Washburn Observatory observed NGC 2362 at the urging of Walter Baade, who pointed out that "there is no discernible nebulosity around the cluster." Johnson's photometric and visual observations, made with the 100- and 60-inch reflectors on Mount Wilson, confirmed this fact. "There is little reason," Johnson wrote, "to suspect the existence of nebulosity in NGC 2362; an examination of the original negative [taken with the 100-inch reflector] likewise shows no nebulosity. The visual appearance with the 100-inch telescope is in agreement with the conclusion that the amount of nebulosity associated with the cluster is negligible." More recently, in a 1991 *Astronomical Journal* paper, David Wilner (University of California, Berkeley) and Charles Lada (Harvard-Smithsonian Center for Astrophysics) referred to NGC 2362 as "a rich young star cluster devoid of natal gas and dust." Other researchers have come to the same conclusion.

Yet there *is* nebulosity around the Tau Canis Majoris Cluster. The confusion, says Lada, stems from the fact that a bubble of gas does surround the cluster, but it is about 5° across — vastly wider than the cluster. Sharpless 310, as the bubble is called, is a very-low-density filamentary HII region that appears on the red-sensitive *National Geographic So-*

ciety–Palomar Observatory Sky Survey prints. This rarefied bubble is centered on the Tau Canis Majoris Cluster and is excited both by it and by 29 Canis Majoris. Lada believes the cloud is the remains of the cluster's natal cocoon, and that it has been blown away by radiation pressure from the cluster's hot, young stars. But that does not mean, he continues, that the nebula should be given an *NGC* designation; nor should Trumpler's original work be altered to note the nebula's presence. Such revisions only pervert the histories of interesting objects like these.

In a 1978 *Astrophysical Journal* paper, Lada described another interesting feature of the bubble. Along its edge are a number of isolated dark clouds and one bright one at the tip of which lies the peculiar variable star VY Canis Majoris 1¼° southeast of Tau. VY Canis Majoris lies at the same distance from Earth as the Tau Canis Majoris cluster and appears to be associated with a vast complex of two colliding molecular clouds. A 10′ × 5′ reflection nebula, vdB 96, lies about 1° north-northeast of Tau Canis Majoris; it may also be illuminated by Tau Canis Majoris and 29 Canis Majoris as well as by the chain of three 9th-magnitude stars embedded within it.

Brent Archinal's database has 60 members in NGC 2362 spread across 6′ (9 light-years). But that number increases substantially if we take into account the several hundred pre-main-sequence stars that the Rosat X-ray satellite revealed in the cluster. The cluster's crown "jewel," of course, is Tau Canis Majoris itself. With a spectral type of *O*9Ib, the magnitude-4.4 star is now known to be a multiple-star system with five components, none of which can be resolved by amateur instruments. The system's core is a close binary, with two stars of nearly equal brightness separated by 0.16″. NGC 2362 also contains three short-period variables; one of them may be the hottest Delta (δ) Scuti star known. Delta Scuti stars are relatively young and hot pulsating variable stars with about 1.5 to 3 solar masses. Though numerous, these stars are difficult to detect because of their very small brightness variations (typically about 0.01 to 0.4 magnitude, or 1 to 4 percent).

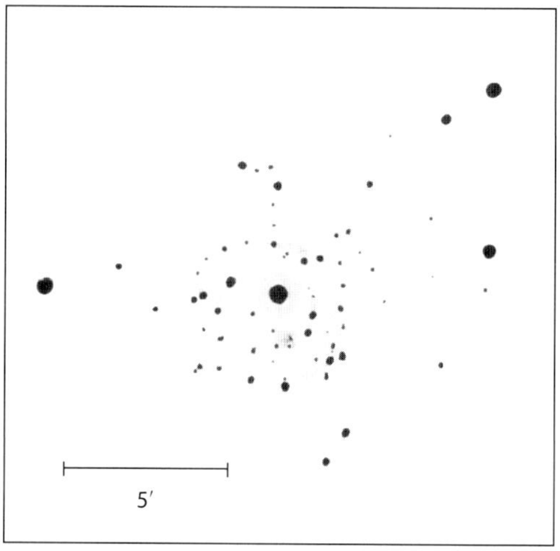

If the star Tau Canis Majoris looks fuzzy to the naked eye, it's because of 5th-magnitude 29 Canis Majoris, which lies a mere 25′ north of it. Our eyes tend to make close pairs of stars look fuzzy, but the fuzziness is just an optical illusion. For those with really young eyes, here is a fantastic test: can you resolve the magnitude-7.3 star about 7′ east of Tau Canis Majoris? The rest of the cluster's members are so tightly packed, and Tau Canis Majoris's brilliance so overpowering, that small-binocular users may be challenged to see any of the cluster's other members. If you do find yourself under a dark sky with binoculars, take a moment to sweep them around this area. The part of the Milky Way that runs through the region is laced with ribbons and arcs of dark nebulosity.

I've given this cluster another nickname: the Liquid Pyramid. Here's why: at 23× in the 4-

inch the cluster shows a pyramid of starlight encompassing an ice blue gem (namely Tau Canis Majoris, which also is known as 30 Canis Majoris). To see the "liquid" you'll first have to let your eyes become dark adapted. Now stare directly at Tau Canis Majoris through your telescope at its lowest power; you should see a bright star surrounded by a half dozen or so faint suns that form a cube of icy starlight. Next, look at the region out of the corner of your eye and a multitude of dimmer suns will ooze into view. In other words, the cube of icy starlight appears to be melting, forming a liquid of faint starlight that surrounds the pyramid. Look directly at Tau Canis Majoris, though, and the "liquid" will evaporate.

Using my imagination, the 72× view of NGC 2362 reminds me of a night flight over a Mayan temple whose sides glitter with decorative lights while a blue beacon shines on top. A mist of unresolved starlight hugs the pyramid's base like fog, though I can see some stray lights trickling away through the "fog" to the north. At high power the view is magnificent. The brightest stars suddenly greet the eye with a visual punch. Fainter stars pop out from the background, though averted vision still reveals a nagging mist of unresolved starlight. Close inspection of Tau Canis Majoris reveals it to be caged by rows of stars (oriented roughly along a north-south axis), while sinuous arms of starlight spiral beyond these "bars" toward the cardinal points. Look for a very faint diamond of stars due south of Tau Canis Majoris. In the 4-inch it appears to have an almost nebulous nature, but that is a visual artifact caused by the closeness of these dim suns.

Bill Arnett shares the following fun experience, which he had during an observing party at California's Fremont Peak Observatory on December 7, 1996. "A nearby observer discovered an interesting phenomenon. While viewing [the Tau Canis Majoris Cluster] through a 14-inch, he commented that when the wind jiggled the telescope, [Tau] appeared to move in a different direction from the [other] cluster stars. This announcement, predictably, was greeted with jeers — until the jeerers looked themselves, and discovered that everyone could see the effect. Most interesting! Presumably the effect has something to do with persistence of vision causing the perceived image of the bright star to remain in place longer than the perceived images of the much fainter cluster stars. Tau [Canis Majoris] was immediately dubbed 'The Mexican Jumping Star.'"

Another open cluster, NGC 2354, lies only 1½° southwest of Tau Canis Majoris. The Genesis offers nearly a 3° field of view at 23×, so placing the Tau Canis Majoris Cluster at the northeastern edge of the field centers NGC 2354 in the eyepiece. Now sweep your scope around, as you did earlier with binoculars, and see if the entire region doesn't look studded with stellar patterns. It's a proving ground for those with fertile imaginations.

65

65

Great Sculptor Galaxy; Silver Coin
NGC 253
Type: Mixed Spiral Galaxy (SABc)
Con: Sculptor

RA: 00^h 47.6^m
Dec: –25° 17′
Mag: 7.6; 7.1 (O'Meara)
Dim: 25.8′ × 5.9′
SB: 13.2
Dist: 9.8 million light-years
Disc: Caroline Herschel, 1783

W. HERSCHEL: [Observed 30 October 1783] Considerably bright, much extended, from south preceding to north following [from southwest to northeast]. Much brighter in the middle. Above 50′ long and 7′ or 8′ broad. Discovered by Caroline Herschel. (H V-1)

GC: *Remarkable*, very bright, very large, very much extended toward position angle 54°, 30′ long, gradually brighter towards the middle, four stars.

NGC: *Remarkable*, very bright, very large, very much extended toward position angle 54°, gradually brighter in the middle.

"STUNNING!" "SPECTACULAR!" "INCREDIBLE!" So many adjectives have been used over the years to describe NGC 253 that it's difficult to find others more appropriate. In photographs the impression is one of magnificent animation, of clumps of starlight and dust stirred together in a stew of galactic vapors. At a glance the galaxy looks like the beginning of universal order emerging from simmering chaos. And in a way it is. NGC 253 is a somewhat normal, somewhat barred spiral, with a complex lens centered on a tight central core that has some of the densest knots of stars ever found. A bar cuts across the lens and is encircled by a bright elliptical ring of stars and dust, off of which fly two spiral arms — one to the northeast and one to the southwest, like two propeller blades.

In the notes that accompanied William Herschel's first catalog of "New Nebulae," the great observer wrote of NGC 253: "This nebula was discovered by my sister Caroline Herschel, with an excellent small Newtonian sweeper of 27 inches focal length, and a power of 30. I have therefore marked it with the initial letters, C.H. of her name." NGC 253 is one of at least 13 objects Caroline discovered while sweeping the heavens for comets in 1782 and 1783. William and Caroline had the disadvantage of trying to study this galaxy close to the horizon. However, William's son, John, observed it high in the sky from the Cape of Good Hope with an 18¾-inch speculum-metal reflector. On November 20, 1835, he noticed that NGC 253 is "somewhat streaky and knotty in its constitution and may perhaps be resolvable."

At a distance of 9.8 million light-years,

NGC 253 is the brightest member of the Sculptor Group of galaxies, the closest such system to our Local Group. NGC 253 travels through space with NGC 247 (Caldwell 62), NGC 300 (Caldwell 70), and NGC 55 (Caldwell 72) and has a recessional velocity of 241 km per second. Handbooks often compare NGC 253 to M31 in Andromeda, but the two really do not have that much in common. NGC 253 measures 54,000 light-years in diameter and has a total mass of about 75 billion Suns, making it about 2½ times smaller and 4 times less massive than M31. As of March 2000, 14 globular clusters had been identified around NGC 253.

Most galaxies form stars slowly but not NGC 253. Like M82 in Ursa Major, NGC 253 is a starburst galaxy, meaning its nuclear region has recently hosted massive and rapid bursts of star birth. Each furious episode generates a "superwind" that sends vast amounts of energy and mass streaming into the intergalactic medium, which it heats up. In a 1994 *Astrophysical Journal* paper, a team led by Yoshiaki Sofue (University of Tokyo) used Anglo-Australian Observatory photographs by David Malin to find 1,000-light-year-long dust features (called "arcs") that appear to be associated with two or more even larger dark "loops." One spectacular bubble may have been formed when a supernova exploded in NGC 253's halo. The entire galaxy's nuclear region appears to be boiling with invisible turbulence.

Midinfrared studies of NGC 253 have found its nuclear region to have three major components: an unusually luminous active nucleus; a "supercluster" of stars; and a large, diffuse envelope containing an older population of supernova remnants and lower-mass stars. The Very Large Array has revealed a family of compact sources of radio waves inside a starburst ring; they are most likely supernova remnants. The VLA also found an H II region that resembles the inner part of 30 Doradus in the Large Magellanic Cloud (see Caldwell 103). This enormous nebula contains approximately 105 million solar masses in the form of stars as well as about 600 million solar masses of ionized gas.

The Hubble Space Telescope imaged NGC 253 in remarkable detail. Hubble's close-up of the galaxy's innermost core shows streaming bands of dark clouds that nearly obliterate the view of a multitude of stellar furnaces below. The entire scene looks gloomy. What Hubble actually imaged is a dusty region of furious star formation; the dim lights peeking through the dust are members of an extremely compact star cluster. Hubble also revealed numerous young star clusters tracing out a 163-light-year-wide ring around the galaxy's active nucleus. Hubble confirmed that starbursts can spawn dense clusters of stars. NGC 253's most recent starburst probably has been going on for 20 to 30 million years.

To zoom in on this extragalactic drama, look 7½° south-southeast of 2nd-magnitude Beta (β) Ceti (Deneb Kaitos) or 4¾° northwest of 4th-magnitude Alpha (α) Sculptoris. That's where you'll find NGC 253 — just 2° northwest of the South Galactic Pole. From more southerly locations the galaxy is a true deep-sky wonder and can be seen with the naked eye on very clear nights. Doing so requires a few minutes of dedication and averted vision. (From Hawaii NGC 253 can be seen this way on the best of nights when on the meridian.) The galaxy is well placed among the stars for naked-eye viewing because it sits between the southern tips of two pronounced asterisms of 5th- to 6th-magnitude stars "beneath" the tail of Cetus. The galaxy is the only object in the gap between the tips, down to at least 9th magnitude, so its light cannot be confused with that of any naked-eye star; you either see the

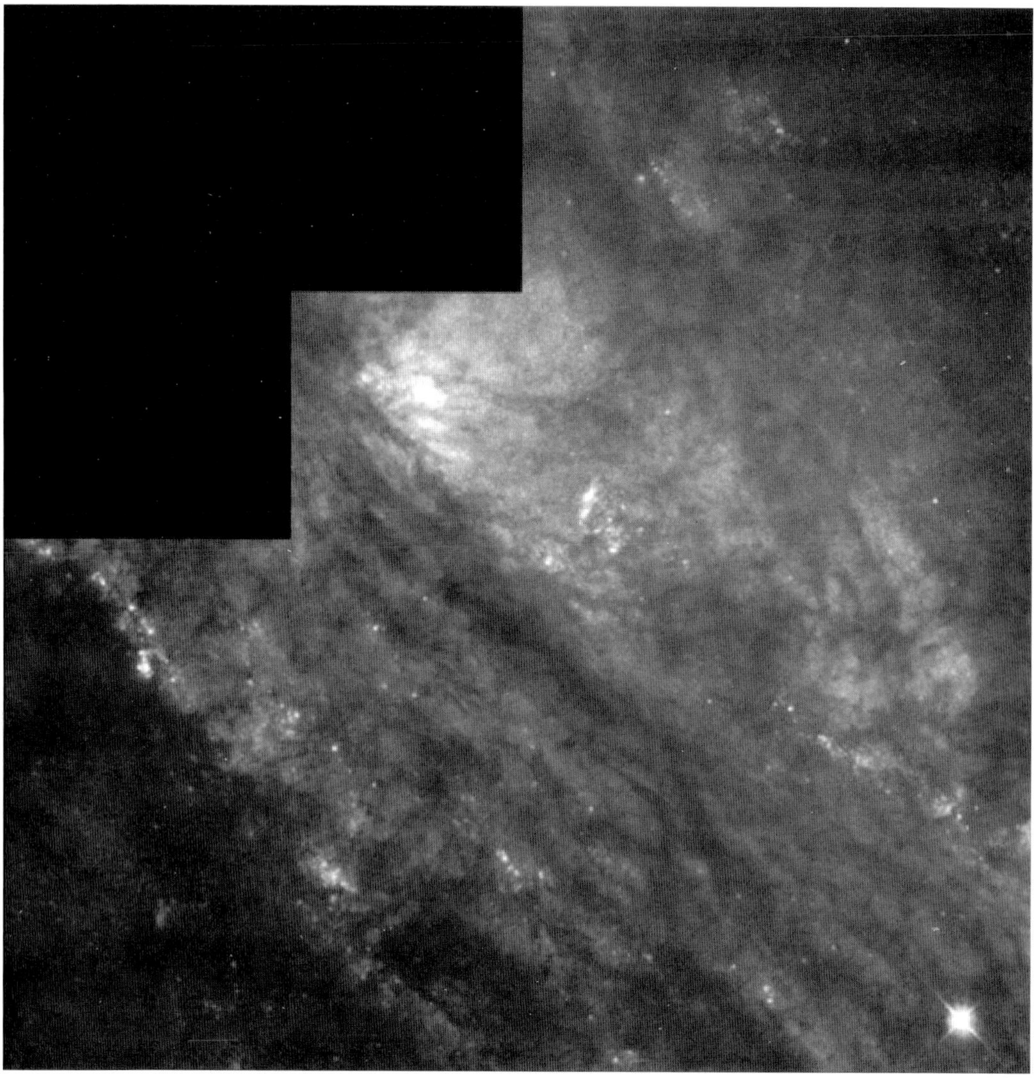

galaxy or you don't. NGC 253's naked-eye appearance reminds me of M33's, but NGC 253 is easier to see. My magnitude estimate of 7.1 was made with 7×35 binoculars. Note that this is 0.5 magnitude brighter than the value published in the *Deep Sky Field Guide*.

NGC 253 is a magnificent binocular object, mainly because the field is shared with the globular cluster NGC 288. The two make quite the dynamic pair; the sight is similar to that of M81 and M82 in Ursa Major. (I was surprised to discover that NGC 288 is not a Caldwell object; I've included it as one of my favorite non-Caldwell targets in the following chapter.) Structure is immediately visible in NGC 253 at 23× through the Genesis. The galaxy's spindle is bordered by a nice pair of roughly 9th-magnitude stars to the immediate south and by one 9th-magnitude star beyond its northeastern edge. Close inspection shows two 12th-magnitude stars, one just to the east of the nuclear region and one farther to the west. With a bit of imagination it looks as if the galaxy was sailing through space until it wedged itself in between these two stars and became forever trapped, like a leaf that was floating down a stream until it got caught between two stones. Seeing the galaxy beyond these foreground stars adds a three-dimensional quality to the low-power scene. (By the way, the Hipparcos satellite did triangulate the 9th-magnitude star to the northeast of the galaxy; it lies 193 light-years from Earth.) The galaxy's highly condensed inner region looks mottled. The brightest patch there is the southwestern edge of the inner nuclear region; it lies well to the

extent at low power; low power emphasizes an apparently sharp edge that makes the eye want to extend the galaxy beyond its true borders. Use moderate magnification (72×) to reliably trace out NGC 253's farthest reaches.

NGC 253 takes high power very well, and high power is needed to make out the complex detail in its center. In fact, at high magnification it's possible to see some of the galaxy's

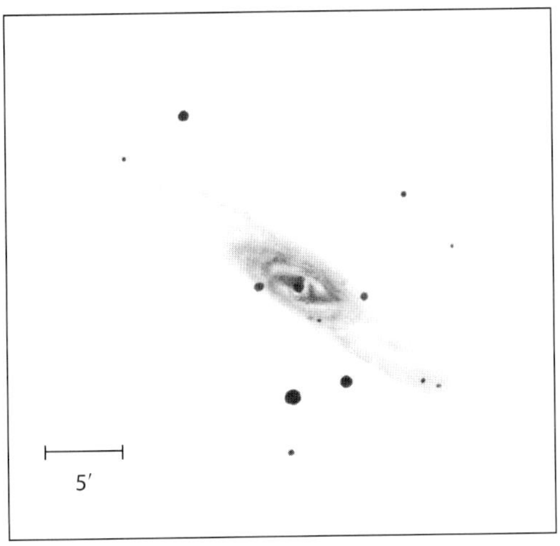

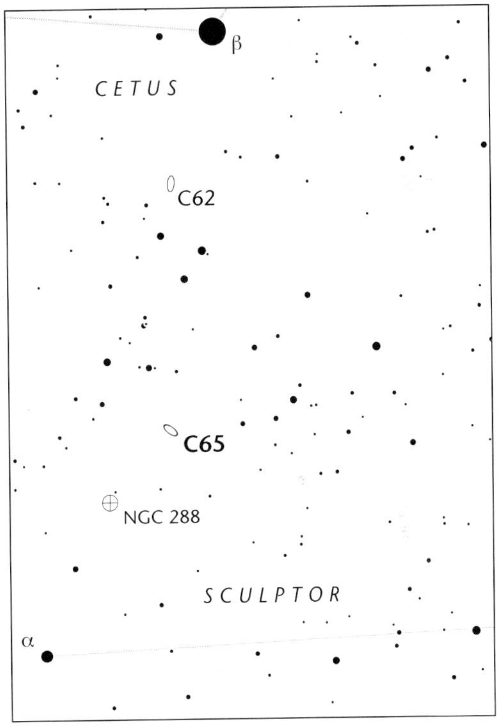

north of the pair of 9th-magnitude stars. If you concentrate, this feature takes on a parabolic shape, with its faintest extension curving gently to the northeast, toward the bright nucleus. Another smaller parabolic patch lies diametrically across from the first one, on the other side of the nucleus. The northeastern side of this patch looks frayed, like a tiny feathery tail. The galaxy's core has a remarkably high surface brightness. It's hard to judge the galaxy's true

compact clumps and stellar associations in a 4-inch telescope. The galaxy's core is also riddled with dark nebulosity, perhaps too much to draw precisely. Moderate power only hints at this detail (or the wealth of it, I should say). High power and a modest amount of hyperventilating will help you study the galaxy's nucleus. I found myself straining to make sense of the vast amount of detail there, but doing so became easier once I relaxed and took strong, rhythmic breaths. In essence, the core of the galaxy looks like two opposing fishhooks. One extends to the northeast, then loops to the southwest; the other extends to the west-southwest, then loops eastward. (These are the parabolas mentioned in the previous paragraph.) The space between the hooks is filled mostly with dark material. The hooks are part of the star-forming ring Hubble imaged, and they appear to be connected by a bar.

The galaxy's two major spiral arms can be traced along its northeastern and southwestern edges, and they are best seen at moderate powers. Not surprisingly, I saw different structures on different nights as atmospheric conditions varied. Supernova hunters should be aware that a smattering of dim foreground stars (13th magnitude and fainter) is superimposed on the galaxy; some of them are shown in my drawing. The drawing on the previous page, by the way, is a composite based on views made with three different magnifications on three different nights. Seeing such detail with a small telescope requires a lot of time, but do not hesitate to try.

The Royal Astronomical Society of Canada's *Observer's Handbook* calls NGC 253 the "Silver Coin Galaxy." Obviously this came about because of the galaxy's appearance in photographs. I wonder who "coined" this name.

66

NGC 5694
Type: Globular Cluster
Con: Hydra

RA: $14^h 39^m 36.5^s$
Dec: –26° 32′ 18″
Mag: 10.2
Diam: 4.3′
Dist: 113,000 light-years
Disc: William Herschel, 1784

W. HERSCHEL: [Observed 22 May 1784] Pretty bright, small, nearly round, bright in the middle, resolvable. (H II-196)

GC: Considerably bright, considerably small, round, pretty suddenly brighter in the middle, resolvable, a single star nearby.

NGC: Considerably bright, considerably small, round, pretty suddenly brighter in the middle, resolvable, a star of magnitude 9.5 south preceding [to the southwest].

NGC 5694 IS THE FINAL OBJECT IN THE LATE Walter Scott Houston's "Hydra Hysteria" — a deep-sky romp through the sky's largest constellation that was published first in *Sky & Telescope* and later in his *Deep-Sky Wonders* book. Hydra spans nearly 100° and contains a cornucopia of splendors. Houston designed the Hydra Hysteria challenge to be "a little different" from the Messier Marathon, for he wanted something that would "sharpen the skills needed to star-hop to small and faint galaxies as well as bright and easy objects." In a way, Houston's Hydra Hysteria mimics Hercules's second labor, which was to conquer Hydra. That was no easy task, as the Sea Monster had nine heads, and every time one was cut off two new ones appeared in its place. But Hercules prevailed; he knocked off eight of Hydra's heads with a club and buried the ninth under a rock. I like to imagine NGC 5694 as this ninth head, for it is buried in the extreme northeastern corner of Hydra, under a southern ledge of Libra and near a western wall of Centaurus.

William Herschel first found NGC 5694 in 1784, but not until 1932 did Lowell Observatory's Carl Otto Lampland and Clyde Tombaugh identify it as a globular cluster. At a distance of 113,000 light-years, the 10th-magnitude globular is one of the most remote in the heavens. Its stars are packed into a region 140 light-years wide. Like M92 in Hercules, NGC 5694 is one of our galaxy's oldest globulars. It is also far (95,000 light-years) from our galaxy's core. It's possible the cluster formed in the inner halo of the galaxy, only to be ejected into the outer halo by a close passage of the Magellanic Clouds. McMaster University astronomer William Harris once wondered if this cluster might be escaping our galaxy. Indeed, some models do show it following a hyperbolic orbit. More research needs to be conducted on this enigmatic globular's space velocity before the truth is known.

In a 1996 *Astrophysical Journal* article, Martha Hazen (Harvard-Smithsonian Center for Astrophysics) recounts how she "blinked" eight reasonably matched pairs of photographic plates of the cluster but found no candidate variable stars. This is not surprising, for other metal-poor globulars share the same trait. Her findings place NGC 5694 into a group of low-metallicity globular star clusters with few (or no) RR Lyrae stars. On average, each member of NGC 5694 has a spectral class of *F*4 and 1⁄70 as much iron as our Sun. CCD photometry also has revealed the presence of bright blue quasi-stellar knots very close to the cluster's core, suggesting the presence of blue stragglers. NGC 5694 certainly displays a high degree of central condensation, with a core measuring a mere 7″.

To find the cluster look a little more than 10° south-southwest of Alpha (α) Librae (Zubenelgenubi), or about 5½° west-southwest of 3rd-magnitude Sigma (σ) Librae. A little less than 2° east of the globular is a lovely 3½°-long chain of 4th- to 6th-magnitude stars; it heads southeast from the star 4 Librae, then bends nearly to the south at 55 Hydrae before ending at 58 Hydrae. These stars form the body of the now-obsolete constellation Noctua, the Night Owl, which appeared in Elijah H. Burritt's *Atlas* accompanying the *Geography of the Heavens*, first published in 1833. The owl stands on a branch that stretches east-west from Pi (π) Hydrae to Sigma Librae, and the branch includes the stars 50, 51, 52, 58, 59, and 60 Hydrae. As Robert Burnham Jr. lamented in his *Celestial Handbook*, the "loss of this little asterism might seem unfortunate to astronomers, since no other celestial creature so appropriately honored their profession."

NGC 5694 is part of a bent arrow of fainter stars that looks like a crooked, 1°-long version of the constellation Sagitta. At 23× in the Genesis the arrow comprises several dimmer asterisms, so the field is quite attractive. The

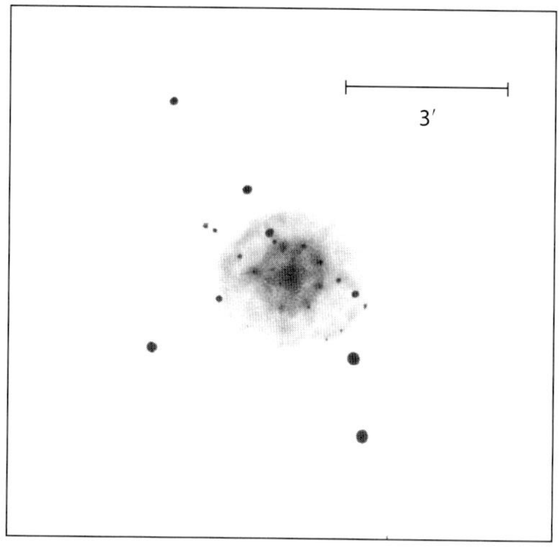

globular is the northernmost "star" in a chain of suns that winds northward from the brightest (middle) star in the arrow's shaft; thus the *NGC* description, "among stars." In fact, the foreground stars near the cluster are so tightly arranged that the area positively glows under averted vision. This glow should draw your eye toward the cluster.

At 72× NGC 5694 reveals itself as a very small, compact glow, and at a glance it looks like a slightly swollen star. It certainly is not a Hydra-Hysteria-stopper — unless, as Houston recommends, it's the last object you view. Its brightest members shine at magnitude 15.5, and its horizontal-branch magnitude is a feeble 18.5. Some field stars do appear to be superimposed on the cluster. Spend some time with NGC 5694 at moderate power. Do

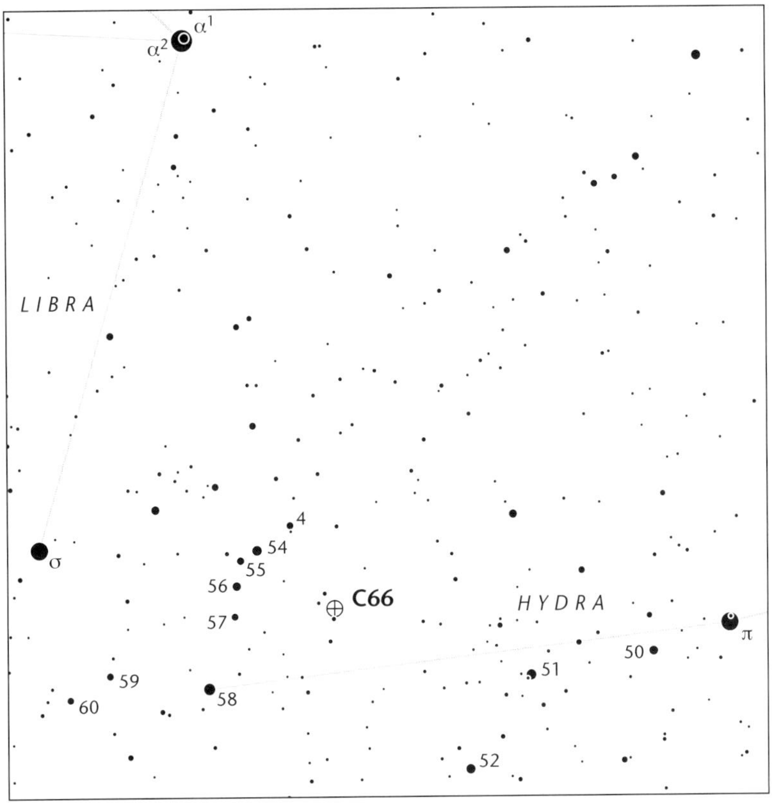

you see its core dancing? Its starlike center seems to jump around. That's because the core is surrounded by a thin shell of light, which is surrounded in turn by a fainter, circular envelope. As your eye moves around the field involuntarily, various parts of these dense halos waltz into view. High magnification does not improve the view much. With time you should make out some concentrations of light forming a cross that marks the main cardinal directions. These glows may be superimposed field stars of roughly 14th magnitude; the brightest lies just northeast of the cluster's core. Look for a dim arc of light at the northeastern edge of the cluster's outer halo.

Larger instruments do not seem to enhance the view much. Christian Luginbuhl and Brian Skiff report that the core and inner halo look granular through a 12-inch. Ernst Hartung failed to resolve NGC 5694's constituent stars, seeing only a "conspicuous round symmetrical haze, well condensed towards the centre." And Barbara Wilson couldn't fully resolve the cluster in a 20-inch f/4 reflector at 72×, though she did resolve about 30 percent of it, reporting some faint 15th-magnitude stars. But I would not give up hope. Like planetary nebulae, tight globular clusters can take magnification well on steady nights, and I believe partial resolution can be achieved in a good 8- to 10-inch telescope (and in a 4-inch under very dark skies). I also suspect that anyone using a moderately long-focus refractor could better resolve this globular cluster if they use higher magnifications — say 800× to 1,000× — on nights of superb seeing. So give it a go.

67

NGC 1097
Type: Barred Spiral Galaxy (SBb)
Con: Fornax

RA: 02^h 46.3^m
Dec: –30° 16′
Mag: 9.2
Dim: 9.8′ × 6.1′
SB: 13.6
Dist: 47 million light-years
Disc: William Herschel, 1790

W. HERSCHEL: [Observed 9 October 1790] Very bright, extended toward position angle 75° from north preceding to south following [from northwest to southeast]. 8′ long. A very bright nucleus confined to a small part, or about 1′ diameter. (H V-48)

GC / NGC: Very bright, large, very much extended toward position angle 151°, very bright in the middle to a nucleus.

A PROMINENT MEMBER OF THE FORNAX Cluster and the Eridanus Cloud, NGC 1097 is a fantastic example of an X-ray-luminous Seyfert galaxy. Photographs show an extremely bright nucleus burning at the center of a broad, dusty bar while two magnificent star-studded arms sweep out in opposing arcs. A magnitude-13.6 elliptical-galaxy companion, NGC 1097A, lies just beyond the bar's northwestern tip, where the spiral's southern arm suddenly frays like an old rope. Three distinct tendrils of stars stretch to the northern tip of this disturbed appendage, while its southern half looks like what happens when someone smears chalk on a blackboard with their hands.

The tiny nucleus of a Seyfert galaxy is an extraordinary energy engine. Seyfert nuclei emit about a thousand times more energy than does our own galaxy's relatively quiescent nucleus. They tend to be in tight binary-galaxy systems, and tidal forces may contribute to their eruptive energy. Two pairs of optical (visible-light) jets have been detected extending as far as 15′ from NGC 1097's nucleus. If these features are indeed jets, they constitute the most extensive system of optical jets known to originate from any galaxy. However, David Carter (Mount Stromlo and Siding Spring Observatory) and his colleagues have their doubts. They investigated one of the most prominent "jets" with the 3.9-meter Anglo-Australian Telescope and concluded that it was actually an unusually thin tidal tail formed during the capture of a low-mass galaxy. The now-defunct Einstein satellite detected strong X-ray sources around these "jets." Six of these sources have been identified as quasars. A total of 31 confirmed quasars populate a 2.8° × 2.8° area around NGC 1097. This number reportedly exceeds normal background levels. Does that mean there is a physical relationship between NGC 1097 and the quasars? Maverick cosmologist Halton Arp (Max-Planck Institute for Astrophysics) thinks so. In a 1998 *Astrophysics and Space Science* paper, Arp notes that the Rosat spacecraft also has found a striking excess of X-ray sources — the brightest of which are quasars — around NGC 1097's optical jets. If, as Arp suspects, the fainter X-ray sources also are predominantly quasars, "at least 10 and possibly considerably more X-ray quasars appear to be associated with NGC 1097." Most cosmologists believe the quasars are the active nuclei of galaxies that are vastly more distant than NGC 1097, but Arp has posited that quasars are ejected from the nuclei of active galaxies.

NGC 1097 is a large galaxy, with a diameter of 123,000 light-years and a mass of 200 billion Suns, making it a good match for our own Milky Way. It is receding from our solar system at a speed of 1,275 km per second. We see NGC 1097's barred form inclined 33° from edge on. High-resolution images of NGC 1097's nuclear region reveal a ring of star formation with a slightly higher metal abundance than our Sun's. If we accept a distance of 47 million light-years to NGC 1097, the starburst ring measures 6,500 light-years across. Outside the ring, H II regions are enhanced where the galaxy's bar and inner spiral arms intersect. No signs of star formation appear inside the ring, though Hubble Space Telescope (HST) images reveal narrow dust lanes intricately winding between the ring and the nucleus. Actually, the ring is incomplete; HST resolved it into a two-armed spiral of H II regions and clumps of bright stars that appear on the leading edges of strong dust lanes. Such two-armed structure may be generic to nuclear rings in barred spiral galaxies. This notion is bolstered by computer simulations that show how gas flows along a galaxy's bar to feed star formation in a ring around the nucleus. In the case of NGC 1097, the burst of star-forming activity along the ring (100 million solar masses' worth) may have occurred nearly instantaneously some 6 to 7 million years ago (in the galaxy's frame of reference, that is; recall that its light takes 47 million years to reach us).

William Herschel did not see the ring when he discovered the "nebula" in 1790, though I wonder if his son, John, did, albeit indirectly, since he saw a pretty large, "round" nucleus. The Rev. T. W. Webb called the object simply "very bright and large." The 9.8′ × 6.1′ galaxy shines at magnitude 9.2, but its surface brightness is rather paltry. I could not see the galaxy with 7×50 binoculars, though it has been spotted in 11×80s with averted vision. The best way to locate NGC 1097 is to star-hop from 3rd-magnitude Acamar, Theta (θ) Eridani, the original end of the River Eridanus. First go 5° north-northwest to the sharp, 1°-wide triangle of Eta^{1,2,3} ($\eta^{1,2,3}$) Fornacis. These stars shine at magnitudes between 5.5 and 6.5. Note that Eta^2 and Eta^3 Fornacis form a near-isosceles triangle with Theta and Iota (ι) Eridani; if you cannot see Eta^2 and Eta^3 Fornacis with your eyes, you can at least point your finder to their location, using Theta and Iota Eridani as guides. With binoculars you can confirm that you've correctly located the field. Continue from Eta^{1,2,3} Fornacis about 3½° north to magnitude-4.5 Beta (β) Fornacis. The galaxy is just another 2¼° north-northwest of Beta Fornacis.

With a glance at 23× in the 4-inch NGC 1097 is not impressive at first, appearing only as a faint glow with a starlike core just 15′ north of a tiny triangle of binocular stars about 5′ across. By the way, if you move your telescope 1½° to the west, you'll see a beautiful dipper asterism of 6th- to 8th-magnitude suns. With north "up," the dipper has its "bowl" on the "table." It also looks a bit like the Pleiades. If you place the galaxy on the eastern edge of a nearly 3°-wide field, the dipper dominates the western and central parts of that field. Now return to the galaxy, but do not increase the magnification just yet. Give your eye some time to adjust to its faint light. With time, some subtle details might start popping out at you. The longer you look the more you should see. For instance, one thing that should become immediately apparent is that the galaxy is not round but slightly elliptical, oriented northwest-southeast. That round section is just the very bright core that John Herschel noticed. Concentrate on the galaxy's core. It should suddenly look like a bright star embedded in a soft oval glow. Stay with it longer and look for an arc of light in the galaxy's northwestern quadrant as well as a knot to the southeast. Few galaxies allow us to study them at such low power and yet see so much detail.

At 72× the galaxy, though tiny, starts to break up into a series of knots and gaps. It looks like it is on the verge of exploding. The knots come and go with the seeing. To make sure I'm not straining my eyes, I'll frequently get up and start walking around my telescope, taking deep breaths and replenishing the oxygen supply to my brain and eyes. Back at the eyepiece, I look at the object for brief periods rather than incessantly. With NGC 1097 your vision needs to be tack-sharp and your mind alert enough to record the most fleeting of details. Once I spot something curious, I immediately take my eye away from the eyepiece and pencil in that detail. By repeating this procedure over and over again, I build up a decent visual record of fleeting glimpses. This is also how I have recorded planetary details at the limit of my vision: by waiting out the atmospheric turbulence and exploiting those fine moments of steady seeing. It takes time and practice, but the rewards are usually satisfying. This is how anyone observing a galaxy

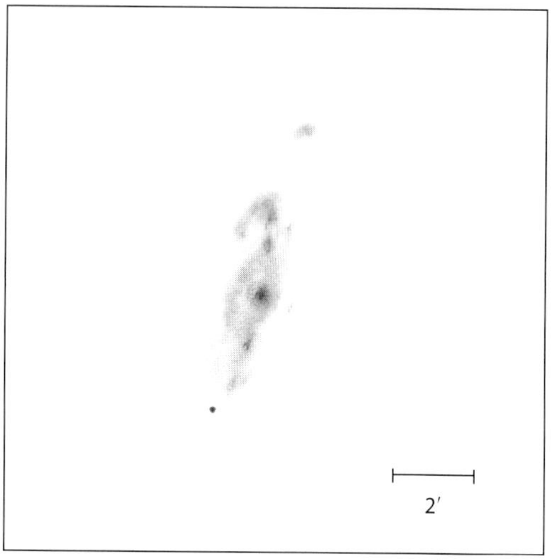

through a small telescope can see details that other observers casually see in larger instruments. I say "casually" because observers with big scopes can see even more if they take the time to do it right. Seasoned amateur Tom Polakis indirectly confirms my suspicions. Of observing NGC 1097 with a 20-inch reflector he writes: "My impression before going out into the field to observe it was that the outer arms would be difficult, and the inner structure spectacular. The object yielded the opposite result. Using a 20-inch at a dark desert site, the outer arms were quite visible even to a novice observer. The bright inner spiral was only a bright, mottled disc. This detail may be more seeing dependent than anything — sort of like looking for detail on Jupiter."

What other details can be seen in NGC 1097 in a 4-inch from a dark Hawaiian site? With moderate power the galaxy's nucleus is a bright pinprick of light surrounded by a tight lens in the middle of a clumpy bar. A dark lane can be seen on the northeastern side of the bar, while the bar's southeastern end has two bright beads in it. A single enhancement of fuzzy light lies just southwest of the nucleus. The southeastern part of the bar looks crooked

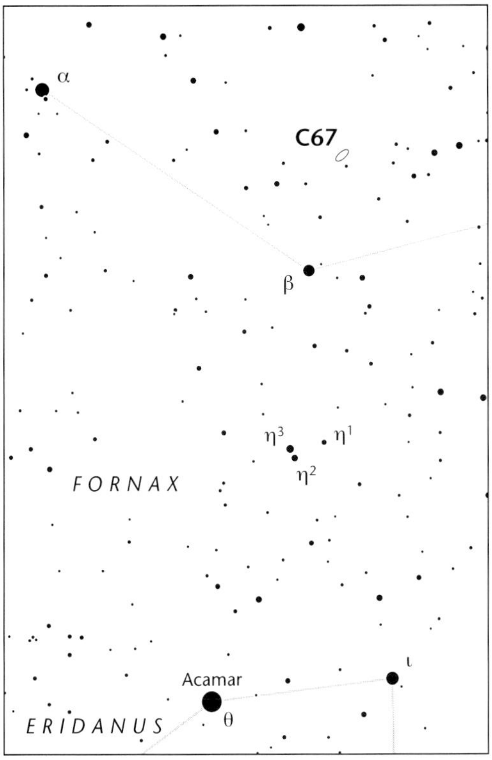

and kinked. All this detail is wrapped in a faint oval halo of light. At this power I see no well-defined spiral arms. With 130× and excellent seeing the northern clumps transform into a knot and a ribbon of nebulosity that curls gently eastward as part of the galaxy's feathery inner spiral structure. A bright arc of light defines the northeastern edge of the outer lens, while an odd fuzzy patch to the southwest may be a dust-disturbed part of the nuclear region. Or is it a faint star?

On the evening of December 30, 1996, I detected a very dim patch of light northwest of the main galaxy at high power. It was in the right position to have been NGC 1097A. But could I have seen that magnitude-13.6 companion galaxy in a 4-inch telescope? Quite possibly. Published galaxy magnitudes in this range are suspect. Christian Luginbuhl and Brian Skiff estimate NGC 1097A's visual magnitude to be around 12.9, though I am certain it's probably a few tenths of a magnitude brighter. Comet hunters should note that the companion is not plotted on the *Millennium Star Atlas*, but it is in *Uranometria 2000.0*.

NGC 1097 is yet another Caldwell object for supernova hunters to keep an eye on. An HST image of NGC 1097 taken on September 30, 1992, captured Supernova 1992bd 12 days prior to its ground-based discovery 2″ east and 9″ south of the nucleus; the supernova shone at magnitude 15 at the time of ground-based discovery. More recently, a magnitude-17.3 supernova appeared 23″ east and 157″ south of the galaxy's center in November 1999.

68

R Coronae Australis Nebula
NGC 6729
Type: Emission and Reflection Nebula
Con: Corona Australis

RA: 19h 01m 54s
Dec: –36° 57.1′
Mag: ~9.5 (O'Meara); variable
Diam: 1.3′; variable
Dist: 424 light-years
Disc: Albert Marth, listed in his *Catalogue of 600 new nebulae*, published in 1876 in the *Memoirs* of the Royal Astronomical Society, Vol. XXXVI

HERSCHEL: None.

GC: None.

NGC: Variable star (11th magnitude and fainter) with nebula.

OUT OF CURIOSITY ONE NIGHT IN SEPTEMBER 1996, I put a low-power eyepiece into my Genesis and swept its 4-inch glass across the northeastern corner of Corona Australis, the Southern Crown, just below the Teapot asterism in Sagittarius. I was pretending to be Charles Messier hunting for comets. I wanted to see if Messier would have picked up the R Coronae Australis Nebula (Caldwell 68). Of course, Messier's instrument would not have been as good as mine, so I took that into account. I also chose to ignore the fact that this region was simply too far south for him to observe. But had he observed from a more favorable latitude, I wondered, what would he have found here? Remember, Messier was not looking for faint nebulae; he was looking for comets. Without question, the father of comet hunters would have noticed the 7th-magnitude globular cluster NGC 6723 in Sagittarius; it lies only 30′ northwest of our target. (Surprisingly, this stunning object failed to make it into the Caldwell Catalog.) It's possible that the great comet hunter would have noticed the close 7th-magnitude double star at the core of the nebula IC 4812, a little more than 10′ southwest of R Coronae Australis; the pair's tightness makes it look fuzzy, like M40 in Ursa Major, so its character probably would have piqued Messier's curiosity. And it's also possible, though unlikely, that Messier would have noticed the faint double nebula NGC 6726-7 just 5′ northwest of our target. But I'm certain Messier would have missed our Caldwell object — a "little" star with a "tail" — because it simply would have been too inconspicuous for this man with a mission.

Not that the nebula isn't visible in a small

telescope. But seeing it takes more effort than one is likely to invest in a random patch of sky during a comet sweep. Actually, NGC 6729's tiny glow escaped the gaze of many great southern-sky observers, including James Dunlop (who discovered NGC 6723) and John Herschel. It took the keen eye of Albert Marth — an "able and hardworking German astronomer who had emigrated to England," as Joseph Ashbrook describes him in his *Astronomical Scrapbook* — to find the diminutive nebula. "Marth," Ashbrook continues, "has never received adequate recognition for his varied contributions to astronomy." In 1861 Marth traveled to the Mediterranean island of Malta with William Lassell, who set up his great 48-inch reflector there. Over the course of the next four years, Lassell and Marth used the telescope to scan the heavens. According to Ashbrook, the "most important result of this survey was the recognition of spiral structure in a number of galaxies. Also, a catalogue of 600 previously unknown nebulae was compiled [by Marth]. Most of the actual observing during this second Malta expedition was not done by Lassell but by his assistant Albert Marth."

NGC 6729 was the 395th object Marth cataloged. It was described as part of a larger, "pretty faint" complex of nebulosity, namely NGC 6726-7. NGC 6729 is but a tiny 1′-long wisp of gas flowing away from R Coronae Australis to the southeast. In photographs it looks like a fan-shaped comet tail. In 1866, Julius Schmidt (Athens Observatory) first noticed that R Coronae Australis's brightness varied irregularly. Generally its magnitude lies between 9.7 and 13.5. The nebula's brightness variations appear to follow the star's. The nebula also changes form rapidly — too rapidly, in fact, to represent true changes in its configuration, since the changes appear at face value to propagate far more rapidly than physically possible. John A. Graham (Carnegie Institution of Washington) and his colleagues documented this phenomenon in the February 1987 *Publications* of the Astronomical Society of the Pacific. (Not that they were the first, as the photographs below, taken from Mount Wilson, attest.) With their CCD images, obtained over a 23-day period in 1984, Graham's team found that the reflection nebula's surface brightness can change

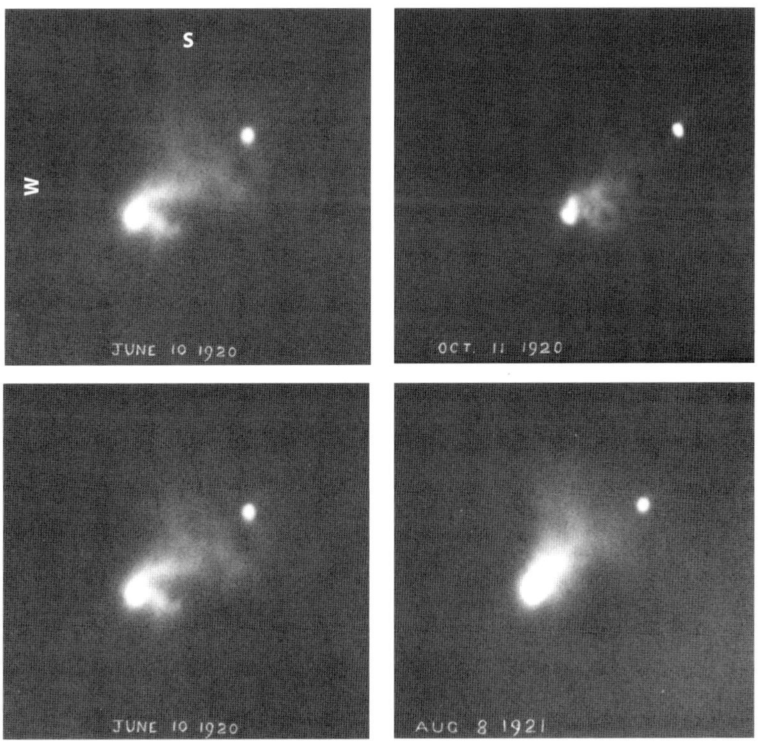

within as little as 24 hours. The researchers also noticed that the variable star brightened by 1.3 magnitudes during their observing period. From these findings they deduced that the nebula's apparent changes were caused by the shadowing effects of moving clouds well within 1 a.u. of the variable star. Similar shadow play occurs in Hubble's Variable Nebula (Caldwell 46).

Our Caldwell target couldn't be easier to find. It is 40′ east-northeast of 5th-magnitude Epsilon (ε) Coronae Australis in the northern arc of the Southern Crown. I use 7×35 binoculars to confirm that I've located Epsilon Coronae Australis, which is 6° south of the midpoint in the base of the Sagittarius Teapot. Then I hop about ½° east to the 7th-magnitude double star associated with IC 4812. (Here's a challenge: can you see this star with the naked eye?) The NGC 6726-7 complex should appear as a fuzzy "star" about 15′ to the northeast of that double star. Our target, NGC 6729, is a fainter "fuzzy" star immediately southeast of NGC 6726-7.

Although we are ostensibly pursuing Caldwell objects here, let's take the time to acquaint ourselves with some of the region's other wonders before rushing on to NGC 6729. Through the 4-inch at 72× the double star associated with IC 4812 is a beautiful pair of soft blue orbs of near-equal brightness; they are separated by 13″. The easternmost star is the brighter of the two at magnitude 6.4 and has a slightly evolved spectral class of *B*9.5IV; its magnitude-6.7 companion has a main-sequence spectral class of *B*7/8V. NGC 6726-7 is a 9′ × 7′ haze surrounding two other stars oriented north-northeast and south-southwest. The brighter member of that pair seems to shine with a slightly warmer (yellower) hue than the blue stars in the 7th-magnitude double. The magnitude-7.2 star responsible for

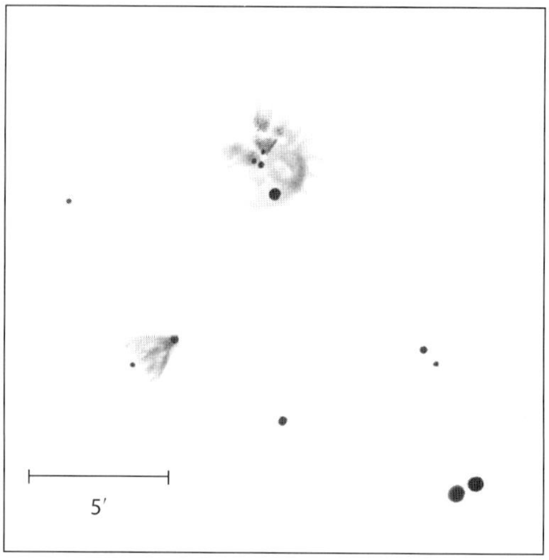

illuminating NGC 6726, the nebula's southwestern portion, is of spectral class *B*9IV. The brightness of NGC 6727, the nebula's northeastern portion, is regulated by TY Coronae Australis, a variable star that fluctuates irregularly between magnitudes 8.7 and 12.4. NGC 6726-7 has a high surface brightness, and if you doubt you're seeing it, just compare it to your view of the 7th-magnitude double star, which should be fuzz-free in a small instrument. With a careful gaze at 72× the nebulae appear mottled. Hints of a ray-type structure can be seen, as well as a large "brow" of nebulosity, which arcs to the northwest over both of the involved stars.

At 130× NGC 6726-7 shatters into several fragments. The two most intense patches form a ten-pin shape, with the brightest portion surrounding the brighter star to the southwest. The "brow" to the northwest comprises two conjoining arcs of light, each of which bristles with faint tendrils that extend radially away from them. Take the time to study the intriguing nebulosity around TY Coronae Australis. In it I see two dense knots, one to the east and one to the north of TY. With a prolonged look and averted vision, dim patches, knots, and

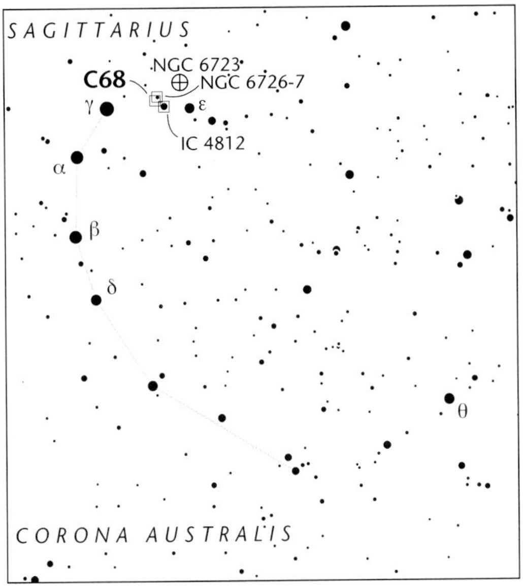

streamers begin to appear, but they will vanish like celestial ghosts under a direct gaze. The knots are the most mysterious features because they waver in and out of view in the 4-inch, leading even an astute observer to question their reality. The million-dollar question is: Does TY Coronae Australis look less stellar than its magnitude-7.2 companion? Perhaps it is immersed in a thicker bed of dust and gas than its companion rests in. What do you think?

Finally, onto our Caldwell target. A very faint petal of nebulosity extends from NGC 6726-7 to the southeast and tapers to a point, like one of those stringy coronal streamers seen during a total solar eclipse. This delicate tendril points directly to our target, NGC 6729, just 5′ away. At 23× NGC 6729 appears fuzzy, but at such low magnification it's hard to tell if this is true nebulosity because its bright "head" (R Coronae Australis) is only 1.3′ northwest of yet another nebulous variable star, T Coronae Australis, and the two stars make a fuzzily close pair. In the 4-inch at 72× NGC 6729 is the least interesting object in the area, at least visually. It shows some definite haziness that extends toward T Coronae Australis to the southeast, but that's all. High power is required to make out any detail. Once you know what you're looking for, however, NGC 6729's comet-like nature becomes apparent. Or so it did in September 1996, when I observed this Caldwell object; the nebula may have changed since. At high power an obvious central bar pointed directly toward T Coronae Australis. A fan of material can be seen on either side of the bar. While NGC 6726-7, with its double star and nebulosity, reminds me of M78 in Orion, smaller NGC 6729 looks like Hubble's Variable Nebula in Monoceros, and photographs of the entire region recall the Pleiades. What a fascinating part of the sky!

Certainly amateurs looking for a long-term project can focus on the variable nature of NGC 6729. CCD owners should first try imaging NGC 6729 each month, then, once a year has passed, create a movie from the successive images. This should show how the nebula changes with time and may well capture the interest of some professional astronomers.

I made a magnitude estimate for NGC 6729 at 23× and got about 9.5. See how this compares to your own estimate. For a visual description of the fantastic globular star cluster NGC 6723 in Sagittarius see page 442 in the next chapter. I searched for the globular with the naked eye but couldn't see it without optical aid. See if you have better luck. I think it can be done.

69

NGC 6302
Bug Nebula
Type: Planetary Nebula
Con: Scorpius

RA: $17^h 13^m 44.1^s$
Dec: –37° 06′ 14″
Mag: 9.6
Dim: 83″ × 24″
Dist: ~5,200 light-years
Disc: Edward Emerson Barnard, around 1883, though he credits James Dunlop, who may have included the nebula in his 1827 catalog.

HERSCHEL: None.

GC: None.

NGC: Pretty bright, extended preceding-following [west to east], planetary nebula.

NGC 6302 IS THE BRIGHTEST PLANETARY nebula in Scorpius and one of the best of its kind — the bipolar kind, that is — in the night sky. It is easily located nearly 4° due west of Lambda (λ) Scorpii (Shaula), the easternmost of the two Stinger Stars in the Scorpion's tail. The nebula now holds the honorable distinction of being the first object imaged by the European Southern Observatory's Very Large Telescope (actually it was imaged by just one of the optical array's four 8.2-meter reflectors).

Edward Emerson Barnard found NGC 6302 around 1883, and he first described and drew its curious form in a 1906 *Astronomische Nachrichten* paper. Barnard, however, credited James Dunlop with the nebula's discovery. It is unclear why, for there is no object at the nebula's position in Dunlop's 1827 catalog. I believe that Barnard — tired, perhaps, from a night's observing — could simply have failed to realize that Dunlop listed his objects by South Polar Distance (the angular distance from the South Celestial Pole). By sheer coincidence, Dunlop's 366th listing has a South Polar Distance similar to the Bug Nebula's declination. What's more, that object's opening description — a "pretty large nebula, extended in the parallel of the equator" — fits the Bug. Dunlop 366, however, is actually NGC 6397 (Caldwell 86), a globular cluster in Ara. (It is conceivable that Barnard observed Dunlop 567. Described by Dunlop as a "very faint, small, ill-defined nebula, with a small star in it," this object has coordinates that, when properly precessed, lie a mere 8′ from the Bug. However, the Bug is both larger and brighter than Dunlop's description seems to imply. Is Dunlop 567 the Bug? Your guess is as good as mine.)

Now consider that neither William nor John Herschel saw the Bug. This seems astounding, because NGC 6302 is a magnificent object that can be seen in 7×35 binoculars. Unfortunately, some sources muddy the waters by listing the nebula's *photographic* magnitude (12.8) in place of its far brighter *visual* magnitude (9.6). Because of their peculiar spectra, which are dominated by a few emission lines, this often occurs with planetary nebulae, and undoubtedly it has dissuaded many observers from trying to find certain ones, NGC 6302 included. Robert Burnham Jr. does not feature the Bug Nebula in his *Celestial Handbook*; Walter Scott Houston never discussed it in his "Deep-Sky Wonders" column for *Sky & Telescope* (which ran monthly for nearly half a century); and the Webb Society does not include it in its *Deep-Sky Observer's Handbook*.

In photographs taken with large telescopes NGC 6302 indeed looks like a squashed bug. Its long body trends east-west and comprises two opposing cones or "wings" of gas. Separated by a dark girdle of dust, these "wings" exemplify the "butterfly" aspect of many bipolar planetary nebulae. Each "wing" is made up of stringy veins and clumps of material whose different elements are being fluoresced by a hot central star. One long tentacle-like extension trails off to the northwest like a squashed bug's guts. Sun Kwok (University of Calgary) believes that the various forms of planetary nebulae — rings; butterflies; and asymmetrical lobes — all can be explained by one generic type of system seen at different angles. In this model a dense torus of dust surrounds the central star and channels its outflowing gas through its open ends. Once free of the ring the gas fans out, giving the planetary a butterfly shape when seen edge-on. Kwok says that bipolar nebulae were once deemed rare, but "they are now believed to be more common than we thought."

Try as they might, as of the year 2000 astronomers had yet to image the Bug Nebula's central star, which ionizes its gases, making them glow. If the star can indeed be seen at all at visual wavelengths, it must be fainter than 20th magnitude. The central star evades visual detection probably because it is very blue and because its light is dimmed by the dense torus of dust separating the nebula's two lobes. The star's presence has been detected, though, in the radio, infrared, and ultraviolet portions of the electromagnetic spectrum. The resulting data imply that the star has an unusually high surface temperature of about 380,000 degrees Kelvin. Distance estimates for the nebula have ranged from 3,900 to 7,800 light-years. In a 1993 *Astrophysical Journal* paper, Yolanda Gómez (Autonomous National University of Mexico) and her colleagues propose a value of 5,200 light-years, plus or minus a third. That would make the object's physical dimensions 2.1 light-years by 0.6 light-year or thereabouts. Gómez's team also estimated the Bug Nebula's total mass to be about 3/10 that of the Sun, placing it among the most massive planetary nebulae.

As I've said, the Bug is easy to spot under dark skies. Again, it's nearly 4° due west of Lambda (λ) Scorpii, so Telrad users have a direct shot at it: place the eastern edge of your Telrad's outermost circle on Lambda Scorpii and the Bug will be on the same circle's western edge. Once I knew where to look, the nebula was seen easily; "no eye strain on this one," I wrote in my field notes. Look for a solitary 9.6-magnitude "star" surrounded by patches of dark nebulosity, which populates this region. The object remains largely stellar at 23× in the 4-inch, though with averted vision it expands ever so slightly. The Bug comes alive at 72×, appearing as a definite high-surface-brightness glow elongated east to west. Even at this modest power the nebula's true bipolar nature can be detected without difficulty. Its brightest section is intense and starlike, with a forked "comet tail"

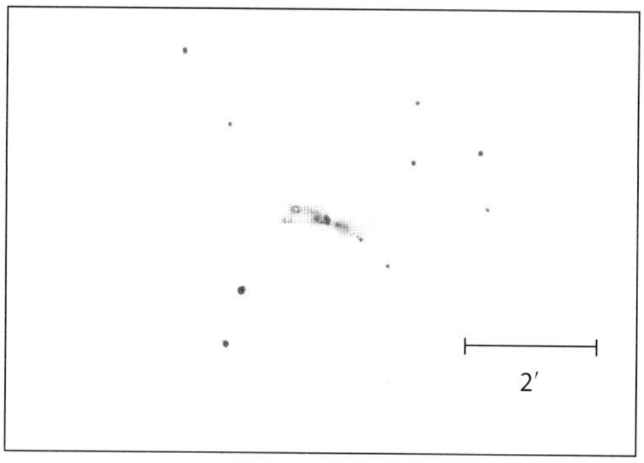

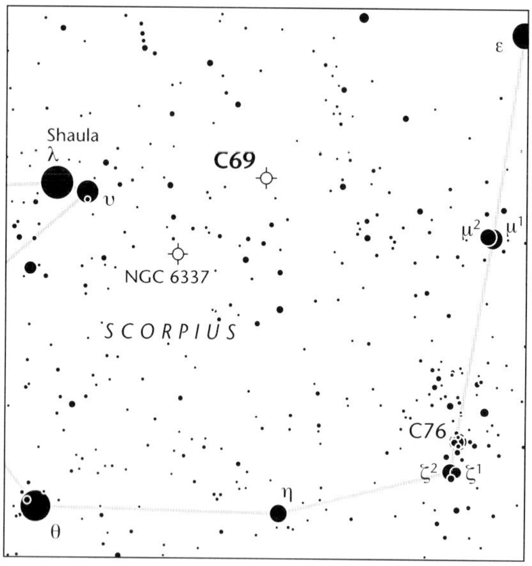

extending to the east; both sides of this parabolic tail have sharp edges. A dark, narrow lane oriented roughly north-south separates that "comet" from a smaller, fainter one to the west; that fainter "comet" shines with a softer and more diffuse glow. The planetary takes high power well, thanks to its high surface brightness, so don't be afraid to use as much magnification as the night will allow. High power should reveal some fine details. In the 4-inch I spotted several knots or enhancements in the eastern lobe: one close to the pseudonucleus; two midway through the lobe; and one near the end. In the western lobe I saw one prominent knot close to the pseudonucleus; Christian Luginbuhl and Brian Skiff noted another one toward the western end of the lobe. (Interestingly, I thought I spied that feature in the 4-inch at 72× but not at high power. Maybe high power reduced the feature's contrast against the background sky.) With averted vision I suspected seeing the entrail-like streamer emanating from the western lobe, but this might have been an illusion created by a string of dim field stars. See what you think. I saw no color in the nebula with the 4-inch Genesis, but Ernst Hartung called NGC 6302 a "bluish nebula" as seen through a 12-inch. Steve Coe found it gray when he saw it through Larry Mitchell's 36-inch reflector at the Texas Star Party.

Just 2° southeast of the Bug is NGC 6337, a 12.3-magnitude planetary nebula, about 48″ across, with a 15th-magnitude central star. It is roughly as distant as the Bug. Finding it with the 4-inch was a real ghost hunt. Start 2½° west of Lambda Scorpii, where you'll see a 1°-long line of three stars (two of 6th magnitude and one of 8th); they are plotted on the finder chart below. Following these stars from north to south will lead you right to the nebula, which is less than 30′ southwest of the southernmost, 8th-magnitude sun. But be careful, because the dim planetary is part of a tight trapezoid of five 9th- to 11th-magnitude stars in a Milky Way region rich in fainter suns. NGC 6337 is at the northern end of that trapezoid. I had to first locate it with high power so its size would give it away; then, after memorizing its position, I was able to pick it out at low power from the rich background of stars.

70

Southern Pinwheel
NGC 300
Type: Spiral Galaxy (Sd)
Con: Sculptor

RA: 00^h 54.9^m
Dec: –37° 41′
Mag: 8.1; 7.2 (O'Meara)
Dim: 20.2′ × 15.1′
SB: 14.0
Dist: 3.9 million light-years
Disc: James Dunlop, included in his 1827 catalog

J. HERSCHEL: [Observed 1 September 1834] Bright; very large; very gradually pretty much brighter towards the middle; very much elongated; irregular figure; 8′ to 10′ long, 3′ or 4′ broad; has subordinate nuclei. (h 2359)

GC/NGC: Pretty bright, very large, very much irregularly extended, very gradually pretty much brighter in the middle.

72

String of Pearls
NGC 55
Type: Barred Spiral Galaxy (SBm)
Con: Sculptor

RA: 00^h 15.1^m
Dec: –39° 13′
Mag: 8.1; 7.5 (O'Meara)
Dim: 34.1′ × 6.5′
SB: 13.6
Dist: 4.2 million light-years
Disc: James Dunlop, included in his 1827 catalog

J. HERSCHEL: [Observed 3 May 1834] Bright; very large; very much elongated in a long irregular train, the preceding end being much the brightest. Whole length = 1.5 [times the diameter of the] field, or 22′. The nucleus is either a double star or a much more sharply terminated nebulous mass, elongated [toward] a different position [angle] (146.5°) from that of the nebula (109.8°). (h 2315)

GC/NGC: Very bright, very large, very much extended, trinuclear.

70 & 72

AT THIS POINT I'VE CHOSEN TO BREAK THE numerical order of the Caldwell Catalog to discuss two objects that are close to one another on the sky. The late-type spiral galaxies NGC 300 (Caldwell 70) and NGC 55 (Caldwell 72) are members of the Sculptor Group of galaxies, which includes NGC 247 (Caldwell 62) and NGC 253 (Caldwell 65). They lie roughly along the same parallel (that is, at nearly equal declinations) and are separated by 8°. NGC 300 is the more northerly of the pair (by 1½°) and the more easterly (by 7⅞°). Although both galaxies belong to Sculptor, they lie near the constellation's southern border, just "across the tracks" from Phoenix. In fact, they are best found by using magnitude-2.4 Alpha (α) Phoenicis (Ankaa) as a guide. NGC 55 lies 3¾° northwest of Ankaa, at the end of a zigzagging chain of binocular stars. NGC 300 lies 7¼° northeast of Ankaa in a more desolate patch of sky. Finding it is a cinch from dark southerly observing sites, though, because it is 1¾° northwest of 5.6-magnitude Xi (ξ) Sculptoris. But from north temperate latitudes it's best to star-hop from bright Ankaa. Begin by moving your telescope 1¾° north-northeast from Ankaa to a pair of 7th-magnitude suns. Another 1¼° to the north-northeast lies an acute triangle of 8th- and 9th-magnitude stars, which is just 1° east of a 5th-magnitude sun. A longer sweep of about 2° to the northeast brings us to a pair of 6th-magnitude stars, Lambda^{1,2} ($\lambda^{1,2}$) Sculptoris. NGC 300 is yet another 2° to the east-northeast of Lambda^{2} Sculptoris, the eastern member of that pair.

In 1834 John Herschel took credit for discovering NGC 300 at the Cape of Good Hope with his 18¼-inch reflector. However, James Dunlop listed the object in *A Catalogue of Nebulae and Clusters of Stars in the Southern Hemisphere, Observed at Parramatta in New South Wales*, which was published in 1827. Dunlop discovered NGC 300 while surveying the southern skies from Australia with a 9-inch f/12 reflector. John Herschel did acknowledge that the object he found was the 530th in Dunlop's catalog, of which Dunlop had written, a "pretty large faint nebula, irregular round figure 6′ or 7′ diameter, easily resolvable into exceedingly minute stars, with four or five

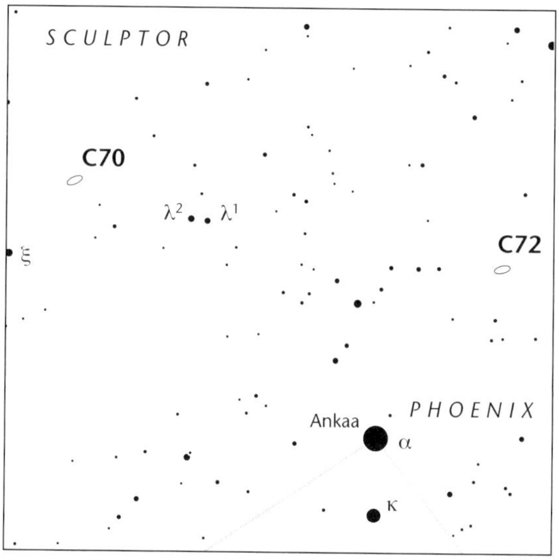

stars of more considerable magnitude; slight compression of the stars to the centre." But during his stay in Africa, the junior Herschel became dissatisfied with Dunlop's survey; he found himself unable to reidentify about two-thirds of the objects in it. In the case of Dunlop 530 Herschel wrote: "Mr. Dunlop's neb 530 is described by him as easily resolvable into very minute stars. Its identity with [NGC 300] is therefore very doubtful."

But *is* it doubtful? The Herschels and many of their contemporaries had imagined that they had resolved many nebulae into stars when there really weren't any stars to resolve. So why would John Herschel suddenly fault Dunlop for doing the same? I cannot find anything wrong with Dunlop's description of the object now known as NGC 300. Johann Louis Emil Dreyer described the galaxy in his *New General Catalogue* as a "complex object with several nuclei." And photographs clearly show the "nebula" studded with field stars and H II regions. Herschel's motion to discredit Dunlop seems even more surprising in light of his own discussion of the object, which placed it among "nebulae of irregular forms having a tendency to several centres of condensation," such as NGC 55 in Sculptor (Caldwell 72) and NGC 1950 in Horologium. That Herschel disagreed with one aspect of Dunlop's visual description of NGC 300 hardly warranted robbing Dunlop of his discovery.

In photographs NGC 300 displays a barless spiral with wide, loosely wound arms and a tiny nucleus; it looks like a miniature, mirror-reversed version of M33, the Pinwheel Galaxy in Triangulum. Like M33, NGC 300 has two dominant S-shaped spiral arms laced with dark lanes and starry clumps; two other feathery arms, forming a weaker S, complete the pinwheel pattern. We see the galaxy's disk 46° from face on. Its true diameter is 22,000 light-years (56 percent smaller than M33's), and it has a total mass of about 30 billion Suns. The galaxy is receding from us at 145 km per second.

NGC 300's disk has an underlying population of old stars with ages between 0.1 and 10 billion years. Deep infrared images have revealed that the brightest stars in the galaxy's central region are massive red supergiants. CCD spectra have confirmed the presence of 28 supernova remnants. Together with a similar number in NGC 7793, these represent the first large sample of supernova remnants identified in a galaxy beyond the Local Group. Thirty-four planetary nebulae, 88 H II regions, and 7 ringlike nebulae also have been identified in the galaxy's disk. The ringlike objects appear to be bubbles blown by winds from unusually hot stars; they span anywhere from 160 to 910 light-years in diameter.

Furthermore, a search for Wolf-Rayet stars with the New Technology Telescope has confirmed the presence of nine such stars within NGC 300. Although NGC 300 is often referred to as the Twin of M33, NGC 300 has twice as much of its mass in the form of cold hydrogen as does M33 and only half the northern Pinwheel's intrinsic luminosity. No nebular emission has been found within the galaxy's bright nucleus, which, most likely, is an unresolved compact stellar cluster similar to the nucleus of M33. X-ray observations of NGC 300 by the Rosat satellite found evidence for a black-hole-bearing binary star, several supernova remnants, and H II regions — but found the galaxy otherwise "unremarkable."

From a dark southern observing site, NGC 300 is visible in 7×35 binoculars as a roundish glow just about 12′ southwest of an 8.5-magnitude star, which could be a variable, and about 2.5′ northeast of a 10th-magnitude star. *Sky Catalogue 2000.0* and the *Observing Handbook and Catalogue of Deep-Sky Objects* list the galaxy's blue (B-band) magnitude as 8.7. Walter Scott Houston found its light to be equal to that of a magnitude-8.5 star, but he also reminded us that since the galaxy's "light is spread over so large an area the surface brightness is low." Thus it would not surprise me if NGC 300 appears fainter from north temperate latitudes, where it can never get far above the horizon. The *Deep Sky Field Guide* lists its visual magnitude as 8.1. Using 7×35 binoculars I got a visual magnitude of 7.2. Try making your own estimate.

In the 4-inch NGC 300 looks very complicated at 23×. Several bright stars are superimposed on the galaxy, which should frustrate supernova hunters. The galaxy's visual impression at low power has convinced me that Dunlop, and not John Herschel, deserves credit for its discovery. Without being aware of the controversy, I penned this field note on December 9, 1996: "The fact that there are so many [stars superimposed on this galaxy] also makes me wonder what someone like [John] Herschel might have thought when he saw this cluster-like haze. In fact, a cluster crossed my mind when I first picked it up." Christian Luginbuhl and Brian Skiff write that "several stars are involved" with the galaxy, including a magnitude-11.5 star about 1.2′ northeast of its center and another brighter star 2.5′ southwest of the center. Dunlop's impression of a nebula that could be resolved into stars hardly seems far-fetched.

NGC 300's nucleus deserves monitoring. The *Webb Society Deep-Sky Observer's Handbook* reports a "star of mag. 11 at centre — possibly a stellar nucleus." In November 1996, Arizona amateur Steve Coe saw the nucleus as "faint, large, and somewhat brighter in the middle" with a 17½-inch reflector at 100×. When I looked at the galaxy a month later than Coe, I found the nuclear region condensed but fuzzy. It was a not-quite-starlike brightening just southwest of a roughly 10th-magnitude star. Curiously, Luginbuhl and Skiff saw no central brightening through a 2.4-inch telescope at low power; nor did they see a central condensation in a 12-inch. I wonder if a flare star is not superimposed on the galaxy near the nucleus. On the evening of December 9, 1996, the star positioned with an X in my drawing appeared bright (roughly 12th magnitude) and obvious to me; then it appeared to fade in front of my eyes. I checked in on the star later that evening and found it to remain faint, around 13th magnitude. The following evening, I returned to the galaxy and saw the star looking about a half magnitude dimmer than it did when I last saw it on the previous night. Now it's your turn to look — what do you see?

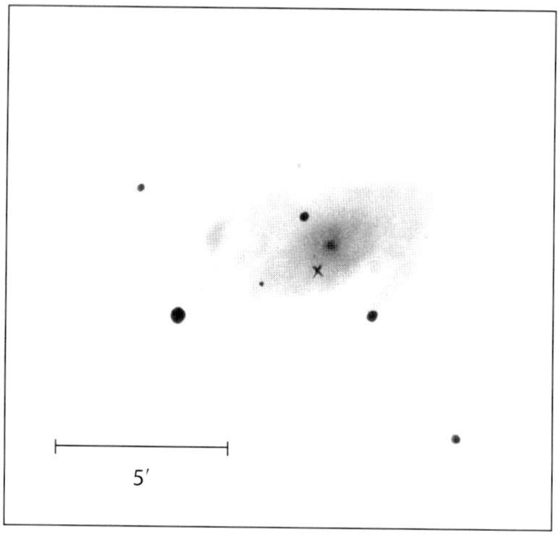

NGC 300's surface brightness is so low that it almost disappears with magnifications higher than 23×. I have glimpsed traces of its two major spiral arms at low power. They're quite a challenge for a 4-inch under a very dark sky. NGC 55, on the other hand, is one of the night sky's finest wonders. It shines two magnitudes brighter than the sky's most famous edge-on spiral, NGC 4565 (Caldwell 38), and its form is just visible in 7×35 binoculars. Of it Houston penned, "Some deep-sky objects offer beautiful, breathtaking visual experiences. NGC 55 is one such object." Luginbuhl and Skiff believed this galaxy to be "one of the most interesting and detailed in the sky." It is the highest on my list of galaxies that show inordinate amounts of detail in small telescopes.

Many sources claim that NGC 55 is the largest galaxy in the Sculptor Group, but this is not so. NGC 55 is a magnificent edge-on barred spiral spanning 30,000 light-years, so it is 27 percent larger than NGC 300 but 45 percent smaller than NGC 253. NGC 55 is receding from us, along with the other Sculptor Group members, at 140 km per second.

Dunlop discovered NGC 55 with his 9-inch f/12 reflecting telescope and cataloged it as his 507th object. "A beautiful long nebula," he wrote of it, "about 25['] in length; position north preceding [northwest], and south following [southeast], a little brighter towards the middle, but extremely faint and diluted to the extremities. I see several minute points or stars in it, as it were through the nebula: the nebulous matter of the south extremity is extremely rare, and of a delicate bluish hue. This is a beautiful object." John Herschel was equally impressed with the galaxy's appearance, seeing it on the night of October 4, 1836, as a "very long irregular crooked ray with 3 nuclei, the second of which appears to consist of stars." In photographs the galaxy appears most unusual. A dense, dusty lens of light dominates the western end of its ½°-long spindle. A lagoon of dark material separates this mass from its faint, splotchy eastern extension, which has a separate IC catalog number. Apparently this dark lagoon caused some problems years ago, as Harold G. Corwin (Infrared Processing and Analysis Center) explains on the NGC/IC Project Web page:

> IC 1537 is the east-southeast[ern] arm of NGC 55. It was first seen, described, and sketched by James Dunlop in the 1820s. [John Herschel] provided a more detailed description and sketch a decade later. Both clearly noted that the south-following [east-southeastern] end of the nebula was much fainter than the north-preceding [north-northwestern], and their estimated sizes (note the typo in Dunlop's paper: in place of 25 arcsec, read 25 arcmin) include the whole galaxy, not just the brighter portion. Furthermore, the fainter following [eastern] part is clearly shown in both published sketches.
>
> In spite of these published observations, [Lewis] Swift claimed this part of [NGC] 55 as his own discovery: "As Sir John Herschel does not mark it [NGC 55] with a sign as being a remarkable object, lends [sic] plausibility to the idea that it [IC 1537] was not seen even by him." And this after implying that Dunlop had certainly not seen the fainter eastern end.

As with NGC 300, deep infrared images have confirmed that NGC 55's disk contains an aged stellar population. Shell structures have also been observed, and their velocities support the idea that diffuse gas is being pushed into the galaxy's halo by supernova explosions and stellar winds. The galaxy is replete with ionized gas features like giant H II complexes, supergiant filaments, and very faint diffuse emission nebulae. Many of these features protrude well above the galaxy's plane, as does one very faint, fragmented shell extending 8,500 light-years above the disk. All of the protruding features are believed to be "chimneys" that allow hot gas from star-forming regions in the disk to funnel into the halo. Several chimneys are "capped" with clumps of ionized gas, one of which appears to be the site of recent star formation. None of this, of course, can be seen with the eye, but this kind of information does help fire our imaginations when we look at this remarkable object.

As noted above, NGC 55 is visible in 7×35 binoculars. But it is not visible to the naked eye (at least not to mine; I encourage others to try). As with NGC 300, I question its published magnitude. Lists often refer to its photographic blue magnitude (7.9), which also matches a 1991 photometric visual-band (V-band) estimate by the late Gérard de Vaucouleurs and his collaborators. My binocular estimate places the galaxy's visual magnitude a bit higher at 7.5. In binoculars the galaxy's spindle is oriented in more or less in the same direction as a 2°-long line of three roughly 7th-magnitude stars, and one roughly 8.5-magnitude star, to its east. I really needed averted vision to make out the spindle in binoculars. With a semidirect gaze the galaxy appears to have a slightly elliptical center with a hint of tapered edges.

At 23× the 4-inch immediately reveals NGC 55's splendor. At a glance the galaxy looks like a beautiful comet, with a graceful dust tail extending to the east-southeast and a small antitail to the west-northwest. The view becomes three-dimensional if you imagine a comet racing toward you at a slightly inclined angle. Make sure you spend time with this galaxy. The longer you look at it, the more you should see. Gradually, the nuclear region should begin to appear more irregular, and faint stars should begin to appear along its length. I like to imagine the galaxy as a squashed comet — like the one Carolyn Shoemaker discovered on photographic plates taken by her husband Eugene and by David Levy. Carolyn called that broken-up comet a "string of pearls" after seeing an image of it taken with the Spacewatch telescope. (That comet, Comet Shoemaker-Levy 9, ultimately smashed into Jupiter.) As I see it, NGC 55 is an extragalactic String of Pearls. Curiously, under averted vision its eastern segment (IC 1537) seems to pulsate with dim, glittering lights. The effect is most dramatic when I look toward the nucleus at 23× but keep my attention on IC 1537. It's like a flashing signal reminding me to use higher magnification.

At medium magnification NGC 55 is stunning. Ribbons of dark material lace its entire nuclear region; the "comet" has been snared in a black net. The scene recalls M82, the fabulous starburst galaxy in Ursa Major. Concentrate on NGC 55's core with averted vision and it should scintillate with faint fuzzes and stars. One phenomenal rift slices through the eastern end of the central bulge before it curls around the

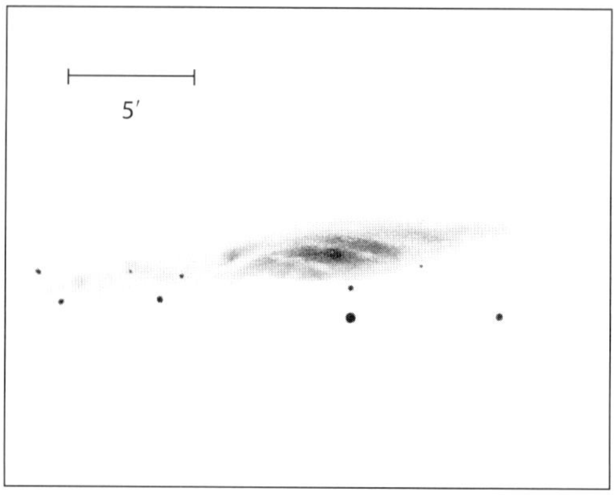

dense core like warped metal. It's as if tiny explosions have blown apart the hull of this extragalactic traveler, which is tipping to the northwest and sinking into a black sea; any turbulence in our atmosphere will cause the dimmest parts to ripple as if against this "ship's" hull. The galaxy's fainter eastern segment is also highly fractured and dappled with clusters of faint stars. This ethereal beauty recalls the words quoted by Strepsiades in Aristophanes's *The Clouds*: "the impetuous tempests, which float through the heavens . . . with aerial wings, loaded with mists."

Despite NGC 55's modest overall surface brightness, its bright nuclear region takes high magnification well. In fact, I think the nuclear region looks best at high power. Here the dark rifts can be seen weaving through the superimposed stars and knots like streams covered with black ice flowing through banks of snow. The sight must be glorious in larger telescopes. Through her 17½-inch f/4.5 reflector Barbara Wilson saw NGC 55 as a "long streak narrower at one end with mottling in the central nuclear region. The galaxy extends beyond the field of the eyepiece at 95×. Incredible."

NGC 55 is without a doubt one of the most spectacular galaxies available to amateurs. But observers at midnorthern latitudes have it tough, since the galaxy is so low in their skies, even when crossing the meridian. Consider Houston's views: "While traveling in Central America I turned my 4-inch rich-field reflector to this galaxy and could see mottling all along its length. At home in Connecticut, however, it's a different story; I feel grateful just to see it with my 4-inch Clark refractor. Even so, on exceptional nights I've glimpsed some mottling with a 4-inch off-axis reflector." If you're ever far south with a telescope at your disposal and Sculptor is available for viewing, do not hesitate; turn your sights immediately to this extragalactic wonder and enjoy the beauty of our hobby.

71

71

NGC 2477
Type: Open Cluster
Con: Puppis

RA: $07^h 52.2^m$
Dec: –38° 32′
Mag: 5.8
Diam: 20′
Dist: 3,700 light-years
Disc: Abbé Nicolas Louis de Lacaille, listed in his 1755 catalog

J. HERSCHEL: Superb cluster, [gradually brighter toward the middle], 20′ diameter, much more than fills the whole field. Stars [of] 10th and 11th [magnitude] all nearly equal. (h 3103)

GC/NGC: Remarkable, cluster, bright, rich in stars, large, little compressed, a star of 12th magnitude.

NGC 2477 IN PUPPIS IS A GLORIOUS OPEN cluster full of subtle detail that should quench the visual thirst of all observers. I listed it as the fifth of "Twenty spectacular non-Messier objects" in *Deep-Sky Companions: The Messier Objects*. The cluster is easy to find. Look 11° southeast of the Greater Dog's tail star, 2nd-magnitude Eta (η) Canis Majoris, and just 2½° northwest of 2nd-magnitude Zeta (ζ) Puppis. But be careful, because our target is also 1½° southeast of the spectacular magnitude-2.8 open cluster NGC 2451. Together NGC 2477 and NGC 2451 constitute one of the most stunning open-cluster pairs in the heavens. Barbara Wilson calls them the "southern M46 and 47." "What is so strange," Wilson says, "is that NGC 2477 and 2451 are visible at the same time [and] in the same constellation as M46 and 47 yet so few people know about them. Stranger yet is that NGC 2477 is brighter, at magnitude 5.8, than M46 is at magnitude 6.1; it just lacks a planetary to show off. [The 11th-magnitude planetary nebula NGC 2438 is superposed on M46.] But NGC 2477 is a richer cluster. I guess it suffers from latitude sickness." What a catchy phrase, "latitude sickness." But it's true. If you live on the north 45th parallel, NGC 2477 can rise no more than 6½° above your southern horizon. Toss in some light pollution and it's a goner. That's why the Winter Star Party is so popular with northerners, for it enables them to dip their telescope tubes deep into Puppis, where these clusters shimmer in all their glory.

The 18th century was a golden age for exploring the far-southern skies. Abbé Nicolas Louis de Lacaille (1713–62) discovered NGC 2477 during an expedition to the Cape of Good Hope in southern Africa. From April 19, 1751, to March 8, 1753, Lacaille cataloged, among other things, 42 "nebulae." His list was by no means complete, though it did suggest what could be accomplished by sweeping the southern heavens with an eye open for nebulous objects. Lacaille placed his discoveries into three classes: "nebulae without stars" (Class I);

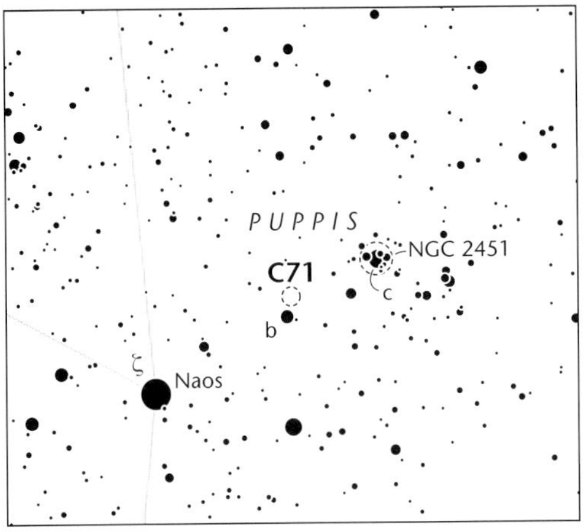

"nebulous clusters" (Class II); and "stars accompanied by nebulosity" (Class III). Of his work Lacaille wrote: "I have found a great number of the three types of nebulosities in the southern part of the sky, but I do not flatter myself to think that I have noticed them all, especially those of the first and third types, because they can only be perceived after twilight and in the absence of the moon. However, I do hope that the list is passably complete in regard to the most remarkable of the three types."

Lacaille's list of nebulae first appeared in 1755, in the *Memoires* of the French Royal Academy of Sciences (16 years before Charles Messier published his first list of 45 objects in the same journal) and was published posthumously as a catalog in 1763. NGC 2477 is the third object in Lacaille's Class I and hence is denoted Lac I-3. This means he did not fully resolve the cluster with his modest telescopes. He simply described NGC 2477 as a "[g]reat nebula 15['] × 20′ in diameter."

In the February 1960 issue of *Sky & Telescope*, Harvard University historian Owen Gingerich related the following report that Lacaille gave to the French Royal Academy:

> The so-called nebulous stars offer to the eyes of the observers a spectacle so varied that their exact and detailed description can occupy astronomers for a long time and give rise to a great number of curious reflections on the part of philosophers. As singular as those nebulae are which can be seen from Europe, those which lie in the vicinity of the south pole concede to them nothing, either in number or appearance. I am sketching out this description and list to serve as a guide for those with the equipment and leisure to study them with larger telescopes. I would have greatly desired to present something more detailed and instructive in this article, but with ordinary refractors of 15 to 18 inches [in length] such as I had at the Cape of Good Hope, I had neither adequate nor convenient enough instruments for this kind of research. Those who do take the trouble to see what has occupied me during my foreign sojourn will see well enough that I did not have time to make that sort of observation.

Lacaille's work did not go unnoticed. Magnanimous Messier credited Lacaille for his discoveries in his now-famous 1771 catalog (which wasn't published until 1774).

James Dunlop also observed NGC 2477 from Parramatta and included it as No. 535 in his 1827 catalog, describing it as a "pretty large faint nebula, easily resolvable into small stars, or rather a cluster of very small stars, with a small faint nebula near the north preceding [northwestern] side, which is rather difficult to resolve into exceedingly small stars."

Observers at tropical or southern latitudes will find NGC 2477 in a naked-eye Cygnus- or Grus-like asterism, with Canopus (Alpha Carinae) serving as the bird's nose and Zeta Puppis the tail; Tau (τ) and Sigma (σ) Puppis are the two brightest stars in the bird's long, crooked neck. Pi (π) Puppis, which is part of the naked-eye open cluster Collinder 135, marks the northern wingtip, while beautiful Gamma (γ) Velorum marks the southern one. Of course, NGC 2451 also is obvious to the naked eye in the northern wing. The challenge, however, is to see NGC 2477, our magnitude-5.8 target, with the unaided eye. From Hawaii I find it *just* visible. The difficulty is that the cluster lies less than ½° north-northwest of magnitude-4.5 b Puppis. (Note that b Puppis is marked as such on the original *Sky Atlas 2000.0* and in *Uranometria 2000.0.* However, the star is not labeled in the second edition of *Sky Atlas 2000.0,* and the *Millennium Star Atlas* tags it as QZ Puppis, an eclipsing binary with a period somewhere between 1 and 9 days.) The cluster also is large (nearly 70 percent as wide as the full Moon) and has a moderate surface brightness.

I find the best way to see NGC 2477 with the naked eye is first to take several deep breaths, then to stare at b Puppis with direct vision while using averted vision to spy the cluster. This is harder than it sounds. It takes time. After several tries, though, NGC 2477 should pop into view. I have no success when gazing directly at the cluster.

NGC 2477 has long been one of my all-time favorite binocular objects because its compact form appears strikingly like that of a tailless comet. The view at 23× in the 4-inch, however, is absolutely gorgeous. I immediately fell in love with the telescopic view of NGC 2477 because even at low power the open cluster looks like a finely resolved globular. In his 1930 book *Star Clusters,* Harvard College Observatory astronomer Harlow Shapley wrote that NGC 2477 is "the richest of galactic clusters; or perhaps it is the loosest of globular clusters," a description that recalls M71 in Sagitta. But unlike M71, which is a true globular, NGC 2477 is an open or galactic cluster with a paltry radial velocity of about six km per second. It's a shame we cannot simultaneously compare and contrast NGC 2477 and M71

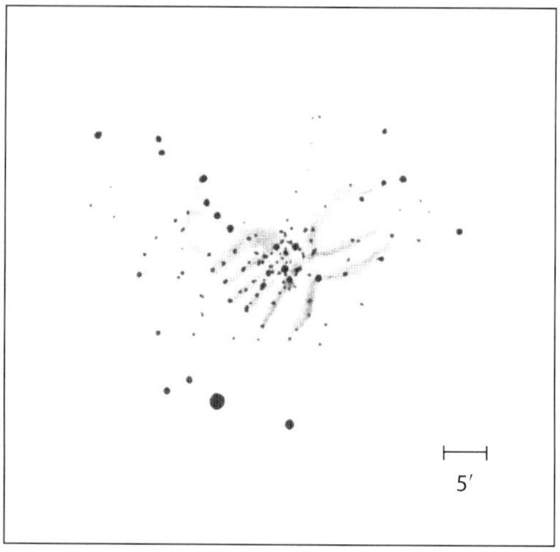

under a dark sky from northern sites, but from temperate southern latitudes the two clusters do appear very briefly just above opposing horizons after sunset in November. As NGC 2477 rises in the east, M71 starts to set in the west — just like the Great Hunter, Orion, and his nemesis, Scorpius.

Brent Archinal says that NGC 2477 has 1,911 known members spread across 22 light-years of space. Since the cluster is seen against a rich Milky Way background, it is difficult to separate its stars from others in the field. The cluster's brightest member shines at magnitude 12, and many others are not much dimmer. NGC 2477 has an estimated 4,400 solar masses and a metal abundance like our Sun's. In 1995 the Hubble Space Telescope was used to image stars as faint as 26th magnitude in the cluster. A study of these stars suggests that the cluster is 1.3 billion years old, about twice as old as the Hyades (Caldwell 41). The Hubble images also revealed four white-dwarf stars, and it's believed that perhaps 120 remain to be discovered. The fraction of a cluster's total mass that is stored in faint white dwarfs is important for determining the evolution of star clusters as well as the masses of our galaxy's disk and halo.

NGC 2477 appears finely resolved under close inspection through the 4-inch at 23×. One linear ridge of stars runs from northeast to southwest, with several spidery arms branching off of it to the southeast and northwest. Each arm is separated from its neighbors by a dark lane. Averted vision reveals ribs of starlight, like an X-ray of a human torso with a spine and a rib cage. Even at 72× the cluster looks magnificent, like brass knuckles studded with diamonds. Some streams form looplike strands. Especially noticeable are a dark hole in the cluster's northeastern quadrant, a dark ellipse on its northwestern side, and a narrow pool of blackness to the southeast. Let your imagination fly at this power. I saw all sorts of fun sights, like a spider's exoskeleton and an aerial view of eroded valleys among hills of starlight. It also looked as if the ebb and flow of dark waters had carved away the southern half of the cluster, leaving in their wake ripples in the stellar sands.

At high power the cluster loses many of its star chains. But the longer I look, the more patterns I see. Near the cluster's core are two stellar concentrations immersed in stellar haziness — the telltale sign of numerous unseen cluster members. Use averted vision and see if you can't make out a tiny "keyhole," a splotch of darkness guarded by three dim stars in the cluster's core. Now relax your gaze and wiggle your telescope tube. Many dark patches and streaks should vibrate into view.

Before leaving this wonderful sight, don't forget to check out NGC 2451 — number 16 in the aforementioned list of spectacular non-Messier objects. This bright open cluster spans a little more than 1½ Moon diameters, and at least 30 of its colorful stars are visible in 7×35 binoculars. These stars seem to burn in adoration of the cluster's brightest member, c Puppis, which shines at magnitude 3.6. I see NGC 2451 as a scorpion approaching me nearly head-on, with two outstretched claws and an upright tail ready to sting. It's fitting that this stinging scorpion travels the sky with NGC 2477, because M71 in the celestial Arrow sets just as NGC 2477 and NGC 2451 are rising. Now we can create our own myth. Ages ago Sagittarius shot Sagitta, the Arrow, toward Puppis, killing our tiny arachnid. Feeling sorry for the innocent creature, Zeus placed the cluster in the company of NGC 2477, as far as possible from the Arrow in the night sky. Thus the Arrow will never again slay the stinging scorpion cluster — just as big brother Scorpius will never again sting the mighty Hunter, Orion.

73

NGC 1851
Type: Globular Cluster
Con: Columba

RA: 05^h 14^m 06.3^s
Dec: –40° 02′ 50″
Mag: 7.1
Diam: 12′
Dist: 39,400 light-years
Disc: James Dunlop, included in his 1827 catalog

J. HERSCHEL: Superb globular cluster; all resolved into stars of 14th mag.; very suddenly much brighter in the middle to a blaze or nucleus of light; [diameter] in RA = 15 seconds of time. Difference of left and right eyes in resolving this cluster very remarkable. Returning from the left to the right eye, the object (in comparison) appears as if glazed over with a kind of dull film. (h 2777)

GC: Globular cluster of stars, remarkable, very bright, very large, round, very suddenly very bright in the middle, resolvable.

NGC: Globular cluster of stars, remarkable, very bright, very large, round, very suddenly very bright in the middle, well resolved, clearly consisting of stars.

COVERING 270 SQUARE DEGREES OF SKY, Columba, the Dove, is the heavens' 54th-largest constellation. It's a sprightly little group of stars that pays homage, in part, to the bird that Noah released from his Ark after the great flood; the dove returned with an olive branch in his beak to alert Noah that the waters were receding. Today the bird stands on the western banks of the river Milky Way, where it has been all but forgotten by Northern Hemisphere observers. Yet the Dove's heart — Alpha (α) Columbae (Phact), a suspected variable star that burns at magnitude 2.6 — lies at a respectable declination of –34°, which places it 11° above the horizon at upper culmination when seen from a latitude of 45° north (and higher still from more southerly locations). The constellation lies directly south of Lepus, the Hare, which is itself directly south of Orion, and three of Columba's brightest stars — Alpha, Beta (β), and Epsilon (ε) Columbae — form an obtuse 4°-long triangle visible to the unaided eye a fist's width west-southwest of 3rd-magnitude Zeta (ζ) Canis Majoris (Furud), the Great Dog's hind paw. In his *Star Names: Their Lore and Meaning*, Richard Hinckley Allen says that Phact was important to Egyptian temple worship, and a dozen temples might have been oriented to this star.

While sweeping Columba with a 9-inch reflector at Parramatta, New South Wales, James Dunlop picked up the fine nebulous glow now known as NGC 1851 nearly 8° southwest of Alpha Columbae and cataloged it as his 508th discovery:

> An exceedingly bright, round, well-defined nebula, about 1½′ diameter, exceedingly

> condensed, almost to the very margin. This is the brightest small nebula that I have seen. I tried several magnifying powers on this beautiful globe; a considerable portion round the margin is resolvable, but the compression to the centre is so great, that I cannot reasonably expect to separate the stars. I compared this with the 68 *Conn. des Tems*, and this nebula greatly exceeds the 68 in condensation and brightness.

Dunlop's mention of the "68 *Conn. des Tems*" refers to the 68th object — a 7th-magnitude globular cluster in Hydra — cataloged by Charles Messier and listed in the French almanac *Connaissance des Temps* in 1783. Like M68, Dunlop's new object is a 7th-magnitude globular cluster, and it is the brightest and most intriguing deep-sky object in Columba. Some 54,500 light-years from the galactic center, NGC 1851 belongs to a breed of old globular clusters in the galaxy's halo. By studying the chemical composition, motions, and other physical characteristics of globular clusters, astronomers have learned much about their evolution. For instance, it appears that the oldest globular clusters appear to populate the halo while the younger ones lie closer to the plane of the galaxy. Globular-cluster ages range from about 12 to 16 billion years. NGC 1851 is 14 billion years old. Its age (comparable to that of the universe) is mirrored in its low metallicity. Each of the cluster's member stars contains, on average, about 1⁄17 as much iron (per unit of hydrogen) as does our Sun, attesting to their formation from primordial gases that had undergone little nuclear processing within previous generations of stars. The cluster's integrated spectral type is *F*7, and its radial velocity has been measured at 320 km per second.

Globular clusters are dominated by aged red-giant stars, and when they are not too poor in "metals" they contain RR Lyrae variables, which astronomers use to measure distances. In a 1998 issue of the *Astronomical Journal*, Alistair R. Walker (Cerro Tololo Inter-American Observatory) revealed that 29 of the cluster's RR Lyrae variables yield a distance of 39,800 light-years, in very close agreement to William Harris's distance for the cluster, which appears in the table on page 288. NGC 1851's true physical diameter, then, is about 140 light-years. Mixed in with NGC 1851's old stars is an apparently younger population, which includes a clan of 13 helium-burning rebels called blue stragglers. (Blue stragglers form either when double-star systems slowly coalesce or when two unrelated stars collide.) NGC 1851 is also the site of an intense X-ray source that could be signaling the presence of a black hole. The source was one of the first discovered in a globular cluster, and is as old as satellite X-ray astronomy. The X-ray source remained unidentified at optical (visual and near-ultraviolet) wavelengths for a quarter century, but a 1996 report in the *Astrophysical Journal* revealed that the Hubble Space Telescope had identified a 21st-magnitude candidate counterpart only 2″ from it.

NGC 1851 resides in a star-poor region of sky, so it can be a challenge to locate. I suggest starting with Phact. Confirm with binoculars that Phact is the northern apex of a triangle with Beta and Epsilon Columbae. Now move your binoculars about 3° south-southwest of Epsilon Columbae and you should see a right triangle of three 5th- to 6th-magnitude stars. Center your finderscope on the southernmost star in that triangle (it is an irregular variable) and look in your telescope's low-power eyepiece. You will see that that magnitude-5.7 star is the brightest member of a 30′-wide triangle with two roughly magnitude-7.5 stars. NGC 1851 is 1¼° southwest of the westernmost star in that triangle (which, together with the brighter star, almost points directly to the cluster). All these stars are plotted on our finder chart below. NGC 1851 should be obvious even in 7×35 binoculars as a slightly swollen "star" of pearly light. From Hawaii the cluster is just visible to the unaided eye with averted vision.

At 23× in the 4-inch the cluster looks like the bright head of a comet. With direct vision a faint coma of light surrounds a dense, round core. Switch to averted vision and traces of arms show in the outer coma. Although the cluster spreads out to 12′ in photographs, visually it's half that size. Now focus your attention on the bright 2′-wide central condensation, which should appear starlike with a glance. The longer you look, the more the core will break down into individual elements; it's like opening packages within packages of stars. The halo immediately surrounding the core reveals definite arms and some resolved stars at the periphery of the halo. Do you see a yellowish cast to the core?

Much can be seen with a 4-inch at 72×. Although the cluster's horizontal-branch stars shine at 16th magnitude, its brightest members are three magnitudes brighter. This means that NGC 1851 can be partially resolved in small telescopes. Try looking at the cluster with your left eye, then your right eye, and see if there's a difference. That's what John Herschel did during his second observation of this object. He also found the stars "barely visible in strong twilight." I find the cluster looks asymmetric to the north, with two fuzzy wings of faint stars, many of which are resolved. The fuzzy wings remind me of those visible in M13, the Great Hercules Cluster. Two spidery arms extend eastward, jutting out of a parabolic envelope of scintillating haze. The western side of the outer halo appears detached from the core, separated by a lane of darkness.

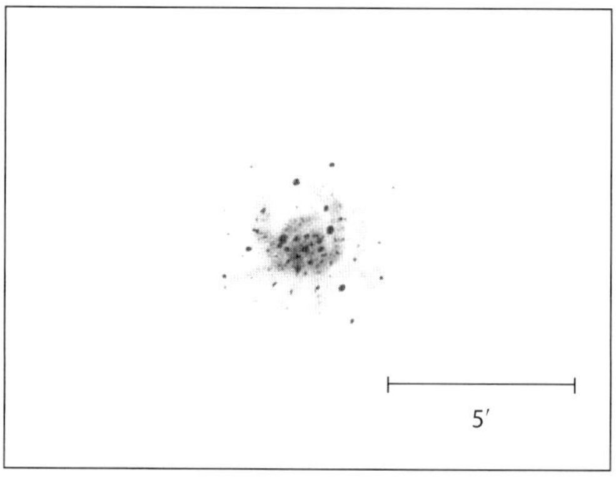

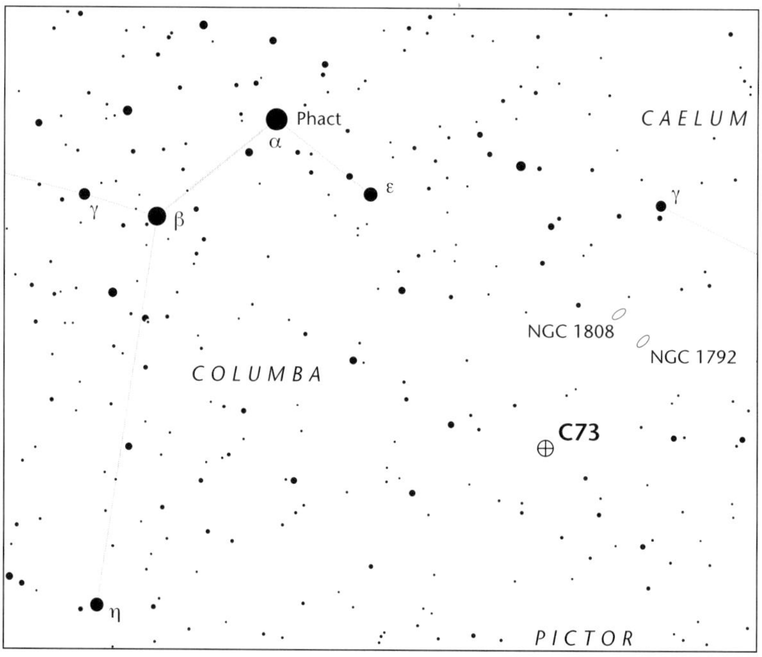

The asymmetry remains at high power. The core shatters into a sharp central pip surrounded by fractured clumps of starlight. Two prominent ripples can be seen to the west along with hints of stellar rays. A spire of mottled starlight extends to the east and is bordered to the south by a ray of unresolved stars. And a pool, empty of starlight, occupies the northern halo. The overall impression is that of the great Hercules globular M13 seen through a fog. At the 1994 Winter Star Party, Barbara Wilson observed NGC 1851 through an 18-inch f/6 reflector and described an intensely bright but mottled core surrounded by a well-resolved and very loose halo.

Just 2¾° northwest of NGC 1851 is a double treat for devotees of extragalactic targets: the twin galaxies NGC 1808 and NGC 1792. These slightly inclined magnitude-9.9 spirals are separated by a mere ¾° and have bright central lenses that should be visible even in binoculars. What do you think?

74

74

Southern Ring; Eight-Burst Nebula
NGC 3132
Type: Planetary Nebula
Con: Vela

RA: $10^h 07^m 01.8^s$
Dec: –40° 26′ 12″
Mag: 9.4; 9.2 (O'Meara)
Dim: 84″ × 53″
Dist: ~2,000 light-years
Disc: John Herschel, 1835

J. HERSCHEL: [Observed 2 March 1835] Planetary nebula, very large, very bright, elliptic; has in it a 9th mag[nitude] star somewhat excentric. Its light is exactly equable, ie. not increasing towards the middle; yet I cannot help imagining it to be closely dotted. It is just like a star out of focus in certain states of the mirror and atmosphere. Three stars near, a = 9th mag[nitude]; b = 9th mag[nitude]; c = 14th mag[nitude]; A very extraordinary object. (h 3228)

GC / NGC: *Remarkable*, planetary nebula, very bright, very large, little extended with a star of magnitude 9 in the middle, 4″ diameter.

NGC 3132 IS ONE OF THE MOST FANTASTIC planetary nebulae accessible to small telescopes. In shape, size, and brightness it is comparable to M57, the Ring Nebula in Lyra, and for that reason it is commonly called the Southern Ring. The main difference between the two objects lies with the brightnesses of their respective central stars: NGC 3132's shines at 10th magnitude, while M57's glimmers around 14th. In truth the 10th-magnitude, type-*A*2 central star in NGC 3132 is not hot enough to saturate the surrounding gases with ultraviolet radiation and make them fluoresce — not without help from a companion, that is. This companion was discovered by Luboš Kohoutek and Svend Laustsen with the 3.6-meter telescope at the European Southern Observatory in Chile. The 16th-magnitude star was found 1.65″ away from its brighter partner at a position angle of 226.3°. This dim partner has an earlier spectral type than its 10th-magnitude companion, and it exhibits an unusually high surface temperature, possibly 100,000° Kelvin. It also lies more directly along the nebula's symmetry axis, so it, and not the 10th-magnitude star, is without question the hot, ionizing dwarf star that blew off the surrounding gas shell. A Hubble Space Telescope image of the nebula clearly shows the two stars near the center of the nebula. The two central stars probably form a true binary pair, and their mutual orbit would explain the nebula's wide and "wobbly" ellipsoidal structure. In the future the brighter star, which is in an earlier evolutionary stage, will probably eject its own planetary nebula, adding more drama to this already glorious spectacle.

In detailed ground-based photographs, the Southern Ring looks like a rosette of overlapping ellipses, each skewed at a slightly different angle. This appearance has led to its other popular nickname, the Eight-Burst Nebula. Astronomers interpret the various forms of planetary nebulae as being different manifestations of one three-dimensional model. In that model, gas flows from a central star and funnels out at right angles to a denser torus of material girdling the star. Ring nebulae like M57 and NGC 3132 manifest themselves when we happen to look down the gas-flow axis. The outer, cooler halos are remnants of the slow wind cast off by the hot central star when it was still a red giant. If we could see ring nebulae from the side, they would probably look much like the planetary nebula M76 in Perseus.

High-resolution images from Hubble also reveal the Eight-Burst features as well as several new ones, including a wide pillar along the nebula's major axis. These features are visible in the light of thrice-ionized oxygen (O III) and hydrogen-alpha. The hydrogen-alpha images also show a thin equatorial band of material girdling the main nebula around its waist, and a fine, jetlike feature extending radially from the girdle. The girdle is made of carbon-rich dust particles that have condensed out of the expanding gases. Astronomers believe that in the distant future the dust could contribute to the formation of new stars and planets. The HST image of all this ethereal splendor sets the imagination free; it's as if we're peering at the Sun (the central star) through a rusted porthole (the ring) of a ship in the shallows (the hot vapors in the center of the ring).

If we accept a distance of 2,000 light-years for NGC 3132, the planetary's true dimensions

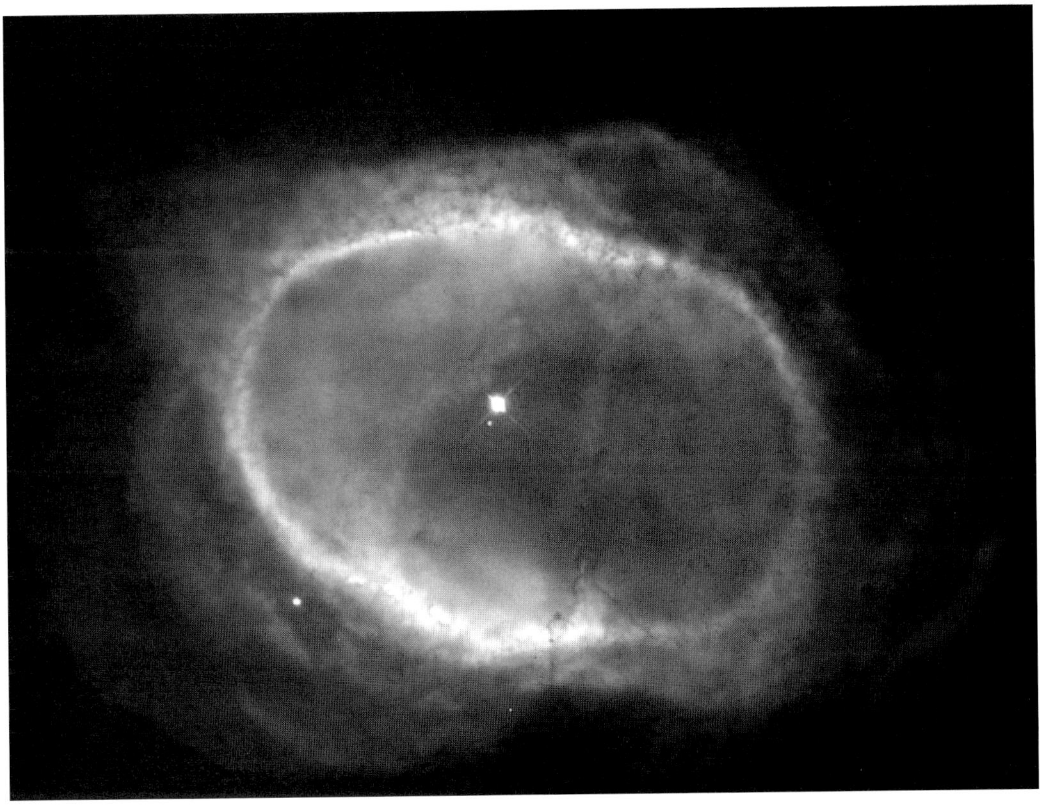

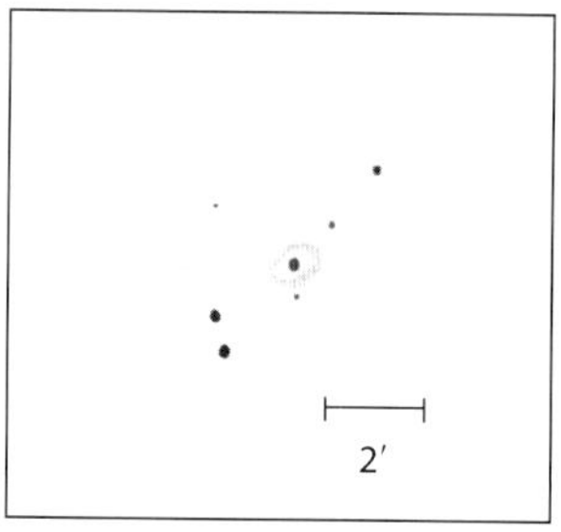

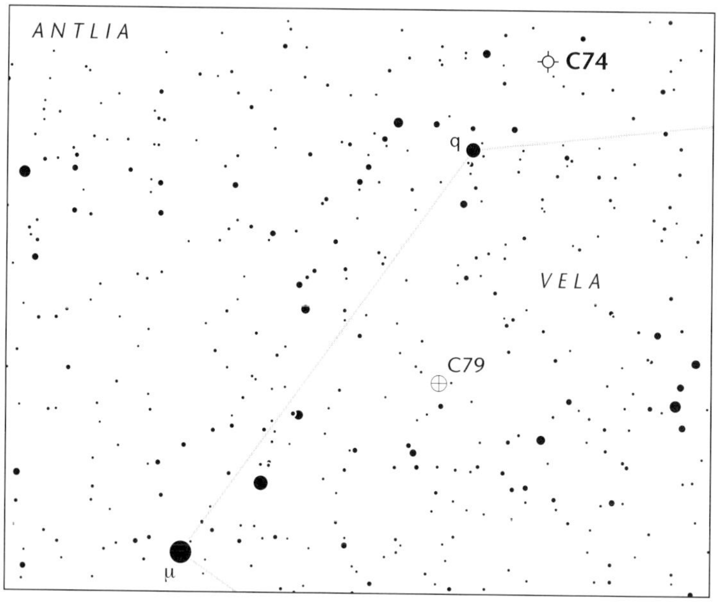

are 0.8 × 0.5 light-year. Its gases are expanding away from the central star at a speed of 14 km per second.

NGC 3132 lies in a rather remote section of Vela, just south of its border with the dim constellation Antlia, the Air Pump. If you live at a southerly locale and have a dark sky, finding the nebula will not be a problem. Start at 2nd-magnitude Lambda (λ) Velorum. A little more than 12° east-northeast of Lambda Velorum is magnitude-3.8 q Velorum, which forms the southwestern corner of a 2°-long equilateral triangle with a 5th- and a 6th-magnitude star. (Confirm this with your binoculars.) NGC 3132 is 2¼° northwest of q Velorum and 1¼° west (and ever so slightly south) of the triangle's northern star.

If your skies are not that great, or if the object is low to the horizon, you can find the planetary with an equatorial mount. First locate 2nd-magnitude Zeta (ζ) Puppis with a wide-field eyepiece and center it in your scope. Zeta is at –40° exactly, so all you have to do is use your setting circles to move your telescope 26.2′ south and $2^h\,3^m$ east of Zeta Puppis. The planetary should be obvious. At 23× in the 4-inch NGC 3132 looks like a 9th-magnitude star and warrants immediate attention at higher powers. Several sources claim the planetary's visual brightness is magnitude 8.2. But this is not true. The error can be traced to Robert Burnham Jr.'s *Celestial Handbook*, which failed to clarify that 8.2 was the nebula's *photographic* magnitude; NGC 3132's apparent *visual* brightness is a full magnitude fainter. (This is precisely the opposite of what one might expect. But there is a reason for this. To show the nebula's faint features, photographic exposures had to be long, and hence they would overexpose NGC 3132's bright central star, exaggerating the nebula's total brightness. The same problem can be seen in NGC 246, which has three stars superimposed on its nebulosity.) This error has led others to assume that the planetary is brighter than it truly is. Most plausible published magnitudes range from 9.2 to 9.7. I observed NGC 3132 in April 1997 using 7×50 binoculars and, blissfully unaware of the controversy, I made an independent visual estimate of magnitude 9.2.

Not much is gained by boosting the magnification of the 4-inch to 72×, though the nebula's nature is clearly revealed by a slightly swollen body like a star just out of focus. With a prolonged look, the object displays a very bright center (the 10th-magnitude central star) surrounded by a tight, tiny halo with an irregular or dappled texture. Depending on how you hold your head or move your eye, spectral glows brighten and fade around the "halo." At high power NGC 3132 becomes immediately intriguing. My first view was of a ring of knots with some skewed outer envelopes surrounding a bright central star. All these features appear to be at the threshold of visibility in the 4-inch. A knot will pop into view, then vanish quickly, just as another curious haze or enhancement gradually materializes, then dissipates. In a sense, this illusory action reminded me of the "rotating" beads of light in M57. Most curious, however, is the appearance of a double shell. The outer shell seems slightly skewed with respect to the inner ellipse. I can also see what appears to be a tiny nebulous hook on the northwestern side of the ellipse, where Christian Luginbuhl and Brian Skiff, looking through a 12-inch reflector, note that the "otherwise ring-like structure is broken." A faint star, perhaps 13th magnitude, borders the ring to the southwest.

John Herschel observed this nebula several times with his 18¼-inch reflector. On one night he saw the nebula as "velvety, or like infinitely fine dust." Ernst Hartung saw no color in NGC 3132 and called it a "bright white annular planetary nebula . . . without any of the bluish tint usual with planetary nebulae." Steve Coe, observing NGC 3132 with a 13-inch f/5.6 instrument, recorded a hue that was "either grey or light green on every occasion I have observed it." Like Hartung, I saw no color.

75

75

NGC 6124
Type: Open Cluster
Con: Scorpius

RA: $16^h 25.3^m$
Dec: –40° 40′
Mag: 5.8; 5.3 (O'Meara)
Diam: 40.0′
Dist: 1,500 light-years
Disc: Abbé Nicolas Louis de Lacaille, listed in his 1755 catalog

J. HERSCHEL: Cluster, bright, large, loosely scattered, not much compressed in the middle, fills nearly a field, consists of about 50 or 60 stars of 9th to 11th mag[nitude]. (h 3626)

GC/NGC: Cluster, bright, large, pretty rich, little compressed in the middle, stars of magnitude 9 to 11.

FEW SIGHTS IN NATURE CAN MATCH THE splendor of the Milky Way stretching across the heavens like moonlit holly. Here is infinite complexity in structure, infinite mystery in design. It is limitless, untouchable, a vision to be grasped by our minds and hearts. And it is here that we continually turn our telescopes to admire a multitude of visual splendors. One aspect of the Milky Way that has continually captured the attention of skywatchers has been the small, cometlike nebulosities that splinter into magnificent clusters of stars with the slightest optical aid. The dazzling naked-eye duo M6 and M7 in Scorpius are two of the finest examples of these easily resolved "clouds." On the other end of the spectrum are clusters too faint to be seen with the naked eye — the ones that manifest themselves with a sweep of the binoculars or a small telescope. If you follow the Milky Way southwest from the Tail of Scorpius toward Norma, you'll find a dozen or more of these tiny clusters in an area only 10° across. But there is one curious object on the western outskirts of the Scorpius Milky Way, straddling this cluster corridor, which seems to exist in a visual limbo.

Open cluster NGC 6124 marks the western apex of an isosceles triangle with $Mu^{1,2}$ ($\mu^{1,2}$) and $Zeta^{1,2}$ ($\zeta^{1,2}$) Scorpii in the Scorpion's Tail. It is also halfway between the Zetas and Eta (η) Lupi. It occupies a dim patch of Milky Way surrounded by blobs of dark nebulosity. The cluster's cataloged magnitude (5.8) seems bright, but it is also deceiving. NGC 6124 is 1.3

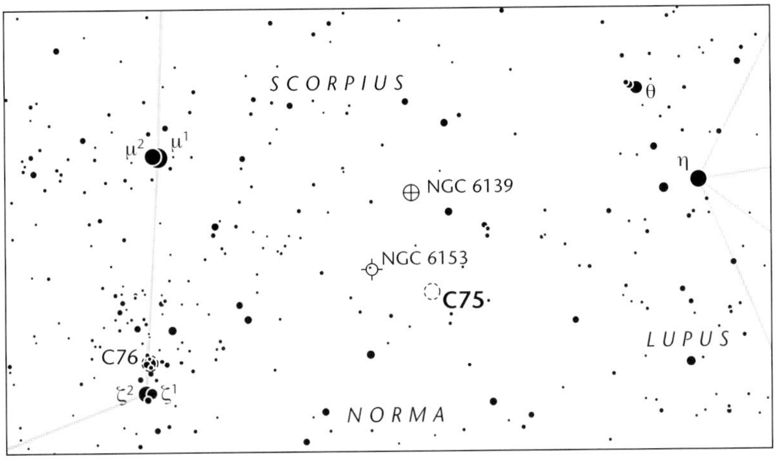

times larger than the full Moon. In other words, take the light of a 6th-magnitude pinpoint of light and smear it out across 40′ of sky — against the Milky Way, no less. Now consider that none of the cluster's 100 known members shine brighter than 9th magnitude. NGC 6124 is one of those objects that can take you by surprise under a dark sky, for it is not immediately obvious to the naked eye, but if you happen to be lying back in a chaise longue and admiring southern Scorpius for aesthetic reasons, you might find yourself springing up convinced that you've just discovered the large, diffuse head of a naked-eye comet. And so it was in the 18th century when Louis de Lacaille came across this object from the Cape of Good Hope — not with his unaided eye but with his ½-inch 8× telescope. "It resembles a big comet without [a] tail," he wrote, adding it to his list of Class I objects ("nebulae without stars"). Some catalogs credit James Dunlop with the object's discovery, since he was the first to describe its true nature: "A round cluster of small stars of nearly equal magnitudes, about 12′ diameter, considerably congregated to the centre, not rich in small stars. . . . [T]here is no nebula." Dunlop listed the cluster as the 514th object in his 1827 catalog, but credit for its discovery should really remain with Lacaille.

When I first saw NGC 6124 I did jump up with a start. While observing the cluster with 7×35 binoculars on September 9, 1996, I noticed not one, but two, objects: a large, round haze some 20′ in diameter and another round haze separated from the first by a narrow lane of darkness. The two glows lay to the northeast and the southwest. I tried to identify the two glows on star charts but found only one: NGC 6124. "A comet!" flashed through my mind. Clouds began moving in, so I went to the 4-inch and confirmed the existence of an ill-defined patch of light northeast of the main cluster before clouds covered the sky. That night many thoughts went through my mind. Is this how Lacaille felt when he first encountered NGC 6124? Much like Lacaille, I was using the rough equivalent of a ½-inch instrument at 7×. And, like him, I had chanced upon a large "nebula" that appeared to lack stars. Perhaps, I thought, this secondary haze was part of NGC 6124. But if that were true, why did Dunlop and some modern-day observers call the cluster "round" instead of "irregular," or "dumbbell shaped," or "a double cluster forming a figure-eight pattern"? I checked various sources for the size of NGC 6124 and all agreed that it was between 20′ and 29′, which placed the secondary fuzzy oval only partly inside the established object's perimeter.

The next night fell clear, so I checked the object for movement but found none. It wasn't a comet. Figuring the entire sky had been covered well enough by now for a new cluster's discov-

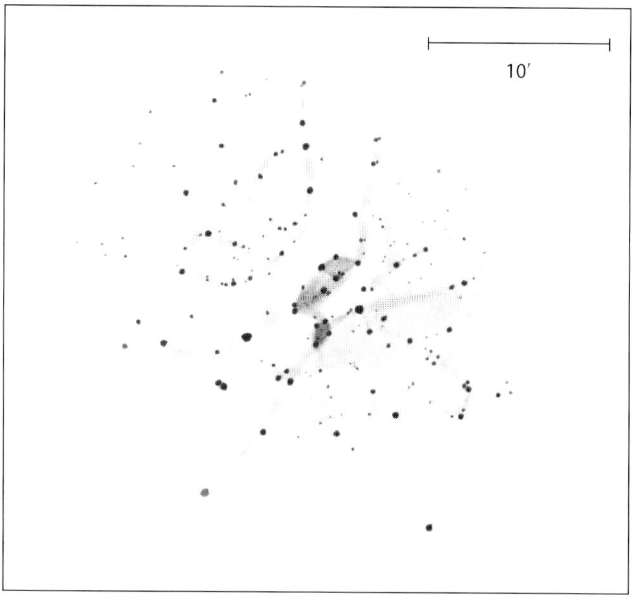

ery to be unlikely, I presumed the glow to be nothing but a spurious patch of bright Milky Way. Through the 4-inch NGC 6124 appeared about ⅓ times larger than this "patch," which was contained by an obvious oval-shaped asterism of 9th- to 12th-magnitude stars. Closer inspection revealed that that single oval was comprised of two fainter ovals side by side, each scintillating with the light of fainter suns. I checked the *Astronomical Journal*, the *Astrophysical Journal*, and the *Monthly Notices* of the Royal Astronomical Society but did not find a reference to this "cluster." Apparently no modern research has been done on it or the surrounding field. The situation remained at a standstill until the year 2000, when Brent Archinal revised the cluster's size from 29′ to 40′. When I plotted his position for the cluster's center (provided in the table below) and placed a circular 45′-diameter template around it, I was pleased to discover that the new dimensions included this curious patch of dim starlight and the bright ovals of stars. But why, then, did I see two distinct "clusters"?

The answer may be simple. On *National Geographic Society – Palomar Observatory Sky Survey* prints I noticed an almost perfect semicircle of darkness (open to the northeast) slicing through the cluster's northeastern quadrant, just where I saw one cluster ending and the other beginning. This dark semicircle outlines the southwestern edge of the large oval of stars I detected in binoculars. Of course, the other possibility is that there really is just a random patch of Milky Way here. Check this object out and decide for yourself. The reason NGC 6124 looks so large and diffuse is because it is relatively nearby (1,530 light-years). If NGC 6124 indeed spans 40′, then its true physical diameter is 18 light-years.

I find that NGC 6124 looks best in binoculars. At a glance it is a circular object with a dense 10′ core. With time, though, the core begins to break up. Soon it's a mottled haze with prominent clumps of stars to the northeast, southwest, and east. Particularly absent is any notable mottling to the west. So the cluster looks somewhat like an arrowhead pointing eastward. At 23× in the 4-inch a pair of parallel and slightly curving bands of stars (oriented northwest-southeast) flows through the center of the cluster's densest part. Look for a tiny gang of knitted suns at the northwestern end of the core, between these two parallel streams. The streams open out into a delta of darkness at the southeastern end of the core; look for a prominent pair of 10th-magnitude suns here. Christian Luginbuhl and Brian Skiff note that the densest portion of the cluster contains about 25 stars in an area 6′ across. All kinds of pairs and groupings dapple the cluster. Using my imagination NGC 6124 looks like the remains of a globular cluster that was torn apart by tidal forces. Sweep your eyes across the field at low power and see if the entire region doesn't look as if you're viewing it through a black fog. That's because the entire region is veiled in dark nebulosity. The extinction of starlight by dust must be substantial along this sightline through our galaxy's disk.

At 72× the cluster's dual aspect is most apparent. Also, look for two asterisms resembling the Southern Cross interspersed among several groupings, pairs, and strings of stars. Ernst Hartung called NGC 6124 a "scattered galactic cluster . . . [with] several orange stars as well as numerous pairs, triplets and small groups." I did not note any colored stars. Do you? If you throw the cluster out of focus it looks like a barred spiral galaxy with a large dust lane slicing through the bar and core, or a shield with a coat of arms.

The magnitude-10.9 planetary nebula NGC 6153 lies just a little more than 1° east-northeast of NGC 6124. The 25″-diameter object is easy to see once you point your telescope in the right direction. It's the southernmost "star" in a 5′-wide diamond-shaped asterism of 9th- and 10th-magnitude points of light. NGC 6153 is an annular planetary nebula with a magnitude-15.4 central star. In the 4-inch the nebula looks like a swollen star at moderate magnifications. And nearly 2° north-northeast of NGC 6124 is the small but bright (magnitude 9.0) globular star cluster NGC 6139, which I found once on a comet sweep. Before leaving the area, do sweep that region I mentioned earlier, southeast of the Scorpion's Tail, toward Norma, and see how many tiny open star clusters and asterisms you encounter. You'll be taking a fantastic voyage through some of the most spectacular Milky Way regions, one that is guaranteed to keep you occupied on many a moonless evening.

76

76

False Comet Cluster
NGC 6231
Type: Open Cluster
Con: Scorpius

RA: 16^h 54.2^m
Dec: –41° 50′
Mag: 2.6
Diam: 14.0′
Dist: 6,000 light-years
Disc: Giovanni Batista Hodierna, before 1654; but known since antiquity

J. HERSCHEL: A fine bright large cluster pretty rich, class VII. 10′, stars [from] 10[th to] 13th mag[nitude]. Place of a double star 5th mag[nitude], the preceding [western] but one of 7 bright stars in the middle. (h 3652)

GC/NGC: Cluster, bright, considerably large, pretty rich, consisting of stars of 10th to 13th magnitude.

HAD THE 18TH-CENTURY FRENCH COMET hunter Charles Messier lived farther south than Paris he most certainly would have noticed the bright 2°-long "comet" that perpetually shines in the Scorpion's Tail. I'm confident Messier would have noticed it, because my wife, Donna, never fails to discover the same "comet" each time she looks upon the stunning Scorpius Milky Way with her unaided eyes from Hawaii. Of course, the object is not a comet, but it is the most cometlike naked-eye spectacle in the heavens that is not a comet. Open cluster NGC 6231 (Caldwell 76) forms the "head" of our pseudocomet and lies only 30′ north of the 4th-magnitude double star Zeta^{1,2} ($\zeta^{1,2}$) Scorpii; the comet's "tail" is the large, elongated open cluster Trumpler 24 to the north. Although these objects have separate catalog identities, they belong to the same group of luminous stars, the Scorpius *OB*1 Association. This widely scattered family of young stars marks the onetime location of a giant H II region. It resides in the Sagittarius-Carina spiral arm, which lies closer to the galactic center than the one containing our Sun. Both of these clusters would have been appropriate for Messier's catalog had he known of their existence.

Which brings us to the next point. Most sources credit Nicolas Louis de Lacaille for discovering NGC 6231 during his exploration of the southern skies from the Cape of Good Hope in the 1750s. But true credit belongs to long-forgotten Giovanni Batista Hodierna, an astronomer at the court of the Duke of Montechiaro, who compiled a catalog of nebulous objects that he observed with a simple 20× Galilean refractor. Printed in Palermo in 1654, his work listed 40 nebulous objects, 19 of which (NGC 6231 included) were of his own discovery. Since Galileo had resolved the misty Milky Way into stars, Hodierna believed that all the heavens' nebulous objects were star clusters, and

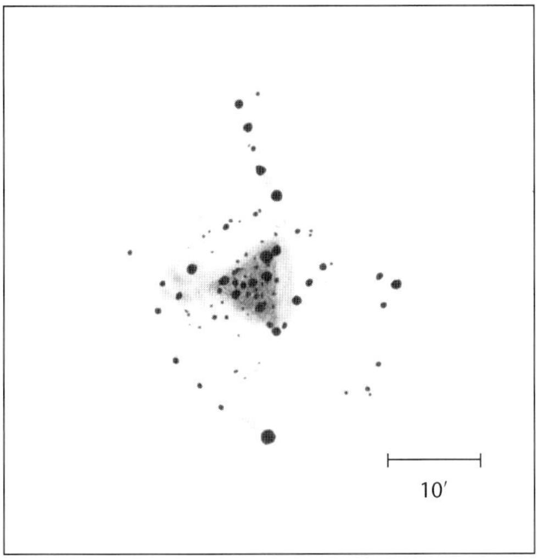

he classified them accordingly. NGC 6231 was the seventh listing under his Class I, *Luminosae* ("stars visible to the naked eye"). Of course, NGC 6231 is so bright and obvious that skywatchers throughout time would have known of its existence, though not necessarily its true nature.

Today we recognize NGC 6231 as one of the youngest open clusters in the heavens. Its precise age, however, appears to be a matter of some debate. Most estimates place its age somewhere between 3.8 and 6.5 million years, though as recently as 1999 Argentine astronomer Gustavo Baume (Astronomical Observatory of La Plata, Argentina) and his colleagues have found stars as old as 10 million years that may be associated with the cluster. Either way, NGC 6231's stars are much younger than those of the Pleiades (70, though some would argue 100, million years). They form the nucleus of the Scorpius *OB*1 Association, which is noted for its preponderance of spectroscopically peculiar stars. Indeed, of the cluster's 93 stars many are luminous *O*- and *B*-type supergiants, including two Wolf-Rayet stars, 6 Beta Cephei stars, and several P Cygni-type eruptive variables. The cluster also hosts several eclipsing binaries. If we accept a distance of 6,000 light-years to NGC 6231, the cluster would span nearly 25 light-years of space while the size of the association would be 10 times greater. Robert Burnham Jr. notes that if NGC 6231 were as close to us as the Pleiades are, the two groups would appear equally large. But NGC 6231's members would outshine the individual Pleiads by a factor of about 50, with the very brightest shining as brightly as Sirius.

From Hawaii NGC 6231 becomes visible in twilight, when 3rd-magnitude stars start to appear. As the sky darkens, the cluster, seen through 7×35 binoculars, seems to gradually energize into a ball of shimmering starlight. Shortly thereafter a nearby asterism of stars resembling a bent Eiffel Tower appears. This is part of the aforementioned Scorpius *OB*1 Association, with Trumpler 24 at the northern end and NGC 6231 and Zeta[1,2] Scorpii at the southern end. The two clusters are joined in celestial longitude by parallel strands of stars, which, with a little imagination, look like strands of mozzarella cheese stretching from a slice of

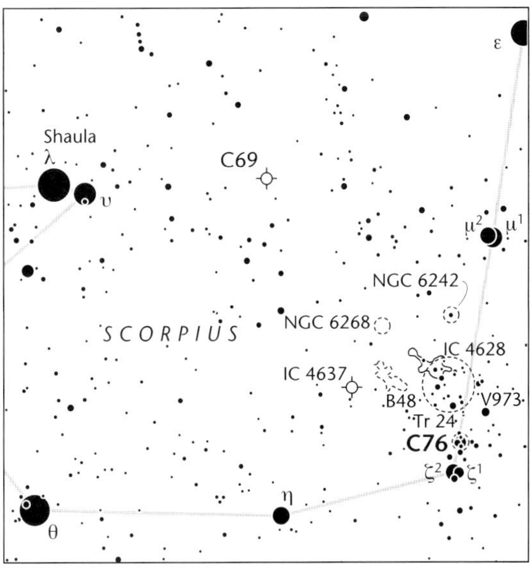

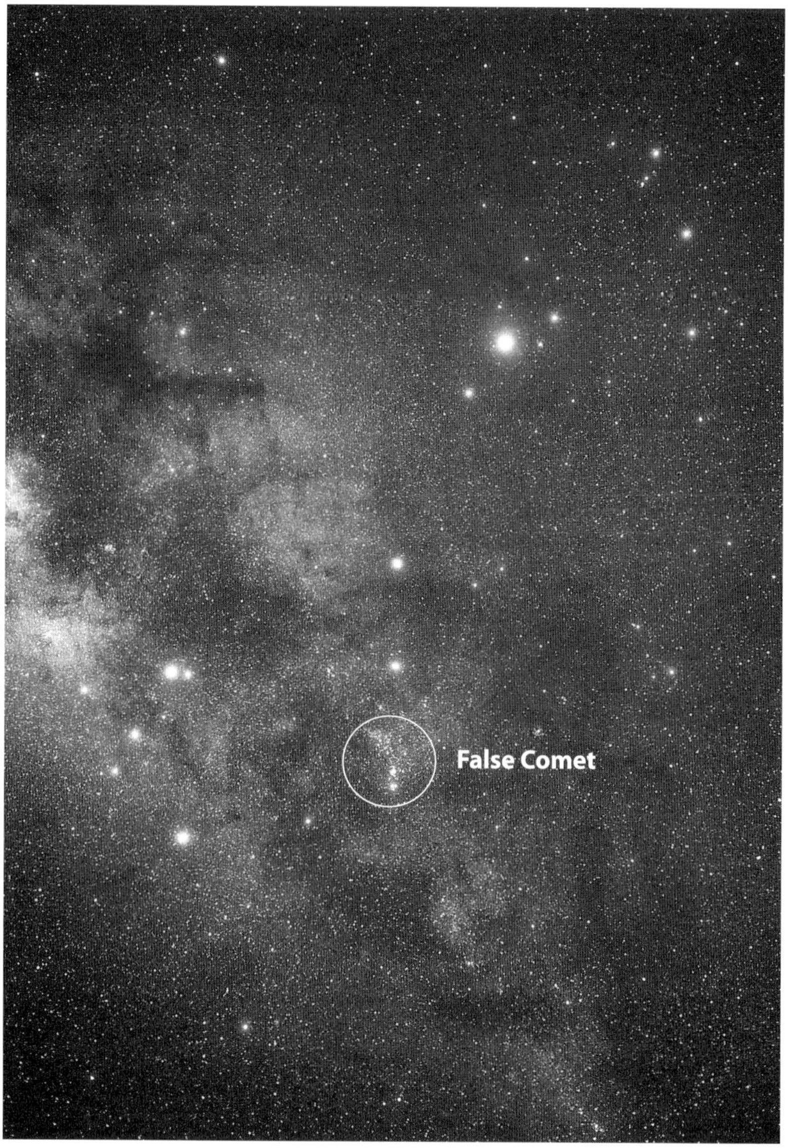

pizza. These strands are probably what James Dunlop saw as "a large branch of very small stars extended on the north side [of NGC 6231]." In a more romantic sense the association can be seen as a diamond bracelet laid on black velvet.

In the 4-inch at 23× NGC 6231 seduces the eye with its fantastic assortment of some two dozen "gemstones" of varying quality and color set in a 10′-wide arrowhead. A bell-shaped mist of dim suns seems to radiate from behind. Attached to the arrowhead's northwestern corner is a northeast-trending shaft of six suns (with magnitudes of roughly 7.5 to 9) that points to Trumpler 24. Several of these stars appear to shine with a topaz light. If you follow the shaft for about 40′ to the northeast you'll see that it encounters a 6th-magnitude star. The line then curves west, dips sharply south (nearly returning to NGC 6231), then loops back north to end at the 5th-magnitude variable star V973 Scorpii. Using my imagination this wavy asterism looks like an inchworm on the move. This may also be Tom Lorenzin's "Baby Scorpion," which "clings to the back of its mother's tail."

At medium and high powers the cluster is most dramatic, taking on a three-dimensional quality. The arrow's shaft seems to be tilted toward us at a 45° angle. It looks as if tiny gems are dripping from an unseen hand and falling into a cache of overflowing jewels (namely

NGC 6231). Under averted vision the cluster appears to lie in the foreground of a galactic mist, an illusion caused by the faint glow of unresolved suns, many of which belong to the rich Milky Way background. Use high power and averted vision, and faint suns will peek out from the dark crevices between the brighter suns. If the arrowhead seems to visually sag to the east, it's because that's the densest part of the cluster, with a miniature assortment of some half dozen suns. Under high magnification many of the stars appear to be paired. The two brightest suns marking the weapon's northern corner are themselves fine doubles; the northernmost of these two pairs shines at magnitude 5.5 with a 7th-magnitude companion 65″ away at a position angle of 13°.

The cluster's radiance is magnified with even a slight increase in aperture. Through a 6-inch telescope Christian B. Luginbuhl and Brian A. Skiff recorded 100 stars (not all cluster members) in a 10′ area, and they saw many blue stars and small geometric patterns at low power. James Dunlop, who used a 9-inch reflector, saw a "cluster of pretty bright stars of mixt small magnitudes, considerably congregated to the centre, about 10′ diameter, with a large branch of very small stars extended on the north side." With a 12-inch telescope Ernst Hartung spied "many bright white and yellow stars, and many pairs and triplets, which sparkle in patterns of lines and small groups." And with her 13.1-inch f/4.5 reflector Barbara Wilson recorded "eight brilliant white stars with dozens of fainter members in a very nebulous region" at the 1986 Texas Star Party; the cluster's center, she noted, "has [a] dimmer streamer curving around it and a smaller streamer [to the] south."

Do not stop with NGC 6231. The entire region is worth exploring. Certainly there is nothing like it in the entire sky for small-telescope users. Just north and east of the cluster, out to about 3°, are three open clusters (Trumpler 24, NGC 6242, and NGC 6268), a planetary nebula (IC 4637), a diffuse nebula (IC 4628), double stars, colored stars, Milky Way patches, and a swath of dark nebulosity (Barnard 48). It's a cornucopia of celestial riches. IC 4628 lies at the northeastern end of Trumpler 24, 1½° northeast of NGC 6231. It is an irregular 90′ × 60′ patch of faintly glowing

fog. I could see it in the 4-inch, but with difficulty, because the region's clustered starlight forms its own pseudohaze. I find a combination of keen averted vision, tube tapping, and back-and-forth sweeping helps to pull the nebula out from the background noise. IC 4628 is actually the brightest part of a 4°-wide circular loop, discovered by Colin Gum in 1952 and centered on NGC 6231. Astronomers once believed that this loop was a supernova remnant, but in 1971 David Crampton and Andrew David Thackeray revealed that the nebula is a normal H II region. As Burnham adds, it is "one of the great 'H II' regions which we often see in the external galaxies, outlining the course of the spiral arms like so many 'pearls on a string.'"

77

Centaurus A
NGC 5128
Type: Peculiar Galaxy (S0p)
Con: Centaurus

RA: $13^h\ 25.5^m$
Dec: –43° 01′
Mag: 6.7; 6.6 (O'Meara)
Dim: 16.7′ × 13.0′
SB: 13.7
Dist: 16 million light-years
Disc: James Dunlop, included in his 1827 catalog

J. HERSCHEL: A most wonderful object; a nebula very bright; very large; little elongated, very gradually much brighter in the middle; of an elliptic figure, cut away in the middle by a perfectly definite straight cut 40 arcsec[onds] broad; pos[ition angle] = 120.3°; dimensions of the nebula 5′ × 4′[.] The internal edges have a gleaming light like the moonlight touching the outline in a transparency. (h 3501)

GC: *Remarkable*, very bright, very large, very much extended toward position angle 122.5°, bifurcated.

NGC: *Remarkable*, very bright, very large, very much extended toward position angle 122°, bifurcated.

NGC 5128 IN CENTAURUS IS THE MOST dynamic and intriguing galaxy in the heavens. It is virtually a superlative in every region of the electromagnetic spectrum. It is one of the brightest naked-eye galaxies; by far the nearest and most violent Seyfert-type galaxy known; one of the most intense radio sources in the heavens; and a wellspring of infrared, X-ray, and gamma-ray radiation. The very sight of its photograph tells us it is the epitome of mystery, a rebel against extragalactic conformity. Its background glow resembles that of an elliptical (E0) galaxy, but its face is masked, Zorro-like, by a severely warped, highly opaque band of dust — one much wider and more chaotic than that found in any edge-on spiral. Compare this celestial Fabergé egg with any other galaxy in the Caldwell and Messier catalogs and you will see why NGC 5128 has been the nerve center of astronomical debate ever since James Dunlop discovered it almost two centuries ago.

Dunlop first encountered this visual oddity while observing from Parramatta, New South Wales. His description of it is rather lengthy:

> A very singular double nebula, about 2½′ long, and 1′ broad, a little unequal: there is a pretty bright small star in the south extremity of the southernmost of the two, resembling a bright nucleus: the northern and rather smaller nebula is faint in the middle, and has the appearance of a condensation of the nebulous matter near each extremity. These two nebulae are completely distinct from each other, and no connection

> of the nebulous matters between them. There is a very minute star in the dark space between the preceding extremities of the nebula: they are extended in the parallel of the equator nearly.

Several years later John Herschel would call NGC 5128 "a very problematic object" that "must be regarded at present to form a genus apart, since it evidently differs from mere 'double nebulae,' not only in the singular relation of its two halves to each other, (having each a well and an illdefined side, their sharply terminated edges being turned towards each other and exactly parallel) but also by the intervention of the delicate nebulous streak intermediate between them and lying in exactly the same general direction."

How bizarre this new nebula must have seemed to John Herschel, who, like his father, believed that the Milky Way and all its nebulae were made up of nothing but stars. What was the dark band that had seemingly ripped apart this cloud? In his 1849 "Outlines of Astronomy" Herschel prophetically decided that it was a "similar but much more strongly marked case of parallel arrangement than that noticed by Mr Bond (in the Andromeda Nebula) . . . one in which two semi-ovals of an elliptically formed nebula appear cut asunder and separated by a broad obscure band parallel to the larger axis of the nebula, in the midst of which a faint streak of light parallel to the sides of the cut appears." Herschel's reference to George P. Bond was critical, for Bond had observed in 1847 two parallel dark "canals" in the Andromeda "Nebula" — one of the last strongholds of William's Herschel's nebular theory, which stated that these masses of nebulous matter were in the process of becoming new solar systems. By 1888 Isaac Roberts had successfully photographed the dark lanes in the Andromeda "Nebula," as did Edward Emerson Barnard two years later. Soon these features would be recognized as dust lanes, part of the visible framework not of spiral *nebulae* but of spiral *galaxies*.

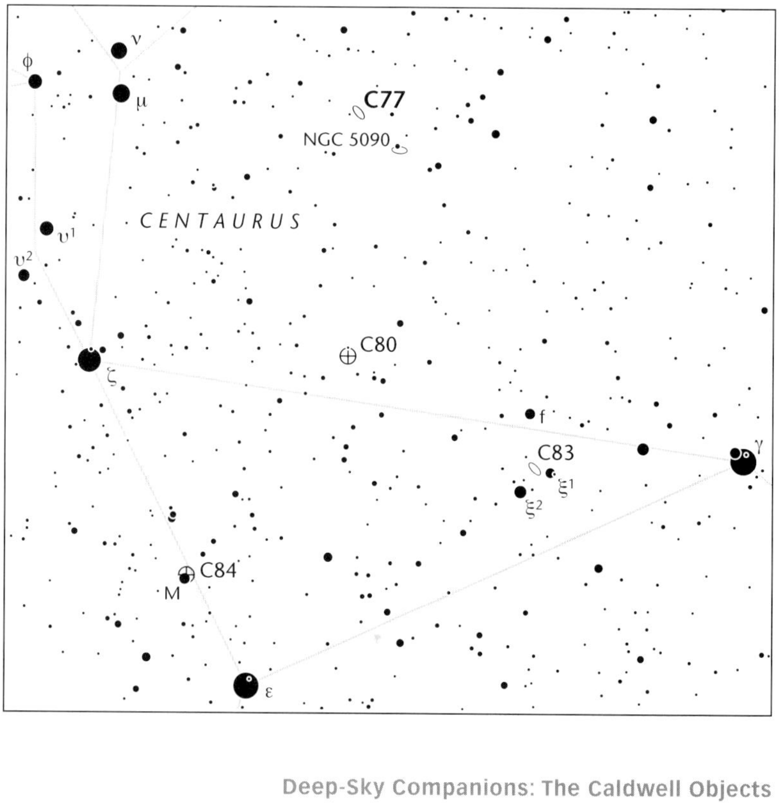

Still, NGC 5128's appearance and nature remained a mystery. In 1918, Heber D. Curtis (Lick Observatory) classified it as an "edgewise spiral with a dark lane." Five-hour exposures taken of it from Arequipa, Peru, with the

Harvard College Observatory's 24-inch Bruce refractor, and a 10-hour exposure taken with Harvard's 60-inch reflector in South Africa in 1934, led John S. Paraskevopoulos to conclude that NGC 5128 could not be an external galaxy. Indeed, Edwin Hubble himself detected gaseous emission lines emanating from the object, and for this reason he classified it in 1922 as a local nebulosity. But 10 years later Harlow Shapley and Adelaide Ames listed NGC 5128 as an irregular galaxy in their now-famous catalog. Shapley explained his reasoning in his 1947 work, *Galaxies*. NGC 5128, he said, is "a 'pathologic' specimen — one of the external galaxies with [a] peculiar spectrum. Any fully successful theory of galactic structure must take into account such abnormal forms." By then even Hubble had conceded that NGC 5128 was truly abnormal.

Two years later, in 1949, it was suggested that NGC 5128 was identical with the strong radio source known as Centaurus A, which blasts out 1,000 times as much radio energy as our Milky Way emits. The galaxy's strongest radio emission originated in its dust lane, while a weaker source discovered in 1954 radiated from a nearly circular area, 2° in diameter, centered on the galaxy's core. Then in 1958, during a Paris Symposium on Radio Astronomy, C. A. Shain announced that this unusual radio galaxy had an X-ray jet and large radio lobes shooting out in opposite directions at large angles to the dust lane. Walter Baade and Rudolph Minkowski explained all these outstanding discoveries by proposing that NGC 5128 was the violent aftermath of a collision between an elliptical galaxy and an edge-on spiral galaxy.

That theory fell in and out of favor over the years, with some arguing that the galaxy is but a single massive object — a link between elliptical and spiral galaxies. But recent Hubble Space Telescope (HST) images have helped astronomers zero in on a new solution. We now know that NGC 5128 is truly a giant, containing more than a trillion solar masses. R. Brent Tully determined that the galaxy lies about 16 million light-years distant and spans 80,000 light-years, though a NASA press release places the galaxy merely 10 million light-years off, implying a reduced diameter of 50,000 light-years. The HST data support the galactic cannibalism scenario, though with a twist. Apparently a massive black hole lies hidden at the center of this Seyfert 2 galaxy, which is voraciously consuming a smaller spiral galaxy that it collided with between 160 and 500 million years ago. The spiral will be fully consumed in a few hundred million years.

Whenever one galaxy plows into another, the interaction triggers a wave of star formation. Indeed, in NGC 5128 HST has found 21 candidate globular clusters, half of which are old, like those in our Milky Way, while the other half are young, implying they might have formed during the recent merger. HST also resolved open clusters of hot, blue stellar newborns along the fringes of the galaxy's dust lanes, which astronomers penetrated with HST's infrared camera. Behind that ring of dust lies a twisted disk of hot gas swept up into what appears to be a gravitational vortex surrounding a black hole. The evidence suggests that the black hole contains the mass of perhaps a billion stars squeezed into a region of space about the size of our solar system. Radio jets have also been detected streaming out of the black hole. These jets reveal the orientation of the black hole's spin axis. They can be seen forming about 16,000 light-years from the galaxy's nucleus, then expanding into plumes that extend 800,000 light-years into intergalactic space. Interestingly, the maelstrom of gas surrounding the black hole is not perpendicular to the jets. Astronomers conjecture that the gas disk might be so young that it hasn't yet become aligned to the black hole's spin axis. Or it may simply have been influenced more by the galaxy's gravitational tug than by the black hole's. The hot gas disk is perpendicular to the galaxy's outer dust belt, while the black hole's spin axis is tilted approximately diagonally to these axes. The black hole's origin remains a mystery. Was it always a part of the giant elliptical? The small spiral? Or was it created during the merger?

You won't solve that puzzle when you visually explore this orb of cosmic wonder with your telescope, but you can contemplate it nevertheless. To find this celestial masterpiece, all you have to do is look 4½° essentially due north of another significant astronomical treasure — the great naked-eye globular cluster Omega (ω) Centauri (Caldwell 80). You could also spot it by sweeping a similar distance west of 3rd-magnitude Mu (μ) Centauri. In 7×35 binoculars NGC 5128 is a large diffuse glow, nearly 60 percent the size of the full Moon. From dark skies there's no mistaking this remarkably bright (magnitude-6.6) glow, which can be seen with the unaided eye. Indeed, NGC 5128 is 0.3 magnitude (32 percent) brighter than M81, 0.9 magnitude (2.3 times) brighter than M83, and 0.5 magnitude (58 percent) brighter than NGC 253 (Caldwell 65) — all of which have also been dimly detected with the naked eye. That brings to six the number of galaxies that I know to be within the reach of human vision. I believe there could be more. The standard of achievable naked-eye limits in amateur astronomy has progressed well beyond the rudimentary boundaries of M31 and M33. We have now journeyed beyond a safe 2.3 million light-years to the razor's edge

at 16 million light-years. If you succeed in seeing Centaurus A with the unaided eye — and I think every amateur south of the 30th parallel should try — you will join an elite club of fellow Earthlings who have helped to expand our natural visual envelope.

Countless millions of skywatchers witnessed a rare sight in mid-April 1986 as Halley's Comet brushed past NGC 5128. The view opened a window on the imagination, allowing us to "see" the awesome dimensions of space. Here was a solar-system snowball millions of *miles* away dwarfing a pair of colliding galaxies millions of *light-years* away. And with a pair of binoculars held just right, Omega Centauri's glittering globe joined the view. Fantastic! Truly an amazing sight, one that anyone can still enjoy today, for photographs of that passage abound. An even greater spectacle occurred several days later. On May 3rd, 1986 (Universal Time), the Rev. Robert Evans of Hazelbrook, New South Wales — history's greatest visual supernova discoverer — became the first to spy an incredible stellar explosion in NGC 5128. The visual blast had achieved an impressive apparent magnitude of 12.5 at the time of its discovery, and it appeared smack-dab on the galaxy's dust lane, 2′ east and 1′ south of the galaxy's nucleus. It also appeared just in time for countless amateur astronomers to witness the aftermath through the monster Dobsonian telescopes set up at the Texas Star Party (which is held each spring on the Prude Ranch near Fort Davis, Texas, in the shadow of McDonald Observatory). Supernova 1986G was the first supernova discovered in Centaurus A.

While comet hunting from Hawaii one evening with the 4-inch at 23×, I encountered NGC 5128 (without knowing what it was) and my heart nearly stopped. My gut feeling was that I had discovered an enormous comet with a parabolic hood and a dark shadow cast by its nucleus onto its broad dust tail. But once I looked up at the sky, I realized where I was, and what it was. Still, I was utterly amazed at how dark and obvious the dust lane appeared with so small an instrument. So I grabbed a pair of

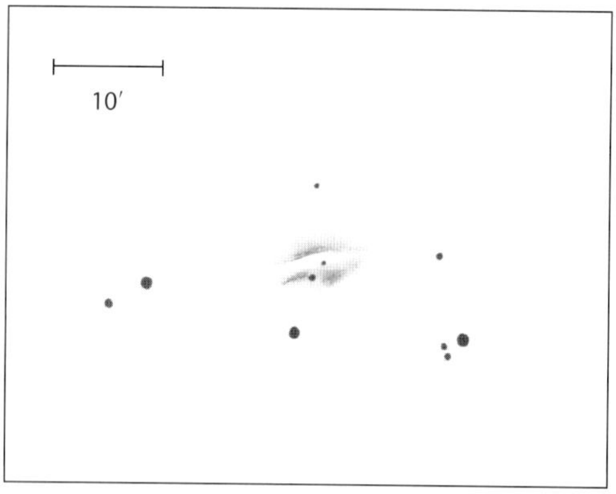

7×35 binoculars and tried to see it in those. With averted vision and a little time, the dust lane indeed was apparent in my binoculars.

At 23× in the 4-inch NGC 5128 looks like a cell undergoing mitosis. Both halves of the galaxy are immediately visible, as is the dust lane. The galaxy's brightest section is oriented east-southeast to west-northwest, and the lane runs fully across its face in that direction. At low light levels, however, the galaxy's long axis extends from northeast to southwest. But at low power, this all blends into a slightly elliptical fog that is perhaps only one-third as large as the galaxy's published angular size. The galaxy's elliptical envelope is so dim that high magnification only diminishes its appearance. So in the 4-inch at 72× and higher powers, the galaxy looks as if it is oriented from east-southeast to west-northwest. With careful scrutiny the dust lane appears wavy and widens on the eastern end. The galaxy's southern half appears brighter than its northern half, and a roughly 9th-magnitude star punctuates the southeastern rim of the outer envelope. (Again, this is visible only at low power.)

At higher power the galaxy is gracefully defined, the dust lane looking like a silken veil. The reason the southern half looks brighter than the northern half also becomes clear — two prominent foreground stars are superposed upon it. One, shining at roughly 12th magnitude, lies dead center, while a dimmer (magnitude-13.5) star lies on the south side of the dust lane to the west. Together the 9th-, 12th-, and 13.5-magnitude stars form an arc of diminishing brightness. Two concentrations of light (or even dimmer suns) lie to the east and west of the 12th-magnitude star. Supernova hunters should take note of these stars, because they do not appear in many photographs, which tend to burn out the galaxy's bright hemispheres. These stars have been mistaken for supernovae in the past. The dust lane's entire northern lip appears bright. Interestingly, this is where the Hubble Space Telescope imaged many of the galaxy's hot young star clusters. The galaxy's uniformly bright northern hemisphere looks something like a German World War II helmet. The dust lane widens to the east, as noted above, and fades into a diminishing sea of delicate vapors, like a slightly submerged walkway. In a more comical vein, Barbara Wilson likened the overall view to that of a hamburger. Through her 13.1-inch reflector, she saw the dust lane as mottled.

If you own a medium-size or large telescope, try swinging it a little more than 1° to the southwest, where a gaggle of six dim galaxies lies in a field only ½° wide. These are NGC 5082, NGC 5086, NGC 5090, NGC 5090A, NGC 5090B, and NGC 5091. Most of them lie immediately southwest of a 7th-magnitude star. Small-telescope users should at least try for the brightest of these, NGC 5090, which shines at magnitude 11.6 from a distance of approximately 150 million light-years. Good luck.

78

NGC 6541
Type: Globular Cluster
Con: Corona Australis

RA: $18^h\ 08^m\ 02.2^s$
Dec: –43° 42′ 54″
Mag: 6.3
Diam: 15′
Dist: 22,300 light-years
Disc: James Dunlop, listed in his 1827 catalog

J. HERSCHEL: Globular, bright, round, extremely compressed, very fine; diameter of most compact part equal to 11 seconds of time in right ascension; stars of magnitude 15 to 16. The scattered stars extend to three times the diameter and die away very gradually. (h 3726)

GC/NGC: Globular, bright, round, extremely compressed, gradually brighter in the middle, well resolved, stars of magnitude 15 to 16.

CORONA AUSTRALIS, THE SOUTHERN CROWN, is an attractive laurel of similarly bright suns close to the waist of Sagittarius, just south of the Teapot asterism. The constellation adorns the southern rim of our galaxy's great central bulge, and though its brightest stars shine at around 4th magnitude they stand out prominently against the Milky Way, like distant campfires dimmed by morning mist. Ptolemy called it the Southern Wreath, perhaps a victor's garland, while classical poets saw it as the crown that Bacchus placed in the sky to honor his mother, Semele. Like its northern counterpart, Corona Borealis, the stars in the Southern Crown have a symmetry that mystically lures our gaze to them despite their modest sheen. Perhaps Ptolemy, who leased his mind to the world of the occult, saw this crown as a befitting decoration to honor the heavens for its victory over human affairs.

To see our next Caldwell target we must shift our gaze 5° west of the Crown, to SAO 228708, a 5th-magnitude sun halfway between magnitude-4.6 Theta (θ) Coronae Australis and 2nd-magnitude Theta (θ) Scorpii in the Scorpion's Tail. SAO 228708 also makes a near-equilateral triangle with Theta Coronae Australis and magnitude-3.5 Alpha (α) Telescopii to the south. If you can locate SAO 228708, even in binoculars, you will have found the equally bright globular cluster NGC 6541, because it lies less than ½° to the southeast of that star. Although the cluster's full extent is 15′ (half the size of the full Moon), its core is greatly condensed and easy to see. In 7×35 binoculars from Hawaii, the globular appears as a small, tight, and concentrated glow. Christian Luginbuhl and Brian Skiff found NGC 6541 visible even in a 12-inch telescope "as a well-concentrated spot with a substellar center." Sharp-eyed observers under decent skies should have no problem detecting this cluster with the unaided eye; the trick is to separate it from its slightly brighter stellar companion. Try keen averted vision on a crisp night just as astronomical twilight is drawing to an end.

James Dunlop discovered NGC 6541 while exploring the southern heavens from his home in Parramatta, New South Wales. It is the 473rd object listed in his 1827 *A Catalogue of Nebulae and Clusters of Stars*. Of its appearance through a 9-inch reflector he wrote: "A very bright round highly condensed nebula, about 3′ diameter. I can resolve a considerable portion round the margin, but the compression is so great near the centre, that it would require a very high power, as well as light, to separate the stars; the stars are rather dusky." John Herschel found it "easily resolvable" with an 18¼-inch reflector, adding: "The left eye resolves it." Here we get a taste of an early experiment in astronomical perception. Which eye works best for you?

It is tempting to think that NGC 6541 looks so compressed because it is one of the 20-odd known globular clusters, like M30 in Capricornus and M15 in Pegasus, that has experienced a core collapse. In dense gravitating systems like globular clusters, stars can swarm toward the cluster's center like bees to a hive. This swarming episode can last a few million years, until the stars get too close. At that point two things can happen. The stars either continue collapsing until they form a black hole (which is unlikely, given the observed properties of globular clusters), or more likely they will rebound back out until they reach a point where they remain in globular-cluster limbo, being both attracted and repelled by one another. In a sense, when we peer into the heart of NGC 6541, we are looking at a snapshot of this drama at some critical point in the globular's evolution. However, our imaginations must be tempered by the fact that the effects of core collapse on a globular cluster almost certainly cannot be resolved visually.

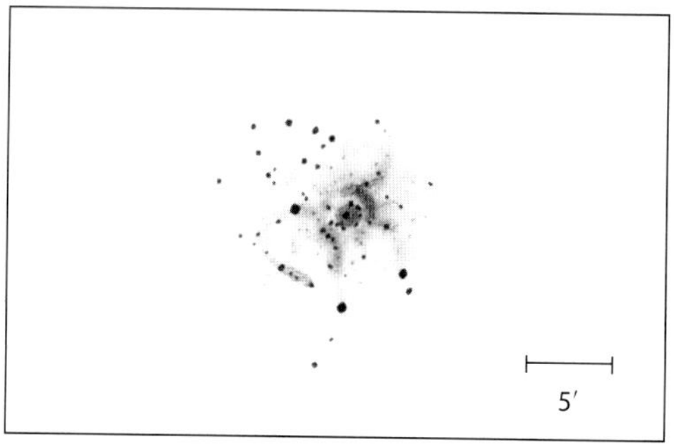

NGC 6541 is a poorly studied cluster. Its age is estimated at 16 to 17 billion years. The cluster's metallicity places it among the metal-poor globulars; each of NGC 6541's stars, on average, contains less than 1/70 as much iron as does our Sun. This is yet another clue that the cluster is one of our galaxy's senior citizens. The cluster's overall spectral type is *F*6, and it is moving away from us at 155 km per second.

At 23× in the 4-inch Genesis, which provides a stunning nearly 3°-wide view of this rich Milky Way region, the globular glows to the southeast of SAO 228708, which is actually an extremely challenging double star whose equally bright components were separated by a mere 1.7″ in 1999. I see these type-*A*5V stars as a single sun burning with a pale yellow light. The tiny (6.9′) magnitude-8.5 globular star cluster NGC 6496 can also be seen in the same field, about 1¾° west-southwest of NGC 6541. There is no mistaking the two globulars for one another; NGC 6496 is but a mere fleck of dandruff in the Milky Way compared to the magnificently crisp form of NGC 6541. When I use averted vision on NGC 6541 at low power, the cluster's outer halo

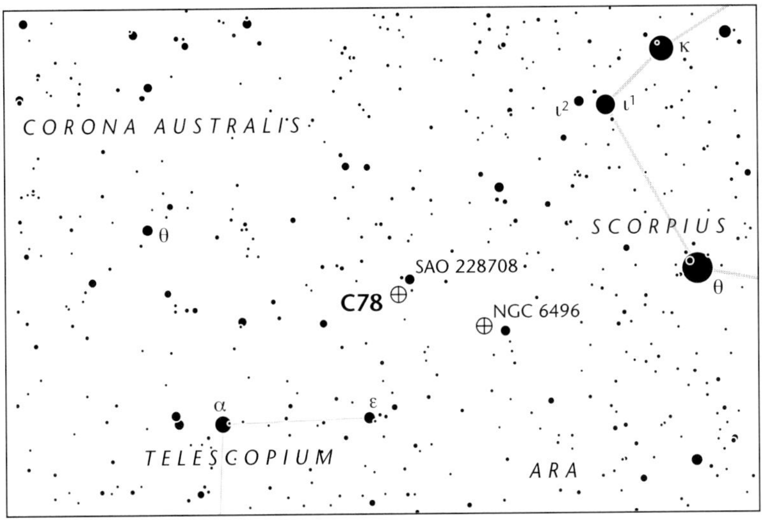

seems to fragment into tiny stars, like snow shaken from a fur-trimmed parka. Do you see the cluster shining with a soft aqua hue?

At 72× NGC 6541 is a wonderful object sporting a very extensive halo glittering with starlight. Several fuzzy arms can be seen, especially to the south, southwest, and west, but only stubs jut out to the east and north. The core remains tight. At high power the cluster is a most remarkable sight. The core is undeniably small, perhaps 2′ across, and it is the site of much visual chaos. A crescent of clustered suns outlines the innermost core's northern rim. Just beyond it to the northwest is a dark lane, which separates the core from a broader crescent of variegated starlight with two jutting spikes of stars -- one to the northwest and one to the west. A bright lash of suns lies on the southeastern side of the core, and it, too, is separated from the core by a dark lane. A roughly 13th-magnitude star blazes to the east of the core and a similarly bright double star resides just outside the inner halo to the southwest. Much of the halo's midsection remains irresolvable, but its outer layers appear peppered with strings and pairs of suns oriented in random directions. Look for a prominent river of darkness coursing through the outer halo to the southeast; the river runs from southwest to northeast and is best seen by defocusing the scope ever so slightly, tapping the tube, and using averted vision.

If your telescope can get you to 15th magnitude, you're well on your way to nearly resolving this enigmatic jewel of a cluster. Partial resolution is attainable even with small apertures, since the cluster's brightest member stars shine at magnitude 12.1, which is within the range of a 2-inch telescope. With a 12-inch instrument Luginbuhl and Skiff saw the core elongated along a north-south axis. Barbara Wilson was most excited with the view of NGC 6541 she had through her 20-inch f/4 behemoth: "Totally resolved at 72× [except for the core, which] is very brilliant. Huge globular; easy in [a] 10×70 finder." It's that core that will keep your eye transfixed. And if you take the time to relax and peer into that stellar vortex, you should see why, because at the very heart of the core is a "star" shining with the luster of burning steel. It is there, in that tiny little pip of light, that countless thousands of suns are battling each other with their gravity. It's a point of imagination, a visible "white hole." What we see there in the mind's eye is limited only by our sense of reason.

79

NGC 3201
Type: Globular Cluster
Con: Vela

RA: $10^h\ 17^m\ 36.8^s$
Dec: –46° 24′ 40″
Mag: 6.7
Diam: 20′
Dist: 17,000 light-years
Disc: James Dunlop, included in his 1827 catalog

J. HERSCHEL: Globular cluster, irregularly round, gradually brighter in the middle, not very much compressed, 6′, resolved into stars of magnitude 13 to 15. (h 3238)

GC: Globular cluster, very large, irregularly round, little compressed in the middle, gradually brighter in the middle, stars of magnitude 13 to 16.

NGC: Globular cluster, very large, irregularly round, little compressed in the middle, stars of magnitude 13 to 16.

THERE IS NO UP OR DOWN IN SPACE. THE celestial equivalents of north, south, east, and west are useful conventions that help us "see" the gears of our clockwork universe. Adopting these conventions has enabled astronomers to find some patterns in the cosmic scheme of things. For instance, most planets spin on their axes from west to east. But Venus has a "retrograde" (backward) rotation, as it spins from east to west. If we look "down" at the solar system, with the Sun's north pole pointing "up" at us, we can see that all of the major planets (and many minor ones) orbit the Sun in a counterclockwise direction, while some minor bodies move around the Sun in retrograde (clockwise) orbits. Likewise, the moon Triton travels "backward" around its parent planet, Neptune. Retrograde renegades appear to be a part, albeit a small part, of our solar system. But what about the Milky Way galaxy? It too has a disk whose constituents (mostly stars) orbit a shared center in the same direction, like skaters in an ice rink. Do we see any renegade objects going against the flow in the Milky Way? The answer is yes, and one of the most remarkable of these is the globular cluster NGC 3201 (Caldwell 79).

NGC 3201 is a roughly 16-billion-year-old globular residing in our galaxy's halo about 17,000 light-years from our Sun and 30,000 light-years from the galactic center. The cluster has an exceedingly fast radial (line-of-sight) velocity, approaching us at 490 km per second — the greatest value, in an absolute sense, evinced by any known cluster. If you remove the effect of the Sun's motion through the galaxy, what remains is an excess velocity of about 240 km per second. This isn't quite fast enough for the globular to escape the bond of our galaxy's gravity. But it does move the cluster through space in a direction that opposes the general rotation of the galaxy's disk. Yet, just as a comet's retrograde orbit doesn't change its physical characteristics, NGC 3201 remains a bona fide globular cluster. Guillermo Gonzalez (University of Washington) reported the following in the *Astrophysical Journal* in 1998: "Although NGC 3201 has an unusual Galactic orbit, we find no new evidence . . . that it is a captured cluster." In an earlier study Gonzalez also noted that "the composition of the stars in NGC 3201 is similar to that of typical halo globulars despite its unusual Galactic orbit."

Typically, a globular star cluster spends most of its time moving slowly through the halo, at the outer extreme of its orbit, like a comet orbiting the Sun; only briefly does it whip in and buzz the galaxy's core. It's possible that at some time in its history NGC 3201 was sent into a new orbit by a close encounter with another globular. It's also possible that it may have been part of an ingested dwarf galaxy; some modern theories suggest that the galaxy's entire halo was built up of such remains.

NGC 3201's discovery is usually attributed to Sir John Herschel, but James Dunlop listed it as the 445th object in his 1827 catalog, which was published six years before Herschel arrived at the Cape. Of it Dunlop wrote: "A pretty large pretty bright round nebula, 4′ or 5′ diameter, very gradually condensed towards the centre, easily resolved into stars; the figure is rather irregular, and the stars are considerably scattered on the south preceding [southwestern] side: the stars are also of slightly mixed magnitudes."

The cluster is easy to find. First locate the enormous naked-eye nebula Eta (η) Carinae halfway between the Southern Cross and the False Cross. Some 10° due north of Eta Carinae is the magnitude-2.7 star Mu (μ) Velorum. Mu Velorum marks the southeastern apex of a 10°-wide naked-eye acute triangle, with 5th-magnitude q Velorum marking the northwestern

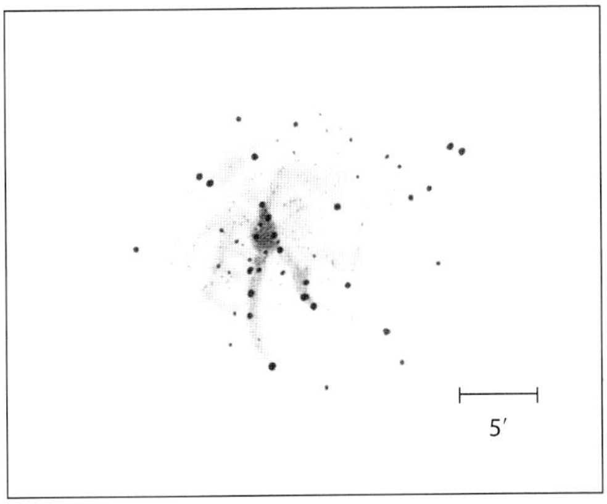

apex and the equally bright stars m and u Velorum occupying the west-northwestern corner. NGC 3201 is near the heart of that triangle, about 4½° due east of the midpoint between m and u Velorum. From Hawaii I find its magnitude-6.7 glow just at the limit of naked-eye visibility. Look for a "double star" in the globular's position, because about 30′ due south of the cluster is a star of essentially equal brightness (magnitude 6.6). The trick is to separate the two objects with averted vision. It's not difficult; the separation is 10 times greater than that of the naked-eye double star Epsilon (ε) Lyrae. Seeing the magnitude-6.6 star should be no problem under a dark sky. The difficulty will lie with seeing the globular, whose light is spread across a much larger area (though in practice you'll be looking for the light from its innermost region, just a few arcminutes across). Plan on spending some time looking. You must be comfortable and relaxed; neck strain will

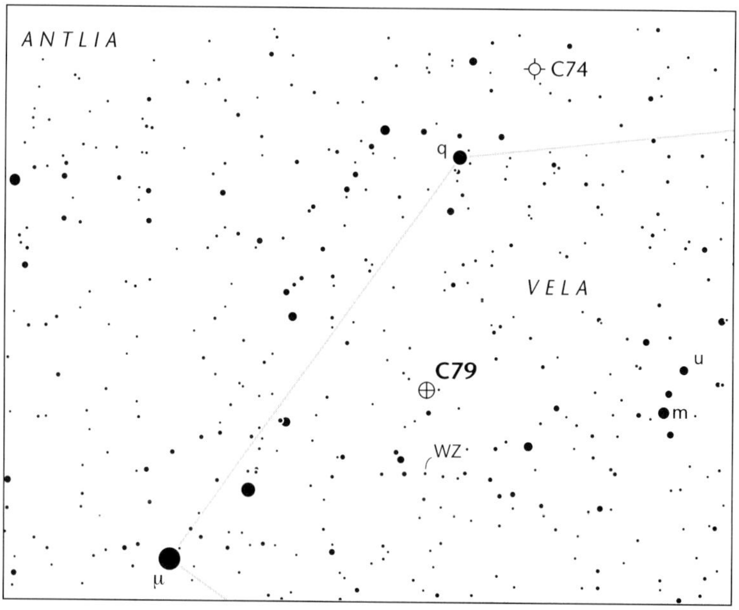

thwart your efforts. Be sure to take frequent deep breaths and a series of quick looks rather than one prolonged gaze, which will just drain your "eye batteries." I found NGC 3201 a bit more difficult to see with the naked eye than the similarly bright and large globular cluster M2 in Aquarius. If you fail, try again on another night.

In 7×35 binoculars the cluster is a cinch, and at 23× in the 4-inch it is a fabulous globular with an intense core surrounded by a wobbly envelope of stars. The apparent direction and extent of the envelope depend on how you hold your head or use your averted vision; this is probably what Herschel meant when he called the cluster "irregularly round." With averted vision the halo is definitely mottled, and some of its outliers can be resolved at low power. The brightest members of the cluster shine at magnitude 11.7, and its horizontal-branch magnitude of 14.8 should enable anyone with an 8-inch Schmidt-Cassegrain to look deep into the cluster's maze of suns. Even in the 4-inch at low power I can make out hints of "spiral" structure and an arc of detached suns to the west-northwest. Before moving on to higher magnification, sweep the cluster's general area at 23× and see if you don't encounter a copse of double stars, many of which are so tight they appear fuzzy. Many sources complain that the cluster's cataloged size is not realistic, but I disagree; when viewed under dark skies with low power the cluster's full 20′ span can be visualized. In physical terms this translates into nearly 100 light-years.

At 72× much of the cluster is immediately resolved. Most fantastic is a bright V of stars. Its apex starts at the north-northeastern side of the cluster's main "starcloud," which suddenly branches into two arms that extend south and southwest. The stars in the branching arms appear remarkably obvious, and most are of similar brightness. This is one of the cleanest visual structures I've ever seen in a globular. A pompadour of dim suns curls off the main cloud to the northwest, and a tail of starlight sweeps in the same direction from the southern side to the outer halo before ending in a prominent

12th-magnitude double star. The cluster's eastern hemisphere is broken by what appears to be a large, dark lane, past which a wedge of starlight continues eastward. The longer I look, the more complicated the view becomes. As my eye becomes accustomed to the view, all manner of star strings, dark halos, faint extensions, and tangles of interacting starlight take shape and follow seemingly random paths. The cluster becomes even more dramatic at high power, when swarms of 13th-magnitude suns sparkle forth from the background haze. A series of at least three dark patches can be seen to the east, each separated by a narrow bar of starlight. A dark patch also stains the northwestern quadrant of the inner core.

Well-resolved globulars can be visually maddening. The eye-brain combination tends to work overtime trying to connect streams of starlight, which come and go with the seeing. It's like watching snow fall through an open window in a heated room. Whenever I get overwhelmed with the view, I gently place the cluster out of focus and look for the strongest concentrations of starlight. Doing this also helps the dark areas to stand out more clearly from the stellar quagmire.

Although regarded as a fine "southern" object, NGC 3201 is theoretically visible as far north as London, Ontario, though from there it would be a horizon-kisser at best. If you return to low power and look 1½° to the south of the cluster, you'll see a red ember glowing just west of a gently curving stellar trio trending northeast to south. That's the light of the type-*M*3 semiregular variable star WZ Velorum, whose brightness oscillates between photographic magnitudes of 9.0 and 10.0 every 130 days.

80

80

Omega Centauri
NGC 5139
Type: Globular Cluster
Con: Centaurus

RA: $13^h\ 26^m\ 46^s$
Dec: –47° 28′ 37″
Mag: 3.7; 3.9 (O'Meara)
Diam: 55′
Dist: 17,300 light-years
Disc: Listed as a star in Ptolemy's *Almagest* of 150 A.D.; probably recognized as a "star" by sky-watchers throughout history

J. HERSCHEL: Diameter full 20′. It much more than fills the field. When the centre is on the edge of the field, the outer stars extend fully half a radius beyond the middle of it. The stars are singularly equal, and distributed with the most exact equality, the condensation being that of a sphere equally filled. Looking attentively, I retract what is said about the equal scattering and equal sizes of the stars. There are two sizes 12th mag[nitude] and 13th mag[nitude], without greater or less, and the larger stars form rings like lace-work on it. One of these rings, 1.5′ in diameter, is so marked as to give the appearance of a comparative darkness like a hole in the centre. There must be thousands of stars. To the naked eye it appears as a star of 5th mag[nitude] or 5[th to] 4th mag[nitude], rather hazy. There is a 9th mag[nitude] star on the S.p. [southwestern] border of it, about 4′ or 5′ south of centre, and several [of] 8th mag[nitude] are scattered far away. My attendant [(John South)] called up, who saw the hole and darkness, and described it as I have done above. On further attention the hole is double, or an oval space crossed by a bridge of stars. Position of axis = 150°. (h 3504)

GC/NGC: *Remarkable*, globular cluster, ω Centauri.

DROP A VISUAL PLUMB LINE 36½° DUE SOUTH of brilliant Spica in Virgo, and you will encounter a soft and solitary 4th-magnitude glow that looks slightly "hairy," like the fuzzy patina of a fading comet — even under a direct gaze. That's Omega (ω) Centauri (Caldwell 80). Omega Centauri also marks the northwestern corner of an isosceles triangle that also comprises the roughly 2nd-magnitude stars Zeta (ζ) and Epsilon (ε) Centauri, and it is 13° northeast of Gamma (γ) Crucis, the northern tip of the Southern Cross.

How is it that the *Alpha* of globular clusters became known as *Omega* Centauri? The answer lies with the way early astronomers assigned Greek letters to the stars.

The Bavarian lawyer and astronomer Johann Bayer (1572–1625) is usually credited with the idea of attributing letters to stars, and the Greek-letter star labels used today are known as Bayer designations. However, as the late *Sky & Telescope* editor Joseph Ashbrook writes in his *Astronomical Scrapbook*, the "idea of using letters to designate stars was . . . not original with Bayer, but goes back to the Italian scientist Alessandro Piccolomini (1508–1578), who was Archbishop of Patras. His book *De le Stelle Fisse,* which went through several editions between 1540 and 1579, contains 48 woodcut constellation maps in which the stars are labeled with [Roman] letters. . . . This parallelism between Bayer's sequence of Greek letters and Piccolomini's Roman ones suggests that the German cartographer sometimes borrowed from his Italian predecessor." His lack of originality aside, Bayer's Greek-letter designations are in universal use today, and in most cases they reflect how each constellation's stars are ranked in brightness. When mistaken for a star, Omega Centauri is outshone by 16 others in Centaurus, and thus it is labeled with the last letter in the Greek alphabet. (Of course we wouldn't call this cluster a "star" today, but pretelescopic astronomers understandably did so.)

Bayer included Omega Centauri in his *Uranometria* of 1603 because Ptolemy had cataloged it in his *Almagest* of 150 A.D. Ironically, there is some question as to whether Ptolemy had even made the observations that appear in that latter work; instead, he may have "borrowed" from Hipparchus and others in order to advance his theories. That aside, Omega Centauri is bright and obvious enough that aboriginal skywatchers throughout time must have seen its luminescence, though they presumably wouldn't have known just what kind of celestial magnificence blazed before them.

Omega Centauri remained a "star" until 1677, when Edmond Halley rediscovered it from St. Helena, recording it as a nonstellar object "in the horse's back." Halley later included Omega Centauri in a list of six "luminous spots or patches" published in 1716 in the *Philosophical Transactions* of the Royal Society. Halley's contribution was titled "An Account of several *Nebulae* or lucid Spots like Clouds, lately discovered among the Fixt Stars by help of the Telescope." Of this new object he writes (in the third person):

> The fourth [luminous spot or patch] was found by M. *Edm. Halley* in the Year 1677, when he was making the Catalogue of the Southern Stars. It is in *Centaur,* that which Ptolemy calls [the one who emerges from the (horse's) back] which He names *in dorso Equino Nebulus* [Nebula on the back of the horse] and is Bayers Omega; It is in appearance between the fourth and fifth Magnitude, and emits but a small Light for its Breadth, and is without a radiant Beam; this never rises in *England,* but at this time its Place is [Scorpio] 5 deg ¾ with 35 deg ⅕ South Lat.

Many sources erroneously credit Halley with first discerning Omega Centauri's cluster nature, but he never did so. Nor did the Swiss astronomer and mathematician Philippe Loys de Chéseaux (1718–51). De Chéseaux included Halley's observations in his list of 21 nebulae, which were presented to the French Academy of Sciences in 1746, investigated by Guillaume Bigourdan in 1884, and published as part of a review of nebulous objects by Bigourdan in the 1892 *Annales de l'Observatoire de Paris.* Of the first 14 objects in de Chéseaux's list Bigourdan writes: "I begin with those which, viewed by telescope, are found to be simple clusters of stars." But Omega Centauri does not appear in that section. Rather, it is listed in the second part of Bigourdan's list, which contained "the nebulae properly so called . . . which, when observed with much larger telescopes, always appear like white clouds." Bigourdan described Omega Centauri as "[that nebula] in Centaurus, discovered by Mr. Halley." Although Bigourdan made an effort to observe most of the 21 neb-

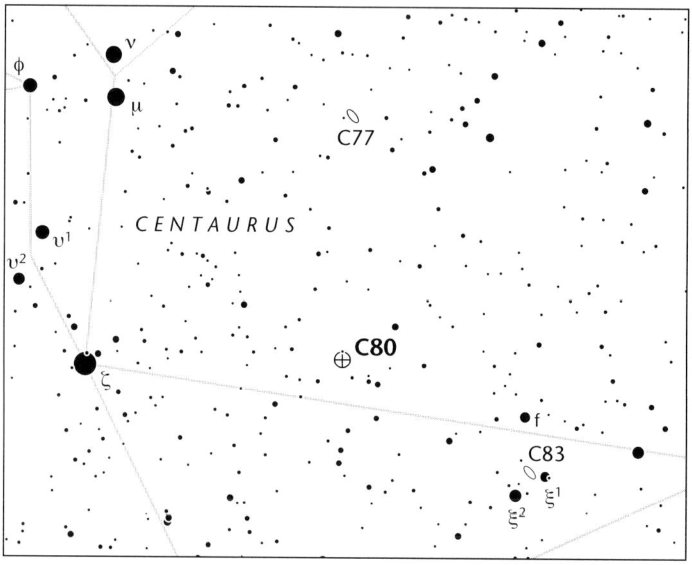

ulae in de Chéseaux's list, he could not have scrutinized Omega Centauri because it was always below his horizon. Clearly, then, his description was based solely on Halley's impressions. Omega Centauri's true nature also eluded the keen eye of Louis de Lacaille, who listed the object in his 1755 catalog under the category of "nebulae without stars." Mind you, Lacaille observed with a simple ½-inch 8× telescope — an instrument inferior to most of today's binoculars. He described it as a "[nebula] in Centaurus; with simple view, it looks like a star of 3rd magnitude viewed through light mist, and through the telescope like a big comet badly bounded."

Some sources imply that Sir John Herschel first identified Omega Centauri as a star cluster. His famous account makes the notion plausible: "The noble globular cluster ω Centauri, beyond all comparison the richest and largest object of its kind in the heavens. The stars are literally innumerable, and as their total light when received by the naked eye affects it hardly more than a star of the 5th or [5th to 4th] magnitude, the minuteness of each may be imagined: it must however be recollected that as the total area over which the stars are diffused is very [considerable] (not less than a quarter of a square degree), the resultant impression on the sensorium is doubtless thereby much enfeebled, and that the same quantity of light concentrated on a single point of the retina would very probably exceed in effect a star of the 3rd magnitude."

But alas, even this claim is wrong. Who, then, first resolved Omega Centauri into stars? None other than James Dunlop, that "gentleman astronomer" from Parramatta, New South Wales. It is the 440th object in his 1827 catalog. Omega Centauri, wrote Dunlop, "is a beautiful large bright round nebula, about 10′ or 12′ diameter, easily resolvable to the very centre; it is a beautiful globe of stars very gradually and moderately compressed to the centre; the stars are rather scattered preceding and following [west and east], and the greatest concentration is rather north of the centre: the stars are of slightly mixed magnitudes, of a white colour. This is the largest bright nebula in the southern hemisphere."

Omega Centauri is generally regarded as the night sky's preeminent globular star cluster. But even that is debatable. Okay, now pick your jaw up off the floor. I haven't lost my mind. Let's look at the statistics. Omega Centauri *is* the brightest globular star cluster, but just barely so. Although many sources show it as bright as magnitude 3.7 or 3.5, my naked-eye estimate of magnitude 3.9 is not so optimistic; my estimate also agrees with the value listed in Brian Skiff's *Observational Data for Galactic Globular Clusters*, which was published in the *Webb Society Quarterly Journal* for January 1995 and has since been updated on the World Wide Web. If we accept a magnitude of 3.9 for the cluster, Omega Centauri has the same apparent brightness as 47 Tucanae (Caldwell 106). So which is the greater globular? Skiff says, "Some prefer the sheer richness of Omega Cen, and others the star-density and remarkable (indeed unique) structure of 47 Tuc. . . . Both are luminous objects with large numbers of stars, as anyone who has observed them can confirm, and their combined light sends them to the top of the total magnitude list. But because of their distance, the brightest [individual] stars [in these clusters] are not as bright as [stars in] nearer clusters."

Indeed, when ranked by the brightnesses of their constituent stars, both Omega Centauri and 47 Tucanae fall short of the top. Omega Centauri's brightest stars shine at an impressive magnitude 11.5, but five other globular clusters beat it in that category, while seven clusters beat out 47 Tucanae. When ranked by horizontal-branch magnitude (an indirect measure of the cluster's resolvability), three clusters beat out 47 Tucanae, while seven clusters are more readily resolved than Omega Centauri in small telescopes. Of course, a cluster's greatness is highly subjective; it's all in the eye or the heart of the beholder. Observing is a highly personal experience, and we all react differently to what we see through a telescope. And even one observer's reaction will differ with telescope aperture. But if we were to magically place all the globulars at the same distance and look at them side by side, we'd find that Omega Centauri is the largest and most massive of them all. It is, in fact, the biggest globular in our Milky Way galaxy. The cluster spans more than 280 light-years of space, contains several million stars, and has a mass of about 5 million Suns, making it about 10 times more massive than other big globulars and a near match for some dwarf galaxies.

Omega Centauri is also one of the oldest objects in the Milky Way; its age is comparable to that of the universe itself. On average, each of the cluster's member stars contains about 1⁄40 as much iron as does our Sun — a pretty typical metallicity for a globular cluster. Its overall spectral type is *F*5, and its radial velocity is 230 km per second.

In 1999 Young-Wook Lee (Yonsei University, South Korea) and five colleagues studied 50,000 of Omega Centauri's stars with the 1-meter reflector atop Cerro Tololo in Chile. They found several stellar populations that had formed in distinct bursts over a 2-billion-year period. The scientists speculate that the prolonged starburst activity would make sense if Omega Centauri were the remnant nucleus of a small galaxy that merged with our Milky Way. If true, this would make Omega Centauri akin to the globular cluster M54, which appears to be the nucleus of the Sagittarius Dwarf Galaxy now being consumed by the Milky Way.

In 7×35 binoculars Omega Centauri is magnificent, a bright, distended cloud of pseudocometary fluff. It lies immediately northeast of a bowl of 7th-magnitude suns in a 2°-long dipperlike asterism. The cluster swells dramatically when one switches from naked-eye viewing to binoculars. As Walter Scott Houston once said, "[the] apparent size of Omega Centauri depends not only on the size of your telescope but also on your reaction to seeing scattered light. It's rather like the better-understood 'cocktail-party effect' — the more sensitive you are to hearing, the more conversations you will overhear." Under averted vision Omega Centauri grows faint radial arms that project against a halo of teasingly faint suns.

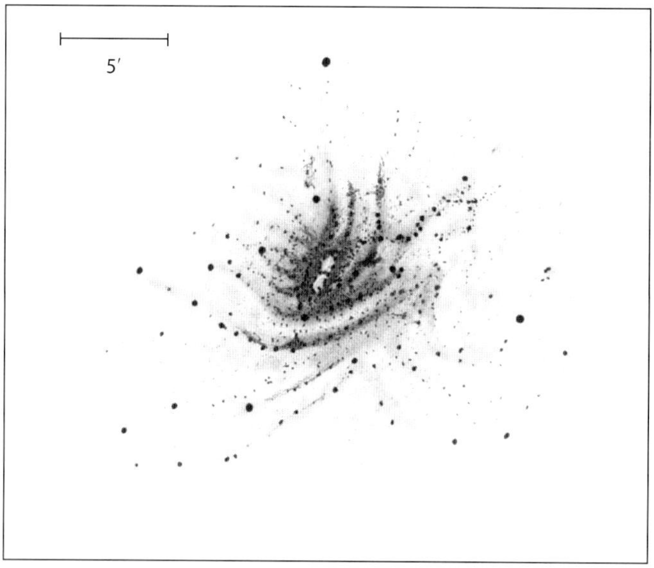

At 23× in the 4-inch the cluster looks like a snowball illuminated from the inside. The east-west trending core seethes with glowing globs of melded starlight. Two hazy beams stretch to the northeast and southwest, respectively (the northeastern one being the more prominent), while claws of stars jut out from the southern rim of the core. A 9th-magnitude star burns about 4′ north of the loose and visually serene center. Omega Centauri's core lacks the intense central concentration of light that its rival 47 Tucanae displays; its texture is softer, recalling salt piled onto a wad of cotton. When the cluster is high in the sky, use low power to look for color. I see a slight yellow tinge to the core while the outer halo has a pale blue luster.

Increasing the magnification to 72× transforms the cluster into a web of gossamer threads surrounding a sizzling skillet of starlight. Look for a pair of 12th-magnitude suns directly at the center of the core, which is bordered by east-west trending arcs of stars to the south. At high power the globular dissolves into innumerable beads of quicksilver. The northernmost section of the inner core has a dark and prominent keyhole or "footprint" stamped on it. (I put "footprint" in quotes because the feature was first pointed out to me by an anonymous observer in the darkness of a Texas Star Party. This also may be the "hole" John Herschel noted in his description on page 318.) Running from the north-northwest to the south-southeast, this footprint lies on the eastern side of a ribbed wedge of stars; look for several thin "comet tails" of fuzzy starlight jutting northeastward from the easternmost rib. The densest part of the cluster core is asymmetrical. It lies on a bed of fainter suns that scintillate in the background like stellar static. A long curving lane of darkness sweeps though the southern half of the dense core like a big smile; its southern levee is lined with stellar filigree. Undulating waves of stars flow away from this black wall to the south. All kinds of clumps, stellar strings, and dark patches scar the globular's face. The arms on the western side are more flamboyantly curled, while those on the eastern side form long wisps that gradually decrease in intensity, like fading smoke trails from skyrockets. Visions of evaporating ice chips, crystal flames, and light-snapping glitter all come to mind as the cluster assaults the eye-brain system. Observing Omega Centauri is like peering into the working mind of the Creator.

A naïve calculation suggests that Omega Centauri is theoretically visible only from latitudes south of N 42° 31.4′. But, as Houston reminded readers of his "Deep-Sky Wonders" column, atmospheric refraction can raise the celestial "limbo stick," making the cluster visible from points ever so slightly farther north. David McCarter and Steve Gauthier of the London Centre of the Royal Astronomical Society of Canada sighted the cluster ½° above the horizon with a 10-inch reflector from a cliff overlooking Lake Erie in southern Ontario (42° 32′ north). And Massachusetts amateurs John Gall and Ralph Pass have seen Omega Centauri with a 10-inch Schmidt-Cassegrain telescope from the state's seaside village of Magnolia (42° 34.5′ north). See if you, too, can match or exceed the theoretical limit.

81

NGC 6352
Type: Globular Cluster
Con: Ara

RA: $17^h 25^m 29.2^s$
Dec: –48° 25′ 22″
Mag: 8.0
Diam: 7.1′
Dist: 18,600 light-years
Disc: James Dunlop, included in his 1827 catalog

HERSCHEL: None.

GC: None.

NGC: Cluster (not nebula), pretty faint, large.

SINCE GLOBULAR CLUSTERS ARE THE OLDEST objects in the Milky Way, they have become proving grounds for theories about the evolution of our galaxy and the age of the universe. By studying the ages, positions, motions, and chemical compositions of these fantastic stellar agglomerations, astronomers have determined that two groups of globular clusters exist in our galaxy. One is huddled near the flat galactic plane; the other is a spherical-halo clan. The "thick disk" clusters that constitute the first group typically have more "metals" (heavy chemical elements) than those in the halo. Yet NGC 6352, an inner-halo object that lies 10,800 light-years from the galactic center, is one of the most metal-rich globulars known. Typically a halo globular might have about 1/100 as much iron (relative to hydrogen) as does our Sun, but the stars in NGC 6352 have one-fifth as much. Clusters like NGC 6352, then, might be connecting links between halo globulars and old open clusters like M67 in Cancer and NGC 188 (Caldwell 1) in Cepheus.

As reported in a 1995 *Astrophysical Journal* paper, Hubble Space Telescope observations of NGC 6352 have revealed it to be very similar to the heavens' prototypical disk globular cluster, 47 Tucanae (Caldwell 106); the clusters are similar in age (about 14.5 billion years) and metallicity. In the opinion of the HST study's authors, this suggests that the thick-disk system of globulars is older than the system populating the halo. This contradicts a widely held consensus that the halo clusters are older than those in the thick disk (see the Caldwell 73 discussion starting on page 288). If it is true, however, the thick-disk globulars probably formed along with the rest of the Milky Way's disk, while the halo clusters formed in satellite galaxies that the Milky Way later cannibalized. Dramatic evidence of this kind of merger was provided in 1994 by the discovery of a dwarf spheroidal galaxy in Sagittarius. That galaxy appears to be undergoing tidal disruption and is depositing its stars and globular clusters into the Milky Way's halo. More likely is a hybrid scenario in which some halo globulars are primordial while others have been accreted along with the

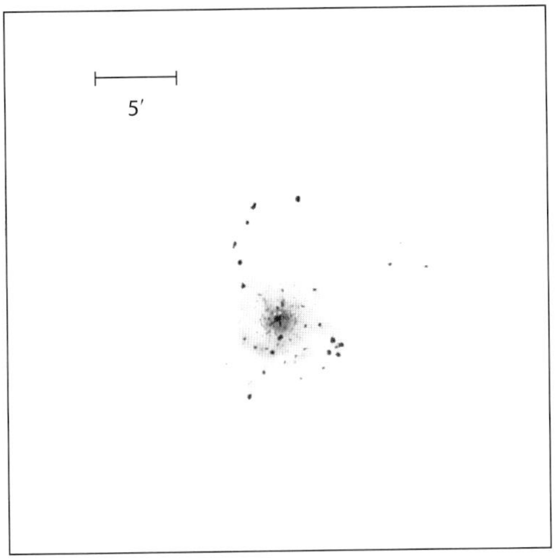

shattered remains of their diminutive host galaxies.

To find NGC 6352 first locate 2nd-magnitude Theta (θ) Scorpii and 3rd-magnitude Eta (η) Scorpii in the southernmost part of the Scorpion's Tail. Now look for 3rd-magnitude Alpha (α) Arae, which forms the southern tip of an acute triangle about 7° south-southeast of the midpoint between Theta and Eta Scorpii. NGC 6352 is 1¾° northwest of Alpha Arae. There's a fine stellar pair nearly 15′ to the southeast of our 8th-magnitude cluster, and the two stars seem to point to our target. Look also for a beautiful star chain some 15′ or 20′ to the southwest; it spans 20′ and comprises six 8th- to 11th-magnitude stars. (Be sure to not confuse NGC 6352 with the brighter, more conspicuous open cluster IC 4651 just 1° west of Alpha Arae.) NGC 6352 can be seen in 7×35 binoculars, but I doubt that this can be achieved from the city or from some suburbs; the cluster's surface brightness appears very low, and I needed averted vision to clearly make it out from Hawaii's dark skies. At 23× NGC 6352 spans 7′ (38 light-years) and appears round; at this magnification, it is more conspicuous than some of the fainter Messier globulars in Sagittarius. I can see a distinctly mottled texture with averted vision but no *bright* central condensation, though the cluster does have a slight enhancement toward the center. With a glance the cluster appears perfectly round, but spend a few minutes with it using averted vision and see if it doesn't look slightly oval. The cluster also possesses a gray sheen, as if it has silvered with age.

At 72× the cluster is a terribly soft glow — but wait a while and don't give up. With time I spotted a very curious detached bar of light to the southwest of the main milky globe, then a very weak and curved arm to the south. Hints of other arms jutted out from the core with filigreed stellar clumps, especially a Crux-like gathering in a dim arm in the cluster's southwestern quadrant. When James Dunlop discovered this cluster from the Southern Hemisphere, he recorded a similar appearance through his 9-inch telescope, seeing a "rather faint nebula, of an irregular round figure, 4′ diameter, slightly branched; easily resolvable into stars, with slight compression of the stars to the centre." The cluster's entire central body is surrounded by a soft glow. The view is disturbingly asymmetrical, and the asymmetry shifts with your gaze; the more you concentrate on one section of emerging detail, the more lopsided that part of the cluster appears. It's like watching a kitten playing under a sheet. Ernst Hartung saw this cluster as "irregularly round . . . and resolved into very faint stars, some of which can be seen with [a six-inch] scattered through the haze and the outlying region." "It is surprising," he added, "that John Herschel's thorough sweeping missed this conspicuous object."

A more coherent picture seems to emerge if you sketch details as you see them. Do we record the same features? For instance, to my eye the cluster's eastern side looks smooth and gracefully curved, while the western half looks

81

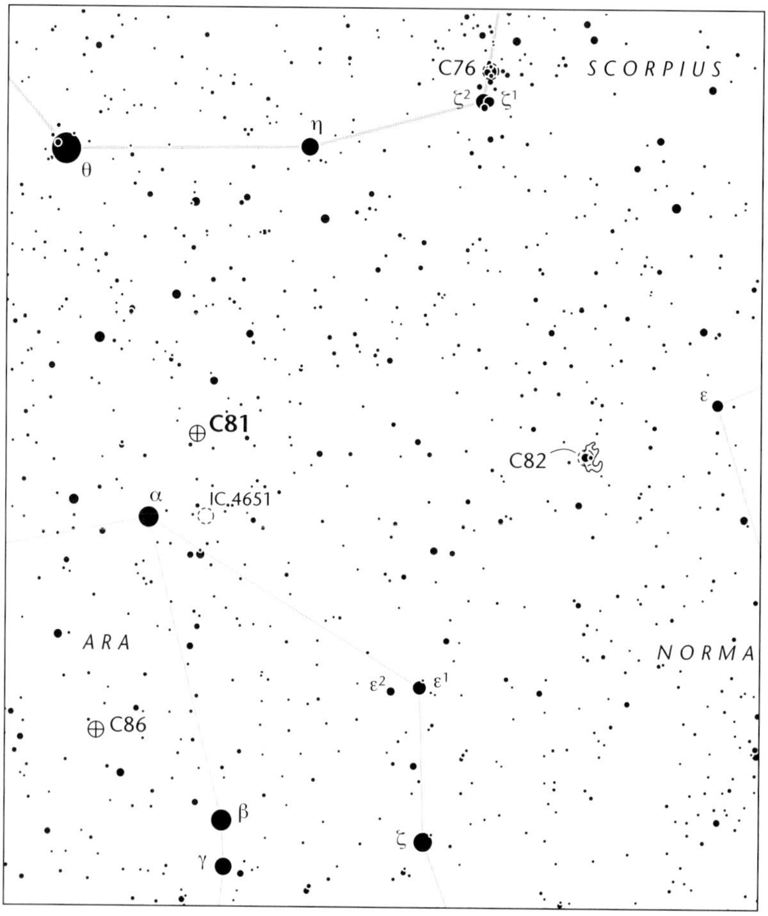

more erratic and ragged. How does it seem to you? Put in high power and tap the telescope tube. Take a few deep breaths, then tap it some more. Doing so will help to bring out faint suns. In fact, when the atmospheric seeing is superb, and my oxygen levels are up, the cluster makes me think of a white volcano experiencing a lateral blast to the southwest. The cluster's horizontal-branch magnitude is 15.1, meaning that you need to be able to see stars that faint to resolve the cluster into a swarm of stars. The cluster's brightest stars shine around magnitude 13.4, and these are the ghostly glints that I follow with my 4-inch, while the fainter suns form a nervous background light. The cluster's spectral class is *G*4, and it is receding from us at 120 km per second.

NGC 6352 is a grass-grazer for many sites across the continental United States. Theoretically it can be seen from latitudes south of 42° N. It is only about 1° farther south than Omega Centauri (Caldwell 80), though it's much fainter, and therefore a greater challenge, than that singular cluster. Through her 20-inch Barbara Wilson saw NGC 6352 from Columbus, Texas, "through the tree limbs in the distance." What's the most northerly site from which one can see this globular? Northern-hemisphere clubs might want to take on this fun challenge.

Before leaving the area, return to IC 4651.

It's a nice even disk of 10th-magnitude suns punctuated to the north-northeast by a 9th-magnitude "beacon." The cluster's core has a dark central hole (dust?) and a thick northwest-southeast-trending dark lane to the west. Together, these two dark masses look like the number "10" when viewed with south up, as in an equatorially mounted Newtonian sited in the Northern Hemisphere. In the 4-inch I also noted an uncataloged cluster or asterism of 10th- to 12th-magnitude suns in a line (oriented northwest-southeast) halfway between Alpha Arae and NGC 6352. Can you see this? Another brighter 20′-wide asterism lies 1¾° northeast of Alpha Arae; it, too, forms a slightly curved line (also oriented northwest-southeast) with about eight suns ranging from 6th to 10th magnitude.

82

NGC 6193
Type: Open Cluster
Con: Ara

RA: $16^h\ 41.3^m$
Dec: –48° 46′
Mag: 5.2; 4.8 (O'Meara)
Diam: 14.0′
Dist: 4,300 light-years
Disc: James Dunlop, included in his catalog of 1827

J. HERSCHEL: Cluster VIII; consists of about a dozen stars [of 10th to 11th] magnitude, and perhaps as many [fainter], with stragglers, which fill field. In its preceding [western] part is a fine double star . . . and yet more preceding [farther west] is a very large, faint nebula, in which the preceding [western] part of the cluster is involved. (h 3642)

GC: Cluster, bright, considerably large, pretty rich, stars of magnitude 10 to 13.

NGC: Cluster, very large, little rich, little compressed, well resolved, clearly consisting of stars, faint nebula involved.

GENTLY STIR CREAM INTO COFFEE AND THE cream will form a spiral pattern before it blends. Can the spiral structure of a galaxy be viewed in much the same way? Only up to a point. As compression waves course through a disk galaxy's plane, matter piles up at the crests of the waves. These density enhancements become the spiral arms as they are curved backward by radial differences in the galaxy's angular rotation speed. But the waves don't blend themselves out of existence, as the coffee-cup analogy would suggest. In the Milky Way and spirals like it, the disk's stars, gas, and dust catch up with these so-called density waves from behind. When a dense interstellar cloud encounters a density wave, it compresses, triggering star formation. Things get really interesting when the compressed cloud's self-gravity becomes a dominant force. Once this happens, the cloud splinters into swirling ribbons, fingers, and globs of cold, opaque gas that can manifest themselves as dark animate shapes; the Horsehead Nebula in Orion and the Eagle Nebula (M16) in Serpens are excellent examples. As the cloud continues to collapse, sections squeeze together so tightly that their interiors start to heat up. Then, suddenly, nuclear reactions fire up within the densest, hottest concentrations. Radiant energy pours out from each newborn star. Some of that radiant energy scatters off the surrounding dust particles, which we can see as icy blue reflection nebulae. The energy also can boil away other sections of the cloud, causing it to glow

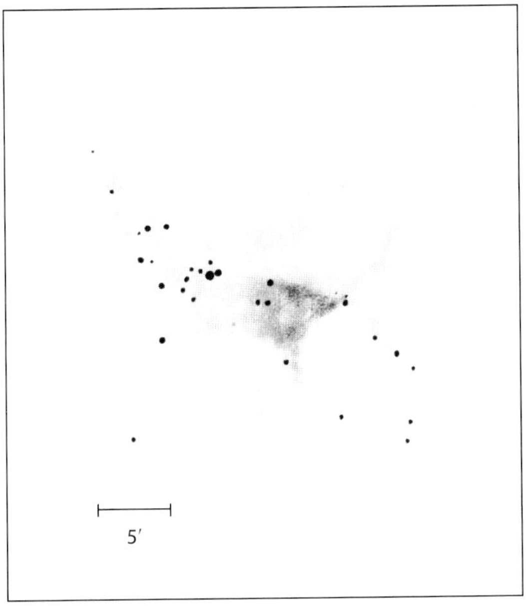

with the fiery reddish light of hydrogen-alpha emission.

Few places in the night sky reveal this wave-driven drama in details large and bright enough to be seen in amateur telescopes. But one region places all the performers on stage, ready for you to critique: it's the Ara *OB*1 Association, a 1°-wide swath of cosmic splendor centered on the naked-eye open cluster NGC 6193 (Caldwell 82). You can find this sparkling cluster nearly 7° southwest of Zeta [1,2] ($\zeta^{1,2}$) Scorpii, itself a member of yet another stellar grouping, the Scorpius *OB*1 Association, whose nucleus is the open star cluster NGC 6231 (Caldwell 76). NGC 6193 is still shedding its swaddling clothes, which can be seen best in photographs as the part emission, part reflection nebula NGC 6188. Cold, opaque clouds of dusty gas make NGC 6193's neighborhood one of the most visually intriguing in the Milky Way. In color photographs the area becomes an elaborate mosaic of cosmic clouds and starlight, replete with baby blue suns, turbulent black eddies, red walls of glowing gas, and diaphanous veils of reflected starlight. Recent studies have shown that the 18-light-year-wide cluster is also on the border of a vast (310-light-year-wide) expanding cloud of atomic hydrogen gas. The dark clouds surrounding NGC 6193 appear to be part of two molecular complexes, one of which might be associated with the cluster.

If you live under dark southern skies, finding NGC 6193 is no chore. It's simply 2½° east-southeast of magnitude-4.5 Epsilon (ε) Normae. Most catalogs list the cluster's magnitude as 5.2, about the same as that of M35 in Gemini. But NGC 6193 is 44 percent smaller than M35, so it's more condensed and thus easier to see. From Hawaii I found NGC 6193 obvious to my naked eye as a magnitude-4.8 "star"; thus my brightness estimate is 0.4 magnitude (44 percent) greater than the commonly published value. The cluster's brightest star, HD 150136, shines at magnitude 5.7 and is also visible to the unaided eye. Through 7×35 binoculars the cluster looks like a bright wedge of Milky Way with a pair of stars marking its southeastern apex. At 23× an irregular scattering of suns begins to reveal itself, though it's hard to determine which belong to NGC 6193 because the region is so rich with stars. (Brent Archinal lists only 14 stars as true cluster members.) The cluster's dominant feature is a small dipperlike asterism with a short handle to the east of HD 150136. A 7th-magnitude star lies about 5′ west of the cluster's dominant star, and a pair of roughly 11th-magnitude suns lie about 4′ due south of that 7th-magnitude sun. The bright Milky Way wedge dominates the area, however, and the reflection nebula is superimposed on it. With a sweeping glance I noticed that the Milky Way region around that wedge has a very peculiar shape, and it soon became apparent that one section of it appears to follow a long

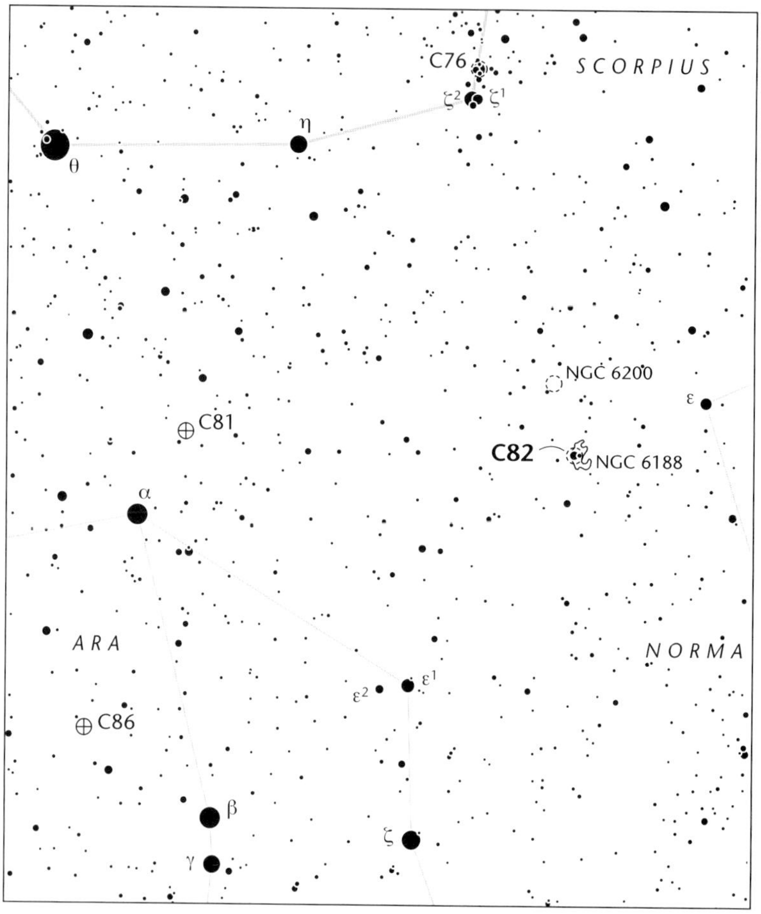

tail of stars that stretches nearly 1½° to the north-northeast of HD 150136, where it connects with another cluster, NGC 6200. Interestingly, NGC 6193's discoverer, James Dunlop, also noted this feature: "A very considerable branch or tail proceeds from the north side, which joins a very large cluster." Another swath of Milky Way curves away to the north-northwest from the bright wedge's western edge. My mind pieced everything together to see an enormous "comet" with a parabolic hood smashing into NGC 6193.

The Milky Way looks very peculiar in this region, especially since its boundaries seem to conform so sharply to specific stellar patterns. One night, as I pondered the view, I noticed a dark pool of nebulosity southwest of the bright wedge; its edges were filigreed with sparkling suns. This pool of cool, opaque matter turns out to be a bright source of infrared radiation and thus a likely site of star formation — and it is readily visible in a 4-inch telescope. It's very difficult (at least visually) to discern any emission or reflection nebulosity in NGC 6193 with a small telescope. Certainly the glow from the bright wedge at the cluster's core is obvious in the 4-inch. But how much of that glow comes from faint, unresolved stars rather than from heated gas or reflective dust? I do not know. Dunlop did not suspect the presence of any nebulosity when he viewed the region with a 9-inch telescope. With his 18¼-inch reflector John Herschel noted a "faint nebula, in which the preceding [western] part of the cluster is involved." Ernst Hartung has definitely seen a nebulous glow in his 12-inch reflector; he recorded "faintly luminous nebular material ... about 100″ across which extends irregularly towards the cluster." He also noted that "[a] clear dark night and good aperture are needed for this object." I know of no other visual reports of NGC 6193's bright nebulosity. How small a telescope will show it? And how will it look if a nebula filter is employed?

Under medium or high magnification HD 150136 can be resolved into a fine double star with a magnitude-6.8 companion that was 9.6″ to the west of the magnitude-5.7 primary star in 1991. James Dunlop discovered that companion in 1836, and the pair has shown little change in separation or position angle since that time. Also in 1836, John Herschel discovered a 10.5-magnitude companion (13.4″ at 160° in 1938) as well as a magnitude-11.4 companion at 14″ and 15° (which also date from 1938; neither pair has changed much since its discovery). The real challenge, however, is spying HD 150136's magnitude-8.5 companion, which in 1938 was a mere 1.6″ off at a position angle of 15°; I did not see this when I looked. Finally, HD 150136 itself is a spectroscopic binary star, one of many in the cluster. The cluster is receding from us at an overall speed of 32 km per second.

83

The Tweezers Galaxy
NGC 4945
Type: Barred Spiral Galaxy (SBcd)
Con: Centaurus

RA: $13^h\ 05.4^m$
Dec: –49° 28′
Mag: 8.8
Dim: 18.6′ × 3.5′
SB: 14.0
Dist: 17 million light-years
Disc: James Dunlop, included in his 1827 catalog

J. HERSCHEL: Bright; very large; very much elongated; very gradually a little brighter in the middle. Length much more than a diameter of the field, or than 15′. Its light extends to a star [of] 14th mag[nitude] beyond the parallel of Brisbane 4299. Position [angle] of elongation 38.7°. (h 3459)

GC: Bright, very large, very much extended toward position angle 38.7°.

NGC: Bright, very large, very much extended toward position angle 39°.

THE SKY IS FULL OF ILLUSIONS. FOR INSTANCE, on spring evenings we can see the Milky Way encircling us along the horizon, making us appear to be at the center of the galaxy; in reality we are well off-center but see several spiral arms meshed together into an apparent ring. Another common illusion relates to double stars. Not all double stars are pairs of physically related stars; the stars in some doubles are many light-years apart but happen to lie almost directly along the same line of sight. This same "coupling" phenomenon occurs with galaxies. Indeed, galaxies are master illusionists. Most curious are those bizarre, apparently disturbed galaxies whose photographs were collected in the 1960s by one of Edwin Hubble's students, Halton "Chip" Arp. Arp investigated relationships between these disturbed galaxies and quasars found nearby. As Dennis Overbye shares in his *Lonely Hearts of the Cosmos*, Arp "turned out to be a genius at finding mystery. Every funny galaxy he inspected turned out to have a quasar tucked under an arm or at the end of a tendril of gas, or lines of them nearby. He photographed luminous bridges of gas that appeared to link galaxies whose redshifts projected them to be billions of light years apart."

Arp also proposed physical relationships between several radio galaxies and other more "normal" galaxies that, in most astronomers' views, simply happen to lie in roughly the same direction on the sky. One of the more exotic and intriguing radio galaxies he collected was NGC 5128 (Caldwell 77), the peculiar radio galaxy Centaurus A. In 1968 Arp reported that one of

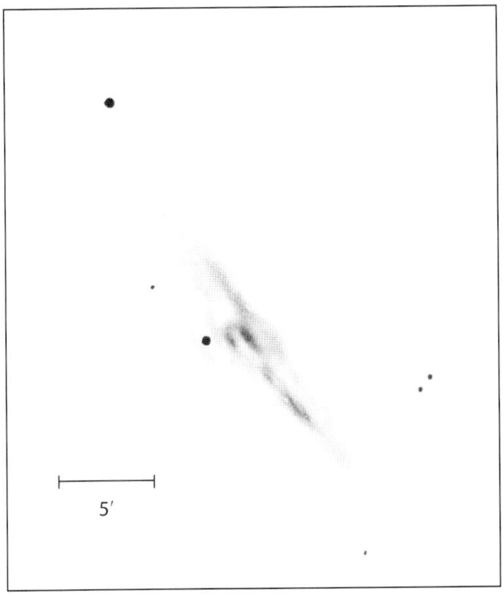

the two jets of radio-wave-emitting plasma streaming out of Centaurus A pointed right to our next target — the nearly edge-on barred spiral galaxy NGC 4945 (Caldwell 83). Although NGC 4945 is about 7½° due southwest of Centaurus A on the sky, the two belong to the Centaurus Group of galaxies and lie at similar distances from Earth. Does the radio jet connect the two galaxies? Maybe, maybe not. "Most astronomers dismissed Arp's diagrams and his computer-processed photographs as coincidences," Overbye writes. But the notion of "intergalactic pipelines" is not totally absurd, especially since HST has imaged a dark string of material flowing between two battered galaxies that bumped into each other about 100 million years ago. That pipeline begins in NGC 1410 and wraps around NGC 1409 like a ribbon around a package. Astronomers used HST to confirm that the 20,000-light-year-long pipeline is a continuous string of material linking these compact galaxies, and it probably was created during a tussle between them.

In photographs NGC 4945 is a similarly chaotic brew of stellar clumps and dusty patches seen through a scrim of foreground stars. The galaxy is quite large, spanning 77,000 light-years, and is receding from us at a speed of 560 km per second. It has the luminosity of 28 billion Suns and is the third-brightest galaxy in the IRAS Point Source Catalog (a list of objects whose brightnesses were measured at far-infrared wavelengths by the Infrared Astronomical Satellite). Most of the galaxy's infrared radiation comes from a compact region around the nucleus, which is surrounded by a dense, roughly 3 light-year-wide torus of dust. Seyfert galaxies, like NGC 4945, are characterized by small, bright nuclei that emit intense and variable X-ray, infrared, and radio radiation while looking comparatively normal at visible wavelengths. Unlike the spectra of most spiral galaxies, which display a continuum and a few absorption lines, those of Seyferts are dominated by emission lines. These lines probably come from interstellar gas heated by X-rays streaming out of the galaxy's nucleus.

There are two types of Seyfert galaxies. Type 2 Seyferts (like NGC 4945) have narrower emission lines than Type 1 Seyferts, indicating gas velocities of a few hundred km per second (or less) as opposed to thousands of km per second. The difference between the two Seyfert types is a consequence of the galaxy's apparent tilt. Type 1 emissions are associated with face-on galaxies, which allow us a clear view of the rapidly moving gas near the core. Since the cores of edge-on galaxies, like NGC 4945, are obscured by dust, we get to see only the narrower emission lines coming from slow-moving gas farther away from the center.

Next to NGC 4151 in Canes Venatici, NGC 4945 is the brightest Type 2 Seyfert galaxy at high X-ray energies, leading astronomers to believe that its nuclear activity may be powered by a supermassive black hole. In a 2000

83

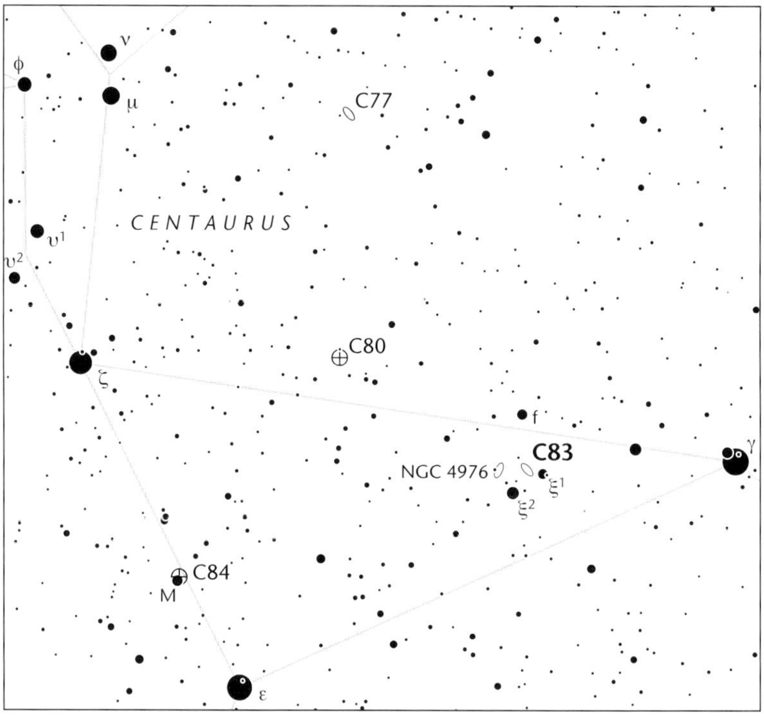

Astronomy and Astrophysics article, Alessandro Marconi and his colleagues discuss how they used the Hubble Space Telescope to detect a roughly 300-light-year-wide starburst ring as well as a conical "super bubble" blown by supernova explosions. Yet not even HST could penetrate NGC 4945's opaque central torus to reveal the active nucleus. With the currently available data it is still not possible to establish whether the intense energy streaming from the galaxy's heart is powered by the active nucleus or by the starburst. If the active nucleus dominates, it must be fully obscured.

You'll find this extragalactic treasure tucked away in a dense nub of Milky Way 4° southwest of Omega Centauri, where background galaxies are dimmed by as much as 0.8 magnitude (a factor of 2) by our galaxy's intervening dust. Were it not for that dust, NGC 4945 would shine at 8th magnitude. And had NGC 4945's disk been tipped a little more our way, it would shine brighter still. To locate this enigmatic galaxy, look halfway between Omega Centauri and 2nd-magnitude Gamma (γ) Centauri, where you'll see a 1½°-long triangle (oriented north-south) of 4th- and 5th-magnitude stars: Xi^1 (ξ^1), Xi^2 (ξ^2), and f Centauri. (Confirm that you can find these stars with your binoculars.) NGC 4945 lies inside that triangle 17′ due east of Xi^1 Centauri, the westernmost member of the triangle. The galaxy is a 9th-magnitude, 19′-long needle of light oriented northeast-southwest. From Hawaii I find it very challenging to see in 7×35 binoculars. But with time its silvery haze glints in and out of view. Seeing NGC 4945 this way requires keen averted vision and patience. Look for a fine thread, sharp yet fuzzy, like a dew-laden strand of cobweb reflecting the first glimmers of dawn.

In the 4-inch at 23× NGC 4945 is very apparent as a slender and diffuse shaft of light, with some patchiness and a hint of a broad central region. The view begs for more power. At 72× the galaxy begins to crumble into tiny details, each requiring time and averted vision to decipher. Most prominent is a bright nub at the galaxy's core, which has a faint skirt of light extending on either side of the major axis. On both ends the galaxy's disk looks gutted, leaving the northeastern and southwestern rims sharply defined and bright, like the prongs of a pair of tweezers. A fairly bright star or clump punctuates the galaxy's northeastern end, and a slightly fainter, though prominent, clump lies due east of the bright hub on the southeastern rim. The galaxy appears to bulge toward the southeast, making the northeastern section of the disk appear narrow. But photographs show this to be an illusion created by a dense ribbon of dust blocking the light. A series of clumps in an arc bowing to the southwest surrounds the nuclear region on the southwestern side, giving the nuclear region a slightly parabolic look. The northern prong on the southwestern limb of the disk appears kinked.

The view at high power is quite interesting. The galaxy no longer appears uniformly bright along its full extent. Rather, its southwestern end is dramatically more obvious. Interestingly, this is exactly what the galaxy's discoverer, James Dunlop, observed:

> A beautiful long nebula, about 10′ long, and 2′ broad, forming an angle with the meridian, about 30° south preceding and north following [from southwest to northeast]; the brightest and broadest part is rather nearer the south preceding [southwestern] extremity than the centre, and it gradually diminishes in breadth and brightness towards the extremities, but the breadth is much better defined than the length. A small star near the north, and a smaller star near the south extremity, but neither of them is involved in the nebula. I have strong suspicions that this nebula is resolvable into stars, with very slight compression towards the centre. I have no doubt but it is resolvable. I can see the stars, they are merely points.

You can sense the enthusiasm of the moment in that passage. Clearly Dunlop was picking out the clumps of starlight shining through the long dust lanes and patches that run the length of the galaxy, as well as a number of foreground stars in that rich section of the Milky Way. In his 12-inch telescope, Ernst Hartung found the northwestern side slightly convex and fairly uniform in brightness except toward the fading ends. And in her 13.1-inch reflector at 105× Barbara Wilson saw NGC 4945 as a "long wispy streak of light that filled the eyepiece [and] looks like a flying saucer."

Overall NGC 4945 reminds me of M82, what with the interplay between its dark lanes and its illuminated portions. The real challenge I found at high power was to resolve the dark notch or "eyelash" between the eastern clump and the nublike core. Much of the dark material in this galaxy can be inferred visually from the patchiness of the luminous sections. Take your time and be patient. I would suggest spending several nights on this galaxy. Keep good records, make plenty of drawings, and your observations should yield more and more detail.

Before moving on, try for NGC 4976, a 10th-magnitude spiral galaxy about 30′ due east of NGC 4945. It's a nice compact (5.4′ × 3.3′) object that can be mistaken for a star. A star of roughly magnitude 8.5 abuts it to the east.

84

84

NGC 5286
Type: Globular Cluster
Con: Centaurus

RA: $13^h 46^m 26.5^s$
Dec: –51° 22′ 24″
Mag: 7.3; 7.2 (O'Meara)
Diam: 11′
Dist: 36,000 light-years
Disc: James Dunlop, included in his catalog of 1827

J. HERSCHEL: Very bright; gradually much brighter to the middle; 2.5′ or 3′ diameter; resolved into 15th mag[nitude] stars; has one star [of] 12th magnitude south following [to the southeast]; the centre near the edge. It is in the field with Brisbane 4618 a star of 6th mag[nitude]. (h 3533)

GC: Globular cluster, very bright, pretty large, round, well resolved, stars of magnitude 15 and fainter.

NGC: Globular cluster, very bright, pretty large, round, well resolved, stars of magnitude 15.

NGC 5286 IS ONE OF THE SKY'S BRIGHTEST, yet most neglected, globular clusters. It is easily located just 5° southeast of the great globular cluster Omega Centauri (Caldwell 80), where it lies directly between magnitude-2.6 Zeta (ζ) and magnitude-2.3 Epsilon (ε) Centauri, though a bit closer to the latter. The cluster literally hides in the blinding topaz glare of 4.6-magnitude M Centauri just 4′ to the southeast. (M Centauri is a rather complex star, being a double with a magnitude-4.6, type-*G*8III primary and an 11th-magnitude secondary — a suspected variable star; and a spectroscopic binary.) NGC 5286 is also part of a 2½°-long binocular asterism that looks like the constellation Scorpius. Shining at magnitude 7.3, the cluster is as bright as, or brighter than, about half of the Messier Catalog's 29 globulars. Is it any wonder, though, that Omega Centauri steals all the attention in this region of sky?

Aside from some of its physical parameters, little is known about the nature of NGC 5286. It has a normal globular-cluster metallicity, with each of its member stars having, on average, nearly 1/50 as much iron (per unit of hydrogen) as does our Sun. The cluster has an integrated spectral type of *F*5. Helen Sawyer Hogg listed eight variable stars in NGC 5286 in 1973, and William Liller increased that number to 16 five years later. All are apparently RR Lyrae types. A 1995 study of the cluster by Nikolai Samus (Institute of Astronomy of the Russian Academy of Science) and his colleagues determined an age between 14 and 18 billion years for the cluster, with the likeliest value being 17 billion years. This finding remains quite controversial, however, both because the cluster has been insufficiently studied and because recent estimates of the Hubble Constant and other cosmological parameters

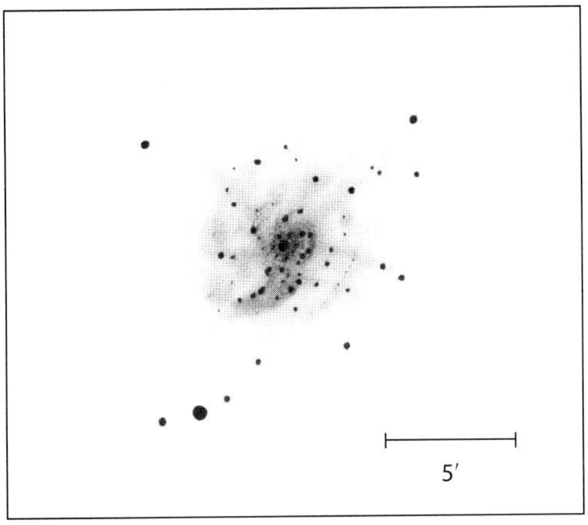

suggest that our universe may be only 12 or 13 billion years old. NGC 5286 lies 28,000 light-years from the galactic center and 36,000 light-years from the Sun. In brightness, apparent size, and distance, NGC 5286 resembles M68 in Hydra. The cluster is moving toward us at 58 km per second, a quarter of Omega Centauri's radial velocity. Interestingly, if you were to reel NGC 5286 in to Omega Centauri's distance, NGC 5286 would still appear about 2.5 times smaller, and 10 times fainter, than that great cluster.

James Dunlop discovered this cluster from Australia with a 9-inch reflector. It is listed as the 388th object in his 1827 catalog. Of it he writes: "A bright exceedingly well-defined rather elliptical nebula, about 1′ diameter, exceedingly condensed almost to the very edge, and gradually a little brighter to the centre. This is about 6′ north of M Centauri. — I have strong suspicion that this is resolvable into stars."

From Hawaii I can see M Centauri with the unaided eye but I could not resolve the cluster next to it. I do not think it impossible to do so, however, especially for someone who observes from the Southern Hemisphere with younger eyes than mine. The globular reveals itself with the slightest of aperture. In 7×35 binoculars it looks like a tiny ghost image of M Centauri, a dim puff of smoke wafting away from that alluring flame-colored star. At 23× in the 4-inch, the color contrast between the two objects is remarkable. M Centauri looks strongly yellow with a tinge of orange — like a dim Arcturus — while the globular cluster glimmers with a very pale ashen blue. Still, at this low magnification little else can be seen; the cluster looks like a circular wafer of uniform light, a breath spot on a mirror. The late, great comet discoverer Jack Bennett of South Africa included NGC 5286 as the 64th entry in his list of objects that looked cometlike through a 5-inch rich-field refractor. Indeed, I have spotted NGC 5286's misleadingly cometary form several times in my own comet sweeps of the Centaurus area.

At 72× NGC 5286 appears as a moderately condensed glow with stumps of starlight poking out at the cardinal directions in a cruciform shape. These are probably the "faint star points" that Ernst Hartung glimpsed with his 4-inch telescope. The cross lies on an elliptical bed of fainter suns. Under a dedicated gaze, star patterns seem to spiral out from the cluster's condensed core, which appears very mottled or patchy, as it is interspersed with dark veins. This view is rather reminiscent of Dunlop's discovery impression. And Dunlop's "strong suspicion" that this cluster would resolve into stars was most certainly justified. Although NGC 5286's horizontal-branch magnitude is 16.5, the cluster's brightest stars shine at a respectable magnitude of 13.5, so partial resolution can be obtained in good telescopes with apertures as small as 3 inches.

At high magnifications the cluster sports some seven arms. Two of them — one to the south and one to the east — form a counterclockwise spiral. The other arms look more rigid. Each is dappled by a few 13th- and 14th-

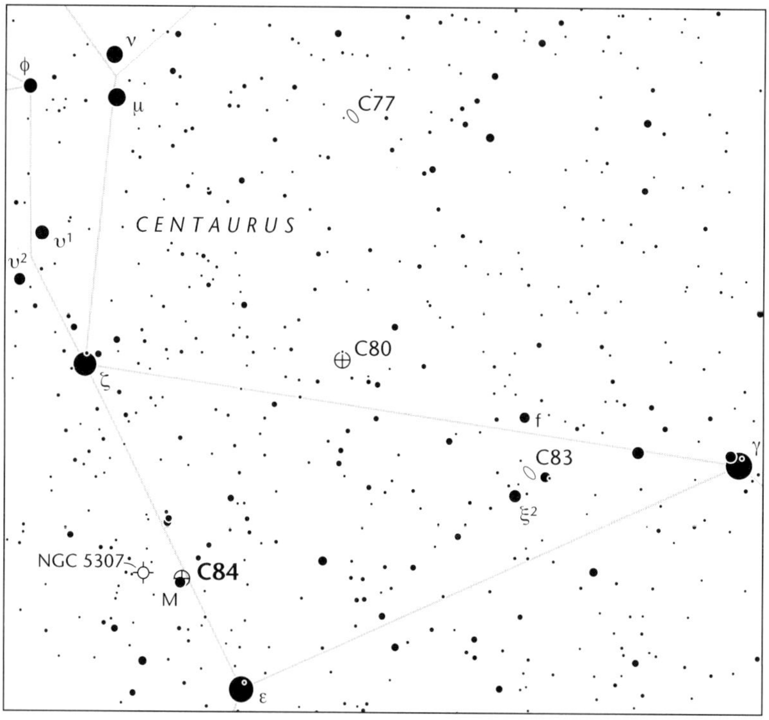

magnitude suns. A dense kite-shaped clustering of roughly 14th-magnitude stars can be seen in the southern arm just beyond the 1.5′-diameter core (which has a very starlike center). The "kite" is cleanly separated from the core by a dark lane running roughly east-west. A few foreground stars, 10th magnitude and fainter, occupy the 11′-diameter cluster's outskirts.

From Columbus, Texas, Barbara Wilson saw NGC 5286 as a "[b]right, high surface brightness cluster" through her 20-inch f/4 reflector; she found it "[c]ondensed and large with a bright center" and "fuzzy" at its low declination. "Must be nice from a southern location," she concluded.

Moving ¾° to the east-northeast of NGC 5286 we find the magnitude-11.2 planetary nebula NGC 5307. It spans a mere 13″ and appears starlike, so high magnification and averted vision (or an Oxygen III filter) are needed to see its uniformly bright elliptical disk. For big-telescope owners, the nebula's central star shines at magnitude 14.6.

One more ancillary note. While observing this cluster and dodging clouds one night, I noticed a peculiar phenomenon. When the clouds rolled over me at Kilauea's 4,200-foot-high summit, I could feel the temperature and humidity rise. I also noticed the crickets and cicadas, who had been chirping beforehand, bring their cacophonous chorus to a halt. When the wind blew the temperature dropped and the insects resumed their singing. Actually, this phenomenon has been known for ages; in fact, a table correlating insect song and temperature has appeared in the *Old Farmer's Almanac.* My point? Though our eyes may be glued to our telescopes, we can keep our other senses open for equally fascinating natural occurrences right here on Earth. Observing is a fully sensual experience.

85

Omicron Velorum Cluster
IC 2391
Type: Open Cluster
Con: Vela

RA: $08^h\ 40.3^m$
Dec: –52° 55′
Mag: 2.5; 2.8 (O'Meara)
Diam: 60′
Dist: 500 light-years
Disc: Abd al-Rahman al-Sufi, about A.D. 964

HERSCHEL: None.

GC: None.

IC: Cluster, compressed, includes star of magnitude 3.7 (o Velorum).

THREE MAJOR "SUPER-clusters" of relatively young stars dominate our region of the Milky Way: the Hyades supercluster, which includes the Beehive Cluster, or Praesepe (M44); the Sirius supercluster, which contains the loose gathering of bright stars in Ursa Major; and the Pleiades supercluster, containing the Pleiades, the Alpha Persei Moving Cluster, and IC 2602 (Caldwell 102). But there are other stellar superclusters in our galactic neighborhood, one of the more conspicuous being the IC 2391 supercluster, which comprises IC 2391 (Caldwell 85) — a large naked-eye cluster in Vela — and some 60 field stars. This supercluster's members are nearly moving together through space, and at least 95 percent of their collective motion is directed toward a spot in Lepus at right ascension $5^h\ 49^m$ and declination –12° 26′.

IC 2391 is commonly known as the Omicron Velorum Cluster because the brightest of the cluster's 30-odd members is the β Cephei-type variable star Omicron (o) Velorum, which

85

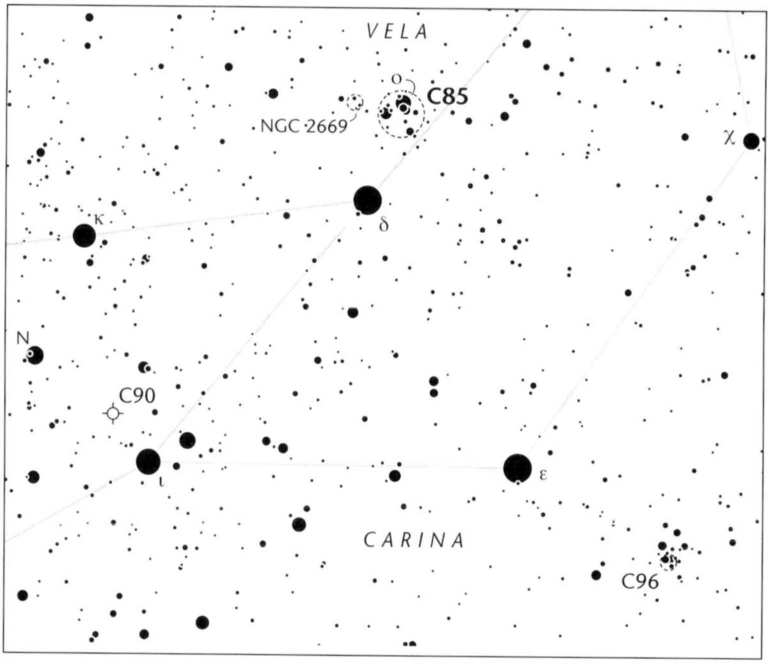

varies between magnitude 3.56 and 3.67 every 3 hours and 7 minutes. The cluster has long been estimated to be about 35 million years young, but in 1999 a team led by David Barrado y Navascués bestowed a new age of 53 million years (plus or minus 10 percent) on it. "While this is considerably older than the age most commonly attributed to this cluster," they wrote in *Astrophysical Journal Letters*, "this [age increase] for IC 2391 is comparable to those recently derived for the Pleiades and Alpha Persei clusters." The scientists had looked for lithium in the spectra of each cluster's faintest stars. They then assessed just how bright a star (or brown dwarf) could be without consuming this fragile chemical element in its central nuclear furnace. In each case, this lithium-depletion measurement yielded a much higher cluster age than had been deduced from more classical analyses.

Like the Pleiades, IC 2391 is a fertile hunting ground for faint, red brown dwarfs — not-quite-stars whose masses (80 Jupiters or less) can't sustain nuclear fusion reactions. As of 1999, extensive photometry had revealed 24 possible brown dwarfs in IC 2391, but more data (particularly near-infrared photometry and spectroscopy) were needed to determine whether or not these candidates truly are substellar brown dwarfs.

Under a dark sky IC 2391 is visible to the naked eye about 1¾° north-northwest of magnitude-1.95 Delta (δ) Velorum, the star marking the northwestern arm of the False Cross. The object is quite bright, and my naked-eye estimate of magnitude 2.8 is reasonably comparable to Brent Archinal's value of 2.5. Persian astronomer Abd al-Rahman al-Sufi (who is also credited with discovering M31, the Andromeda Galaxy) first noted IC 2391 on or around 964 A.D. Of course, the object is bright enough that anyone, from poet to pig farmer, would have seen its light if they happened to gaze toward that region of sky. Abbé

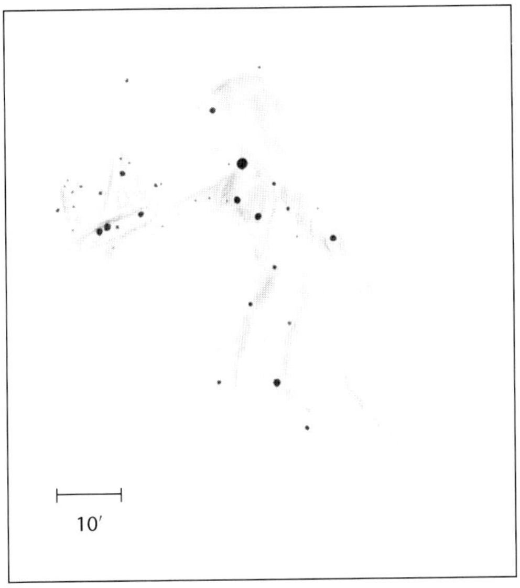

Nicolas Louis de Lacaille, however, is credited with first describing it as a star cluster. It is the fifth object listed in his 1755 catalog's Class II (Nebulous Clusters). In his ½-inch 8× telescope he saw it as a "small star group."

Whenever I take my first naked-eye glance at this cluster, I immediately see a double star, oriented northwest-to-southeast, whose components are separated by about 20′; these "stars" are Omicron Velorum and what is actually the combined light of a magnitude-5.5 star and a magnitude-4.9 star, themselves separated by less than 2′. With a little more time, a magnitude-5.5 star pops into view about ½° south-southwest of Omicron Velorum. These stars form a tight naked-eye triangle. Now return to Omicron Velorum with averted vision. Does it look fuzzy? It should, because it is the brightest of three naked-eye stars that form a tiny (10′-long) arc running from northeast to southwest. Can anyone resolve these stars with the unaided eye? Try to do so in the twilight.

In 7×35 binoculars the cluster's main body comprises about a dozen suns. These stars form an asterism that, to me, looks like the stick figure of a woman extending an arm to the southeast. In her hand is a pillow (the close pair of magnitude-5.5 suns and others). And atop that pillow is a diamond tiara formed by a tight sprinkling of 8th- to 11th-magnitude suns. It looks like a tiny cluster-within-a-cluster. The view is better yet in the Genesis at 23×. The star immediately south of aqua-tinted Omicron Velorum burns with a lemony hue. Among the roughly two dozen suns readily visible in the 1°-wide cluster, several pairs can be seen, and many of these make up the tiara. If we accept the distance tabulated on page 339, this loose cluster is spread across 9 light-years of space. The stars are scattered unevenly, and no point forms a distinct visual center. In fact, Archinal's latest data shift the cluster's center northward to Omicron Velorum, creating an even more disturbingly asymmetric picture. Imagine how different the view of the night sky would be from hypothetical planets orbiting various stars in this lopsided cluster.

Contrary to some reports, IC 2391 does have appeal under higher magnification. At 72×, for instance, about a dozen suns burn forth amid a bright trapezoidal asterism, inside of which lies a smattering of various scintillating suns. It looks as if Orion's Trapezium has been magnified 1,000 times over without any loss of image quality. Omicron Velorum and its two bright counterparts each sport one companion or more. The pillow-and-tiara region also stands apart, surrounded by blackness.

Before moving on, hop about 1° due east of Omicron Velorum to the 12′-wide open cluster NGC 2669. Also known as Harvard 3 (H3) on some charts, it's a neat little 6th-magnitude cluster of about a dozen or so suns, with four prominent pairs forming a mirror-image Greek letter lambda (λ). Can you see it with the unaided eye?

86

NGC 6397
Type: Globular Cluster
Con: Ara

RA: $17^h 40^m 41.3^s$
Dec: –53° 40′ 25″
Mag: 5.7; 5.3 (O'Meara)
Diam: 31′
Dist: 7,500 light-years
Disc: Abbé Nicolas Louis de Lacaille, listed in his 1755 catalog

J. HERSCHEL: Globular cluster, fine, large, bright, round, gradually brighter to the middle, not very compressed, 5′ diameter, but stragglers extend a great way. In the middle is a more compact group of much smaller stars. The stars at circumference are larger than in the middle, at [the northeastern] border is a double star. (h 3692)

GC: Globular cluster of stars, bright, very large, rich, stars of 13th magnitude and fainter.

NGC: Globular cluster of stars, bright, very large, rich, stars of 13th magnitude.

GLOBULAR CLUSTER **NGC 6397** IN ARA HAS an impressive résumé. At magnitude 5.3 it's the fourth-brightest globular cluster in the heavens; only Omega Centauri (Caldwell 80), 47 Tucanae (Caldwell 106), and M22 in Sagittarius are brighter. It is also the second-closest globular to our Sun; only M4 in Scorpius is closer, and by a statistically insignificant 700 light-years at that. What's more, NGC 6397 ranks first among globulars for brightest member stars (10th magnitude) and resolvability (its horizontal-branch magnitude is 12.9, so even a good 2½-inch refractor can penetrate its inner sanctum).

Unfortunately for Northern Hemisphere observers, NGC 6397's declination places it below the horizon from midnorthern latitudes. Even from more southerly locales in the continental United States it hardly clears the horizon. Yet Texas amateur Barbara Wilson calls it a "gorgeous cluster even at this low altitude, only 3° above [the] mountaintops." If you are among the lucky ones who can reach a declination of –54° from home, or if you travel, say, to the Texas Star Party, you'll find the cluster easy to locate. Look 11° south and slightly east of 2nd-magnitude Theta (θ) Scorpii in the Scorpion's Tail. There sits the cluster on the eastern fringe of the Milky Way. The apex of a bright acute triangle comprising Theta and Eta (η) Scorpii and Alpha (α) Arae points right to it. From Hawaii the cluster is easily visible to the naked eye as a 5th-magnitude glow 2° east and a tad south of the midpoint between Alpha and Beta (β) Arae. It lies just 1° north-northeast of 5th-magnitude Pi (π) Arae. In deep photographs the cluster is quite large, covering as

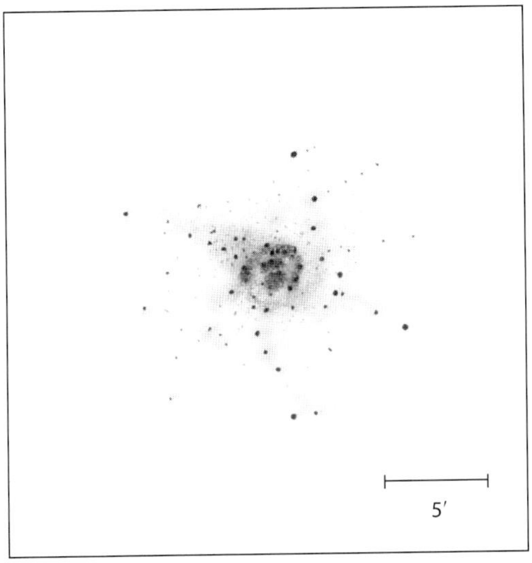

much sky as does the full Moon, though it's the dense inner portion you'll be spying visually.

It's possible that keen-eyed skywatchers in the Southern Hemisphere had detected this cluster's glow many years before Lacaille found it with his ½-inch 8× telescope. But Lacaille was the first to list it. In his 1755 catalog it's the 11th entry under his Class III (stars accompanied by nebulosity). His description reveals little else. James Dunlop was the first to describe the object's true nature, writing, the "nebula is resolvable into stars." NGC 6397 is the 366th entry in his 1827 catalog.

Like 47 Tucanae, NGC 6397 resides close to the plane of our galaxy and is about 20,000 light-years from the galactic center. It probably formed in the galactic halo some 16 billion years ago. Evidence of its birthplace is its paucity of "metals" (chemical elements heavier than helium); on average, each of the cluster's stars contains only 1⁄90 as much iron (per hydrogen atom) as does our Sun. The cluster's spectral class is a yellowish *F*4.

NGC 6397 is so close, so large, and so loose that the Hubble Space Telescope had no trouble seeing right through it to far more distant galaxies beyond. This was not what astronomers expected. Star-formation theory had predicted that red dwarfs with only 1⁄5 the mass of the Sun (or less) should abound, and that's what astronomers expected to see when HST trained its powerful eye toward NGC 6397. But while HST did discover many previously unseen stars, their brightnesses and colors implied masses of 0.2 Sun and above. This showed that low-mass red dwarfs are much rarer than expected, and in doing so, ruled out a leading candidate for dark matter. HST has also discovered 100 white-dwarf stars, four cataclysmic variables, and six blue stragglers in this dense core-collapse globular. Two of the latter are excessively blue and hot even by blue-straggler standards. This is probably the result of either yet-undetected companions or mass accretion within these binary-star systems. HST also identified a nonvariable blue star that may represent a new class with properties reminiscent of white dwarfs. Such stars could result from the stripping of red-giant envelopes in binary systems. Overall, the HST results spotlight the exotic outcomes of stellar interactions and binary evolution in the cores of densely populated globular clusters.

But how serene is the view on dark starlit nights, when through 7×35 binoculars NGC 6397 stands out not as a stellar dynamo but as a soft and gentle glow. Even in binoculars you can feel its closeness, the cluster is so large and conspicuous. With averted vision look for a gradual brightening toward the center. Many sources say the telescopic view is similar to that of M4, but this is not the case. Yes, the cluster is at essentially the same distance as M4, and both shine at about 5th magnitude, but no other globular has M4's brilliant spine of suns, which seems to sizzle with unbridled energy. In the 4-inch at 23× NGC 6397 appears as a clean wash of loosely packed starlight.

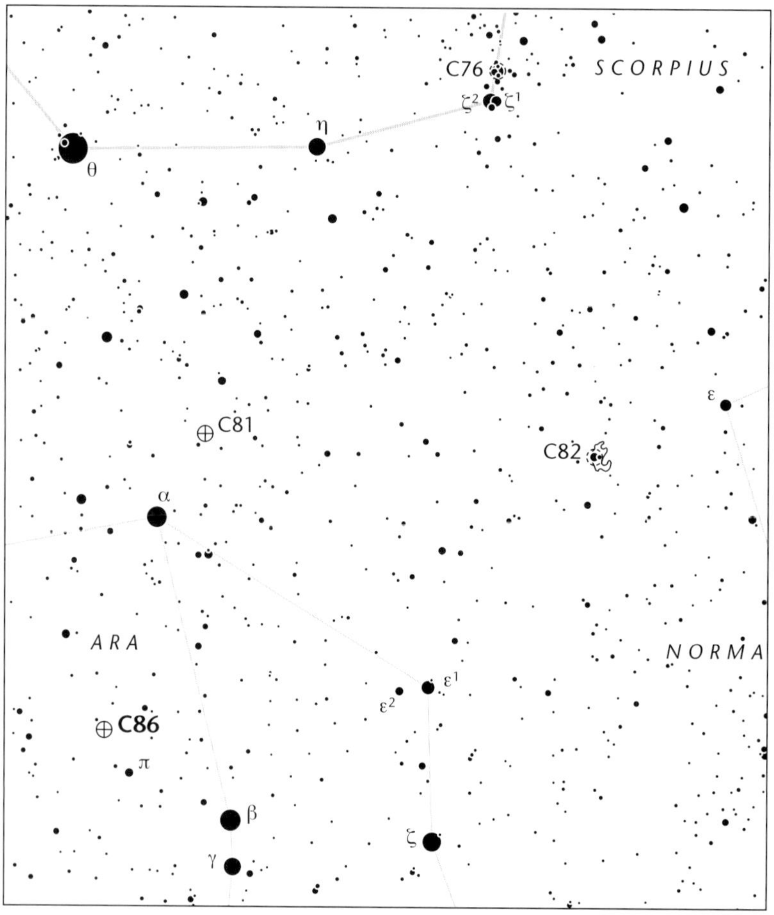

With averted vision it immediately resolves into a granular sphere of crushed diamonds with brighter jewels seemingly tossed helter-skelter about its periphery. About two dozen of the cluster's suns shine between 10th and 12th magnitude. At 72× the cluster looks more cometlike; though the core is circular, a bright tail of mixed suns tapers off to the east-northeast. Aside from the round center, this tail is the single most prominent stellar gathering in the entire cluster. An arrow of four 11th- to 12th-magnitude suns appears opposite the tail to the west-southwest. One meandering arm courses south to a 10th-magnitude sun, while a weaker string of stars points northwest. The remaining suns in the periphery form all manner of shapes: arcs, right angles, Ts, hooks, circles. What appears is a dizzying array of chaos. It looks as if the cluster's center of gravity suddenly ceased to exist, allowing it to fly apart. High magnification will help you plumb the 2.5′-wide core.

87

NGC 1261
Type: Globular Cluster
Con: Horologium

RA: $03^h\ 12^m\ 15.3^s$
Dec: –55° 13′ 01″
Mag: 8.3
Diam: 6.8′
Dist: 55,500 light-years
Disc: James Dunlop, included in his 1827 catalog

J. HERSCHEL: Bright; large; irregularly round; 2.5′ diameter; all resolved into equal stars [of] 14th magnitude. Has a star [of] 9th magnitude [at a position angle of] 45 [degrees] north following [to the northeast]. 3′ distant. (h 2517)

GC/NGC: Globular cluster, bright, large, round, partially resolved, some stars seen.

OUR NEXT CALDWELL OBJECT, **NGC 1261,** is an unobtrusive globular cluster inhabiting an inconspicuous swath of celestial real estate in the unassuming constellation of Horologium, the Clock. Aside from its 4th-magnitude lucida, Alpha (α) Horologii, that constellation is but a dim sprinkling of 5th- and 6th-magnitude suns paralleling the southeastern bank of the celestial River Eridanus. The Abbé Nicolas Louis de Lacaille created the constellation during his famous exploration of the southern skies from the Cape of Good Hope between 1751 and 1753. But, as Julius Staal reminds us in *The New Patterns in the Sky,* Lacaille named the constellation Horologium Oscillatorium, the Pendulum Clock. Lacaille wanted to commemorate the technological achievement of not the *clock* but the *pendulum,* which was designed to make the clock run more accurately. Unfortunately, as time marched on, it became popular to refer to the constellation as simply Horologium. In doing so we have denied ourselves a vital piece of astronomical history.

The best way to find NGC 1261 is to start at 1st-magnitude Achernar (α Eridani). Some 6° to the north-northeast are the roughly 4th-magnitude stars Chi (χ) and Phi (φ) Eridani. Now use your binoculars and look 3½° southeast of Phi Eridani for a 2°-long acute triangle comprising 5th-magnitude Eta (η) and Zeta (ζ) Horologii and a 6th-magnitude star. (Zeta Horologii marks the southern tip of that triangle.) Next, look a little more than 2° southeast of Zeta Horologii for a 1°-wide right triangle of magnitude-6.7 stars. (Finding these triangles should not be a problem because they contain the brightest stars in their respective sky regions.) At this point it may be best to switch to your telescope and put in a wide-field eyepiece, because NGC 1261 shines weakly at magnitude 8.3 and its disk is small and compact. You'll find it just shy of 2° due east of the 1°-wide triangle. TW Horologii, a semiregular variable star that fluctuates between magnitudes

5.25 and 5.95, lies 2° south of the globular; it shines with a distinctive orange hue.

NGC 1261 does not pack a visual punch at low power in the 4-inch. This is not surprising given the object's great distance. NGC 1261 is a halo object 55,500 light-years from our Sun and 59,400 light-years from the galactic center. Although the cluster is far from the plane of the Milky Way, the field at 23× is busy with stars congregating so closely together that they look like dim deep-sky objects or minicomets, giving rise to many false alarms that something new and wonderful lurks near the cluster. I wonder if any of these asterisms fooled John Herschel, since during his second observation of the cluster he penned: "A very faint nebula (??) precedes [lies to the west]." The cluster's small form all but hides among these fuzzy apparitions. At low power you'll find NGC 1261 marking the eastern arm of a 40′-long "warped Northern Cross" asterism of 8th- and 9th-magnitude suns. It manifests itself as a soft but compact 8th-magnitude glow immediately southwest of a roughly magnitude-9.5 star. Once you fix your gaze on it, the cluster is obvious. James Dunlop listed it as the 337th object in his 1827 catalog, describing it as a "very bright round nebula, about 1½′ diameter, pretty well defined and gradually bright to the centre."

At 72× the globular offers the 4-inch a hint of resolution. With averted vision its nearly 3′-wide central condensation appears as a swollen mass punctuated by several stellar clumps. The outer envelope looks fractured, especially to the north. At high power the cluster's core starts to resolve well, with many similarly bright stars in geometrical patterns separated by apparent dark lanes or absences of stars. In this way NGC 1261 appears very similar to NGC 5286 (Caldwell 84), and indeed, each cluster has a horizontal-branch magnitude of 16.7 and a smattering of 13th-magnitude stars. But NGC 1261 "looks" older, as if it's decaying. In fact, high-resolution hydrogen-alpha images of the cluster's core taken with the European Southern Observatory's 3.5-meter New Technology Telescope show a similar effect: the core appears to be melting, with globs of starlight dripping southward from its northern rim. The cluster is about 15 billion years old, according to some es-

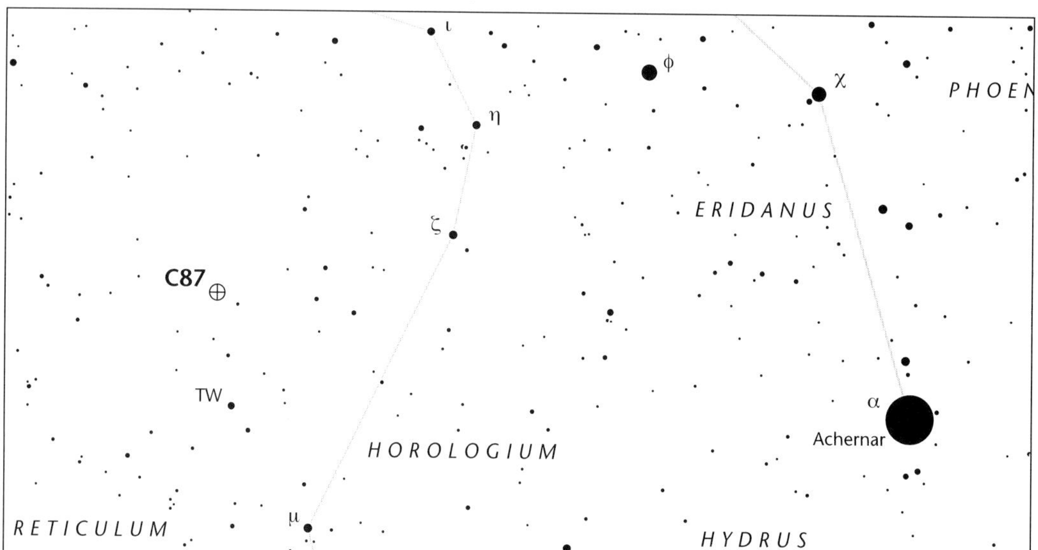

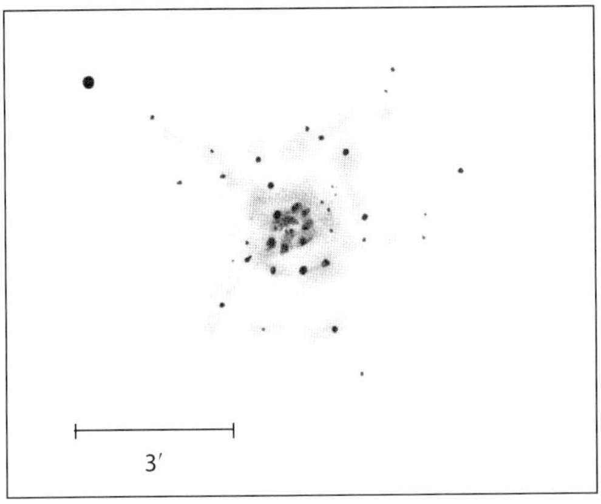

timates, making it 2 billion years younger than the "best-fit" age for NGC 5286 (Caldwell 84).

At high power look for a filigreed arm extending southeastward from the outer rim of the central condensation, and for a prominent arc of 13th- to 14th-magnitude stars on the south side of the cluster's halo. The latter is separated from the main disk by a curving dark rift. Overall, the entire cluster is riddled with rifts, as if it's in the process of being blown apart. The most prominent rift (which rivals the one in M2 in Aquarius) appears in the cluster's northwestern quadrant, just beyond the outer rim of the central blaze. It runs from northeast to southwest. The stars on this rift's northern side taper off to a point, making it appear as if the cluster is wearing some kind of fashionable hat. I find the cluster's multitude of dark lanes as interesting as any of its other features. (Unfortunately they are not visible in most photographs, ours included, because those photographs tend to overexpose the cluster's inner zones.) Be sure to spend much time looking at NGC 1261 with averted vision while playing with the focus knob. Some nicely resolved stars reside in the globular's outer halo. Under averted vision, many of these stars snap into view before fading into the background mist. In this way, the cluster takes on the appearance of a busy beehive. While traveling in Tacna, Peru, Barbara Wilson observed NGC 1261 with a 7-inch f/5.1 reflector and found the cluster to have "an extremely bright little nucleus and fuzzy edges. A few stars are seen but not many, the core is not stellar. . . . Amazing to semi-resolve this cluster with a 7-inch!" And with his 12-inch reflector Ernst Hartung commented that this "well condensed" globular resolves into "crowded stars right to the centre. It is about 2.5′ and the scattered outliers do not extend far; the stars are however very faint and hard to detect with [an 8-inch] although the cluster looks granular."

Each of NGC 1261's constituent stars has 1/20 as much iron as does our Sun, making the cluster moderately metal rich as globulars go, especially since it resides in our galaxy's halo. Its spectral type is *F7*. Since the globular is away from the plane of the galaxy, our sightline to it is virtually free from annoying interstellar dust. Images taken with large telescopes show some very small and dim galaxies on the cluster's immediate outskirts. Noteworthy among them are an open-faced spiral about 4.5′ to the south-southeast of the cluster's core; an edge-on galaxy about 5.5′ to the south; and a fantastic pair of possibly interacting galaxies about 5.0′ to the south-southwest, along a line to a roughly 13th-magnitude sun. The cores of all these galaxies, save for the edge-on one, are apparent in images from the Digitized Sky Survey (one of which appears on page 345). These galaxies are not plotted on the *Millennium Star Atlas*. What's the smallest aperture that will show them to an observer at the eyepiece?

Small-telescope users should be sure to look for one particular Y-shaped asterism of 10th- and 11th-magnitude stars, like a dim version of M73, just 35′ to the southeast of NGC 1261. Asterisms of this type are fairly common; keep your eyes peeled for them.

88

88

NGC 5823
Type: Open Cluster
Con: Circinus; Lupus

RA: 15^h 05.4^m
Dec: –55° 36′
Mag: 7.9
Diam: 12.0′
Dist: 3,400 light-years
Disc: John Herschel, 1836

J. Herschel: [Observed between 15 and 20 April 1836] Cluster class VII; a fine large cluster of separate stars [of] 13[th to] 14th magnitude, 10′ diameter; not much compressed in the middle; nearly fills the field. (h 3589)

GC/NGC: Cluster, considerably large, rich, little compressed in the middle, stars of magnitude 13 to 14.

Open cluster NGC 5823 is one of the few Caldwell objects with dual celestial citizenship. Although its center (as tabulated by Brent Archinal) lies just within the border of Circinus, the Drawing Compasses, its northernmost stars belong to Lupus, the Wolf. John Herschel found it "a fine large cluster." Others have exclaimed that it is "rich" and "exuberant." But observers beware: you may have difficulty finding the cluster or verifying its presence in your eyepiece's field of view. But once that's achieved, the cluster becomes a sight well worth pondering. So first let's get you there.

Finding the area *containing* NGC 5823 is a cakewalk. First draw an imaginary line between zero-magnitude Alpha (α) Centauri (Rigil Kentaurus) — the brighter of the two stars pointing to the Southern Cross — and 3.4-magnitude Zeta (ζ) Lupi 10° to the north-northeast. Now use your binoculars. Locate Zeta Lupi, then move 2½° along that line to the southwest (that is, back toward Alpha Centauri). Do you see a large circular "smoke patch" filled with faint, glittering suns? This is the 5th-magnitude open cluster NGC 5822 in Lupus. It spans an impressive 40′. If you're under a dark sky, lower your binoculars and look for this cluster with the unaided eye; you should be able to detect it. Our Caldwell target, NGC 5823, is 3 magnitudes fainter, lies about 1¼° farther south, and is in a dense, patchy region of the Milky Way. So it's easy to imagine all manner of clusters around it. In binoculars NGC 5823 should appear as if it were a dim, detached segment of

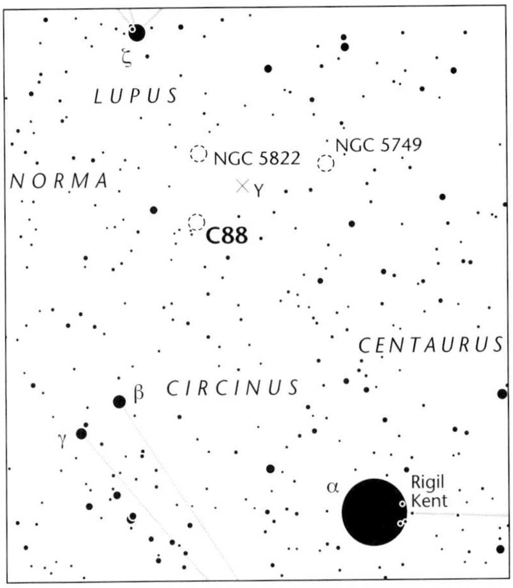

NGC 5822. Don't be surprised if you think you see hints of resolution. This is somewhat illusory, though. While several 10th-magnitude field stars dapple NGC 5823's face, they are foreground stars; NGC 5823's true members shine at magnitude 13 and fainter. The cluster may also look segmented, but this is another illusion created by a pair of roughly 8.5-magnitude stars 15′ to the southeast. The pair looks fuzzy in small binoculars.

At 23× in the 4-inch the region is most complex. NGC 5822 dominates the field. It is so open and patchy and the surrounding swath of Milky Way so rich that it is difficult to fathom its true extent. Furthermore, and most confusingly, a most peculiar gathering of some 20 suns (magnitude 8 to 11) lies 50′ southwest of the core of NGC 5822, where it trends east-west and covers an elliptical area 20′ long. This elliptical cloud's brightest stars favor its west side; included among them is the Mira-type variable Y Lupi, which dips from magnitude 8.2 to 15.2, then brightens, every 401 days. To my knowledge, this "cluster" (RA $15^h 01^m$; Dec. –55° 00′) does not appear on any star atlases as such, so it may be just a dense patch of Milky Way. However, this region of the Milky Way has been poorly studied by professionals; someone should obtain photometry on these stars to see if they are related. A faint tail of dim suns runs northward from our new "cluster" toward NGC 5822, while another travels an equal distance (50′) to the southeast, where NGC 5823 lies. Together these three clusters form a large equilateral triangle and look as if they are but pieces of one massive "supercluster" spanning nearly 1½° of sky.

Actually, that idea is not totally out of kilter. In fact, the visual confusion provided by these clusters and the Milky Way has had professional astronomers scratching their heads, too. In 1968 Mary T. Bruck (University of Edinburgh, Scotland) suggested that NGC 5823 might be physically related to NGC 5822. And while Dennis W. Dawson (Western Connecticut State University) all but disproved that notion 10 years later, he introduced another concept — namely, that NGC 5823 might not be a cluster at all! His study of its red-giant stars showed a wide scattering of physical properties, implying that the stars are not related and, therefore, that NGC 5823 is not a cluster. Furthermore, the density of stars in NGC 5823 seemed to be no different than that of the surrounding Milky Way. Such is the difficulty with open clusters in extremely rich regions of the galactic plane. That being said, the most recent studies of NGC 5822 and NGC 5823 have all but proved they are independent clusters with dissimilar distances and ages. Writing in the August 1981 *Astronomical Journal*, Kenneth A. Janes (Boston University) asserted that Dawson's study was flawed because it included stars across too large an area. Janes's own study stayed within 10′ of the cluster's core, and it showed stronger relationships between the cluster's putative

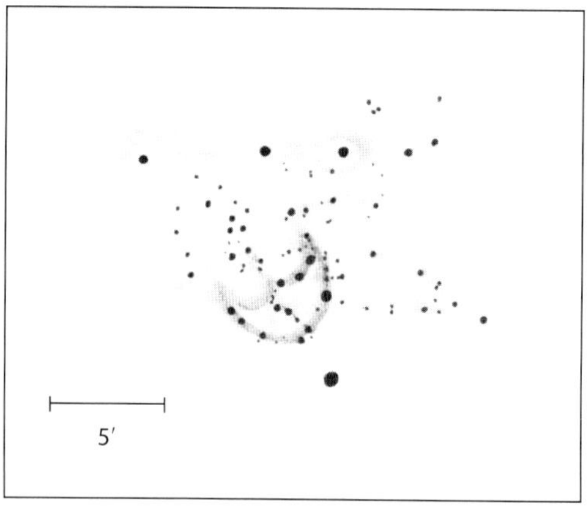

members. Archinal lists 103 cluster members within a 12′-wide area. NGC 5823 extends 12 light-years in space. (Very conveniently, each arcminute equals one light-year at the cluster's distance of 3,400 light-years.) NGC 5822, by the way, is 1,600 light-years closer to us than is NGC 5823, and a bit older; with an age of 600 to 800 million years, it is more or less coeval with the Hyades.

Now that the cluster's identity has been established, we can sit back and enjoy the view. At 23× in the 4-inch NGC 5823 looks boxy, with its brightest stars — actually foreground stars — lining its ragged edges. Inside this box the cluster's main stars trace a distinct tuliplike figure. At 72× the stars outlining the tulip appear most prominent. The base of the tulip is a fine C-shaped curve of equally bright suns whose open end faces northeast. A tiny wave of starlight just northeast of the C completes the flower. Using my imagination I can transform the tulip into a skull half-buried in the stellar sands, while a prominent row of stars to the north (and oriented east-west) becomes a bone lying nearby. Magnifying the cluster will bring out its fainter members, but at the cost of diminishing the cluster's overall appeal. As John Herschel noted, the cluster is "not much compressed in the middle," so magnifying the view just spreads out this already loose aggregation of faint suns. Be sure to look for a tiny clustering of five or six roughly 13th-magnitude suns a few arcminutes due west of the cluster core. It is cleanly separated from the main body of stars and looks rather like a stick figure of a fish.

Of course, the entire region around NGC 5823 and its neighboring clusters deserves attention. A little more than 2° to the northwest of NGC 5823 lies yet another fine open cluster, NGC 5749. This roughly 8th-magnitude glow contains a wishbone-shaped asterism of 10th- and 11th-magnitude stars at its core. Unfortunately, this 7′-wide cluster was omitted from the second edition of *Sky Atlas 2000.0,* but it appears on the chart on page 349. And if you'd like to test your color sensitivity, Zeta Lupi is a nice double star for small-telescope users. Its magnitude-3.5 (*K*0III) primary shines with a brilliant yellow cast, while its magnitude-6.7 (*F*8V) secondary seems to shine with a rust-colored hue. (Of course, star color is highly subjective; the color difference might be the result of the brightness thresholds for stimulating different color perceptions in the eye, rather than intrinsic to the stars. What do you see?) The secondary is 72″ west of the primary (or so things stood in 1991, the most recent date for which the stars' relative positions are tabulated in the *Washington Double Star Catalog*).

89

S Normae Cluster
NGC 6087
Type: Open Cluster
Con: Norma

RA: 16^h 18.9^m
Dec: –57° 54′
Mag: 5.4; 5.2 (O'Meara)
Diam: 15.0′
Dist: 3,300 light-years
Disc: James Dunlop, included in his 1827 catalog

J. HERSCHEL: Cluster VIII class, large, loose, brilliant, irregular figure, fills field, chief star about 7th mag[nitude] taken. (h 3619)

GC/NGC: Cluster, bright, large, little compressed, stars of magnitude 7 to 10.

AS I LAMENTED IN THIS BOOK'S INTRODUCTION, the Caldwell Catalog, as originally published in *Sky & Telescope* magazine, included a serious typographical error. There Caldwell 89 was identified as the open cluster NGC 6067 in Norma. In fact it is NGC 6087 — another open cluster in Norma. As fate would have it, NGC 6067 lies just 3¾° north and slightly west of our true mark, the delightful S Normae Cluster, a dynamic assortment of 349 suns centered on the bright Cepheid variable star S Normae.

James Dunlop first discovered this "group of very small stars of an irregular branched figure," and he listed it as the 326th object in his *Catalogue of Nebulae and Clusters of Stars in the Southern Hemisphere*. The 5th-magnitude cluster can be found in southern Norma, just beyond a fabulously rich and dense part of the Milky Way. To find it with the naked eye, I start with Alpha (α) and Beta (β) Centauri. Just 5° east-northeast of Alpha Centauri (which is also known as Rigil Kentaurus) is a naked-eye double comprising two 4th-magnitude stars, Beta and Gamma (γ) Circini. And a little more than 5° farther east-northeast of Gamma Circini is Iota^1 (ι^1) Circini, the westernmost of four 5th- and 6th-magnitude stars that form a 2°-long, east-west-trending chain. That chain's easternmost "star" is the open cluster NGC 6087.

Under dark skies the 15′-wide cluster is just visible to the naked eye, though it's difficult to separate it from the magnitude-5.6 star some 25′ to its west. Averted vision is required to see the combined glow of the cluster's brightest suns. Most of NGC 6087's light comes from S Normae, a true cluster member. A Cepheid variable, S Normae fades from magnitude 6.12 to 6.77, then brightens again, every 9.75 days. The naked-eye view may take a lot of squinting; try brief, relaxed bouts of averted vision.

In 10×50 binoculars I estimated NGC 6087's magnitude to be 5.2, which is 0.2 magnitude (20 percent) brighter than the published value. Actually, I was quite taken with the binocular view of this cluster. Train your binoculars on it, and you should see the cluster as an east-west trending chain of stars surrounded by a magnificent halo of unresolved suns. S Normae immediately draws attention to itself with its orange color. A 9th-magnitude star pops into view at the extreme northern edge of the halo, and a pair of 8th- and 9th-magnitude suns marks the halo's southern boundary. Overall, the binocular view reminds me of a granite Celtic cross — the kind you'd find glowing in a cemetery, as if moonlit on a foggy night. I cannot see this background glow with the naked eye, only the bright bar of stars; the background swells into view only when binoculars and averted vision are used.

NGC 6087 is a relatively young open cluster, about 65 million years old. It has been of particular interest to astronomers because of its association with its Cepheid lucida. Cepheids have extremely stable periods of variability and very reproducible light curves. In the early 20th century Henrietta S. Leavitt discovered that Cepheids have a very special relationship between their intrinsic brightnesses, or luminosities, and their periods of variability. Basically, the brighter a Cepheid is, the longer it takes to complete one cycle of brightening and fading. By observationally determining the period of a Cepheid variable, we learn its intrinsic brightness. Then, by comparing how bright it appears with how bright it really is, we can deduce its distance. Applying the period-luminosity rule to S Normae, we get a distance of 1 kiloparsec, or 3,300 light-years.

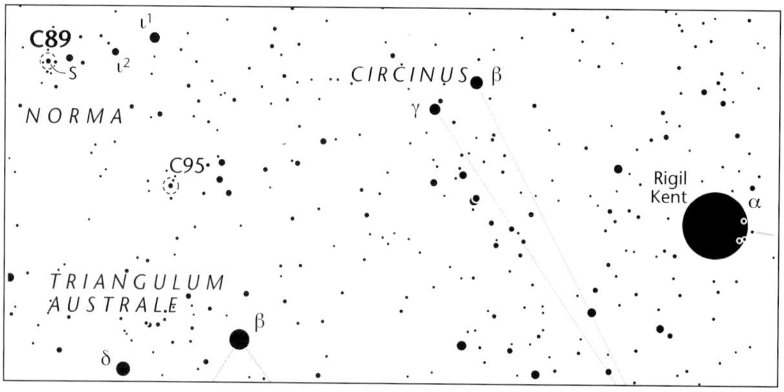

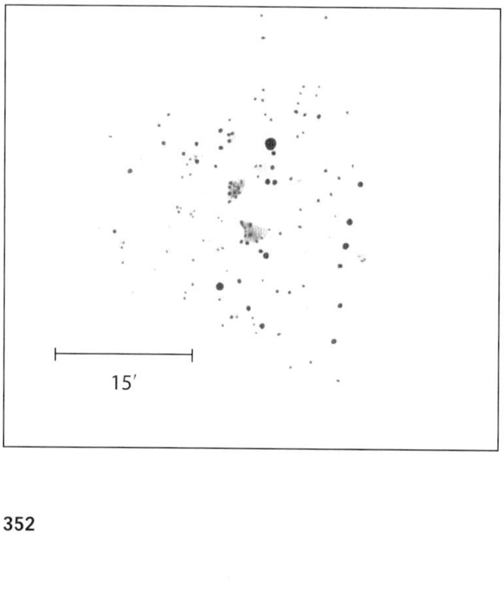

At 23× in the Genesis it's almost hard to believe you're looking at a star cluster with half the angular size of the full Moon, because NGC 6087's brightest stars do not seem to blend together very well. Except for a 5′-wide gathering of suns close to S Normae, the cluster's remaining constituents appear to be scattered like stellar marbles, forming all manner of geometrical patterns — right angles, lines, arcs, diamonds — in a most haphazard way. Add the rich Milky Way background and it's a visual mess. But if you relax your gaze and let your eye roam around the field you should see that the cluster's edges fit within the confines of a boxlike arrangement of field stars. It looks as if several trucks loaded with stars had approached S Normae from various directions, dropped their wares on top of it, and then departed. In fact, if you look about 7′ southwest of S Normae you will see a wonderful 17′-long trail of seven suns (ranging from 8th to 11th magnitude and oriented north-south) that look as if they had fallen out of one of the imaginary trucks as it sped away. Immediately south of S Normae is a tight acute triangle of 9th-magnitude suns; 5′ to the northwest of S Normae is a near-perfect right triangle of 10th-magnitude suns; and 5′ due east of S Normae is a stellar rhomboid. The area immediately surrounding S Normae and the southern triangle glows with the nagging light of stars near the limit of visual resolution, and this background glow continues to the east, tapering along the way, until it reaches the rhomboid. With averted vision, this patch has the shape of a decapitated fish.

At 72× the cluster is a stellar Rorschach test. Its entire 15′ diameter is dappled with at least a dozen obvious stellar groupings and several other seemingly nebulous patches that crumble into stars with any scrutiny. Pale orange S Normae looks like a nova at the tip of a tiny Sagitta-like asterism. (Note that S Normae is not centered in my drawing because, when I was at the eyepiece, the greatest concentration of stars seemed to lie to the south of that star.) The southern end of this arrow is a fine double star; it marks the southeastern corner of the acute triangle I saw at low power. Some 5′ southeast of S Normae lies a small circlet of about a dozen dim suns — one of those nebulous patches that crumble into stars when examined with averted vision. In part it looks nebulous because at least three of its constituent stars are close pairs. Those of you with moderate- to large-sized telescopes should use high power on this circlet and see if you can't "discover" a roughly 13th-magnitude double on its western edge. A really nice double lies 5′ east of S Normae in the rhomboid. And the northern star in that pair is an even closer double that should be scrutinized at high power. These descriptions of doubles could go on forever. The field is littered with them. And it's really futile for me to say what geometric patterns you'll see beyond those I've mentioned, because these are only the obvious ones (to me, anyway).

If you happen to have a visitor at your telescope when you're looking at NGC 6087, share with him or her the fact that each arcminute of cluster spans about 1 light-year (9.5 trillion km) in space. That should enable your companion to fathom the astounding vastness of our Milky Way.

Planetary nebulae are extremely rare in open clusters. NGC 2818 in Pyxis is the only open cluster known to contain one. Other apparent pairings, such as that of the cluster M46 and the nebula NGC 2438, are coincidental alignments of objects at different distances. So you can imagine the excitement Detlev Koester (Louisiana State University) and his colleagues must have felt when they discovered a previously unknown planetary nebula in the field of NGC 6087 (its equinox 2000.0 coordinates are R.A.16^h 19^m 16^s, Dec. –57° 58′ 25″). Alas, in a 1989 issue of *Astronomy and Astrophysics* they concluded that the cluster and the nebula are not physically associated. PN G327.7–05.5, the 19th-magnitude planetary, is 28,000 light-years distant — nearly nine times farther away from us than the cluster. The planetary spans 14.2″, which translates into a shell of gas nearly 2 light-years in diameter. Now there's a wonderful challenge for you CCD users.

90

90

NGC 2867
Type: Planetary Nebula
Con: Carina

RA: $09^h\ 21^m\ 25.3^s$
Dec: –58° 18′ 41″
Mag: 9.7
Diam: 11″
Dist: ~5,500 light-years
Disc: John Herschel, 1834

J. HERSCHEL: [Observed 1 April 1834] The finest planetary nebula I ever remember to have seen for sharpness of termination; 3 arcsec[onds in] diameter; exactly round; no more haziness about them than would be about a star of the same magnitude to-night (which is a favourable one). Light, a pale white = star of 9th mag[nitude]. Position [angle] of companion star = . . . 58.6°; star = 15th mag[nitude]. A very remarkable object. Showed to [John] Stone, who distinctly perceived the total difference of appearance between it and a star [of] 9th mag[nitude] very near it. A second companion star suspected (at about half the distance of the 1st by diagram, and at an estimated position [angle] of 330[°]) among [a] multitude of large and small stars. (h 3163)

GC: *Remarkable,* a planetary nebula equal to a star of 8th magnitude, 59° 8′, 13″ wide.

NGC: *Remarkable,* a planetary nebula equal to a star of 8th magnitude, very small, round, a single star of 15th magnitude, position angle 59°, 13″ wide.

ON APRIL 1, 1834 — OUR APRIL FOOLS' Day — John Herschel was sweeping the southern heavens with his 18¼-inch reflector when he chanced upon a nebula so sharp and round that he believed it to be a new planet. The object measured 3″, as Herschel saw it, and it resembled the planet Uranus, which his father had discovered in 1781. Herschel's heart must have pounded with excitement over the prospect that he might follow in his father's footsteps and contribute to the expansion of the solar system. Anticipating that the object would move, John Herschel meticulously measured its position relative to nearby field stars. He wrote of a 15th-magnitude "companion," which he presumably suspected to be a moon of the putative planet. No motion, however, was detected that night. And though the fol-

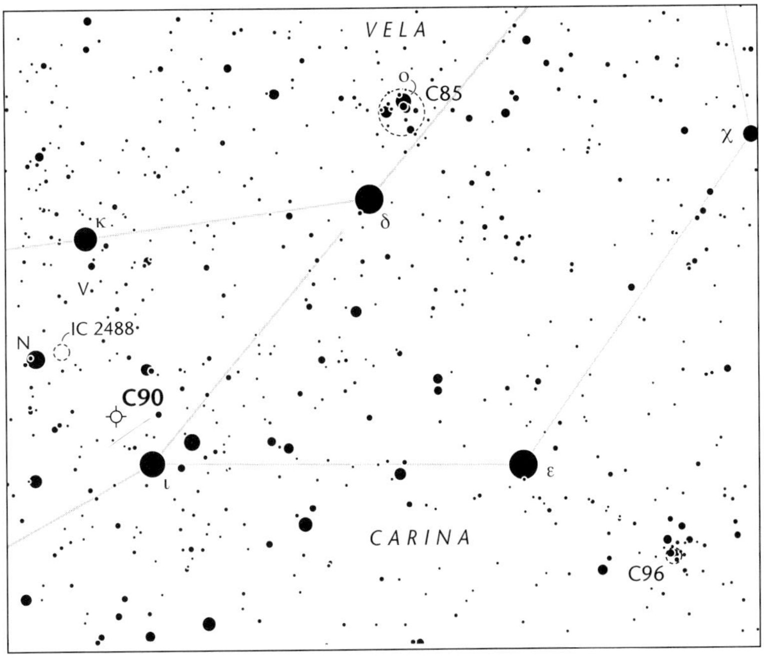

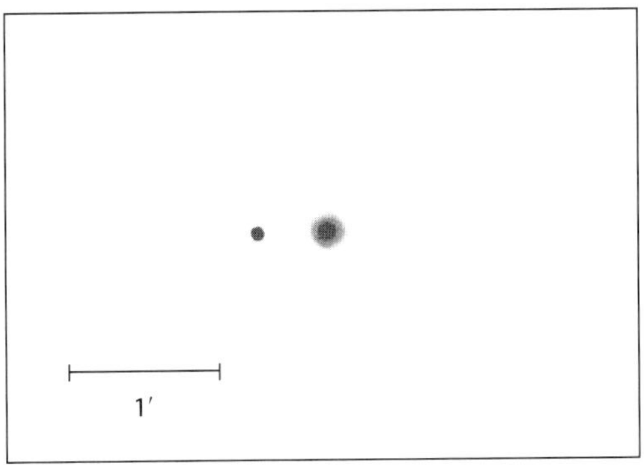

lowing night fell clear, when Herschel returned to his "potential planet" he discovered that "[t]he small star is still 1.5 diam. from edge. It has therefore not moved perceptibly, and is therefore not a planet." We do not need to be at Herschel's side to sense his disappointment. It's the feeling every observer gets after he or she encounters a strange nebulous glow and believes it might be a comet — until its true identity as a star cluster or faraway galaxy is revealed on a star chart.

Today we know Herschel's planetary nebula as NGC 2867, and we know that it is an expanding shell of gas blown off the surface of a dim and dying star. NGC 2867 is one of the easiest objects of its type to spy in the sky. Shining at magnitude 9.7, it is only 0.9 magnitude (roughly a factor of 2) fainter than M57, the famous Ring Nebula in Lyra. But NGC 2867 is seven times smaller than the Ring, so its light is much more concentrated on the plane of the sky. In fact, the nebula has such a high surface brightness that it appears virtually stellar at low powers, a 10th-magnitude "star" among the multitudes in the rich Carina Milky Way. The object's field is a cinch to locate, for it lies a little more than 1° northeast of the 2nd-magnitude star Iota (ι) Carinae (Aspidiske), which marks the southeastern arm of the False Cross. The nebula is the center "star" in a 10′-long chain of three roughly 9th-magnitude, equally-spaced suns oriented northwest to southeast. Almost any instrument will show NGC

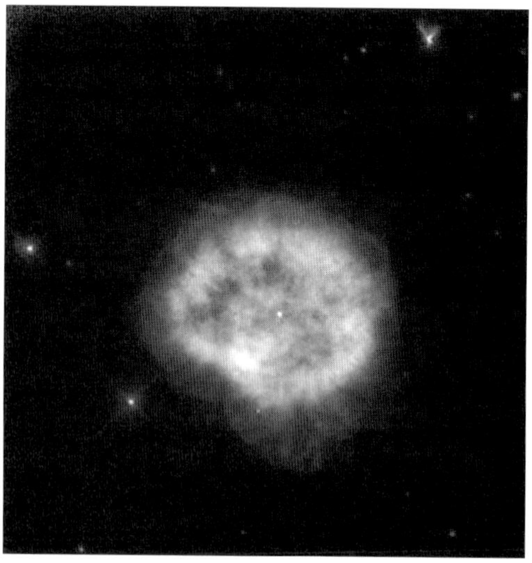

2867 as a "star," but a fair bit of magnification and a stable atmosphere are required to identify it as a planetary nebula. (That is, unless you slip an Oxygen III filter in front of your eyepiece, then out; this will briefly pop the nebula in and out of view.)

Several times I looked at NGC 2867 when it was low to the horizon from Hawaii. Under such circumstances the field stars were so bloated, they *all* looked like planetary nebulae, even at 23× in the 4-inch, and higher powers only magnified their swollen masses. But when the object was higher in the sky, and the stars appeared as tack-sharp points of light, the nebula revealed itself as a slightly swollen disk, especially with averted vision. At 72× its tight inner disk was surrounded by a faint, thin ring of diffuse light. I have seen this feature several times. It shows best with medium magnification; high power tends to magnify the dim light beyond recognition. The inner disk is so condensed, however, that you can crank up the power to ridiculously high levels while viewing it. But don't be disappointed if you still cannot see much more than what I've described here. Even when seen through a 12½-inch telescope, the object has been reported as "small." Andrew Murrell (Ilford, Australia) used a 20-inch f/5 Dobsonian at 200× on NGC 2867 and saw a "small bright planetary with a diameter of 20″." He also said it had an even surface brightness, and he saw no central star. "The planetary appeared round," he continues, "even with high magnification. At low power, the planetary had a very pale aqua colour which was lost when observing above 200×." Ernst Hartung also saw a "pale blue" color. Using 72× with the 4-inch on May 24, 2000, I had momentary impressions of a pale glacial blue cast, like that of the daytime sky reflected in ice. I know of no positive sightings of the 15th-magnitude central star, which may vary in brightness.

Fortunately we have instruments like the Hubble Space Telescope, which can turn the mundane into the magnificent. HST showed NGC 2867 to be a multishelled planetary with a prominent ring (like M57's) that appears "dented" on its eastern edge, as if a car had accidentally backed into it. Close inspection shows that ring to be clotted and irregular. That bright annulus also is surrounded by a turbulent collar of glowing green gas. Inside the ring is a faint ellipsoidal cloud, with the hot central star "pushed" against its western edge. It's as if we are looking down a barrel at the central star from a slight angle. In essence, we are doing just that; the inner oval is what we see when we look down the funnel of a bipolar flow of gas, while the fainter outer halo is the remnant of the slow wind that the central star cast off when it was still a red giant. Recent studies place NGC 2867 about 5,500 light-years distant, making the shell's true physical extent 0.3 light-year. Some estimates, however, place the nebula 1,600 light-years farther out, which would increase the planetary's true diameter by about 30 percent.

NGC 2867 is a nitrogen-rich planetary nebula; its nitrogen abundance is 1.6 times greater than that found in our Sun. Excess nitrogen in planetaries is created when carbon is destroyed in nuclear reactions; this takes place when the nebula's precursor star starts out with 3 to 11 solar masses. Spectra of NGC 2867 also have shown it to have more carbon (per unit of hydrogen) than our Sun contains. The carbon and nitrogen both may have been dredged up by an episode of convection late in the star's life, when carbon, oxygen, and nitrogen were catalyzing the fusion of hydrogen into helium. On the other hand, spectroscopic observations from the Anglo-Australian Telescope and the International Ultraviolet Explorer satellite show no significant chemical-abundance abnormalities. That being said, in 1998 Walter Feibelman did describe a rich variety of previously unreported emission lines in the IUE spectrum of the nebula and its central star. That star, by the way, is a 15th-magnitude Wolf-Rayet star of sorts; it has a spectrum similar to the much more massive Wolf-Rayets like HD 192163 in NGC 6888 (Caldwell 27). NGC 2867's nebulous shell is believed to contain 20 or 30 percent as much mass as our Sun and (as seen from Earth today) to be about 3,000 years old.

Just over the border in Vela, 1½° northeast of NGC 2867 and 30′ west of the star N Carinae, is the 6th-magnitude (my estimate) open cluster IC 2488 — a splash of 9th- to 11th-magnitude stars that look like a drooping bow tie. And just a little less than 1° due south of Kappa (κ) Velorum, the northeastern tip of the False Cross, is the bright binocular variable V Velorum, a classical Cepheid whose brightness fluctuates between magnitude 7.2 and 7.9 every 4.37 days.

91

The Pincushion Cluster
NGC 3532
Type: Open Cluster
Con: Carina

RA: $11^h\ 05.5^m$
Dec: –58° 44′
Mag: 3.0; 3.2 (O'Meara)
Diam: 50.0′
Dist: 1,600 light-years
Disc: Abbé Nicolas Louis de Lacaille (in his 1755 catalog), but probably known since antiquity

J. HERSCHEL: Chief star of a very large, round, loosely scattered cluster of stars [of] 8[th to] 12th magnitude, which fills 2 or 3 fields. A fine bright object. (h 3315)

GC/NGC: *Remarkable*, cluster, extremely large, round, little compressed, stars of magnitude 8 to 12.

MIDWAY BETWEEN THE FALSE CROSS AND the Southern Cross lies one of the heavens' true stellar spectacles, the 3rd-magnitude open star cluster NGC 3532 (Caldwell 91). It is a burst of visual splendor in the swirling tempest of the rich Carina starcloud. Together with the Eta Carinae Nebula (Caldwell 92, just 3° to the west-southwest) and IC 2602, the Southern Pleiades (Caldwell 102), NGC 3532 is part of *the* most magnificent tract of Milky Way in the entire heavens. Doubtlessly it has been a visual landmark ever since aboriginal people first began looking to the heavens. Yet its discovery is credited to an 18th-century European explorer, Abbé Nicolas Louis de Lacaille; he included it in his 1755 catalog as the 10th member of his Class II (nebulous clusters). It seems almost blasphemous to consider this object an 18th-century European revelation, and I'm certain Lacaille himself would have agreed. That being said, this cluster's brightest stars shine at magnitude 8, so they are too faint to discern individually without optical aid. To the naked eye the cluster looks like an enhancement of the Milky Way, a fantastic knot in the fabric of space — but not a star cluster. It took Lacaille's meager ½-inch 8× telescope to reveal, for the first time, that this diffuse glow was actually "a prodigious number of faint stars forming a semi-circle of 20/25 [arc]minutes diameter."

Nearly three-quarters of a century later James Dunlop trained his 9-inch telescope on the southern skies and encountered NGC 3532, which he described as "a very large cluster of stars about the 9th magnitude, with a red star of the 7-8th magnitude, north following [northeast of] the centre of the cluster. Elliptical figure: the stars are pretty regularly scattered." Dunlop's description falls far short of declaring the object's true glory, which John Herschel and, later, William Pickering extolled. Herschel called NGC 3532 a "superb" cluster, "the most brilliant object of the kind I have ever seen." And like Herschel before him, Pickering saw it as the finest irregular star cluster in the sky.

Today NGC 3532 is classified as a very rich, intermediate-age open cluster. Recent estimates date it at about 300 million years; that's halfway between M37 in Auriga (200 million years old) and M44 in Cancer (400 million years). NGC 3532 weighs in with at least 2,000 solar masses, and its brightest stars huddle in its central region. In the 1930s Harlow Shapley found the cluster unusually rich in bright members of the *A*-type spectral class. Surprisingly, the cluster received little professional attention since then, until interest picked up again in the 1980s. Today we recognize 677 stars as cluster members. The vast majority of these are *A*-type stars, though there are also eight *K*-type stars. As Herschel suspected, the cluster is also binary-rich (he noted "several elegant double stars, and many orange-coloured ones"). NGC 3532's metallicity is essentially the same as the Sun's. In a 1993 *Astronomy and Astrophysics* article, Detlev Koester (University of Kiel, Germany) and Dieter Reimers discovered seven white-dwarf candidates and confirmed that three of them were in fact white dwarfs, with temperatures of about 28,000° Kelvin. Two of these white dwarfs have 0.6 solar mass, while the third is a 0.9-solar-mass stellar cinder. White dwarfs are the evolutionary endpoints of relatively low-mass stars like our Sun; more massive stars explode as supernovae. Where exactly is the dividing line between white-dwarf precursors and supernovae-to-be? According to Koester and Reimers, it lies somewhere between 6 and 8 solar masses.

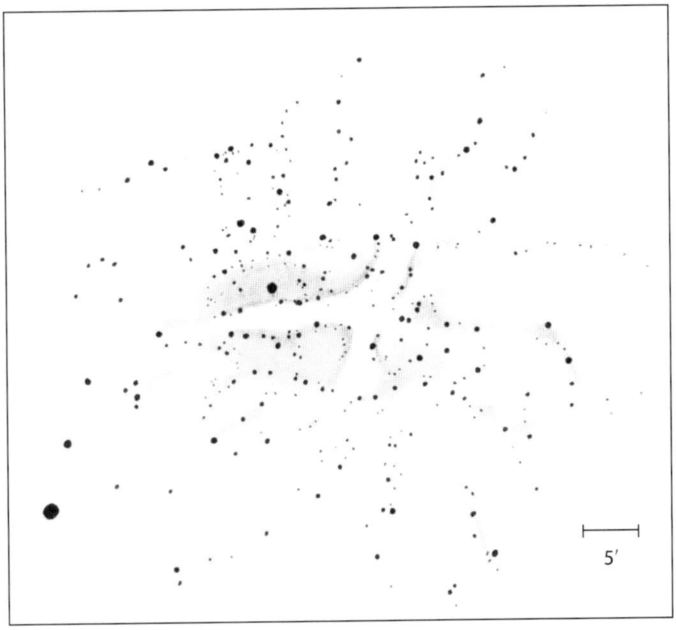

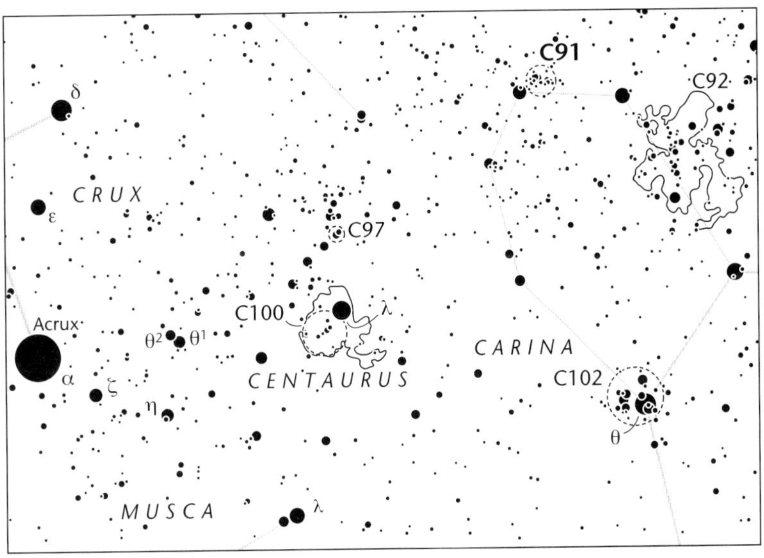

Interestingly, Koester and Reimers found no white dwarfs in the cluster's center, where the brightest stars convene; their finds all lie 15′ to 60′ away. This strengthens the theory that low-mass and high-mass stars become segregated in clusters as time passes. In NGC 3532's case, this process is estimated to have taken about 40 million years to complete.

To the naked eye NGC 3532 appears as an intense cloud more than 1½ Moon diameters across. A 4th-magnitude Cepheid variable (V382 Carinae) lies on the southeastern edge of the cluster, whose dense, elliptical core of bright suns curves away from that star to the west like a tiny comet tail. My naked-eye estimate of the cluster's apparent magnitude agrees fairly well with the published value of 3.0, though I find the cluster 0.2 magnitude (20 percent) fainter. (Note that V382 Carinae isn't included in my estimate of the cluster's apparent magnitude.) Train a pair of 7×50 binoculars on NGC 3532 and *bang!*— five dozen suns come screaming out at you through a blizzard of background starlight. The core itself looks like a 30′ × 15′ pile of shattered glass scattered to the east and west, as if a crystalline globe had just fallen victim to a steamroller. Another yellow star, shining at 6th magnitude, lies on the northeastern fringe of this glassy core.

At 23× in the 4-inch NGC 3532 is a stunning agglomeration of stars packed into a diamond-shaped core crisscrossed by thick dark lanes, giving it the appearance of a diamond brocade set in black gold. Filigreed arms radiate in all directions from this core and some take on macabre forms. Defocus the view slightly and some equally magnificent dark features materialize out of the background fog. Most noticeable (when seen with south up) is a dark numeral 7 or letter T that slices through the cluster's eastern side; an ink-black well lies immediately north of it. Due west of the cluster's elongated core is a double pond of darkness with a river of shadows pouring off to the south. I could go on, but I prefer to let your imagination give form to the cluster's many dark features. High magnification does little to enhance this cluster's beauty. Some black voids show up better at high power, as do the cluster's many doubles and tiny stellar groupings. But NGC 3532, like many of the night sky's treasures, is best admired at low power.

During my January 2000 trip to New Zealand I spent some time with Ian Cooper, that country's leading deep-sky wizard. We were observing what we could with the naked eye and binoculars. The Large Magellanic Cloud and the Carina Milky Way had just peeked above a thick band of clouds. Suddenly Cooper thrust a pair of 11×80 binoculars my way and pointed excitedly to NGC 3532. "There ya go, Stephen," he said with his delightful accent. "Take a look at that . . . it's the Pincushion." I had to laugh. What a grand and appropriate name! I don't know who first pinned it to the cluster, but I hope it sticks.

92

Eta Carinae Nebula
NGC 3372
Type: Emission Nebula
Con: Carina

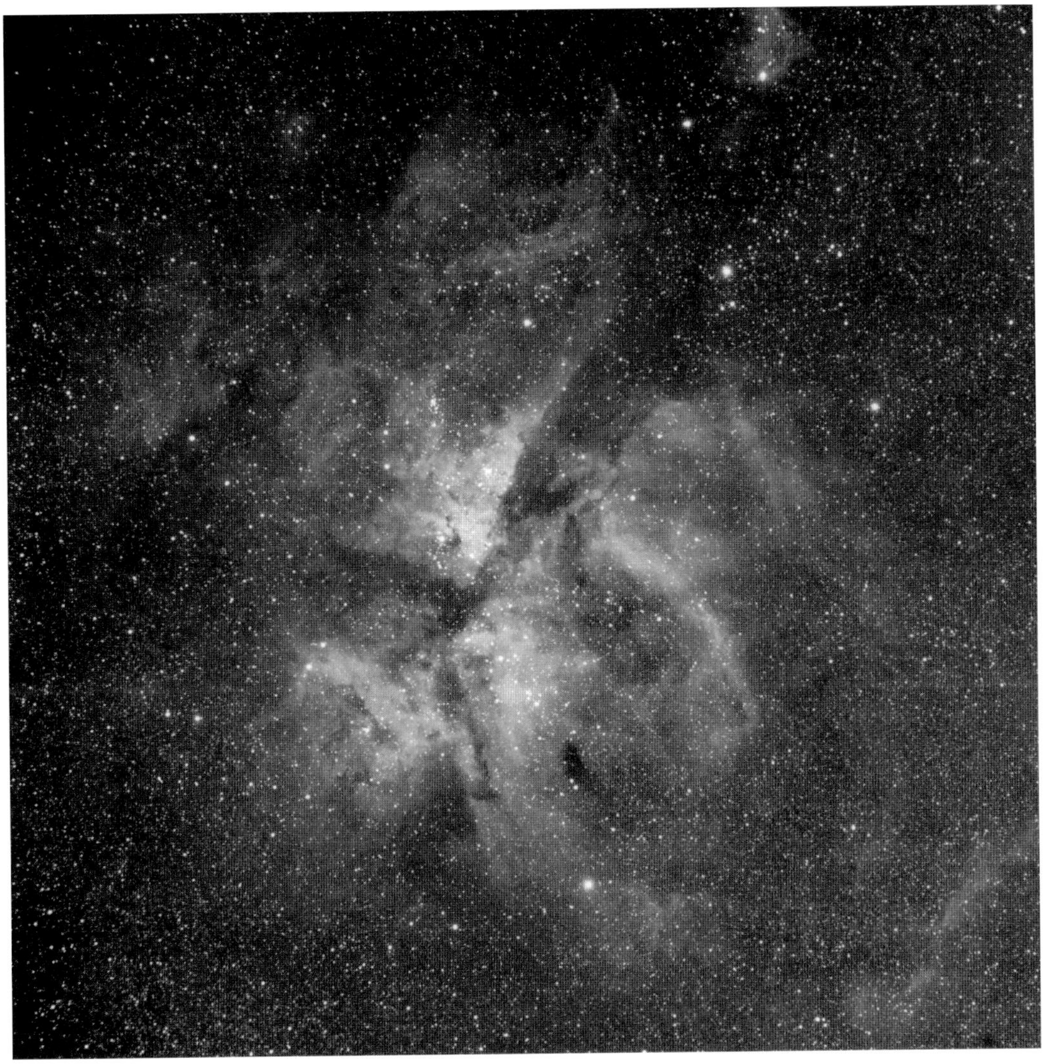

RA: 10^h 43.8^m
Dec: –59° 52′
Mag: 4.8 (variable), nebula; 4.5 (O'Meara)
Mag: ~4 (variable), star
Diam: 120′ (variable)

Dist: ~7,500 light-years
Disc: Known since antiquity (?)

J. HERSCHEL: [Notes below.] (h 3295)

GC: η Argus, the great nebula, monograph published.

NGC: *Remarkable*, great nebula, η Carinae.

92

IF WE COULD PHOTOGRAPH HUMAN IMAGI-nation, the result might look something like the Eta (η) Carinae Nebula — an elaborate tapestry of stellar energy fused into luminous folds of gases and strung together with threads of dust. To the naked eye the complex appears as a silver pendant of nebulous matter in the sparkling collar of the Carina Milky Way. It is the most spectacular example of an emission nebula in the heavens, and it resides in the richest and most dramatic region of the Milky Way, a stormy swath of suns between the Southern Cross and the False Cross. Wide-field photographs show Eta Carinae as a symphony of visual splendor; the heart sees it as the Taj Mahal of galactic wonder.

Some books credit Abbé Nicolas Louis de Lacaille with discovering the nebula. This claim seems ludicrous. Today, the nebula is apparent at a glance. So why wouldn't southern stargazers have known of it throughout time? For a clue, let's look at Lacaille's "discovery" descriptions. Looking at the nebula through a ½-inch 8× telescope — very roughly the equivalent of a small pair of binoculars — he noted "[t]wo small stars surrounded by nebulosity." On another night he described it as a "[l]arge group of a great number of small stars, little compressed, and filling out the space of a kind of a semi-circle of 15 to 20 minutes in diameter; with a *slight nebulosity* widespread in space" (emphasis mine). Would anyone today looking at this region of space in binoculars call the nebulosity *slight?* Equally remarkably, Edmond Halley is said to have observed Eta Carinae, the bright star embedded in the nebula, from St. Helena in 1677, nearly 80 years before Lacaille. Yet he failed to mention the presence of any nebulosity. Not until James Dunlop observed the object in the 1820s do we get a sense that a remarkable nebula can be found here:

> [Eta (η)] Roboris Caroli* . . . is a bright star of the 3rd magnitude, surrounded by a multitude of small stars, and pretty strong nebulosity; very similar in its nature to that in Orion, but not so bright. . . . I can count twelve or fourteen extremely minute stars surrounding η in the space of about 1′; several of them appear close to the disk: there is a pretty bright small star about the 10th magnitude north following [northeast of] the η, and distant about 1′. The nebulosity is pretty strongly marked; that on the south side is very unequal in brightness, and the different portions of the nebulosity are completely detached. . . . There is much nebulosity in this place, and very much extensive nebulosity throughout the Robur Caroli, which is also very rich in small stars.

Is it possible that the nebula has been gradually brightening over time? Today its brightest parts are so intense they can be glimpsed even in astronomical twilight, and the Eta Carinae Nebula is as much a part of the naked-eye fabric of the Milky Way as, say, the Double Cluster in Perseus. So how could its discovery belong to anyone in particular? Before answering that question, let's see what modern science has to say about the nebula and its source of radiant energy.

The Eta Carinae Nebula is a remarkable H II region, the largest known in our galaxy. It measures some 260 light-years across — nearly 7 times the size of the Orion Nebula. The nebula appears severely disjointed, with fantastic blossoms of glowing gas clinging to a twisted

* Robur Carolinum (Charles's Oak) is a now-defunct constellation that Halley designated in 1679. Eta Carinae was briefly known as Eta Roboris Caroli. Eta Carinae also was known, in the 19th century, as Eta Argus, when Carina was considered merely a part (the Keel) of Argo Navis (the Ship Argo), another now-defunct constellation.

branch of cold, dark material. Indeed, no less than eight star clusters lie in or near the nebula, and four of those lie very close to the nebula's heart. The Eta Carinae complex harbors one of the highest concentrations of early *O*-type stars known in the galaxy. And it is these hot, young stars, perhaps just a few million years old, which energize the entire nebula. Furthermore, in the nebula's bright northern fan we find the most intriguing star in the night sky — the enigmatic novalike variable Eta Carinae, one of the most powerful laboratories available for investigating the early evolution of the most massive stars.

Eta Carinae has roughly 100 times the Sun's mass and (at the time of this writing) 4 million times its luminosity. Over the years, though, its brightness has varied most spectacularly. Historical records show the star ranging from magnitude –0.7 (outshining nearly all the night sky's stars) to about magnitude 7.6 (well below naked-eye visibility). Little is known about its intensity prior to Halley's 1677 observation. However, German orientalist Peter Jensen believes this star's fluctuations have been known since almost prehistoric times. The star, he says, is mentioned in Babylonian inscriptions. Apparently Eta Carinae was a temple star associated with Ea, or Ia, otherwise known as Oannes, the mysterious human fish and the greatest god in the kingdom of Eridhu (the Holy City). This implies that Eta Carinae shined magnificently near the dawn of human civilization. In his *Star Names: Their Lore and Meaning,* Richard Hinckley Allen adds that in ancient China, Eta Carinae was known as *Tseen She* (Heaven's Altars) — yet another moniker implying greatness on high. But little else has been uncovered about the star or its nebula. Even Halley's observation has an aura of mystery. John Herschel credited Halley with first recording the star's brightness (4th magnitude). But, as Allen notes, Eta Carinae does not appear in Halley's *Catalogues Stellarum Australium,* the compendium of 341 stars he observed while at St.

Helena. How, then, did Halley's observation become public?

In the 150 years following Halley's observation, Eta Carinae's brightness was to fluctuate erratically, achieving a maximum, at 1st magnitude, in 1827. The star then dipped back to 2nd magnitude, where it remained for about five years. Writing in his 1847 "Results of Astronomical Observations," John Herschel picks up the story from there:

> When first observed by myself in 1834, [Eta Carinae] appeared as a very large star of the second magnitude, or a very small one of the first, and so it remained without apparent increase or change up to nearly the end of 1837 . . . It was on the 16th December 1837 that . . . my astonishment was excited by the appearance of a new candidate for distinction among the very brightest stars of the first magnitude . . . After a momentary hesitation, the natural consequence of a phenomenon so utterly unexpected . . . I became satisfied of its identity with my old acquaintance Eta Argus [Eta Carinae]. Its light was however nearly tripled. While yet low it equalled Rigel, and when it had attained some altitude was decidedly greater. It was far superior to Achernar. Fomalhaut and Alpha Gruis were at the time not quite so high, and Alpha Crucis much lower, but all were fine and clear, and Eta Argus would not bear to be lowered to their standard. It very decidedly surpassed Procyon, which was about the same altitude, and was far superior to Aldebaran . . . . From this time its light continued to increase. On the 28th December it was far superior to Rigel, and could only be compared with Alpha Centauri which it equalled, having the advantage of altitude, but fell somewhat short of it as the altitudes approached equality. The maximum of brightness seems to have been obtained about the 2nd January 1838 . . . it was judged to be very nearly matched indeed with Alpha Centauri . . . After this its light began to fade . . . On the 20th, it was visibly diminished — now much less than Alpha Centauri, and not much greater than Rigel. The change is palpable.

An even more spectacular event occurred in 1843, when the star reached about –0.7 magnitude, outshining every star in the sky with the exception of Sirius. Eta Carinae then slowly faded from naked-eye view, hitting a rock-bottom magnitude of 7.6 in 1968. It then rebounded to 6th magnitude in 1978. The star brightened to 5th magnitude between December 1997 and February 1999. On June 12, 1999, I estimated Eta Carinae's apparent magnitude to be 4.2 — about as bright as it was when Halley saw it in 1677. Kris Davidson (University of Minnesota) believes this latest brightening episode may be Eta Carinae preparing itself for another major outburst.

Today we know that the great 1843 outburst (which became known as Nova Carinae 1843) was the biggest explosion that any star is known to have survived. According to Davidson, several solar masses of matter were ejected at speeds around 1,000 km per second. The explosion shot out two opposing lobes of expanding gas, which squeezed out of a dense, 15-solar-mass torus of cold dust and gas perhaps 0.1 light-year in diameter. The gaseous shell expanded like a balloon inflating in a napkin ring. Also called the Homunculus, that balloonlike portion of the nebula has been expanding ever since; presently it spans about 0.8 light-year. Meanwhile, Eta Carinae and its eponymous nebula together have become the brightest object beyond the solar system at infrared wavelengths.

The Hubble Space Telescope has imaged the star's expanding bipolar lobes in unprecedented detail. Two polyps of gas pulse with veins of dust and strange radial streaks that

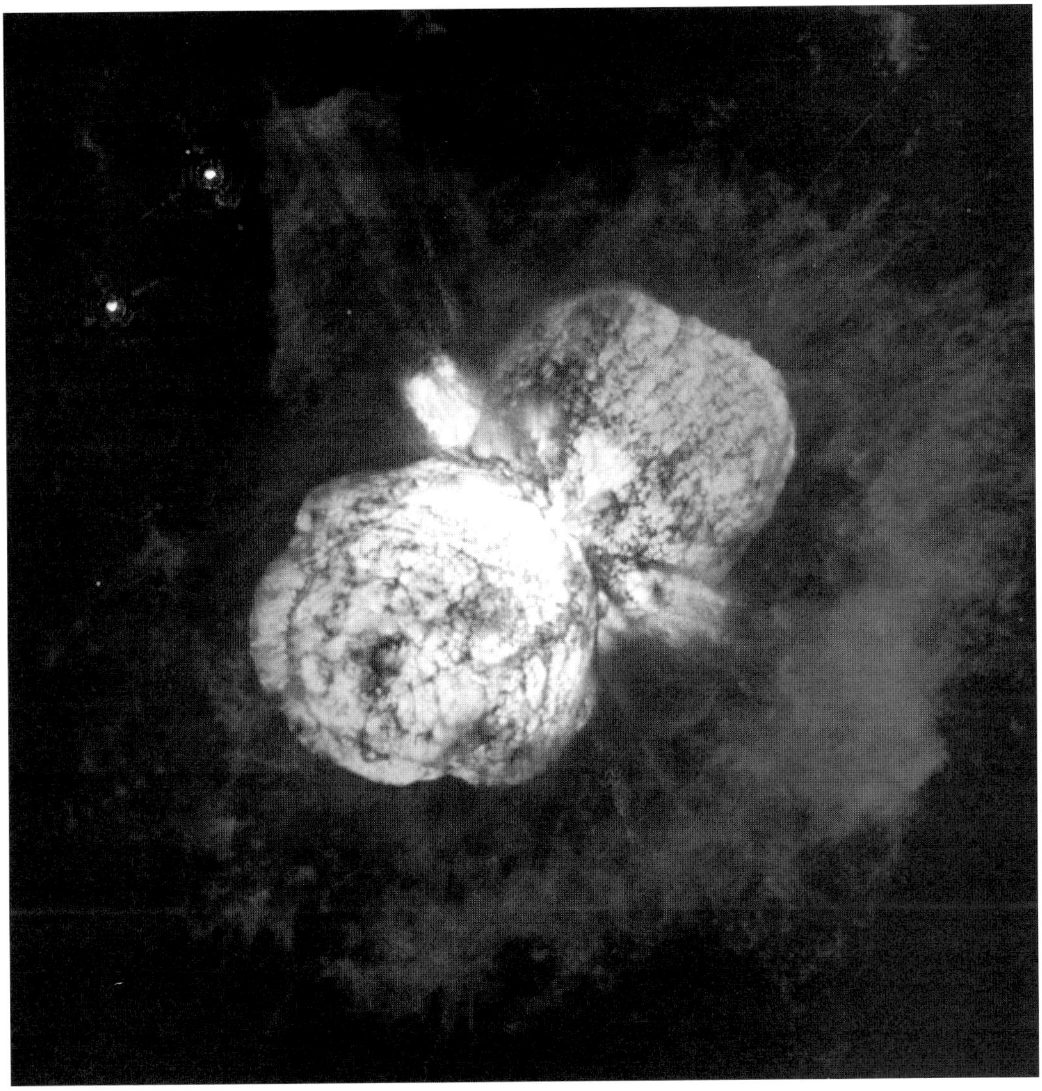

stretch 16 billion km from the central star. HST's high-resolution spectrograph also has resolved individual gas blobs that can fade or disappear in days or weeks. As more data are gathered, a new picture of Eta Carinae is emerging from its shroud of stellar mystery. Eta Carinae is most likely a binary star, and the "napkin ring" is the outer mantle of one of the binary's two stars, which was shed some 1,000 or 2,000 years before the system assumed its present appearance. The material was stripped from the star and yanked away by the gravity of its stellar companion. Thus stripped of some of its mass, this "naked" star became unstable, leading to the powerful 1843 blast. The spectral signature of the torus supports this theory, for it is nitrogen-rich and oxygen- and carbon-poor, implying that it contains processed material from the star's interior.

At radio wavelengths, writes Davidson,

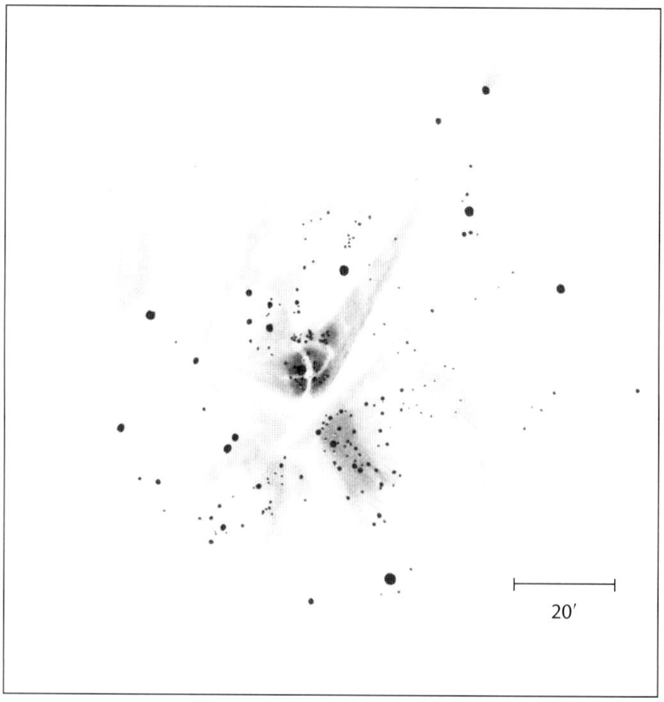

Eta Carinae produces the brightest known stellar wind, which might also explain the star's unique X-ray-emitting sheet of scorching hot gas (60 million degrees Kelvin). The X-ray emission probably arises as stellar winds flowing away from each member in the binary pair collide, or, if Eta Carinae is a solitary star, by a fast wind colliding with an older, slower one. A binary system would also explain why the star's spectrum changes, with clocklike precision, every 5½ years. Augusto Damineli (University of Sao Paulo, Brazil) and his colleagues observed Eta Carinae during a predicted 1998 episode of spectral change. From their data they deduced that the star is indeed a binary system whose members travel around a common center of mass in highly elongated orbits every 5.53 years. However, in a 2000 *Astrophysical Journal* paper Davidson admits that HST could not confirm that model with its high-resolution spectrographs.

As astronomers continue to delve deeper into these many mysteries, most agree on one thing: Eta Carinae doesn't have much longer to live. The star is one of only a few luminous blue variables in the Milky Way. These extremely massive stars live their lives in the fast lane of stellar evolution, and Eta Carinae could demolish itself in a supernova explosion today, tomorrow, or in a few hundred thousand years. (In fact, since it lies thousands of light-years away, it may have done so already. Doesn't all this excitement make you want to grab your telescope, and, if necessary, a seat on a southbound jet?)

Back to our historical conundrum. Considering Eta Carinae's dramatic variability, isn't it possible that the brightness of the Eta Carinae Nebula itself changes with time? Of course it is — especially since erratic Eta Carinae is the most powerful star in what is now the brightest section of the nebula. According to Allen, the "most brilliant portion [of the nebula], as drawn by Sir John Herschel, seems to have disappeared at some time between 1837 and 1871." Impossible? Not really, especially if you consider how other variable nebulae, such as Hubble's Variable Nebula (Caldwell 46) and the R Coronae Australis Nebula (Caldwell 68), behave. (The Eta Carinae case also tells us something about the value of astronomical drawings.) I know of no catalogs that offer brightness estimates for the Eta Carinae Nebula. In June 2000 I estimated the nebula's brightness as magnitude 4.5. Given the nebula's likely variability, it's entirely possible that

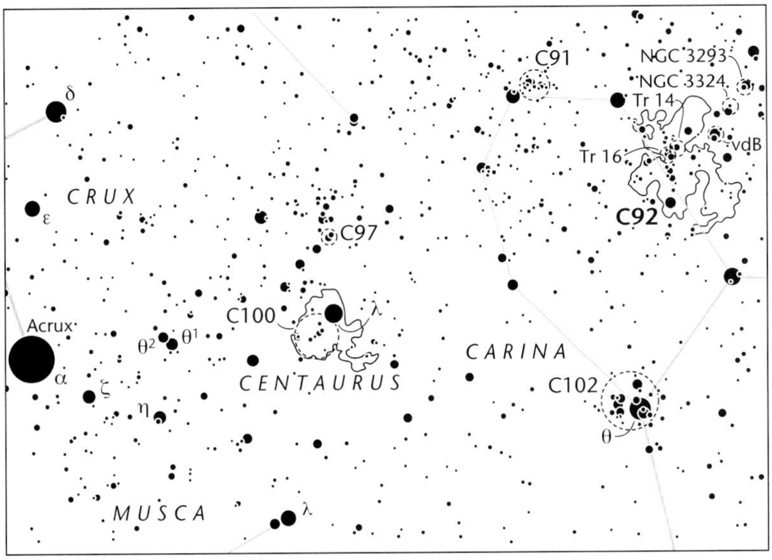

nitude glow. (Strictly speaking, NGC 3293 comprises both clustered stars and nebulosity, but only the former stood out in the twilight.) As darkness fell, more and more of the Eta Carinae Nebula appeared in patches, until its bright central chrysalis of stars and gas emerged as a celestial butterfly with impressively broad wings.

At 23× in the 4-inch the Eta Carinae Nebula displays three tiers of intensity. Its brightest material, which is centrally located, comprises two concentrations of light. The brighter of these two, the northern one, is teardrop-shaped. The teardrop's head contains two open star clusters: 10′-wide Trumpler 16 (with ten stars, including Eta Carinae) and 5′-wide Trumpler 14 (with five stars), 10′ to the north. The nebulosity surrounding these clusters tapers to the northwest, where, 45′ away, it beads up again at the 14′-wide cluster vdB-Ha 99, a loose but rich aggregation of suns collected around a 5th-magnitude star. (For brevity this modest cluster is labeled "vdB" in the chart above.) A narrow vapor trail streams away from vdB-Ha 99 to the northwest, like a comet's thin ion tail. In fact, this brighter northern section reminds me of the remarkable Comet Hyakutake (C/1996 B2). The tip of the tail points to NGC 3324, a bright patch of irregular nebulosity that surrounds a star cluster and the 9th-magnitude double star h 4338 (6″ separation; east-west orientation). And the cluster NGC 3293 (which has its own overlapping nebulosity) lies ½° north-northwest of NGC

the object was known to some civilizations but not others.

Herschel wrote that it "would manifestly be impossible by verbal description to give any just idea of the capricious forms and irregular gradations of light affected by the different branches and appendages of this nebula." But I'll try. My first view of the Eta Carinae Nebula was with the naked eye in February 1982, while flying at 40,000 feet (12 km) to New Zealand. The nebula was a magnificent presence, not the eye-squinting cough of light that some sources imply. And that's the way it looks now from Hawaii. Its bright northern section was so intense in February 1998 that I saw it in diminishing twilight with the unaided eye. And in twilight, with 7×35 binoculars, the object looked like a wedge-shaped asterism of two separate star clusters surrounded by gas. In fact, on the face of it my *twilight* view must've been very similar to Lacaille's *dark-sky* view with a vaguely comparable instrument. In the twilight another object, open cluster NGC 3293, is visible, about 2° to the north-northwest of Eta Carinae, as a compressed 5th-mag-

3324. This section of the Eta Carinae complex is a virtual five-car train of clusters and glowing hydrogen gas.

Now return to the head of the teardrop and study it carefully. It is bisected by two criss-crossing dust lanes. The darker of these two lanes runs from the teardrop's southeastern rim and curves northeastward to the west of orange Eta Carinae. In photographs this lane is a black S-shaped ribbon of opacity, with a dark black knob at its northern end and a bell-shaped bottom. These intensely dark segments resemble a keyhole, which is why it is referred to as the Keyhole Nebula. (Remember that "Keyhole" refers to a tiny black smudge within the bright nebula, and not to the bright nebula itself.) Try as I might, with the 4-inch I could not distinguish the keyhole within this lane of darkness. Another fainter lane of darkness courses east-west through the teardrop just south of Eta Carinae. It does not show well in long-exposure photographs but is very apparent at the eyepiece.

Of course, when we look at Eta Carinae we are not seeing the star itself but the expanding gas lobes — the aforementioned Homunculus. Enrique Gaviola invoked the term in a 1950 *Astrophysical Journal* article, saying the lobes looked like a mannequin or homunculus, complete with "head pointing northwest, legs opposite and arms folded over a fat body." These globs of nebulosity should be monitored very carefully. (It's amazing we can see them at all!) In the first edition of his *Astronomical Objects for Southern Telescopes,* Ernst Hartung writes that "bright orange η Car is surrounded by an orange red nebula about 3″ wide, just visible with a [4-inch], and the spectrum of the star shows numerous bands with the red H α shining like a tiny lamp at one end." With my 4-inch I could not see the famed Homunculus as distinctly as Hartung did with his, but Eta Carinae was brighter when I was observing, so the star's increased brilliance conceivably could have made this delicate feature less distinct. To me Eta Carinae looked like two slightly overlapping orange spots. In 1982 New Zealand amateur Graham Blow took me to the 9-inch refractor at Carter Observatory in Wellington to view Eta Carinae, and it was there that I observed the Homunculus for the first time with amazing clarity. Through that scope the Homunculus looked like two puffs of glowing orange light 1″ on either side of an intense core of light. Tom Clark, the editor of *Amateur Astronomy* magazine, described the Homunculus as two balloons covered with kudzu after seeing it through a 25-inch reflector that he took from Florida to Australia.

A massive, dark checkmark separates the brighter, cometlike northern section of the nebula's heart from its dimmer, more rectangular southern section. Allen notes that this vacancy was called the "Crooked Billet" by Francis Abbott. The southern section is a dense 20′-long aggregation of sparkling starlight immersed in a smooth cotton coffin of glowing gas. The coffin is oriented north-south and contains the scattered "bones" of Collinder 228, a magnitude-4.4 cluster with an indeterminate number of stars. Collinder 228's dominant stars are strung out like a skeleton's groping hand and forearm. The hand, with strings of stars radiating like outstretched fingers, is to the north, while the starlit arm lies to the south.

Thus run the Eta Carinae Nebula's brightest components. Its second tier of nebulosity swirls like four butterfly wings, with the densest to the south. These features are delicate, and in fact they are better seen in binoculars, where their light remains condensed, than in a telescope. The faintest third tier is an even more delicate fan of gas extending from the northern "comet" to the northwest and northeast, and a dim lip of gas to the southwest.

Although the Eta Carina Nebula's visual diameter is 2°, some photographs show nebulosity extending across an area twice as wide, though it is hard to tell if this outer nebulosity is physically related to the Eta Carinae Nebula complex. For instance, NGC 3324 apparently has no association to it. Nor does NGC 3293. But NGC 3293 is a marvel unto itself. I call it the Little Jewel Box. It is diamond-shaped, with a Y-shaped core of about a dozen colorful suns laid on a fainter background of crushed gems.

Incidentally, despite its grandeur, and despite what you may have read about how magnificently the Eta Carinae Nebula "dwarfs" the Orion Nebula, Eta Carinae Nebula does not pack a more powerful visual punch than the Orion Nebula, especially once a telescope is trained on it (though some may beg to differ).

Finally, I can never look at Eta Carinae without recalling David Levy's biography of Bart Bok. Bok had spent a good part of his life studying this enigmatic star and its associated nebulosity. On November 15, 1975, writes Levy, Bok and his ailing wife, Priscilla, were waiting for Senator Barry Goldwater in the University of Arizona's Flandrau Planetarium. The Boks were looking at a panorama of the entire Milky Way. The exhibit had tiny red lights that could be turned on and off to point out specific Milky Way features. As Levy recounts:

> Priscilla pressed one of the buttons, and a faint red light turned on to identify the Eta Carinae nebula. . . . They stood there quietly for a moment, remembering the stunning photographs they had taken of it during years of observing. Finally Priscilla spoke. "You know, Bart," Priscilla said, "when I am gone, that is where I am going. I will ask St. Peter to give me a front row seat right at the center of the nebula. I'll see stars forming right before my eyes!"
>
> Overwhelmed by this profound merging of his two greatest loves, Bart tried to hold back tears as he hugged Priscilla. "Eta Carinae," Priscilla repeated as they walked into the planetarium theater, "that is where I want to be."

93 & 101

93

The Starfish; The Windmill
NGC 6752
Type: Globular Cluster
Con: Pavo

RA: $19^h\ 10^m\ 51.8^s$
Dec: –59° 58′ 55″
Mag: 5.4
Diam: 29′
Dist: 13,000 light-years
Disc: James Dunlop, included in his 1827 catalog

J. HERSCHEL: Globular, bright, round, pretty rich, pretty suddenly much brighter towards the middle, 5′, stars of two sizes, 11th to 12th magnitude and 15th to 16th magnitude, south preceding [to the southwest] is an elegant first class double star. (h 3778)

GC: Globular cluster, very bright, very large, irregularly round, well resolved, clearly consisting of stars, stars of magnitude 11 to 16.

NGC: Globular cluster, bright, very large, irregularly round, well resolved, clearly consisting of stars, stars of magnitude 11 to 16.

THE CELESTIAL HERITAGE OF PAVO, THE Peacock, can be traced to two different mythologies. The first would make a good modern soap opera. It starts with the sexual escapades of Jupiter and the maiden Io. Io becomes pregnant. Jupiter's wife, Juno, discovers the truth and becomes enraged. For revenge, Juno turns Io into a heifer, then orders Argus — a demon-giant with 100 eyes — to "keep an eye" (any eye!) on Io. Enter swift Mercury, who, under Jupiter's urging, slays Argus. Io is freed. Juno is horrified. But she can't let those loyal eyes of Argus go forever unseen. So she removes them from the giant's head and attaches them, one by one, to the tail feathers of her favorite bird, the peacock. And *poof!* — the peacock becomes a sacred symbol of immortality. The second myth is shorter. When Argos (not to be confused with Argus) — the builder of the Ship Argo, which Jason sailed — died, Juno turned him into a peacock and placed him in the sky. I'd buy serial rights to the first myth.

Sometime between 1595 and 1597, Pieter Dirkszoon Keyser and Frederick de Houtman invented the stellar Peacock that we turn our eyes to today. Johann Bayer included the figure in his 1603 *Uranometria* atlas. Pavo has 378 square degrees, making it the 44th-largest constellation. It can be found south of the "glass eye," Telescopium (the Telescope), which, in turn, lies south of Sagittarius, the Archer. Second-magnitude Alpha (α) Pavonis (Peacock) displays its pure blue luster in the constellation's northeastern corner. Peacock's declina-

101

NGC 6744
Type: Barred Spiral Galaxy (SABbc)
Con: Pavo

RA: $19^h\ 09.7^m$
Dec: –63° 51′
Mag: 8.6
Dim: 20.8′ × 13.3′
SB: 13.9
Dist: 34 million light-years
Disc: James Dunlop, included in his 1827 catalog

J. HERSCHEL: Pretty bright, round, [at] first very gradually, then suddenly very much brighter in the middle; total diam[eter] 2′, but that of the bright part 15 arcseconds. (h 3776)

GC/NGC: Considerably bright, considerably large, round, very gradually, suddenly very much brighter in the middle, resolvable.

---

tion (–56° 44′) makes it visible, in principle, from latitude 33° 16′ N (ever so slightly farther north if you consider refraction) and points south. This means it is visible from most of the southern United States.

Globular star cluster NGC 6752 (Caldwell 93), however, is more challenging than Peacock. It lies 10° west-southwest of Peacock, so you basically need to be at, or south of, latitude 30° N to see it. That means the cluster just misses being visible from the grounds of the Texas Star Party near the McDonald Observatory. But if some enterprising TSP attendee were to go and view the lights of Marfa, Texas, or drive south of the 30th parallel on their way, say, to Presidio at the Mexican border — and take a pair of 7×50 binoculars with them — they just might spot the southernmost Caldwell globular visible from the continental United States.

NGC 6752 is the third-easiest globular star cluster to resolve, by virtue of its horizontal-branch magnitude of 13.7, which compares favorably to NGC 6397's (12.9) and M4's (13.4). (A cluster's horizontal-branch magnitude indicates how bright a fair fraction of its stars are; according to deep-sky expert Brian Skiff, a telescope must show stars of that brightness to be able to resolve the cluster well.) Robert Burnham Jr. ranked NGC 6752 among the finest dozen globulars in the heavens. At magnitude 5.4 it can be seen with the naked eye, and it ranks 5th among globulars in total brightness.

And when it comes to its *brightest* stars, NGC 6752 ranks second, trailing behind only NGC 6397 (Caldwell 86) in Ara. Yet how many North American observers have seen it? And how many North American observers have seen another Pavo prize, NGC 6744 (Caldwell 101), one of the largest barred spirals known? Since NGC 6744 lies only 4° south of Caldwell 93, we will throw order to the wind and explore both of these Caldwell gems here.

James Dunlop discovered NGC 6752 from Parramatta, New South Wales, listing it as the 295th object in his 1827 catalog of southern deep-sky objects. He described it thus:

> [A] "pretty large and very bright nebula, 5′ or 6′ diameter, irregular round figure, easily resolved into a cluster of small stars, exceedingly compressed at the centre. The bright part at the centre is occasioned by a group of stars of some considerable magnitude when compared with those of the nebula. I am inclined to think that this may be two clusters in the same line; the bright part is a little south of the centre of the large nebula.

From the Cape of Good Hope, John Herschel noted that the cluster's brighter stars "run out in lines like crooked radii," while the fainter ones "are massed together in and round the middle."

Dunlop and Herschel both noted one of the most singular and fascinating qualities of this cluster: its extremely high degree of central compression. The central condensation appears to be about 2.5′ in diameter, but this globular also hides an extremely tiny inner core only 20″ across. While the cluster's true physical extent is 110 light-years, this tiny inner core spans but 1.3 light-years. NGC 6752 has provided astronomers with one of the best examples of a remarkable phenomenon called *core collapse*. Astronomers reviewed more than 100 Hubble Space Telescope images of the cluster's inner core and discovered that about 20 to 40 percent of all the stars there are binaries. The fraction of binary stars in a globular star cluster provides important insight into the cluster's evolution. The orbital energy stored within binary systems is the main source of a globular's

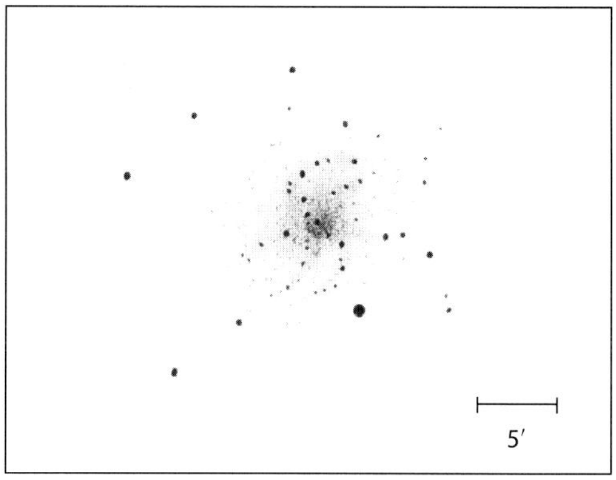

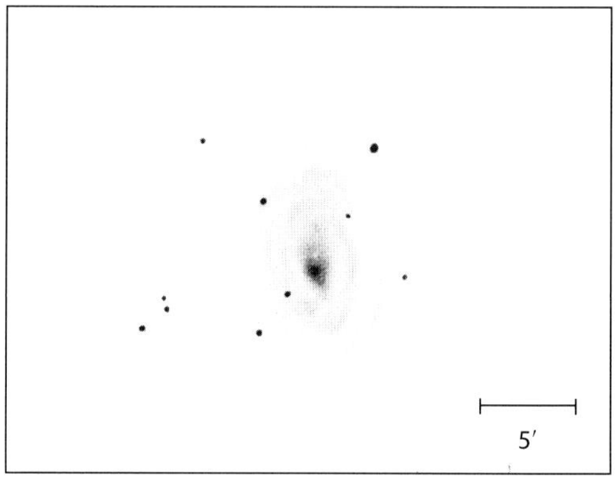

dynamical "heating," which prevents its core from collapsing into a single massive object. But in the case of NGC 6752, the intense gravitational pull of so many stars in such a small volume seems to be causing stars to converge on the cluster's core like bees swarming to their hive. Such a collapse may be the process responsible for creating such exotic objects as X-ray pulsars, blue stragglers, and possibly (though doubtfully) black holes. Blue stragglers have been identified in NGC 6752's core.

Astronomers have had great success in probing the depths of NGC 6752's extremely tight core because it is one of the closest globulars to us. Its computed distance of 13,000 light-years makes it about 1,700 light-years closer than 47 Tucanae (Caldwell 106). NGC 6752 lies 17,000 light-years from the center of our galaxy. With an age of perhaps 15 billion years, the cluster is quite old, and, like most globulars, it is metal-poor; each of its stars has, on average, 1/36 as much iron (per unit of hydrogen) as does our Sun (a fairly typical value for a globular cluster). Expert William E. Harris lists its overall spectral type as *F*4 or *F*5. From 1897 to the 1980s, only two variable stars were known in the cluster. But that number has risen sharply. When the cluster was monitored in 1996 and 1997 with the 1-meter Swope Telescope in Chile, 11 new variables were identified.

From Hawaii NGC 6752 can rise as much as 10° above the southern horizon. It is visible (albeit with effort) to the unaided eye and is splendidly apparent in 7×50 binoculars. It lies just 3° northeast of 4th-magnitude Lambda (λ) Pavonis and 1½° east-northeast of yellow, 5th-magnitude Omega (ω) Pavonis. There is nothing else in the area to mistake it for. Its small, pale disk of white light suddenly swells to nearly ¾ the size of the full Moon when averted vision is used. A magnitude-7.4 star lies 5′ south-southwest of the cluster's core.

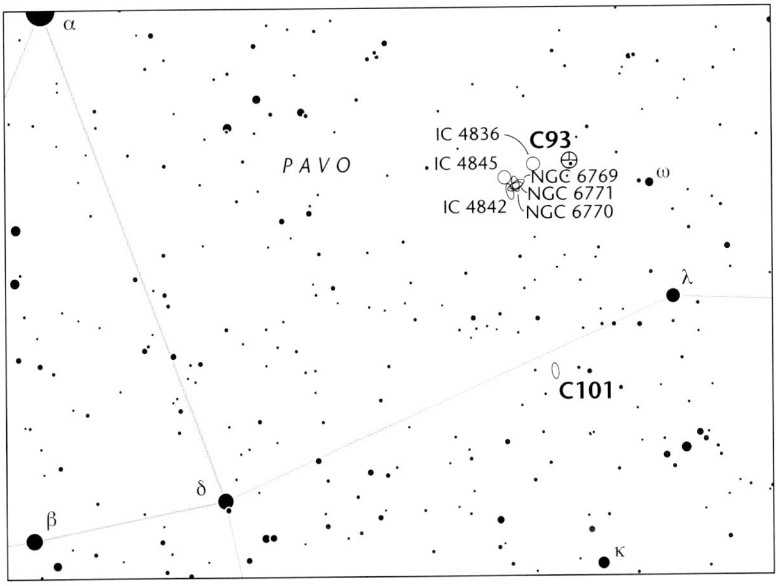

At 23× in the 4-inch Genesis the cluster is amazingly bright and round, and its outer halo seems to sweep across the light of the 7.4-magnitude star like windblown sand. When you look at this part of the cluster with averted vision, then directly, and repeat the process, the "dust" seems to flicker eerily. The cluster's tight inner core gradually extends outward into an irregular shell of scintillating light. At 72× NGC 6752 is most remarkable. The soft glow of countless suns snaps into view. The core has a dense cross of bright members whose magnitudes range from 10.5 to 12. The cross's arms are oriented northeast-southwest and northwest-southeast. The northeastern arm loops to the north, forming a starry atoll that encircles a dark lagoon. Three weaker arms extend toward the south and southeast, forming a letter E. Most curiously, the brighter members of these arms appear to cluster away from the core. Seeing this helped me to understand how Dunlop could have felt that NGC 6752 might be two different clusters along the same line of sight. These impressions don't hold together at high power, though. Rather, high magnification makes the cluster look like an old windmill with 10 broken arms. Or so I see it. Marilyn Head of Wellington, New Zealand, sees these bright, curving arms differently. She calls NGC 6752 "The Starfish," which I rather enjoy, because it fits well with my "atoll" metaphor.

Just 1° southeast of NGC 6752 is a cluster of five *NGC* and *IC* galaxies: IC 4845, NGC 6769-70-71, and IC 4842. Another galaxy, IC 4836, lies about halfway between the globular and these galaxies. The *NGC* objects are all about 12th magnitude, while IC 4845 shines at magnitude 11.6 and IC 4842 at 13th magnitude. These six galaxies are plotted in the *Millennium Star Atlas,* along with others. Can you detect any of them?

Located 4° due south of NGC 6752, one of the sky's most aesthetically stunning galaxies, NGC 6744 (Caldwell 101), glows from behind a rich thicket of Milky Way suns. This enormous barred-spiral system extends 193,000 light-years and has a total luminosity of 56 billion Suns. It belongs to the Pavo-Ara Cloud of galaxies and is receding from us at 842 km per second. Photographs taken with large telescopes reveal a maelstrom of arms clotted with H II regions. These arms spiral out from a lens-shaped bar centered on a fractured inner ring seen 35° from edge on. The area between the prominent bulge and the inner ring is virtually devoid of gas and star-formation activity. The void may be the result of an orbital resonance, like the gaps in Saturn's rings. At least a half dozen major spiral arms project tangentially from the inner ring like the barbs of a feather before winding almost continuously out to the edge of the disk. Two arms display especially intense star formation, though H II regions pepper all the arms. Maps made with the Australia Telescope Compact Array of radio telescopes reveal that the galaxy's invisible atomic-hydrogen gas resides in a ring underlying the outer portion of the galaxy's disk. They also show two spiral arms extending almost 50 percent farther than the visible disk. Astronomers also have found evidence for tidal interactions between NGC 6744 and its two companion galaxies, NGC 6744A and ESO 104-G44.

Alas, NGC 6744 is *not* a retina burner. It is relatively bright (magnitude 8.6) as galaxies go, but its light is spread across 21′ of sky. As a result, its surface brightness is a relatively feeble 13.9 magnitudes per square arcminute. This means that each square arcminute of the galaxy's luminous surface shines with roughly the brightness of Pluto. Dunlop deserves considerable credit for finding this low-surface-brightness object. It is the 262nd object in his catalog, and he observed it on three evenings. He described it as a "pretty large very faint nebula, about 5′ or 6′ diameter, slightly bright towards the centre; a minute star is north of the nebula, and two stars of the 7th mag preceding [to the west]." John Herschel's descriptions of it — "pretty bright, round . . . first very gradually, then suddenly very much brighter in the middle" and "resolvable" — make NGC 6744 sound more like a globular cluster than a galaxy. But, as we have seen, with his 18¼-inch reflector Herschel was undoubtedly resolving the galaxy's innumerable H II regions, some of which are as large as 5″, and not its individual stars, which would challenge even HST.

At 23× the 4-inch Genesis shows NGC 6744 as a ghostly pale oval with a bright central pip burning through a tight, slightly elliptical haze with an enhanced rim. By concentrating I can make out a brightening on the southeastern edge of this rim; it could be a bright row of H II regions in the inner ring. A dark region separates the ring and what appears to be the brightest spiral arm(s) to the north, which arc(s) toward the west. A roughly 10.5-magnitude star on the galaxy's northwestern edge could fool a supernova hunter, so beware! At high power two spiral arms (or groupings of spiral arms) are seen with difficulty: the one mentioned above and another, fainter one to the south, which curves to the east. The brightest H II regions seem to pulse in and out of view to the east and southeast. After observing NGC 6744 with a 12-inch telescope, Ernst Hartung described it "as a large irregularly elliptical faintly luminous haze more than 5′ across with some faint stars involved. The centre about 30″ wide is bright and looks granular . . . In good conditions [a 4-inch telescope] shows the nebula plainly." Others have clearly resolved the galaxy's spiral arms with a 10-inch instrument. Still others have commented on a greenish hue. Has anyone else seen this?

94

94

Kappa Crucis;
The Jewel Box
NGC 4755
Type: Open Cluster
Con: Crux

RA: $12^h 53.6^m$
Dec: –60° 21′
Mag: 4.2
Diam: 10.0′
Dist: 4,900 light-years
Disc: Abbé Nicolas Louis de Lacaille, included in his 1755 catalog

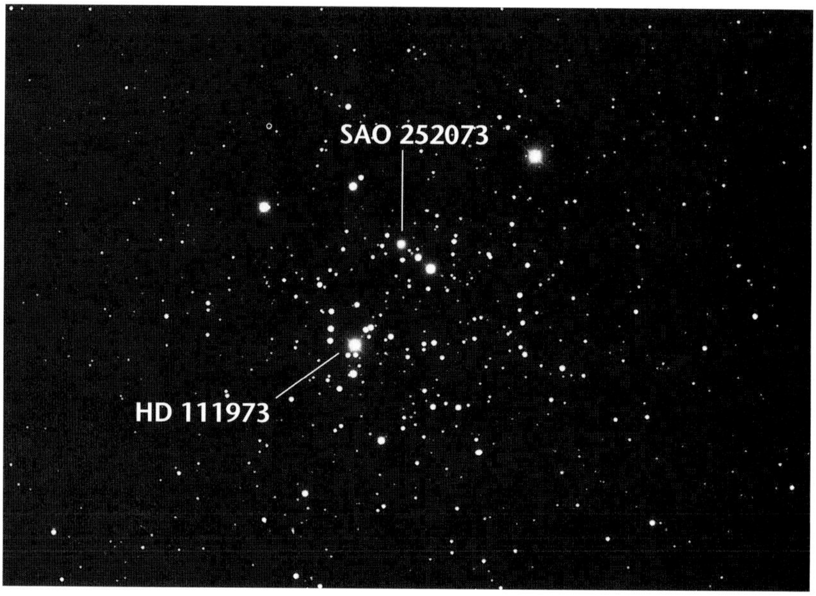

J. HERSCHEL: The central star (extremely red) or a most vivid and beautiful cluster of from 50 to 100 stars. Among the larger there are one or two evidently greenish; south of the red star is one [of] 13th mag[nitude], also red; and near it is one [of] 12th mag[nitude], bluish. (h 3435)

GC/NGC: Cluster, very large, stars very bright (κ Crucis).

THE JEWEL BOX, A TRUE TREASURE AMONG open clusters, adorns the Holy Grail of constellations, novelist Edward Robert Bulwer-Lytton's "great cross of the South." If there is one constellation wholly symbolic of travel and adventure, it is the Southern Cross. Early sailors saw the Cross as a good omen sent to guide them across uncharted seas. But before I continue, a note of clarification. Crux, the constellation's formal name, means "Cross," not "Southern Cross" (which would be Crux Australis). Astronomers often refer to Crux as the "Southern Cross" to differentiate it from the Northern Cross, an asterism in Cygnus, the Swan. The phrase "Southern Cross" has become so entrenched in our vocabulary, however, that most modern references equate Crux with it. But that's like saying "Corona" means "Southern Crown" or "Aurora" means "Southern Lights."

The Cross first appeared as such on celestial globes in 1592. Before that its stars belonged to Centaurus and could be seen on Arabic globes dating to 1225. Somehow Dante Alighieri — "the spokesman of ten silent centuries," as 19th-century Scottish essayist and historian Carlyle called him — learned of the stars' existence, probably through conversations with the great adventurer Marco Polo or other equally well-traveled men. For he alluded to the stars in his *Divine Comedy,* which he completed in 1321. In that tale, after Dante and Beatrice Portiani ascend from Hades on the other side of the world, the first thing they see when looking skyward are four brilliant stars representing the four principal virtues: Justice, Prudence, Fortitude, and Temperance. Among the first true navigators to glimpse this "other side of the world" was Christopher Columbus. It was on

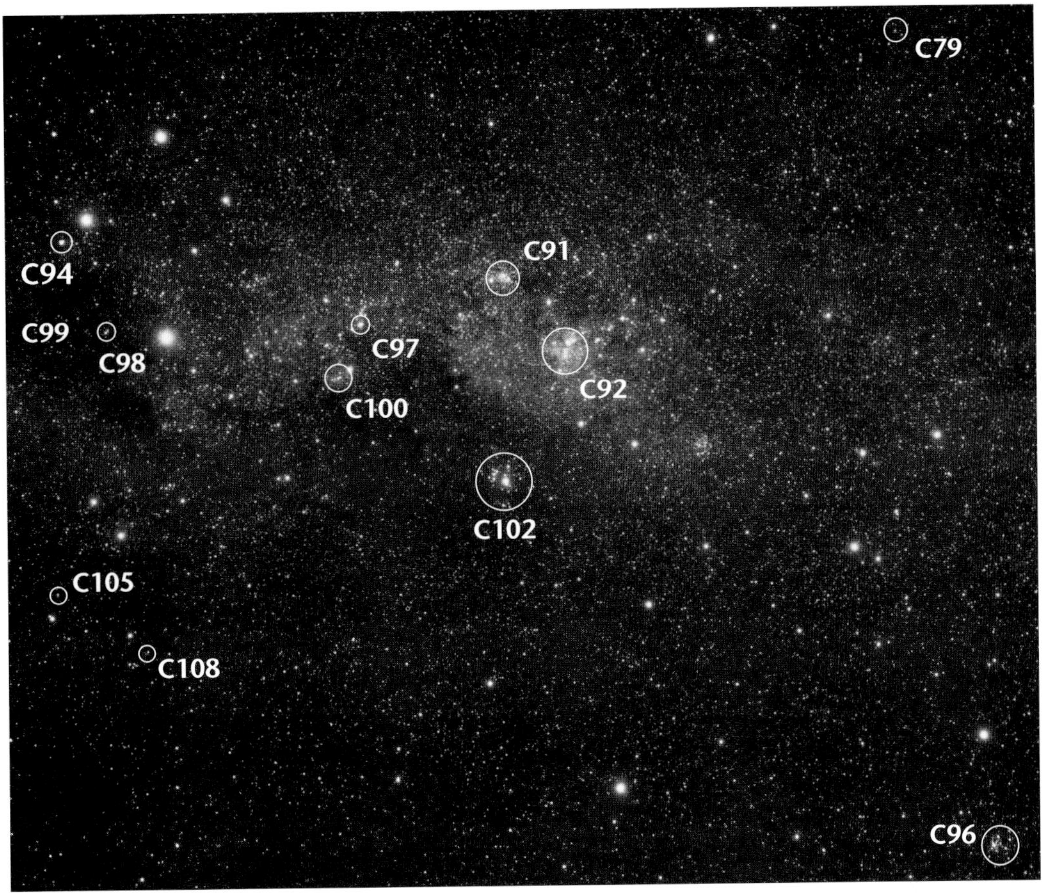

his maiden voyage of discovery in 1492 that Columbus glimpsed four stars in the form of a cross over the West Indies. Surprisingly, he believed that these stars were the ones referred to in Dante's allegory. Richard Hinckley Allen notes that on Amerigo Vespucci's 1501 voyage, Vespucci "called to mind the passages from Dante, insisting that he himself was the first of Europeans to see Four Stars." Alas, Vespucci did not see these stars in the form of a cross but as *Mandorla,* "an almond." But do not despair; Vespucci's reference was not to the tasty treat but to the *vescica piscis,* or "oblong glory," surrounding the bodies of saints ascending to heaven in Italian art. Of course, today many people don't seem to care much about virtues or oblong glories (unless they're chastising politicians or enjoying an *X-Files* episode in which someone is ascending into an alien spaceship). But we all like to listen to music. And what better verse summarizes the spirit of adventure than this one, from "The Southern Cross" by Crosby, Stills, and Nash: "When you see the Southern Cross for the first time you'll understand now why you came this way."

Actually, most people who see the Southern Cross for the first time don't understand why it's so small. In fact, it doesn't even look like a cross (the way we think of one today) but a kite. So much for poetry and romance. I honestly believe the romance of the Southern Cross lies not in its handful of brilliant stars but in the visual majesty of the Milky Way surrounding it. The Southern Cross marks the eastern end of a swath of swirling madness running between 10 and 13 hours of right ascension and from –55° to –65° in declination. This region contains about two dozen open star clusters (including IC 2602, or Caldwell 102, the Southern Pleiades), about a dozen nebulae (including Caldwell 92, the Eta Carinae Nebula), thick star clouds, and eerie lagoons of darkness like the Coalsack (Caldwell 99). No other part of the visible heavens draws more attention to itself than this. The Southern Cross simply happens to be the most obvious asterism in this Van Goghian celestial landscape.

And of all the clusters that dapple this region, none has captured the imagination more than the Jewel Box. Early explorers saw the Jewel Box as a 4th-magnitude star and designated it Kappa (κ) Crucis. Many believe that John Herschel discovered the cluster because his description of it led to the Jewel Box nickname: "this cluster, which though neither a large nor a rich one, is yet an extremely brilliant and beautiful object when viewed through an instrument of sufficient aperture to show distinctly the very different colour of its constituent stars, which give it the effect of a superb piece of fancy jewellery." But credit for the Jewel Box's discovery (if not its name) belongs to Nicolas Louis de Lacaille, who designated Kappa Crucis a "nebulous cluster" during his 1751–53 exploration of the southern skies. The cluster is the 12th object in his 1755 catalog, where it is described as "five to six stars between two mag. 6 stars."

That Lacaille saw the cluster as nebulous is not surprising; he used tiny ½-inch telescopes. Through such instruments any background of faint stars would have appeared as a milky haze. James Dunlop split that haze into individual stars with his 9-inch telescope in Parramatta, New South Wales. His notes: "κ Crucis . . . is five stars of the 7th magnitude, forming a triangular figure, and a star of the 9th magnitude between the second and third, with a multitude of very small stars on the south side."

At the heart of the Jewel Box John Herschel noticed a bright "extremely red" ruby among a smattering of emeralds "of different shades of green." In all, Herschel found eight stars in which the "colour is conspicuous." Many sources identify this ruby star at the cluster's center as the star Kappa Crucis, but it is not. That central star (a type-*M*2Iab red supergiant known as SAO 252073) is too faint,

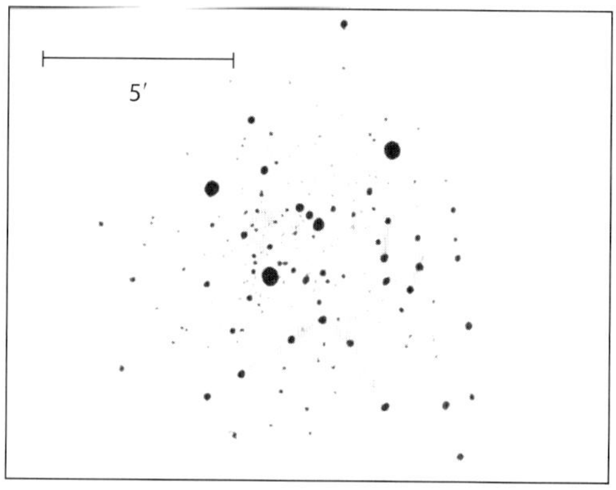

at magnitude 7.2, to have received a Bayer designation. According to the *Yale Bright Star Catalogue,* Kappa Crucis is HD 111973, the magnitude 5.9 star at the southeastern end of a pyramid of four 6th- to 8th-magnitude stars. Kappa Crucis, then, would be a *blue* (type-*B*5Ia) supergiant. Alas, the *Yale Bright Star Catalogue* also is mistaken. The naked-eye observers who first charted Kappa Crucis saw the *entire* cluster as a naked-eye star, just as they did with Omega Centauri. On many nights I have gone out and looked at the Jewel Box with the naked eye, and I have convinced myself that our astronomical ancestors did not isolate HD 111973 from the rest of the compact cluster. It simply takes much too much effort to do so. So, just as no single star in the Omega Centauri cluster is Omega, nor is any star in the Jewel Box Kappa.

The Jewel Box packs 281 known members into an area only 10′ (14 light-years) across. The hypothetical inhabitants of a planet orbiting any of the cluster's dimmer stars would have one of the most glorious night skies

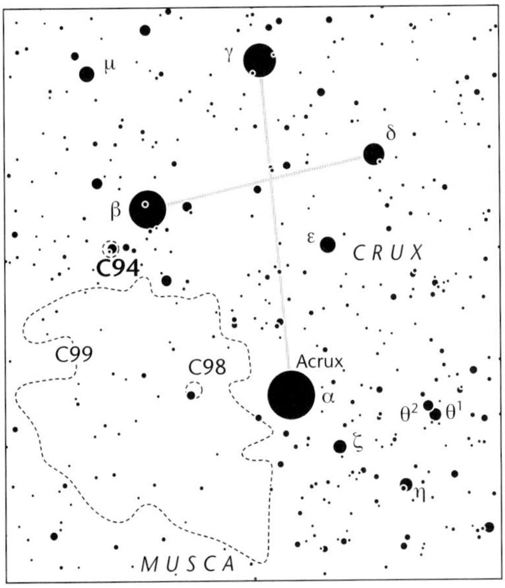

imaginable: four Rigel-like stars, each blazing brighter than a quarter Moon, with hundreds of other stellar companions projected against luminous splashes of Milky Way and dark, dusty voids. The light we see shining from these stars today began its journey across space shortly before the Syrians slew Ahab, King of Israel, at Ramoth-Gilead.

In the May 1995 issue of the *Astronomy and Astrophysics Supplement*, Ram Sagar estimates the Jewel Box's age to be about 10 million years, making the cluster a contemporary of the Double Cluster in Perseus (Caldwell 14) and the Tau Canis Majoris Cluster (Caldwell 64). After sampling 813 of the region's stars, some as faint as 20th magnitude, Sagar and his colleagues determined that obscuring dust is not uniformly distributed across the cluster's face. On average, though, each of the cluster's stars is dimmed by 0.4 magnitude (30 percent). The cluster's stars formed nearly at the same time from a molecular cloud that may have predated the cluster by 6 or 7 million years or more. To date nine Beta Cephei variables have been discovered in the cluster.

To find the Jewel Box, look 1° southeast of 1st-magnitude Beta (β) Crucis, the eastern arm of the Cross. The Jewel Box is visible to the naked eye as a 4th-magnitude "star" even under the light of a quarter Moon. Binoculars will bring out four bright stars against a background of misty light. The four bright stars form a tiny pyramid — the triangular figure seen by Dunlop. (By the way, if you doubt my claim that it would have been difficult for our ancestors to resolve the Jewel Box with the naked eye, give it a try. Its two brightest members shine at 6th magnitude and are separated by about 1.5′. I resolved these two stars with effort from Zambia in June 2001. And though it was not *that* difficult, I did have the advantage of knowing what to look for.)

At 23× in the 4-inch (with south "up") the cluster looks like an ice-cream cone. The bright pyramid of stars is the cone; a faint sprinkling of stars to the south and west is the ice cream; and a roughly 11th-magnitude pair of stars is a fuzzy cherry on top. The Jewel Box also looks like two superposed clusters, one bright, one faint. Under a direct telescopic gaze its brightest stars seem chiseled out of black marble. Look with averted vision at the 6th-magnitude sun in the southeastern corner of the pyramid — the *Yale* Kappa, or HD 111973 — and see if the area around it doesn't sparkle with faint suns. That star is the jewel at the center of a pendant-shaped asterism of 10th-magnitude suns, which is best seen with high magnification.

From a dark-sky site the cluster appears to be snared within a web of dark nebulosity. Throw the low-power image slightly out of focus and see if the region southwest of the pyramid appears especially dark. Perhaps a dust cloud has dimmed starlight here.

At 72× the cluster fractures into smaller groupings, even as the pyramid remains bold and bright. The pendant containing HD 111973 reveals at least seven stars that form a smaller pyramid at right angles to the larger, brighter one. HD 111973 looks yellow at 72×, as does HD 111904, the magnitude-5.8 star at the northwestern corner of the large, bright pyramid. In February 1997 and again in April 2000, HD 111973 appeared at least a half magnitude (60 percent) brighter than HD 111904 to my eyes — even though, according to the second edition of *Sky Catalogue 2000.0,* it should be 0.1 magnitude (10 percent) fainter. An undated drawing by Kenneth Glyn Jones in Volume 7 of the *Webb Society Deep-Sky Observer's Handbook* (published in 1987) shows HD 111973 significantly fainter than HD 111904, and this observation was based on a view through a 12-inch reflector. Most photographs I've seen show HD 111973 fainter than HD 111904, but that may be an artifact of the photographic emulsions. (Note, by the way, that in both Ernst Hartung's and Robert Burnham Jr.'s handbooks the photographs of the cluster have been flopped, with east and west reversed.) Clearly these stars warrant watching.

At 72× in the 4-inch SAO 252073 looks more orange than red to me. It is the northeasternmost star in a tiny line of three similarly bright stars oriented northeast to southwest. SAO 252073, to my eyes, is the faintest of these three stars. Glyn Jones, however, saw it as the brightest. My eyes tend to be blue-sensitive while other observers are more sensitive to red light, so this apparent discrepancy might be related to the way different observers perceive color, and not to any real brightness variations. Yet Edward James Stone, Her Majesty's Astronomer at the Cape from 1870 to 1879, suspected that this red star had altered its brightness since John Herschel had observed it. And the Hipparcos satellite's database seems to support the notion, for on its watch the star's apparent visual magnitude appears to have ranged from 7.08 to 7.52. However, Hipparcos didn't enable astronomers to determine a period of variability or to classify the star's light curve. Clearly it would behoove someone with a photometer to monitor all the Jewel Box's supergiant suns for variability.

Moderate magnification reveals other treasures in the Jewel Box. The brightest stars to the southwest of the large pyramid form two semicircles, like wide crowns, facing in opposite directions: one opens to the north-northwest and the other to the southeast. Immediately south of these crowns is a long chain of fainter suns oriented in the same northwest-to-southeast direction as the two semicircles. The void between these groupings may be a long lane of dust.

At high power the cluster fractures completely and the Jewel Box loses its luster. Most prominent is HD 111973 and the pyramidal pendant. Look for a black cross at the southwestern edge of the pyramid. Look also for a tight double star immediately northwest of HD 111973 and another about 2′ to the southeast. For a slightly greater challenge, see if you can detect a red, 13th-magnitude star south of SAO 252073. It is one that John Herschel first noticed, along with so many others of the Jewel Box's sparkling gems (though he didn't say *how* far south of SAO 252073 it is).

95

NGC 6025
Type: Open Cluster
Con: Triangulum Australe

RA: $16^h 03.3^m$
Dec: –60° 25′
Mag: 5.1; 5.5 (O'Meara)
Diam: 15′
Dist: 2,500 light-years
Disc: Abbé Nicolas Louis de Lacaille, included in his 1755 catalog

J. HERSCHEL: Large, brilliant cluster VII class; fills field, not rich, stars 8, 9, 10, 11th mag[nitude], with smaller [fainter]. Chief star 8th mag[nitude] taken, in the southern part of cluster. (h 3616)

GC/NGC: Cluster, bright, very large, pretty rich, little compressed, stars of 7th magnitude and fainter.

ALPHA (α) AND BETA (β) CENTAURI POINT almost directly to our next Caldwell object, the 5th-magnitude open cluster NGC 6025 in the south circumpolar constellation Triangulum Australe. The cluster literally kisses the border of Norma and is only about 3° southwest of NGC 6087 (Caldwell 89). To find NGC 6025, just point your binoculars 10° (the width of a fist held at arm's length) due east of Alpha Centauri and look for a bright but fuzzy string of starlight. Better yet, hop 3° north-northeast of magnitude-2.8 Beta Trianguli Australis.

The Southern Triangle is a fine constellation. It first appeared in Johann Bayer's 1603 *Uranometria,* but Richard Hinckley Allen notes that Pieter Theodor had formed it nearly a century earlier. Unlike the sky's dim northern Triangle, Triangulum Australe is obvious, as it comprises three stars of near-equal brightness: the "three patriarchs" (probably Abraham, Isaac, and Jacob) of Jansenius Caesius, a 17th-century Dutch globe maker. Beta Trianguli Australis marks the Southern Triangle's northern tip. Years ago, astronomers believed that the star was 10 parsecs (32.6 light-years) distant — precisely the distance at which a star's apparent magnitude equals its absolute magnitude. This made the star a standard for determining other stars' absolute magnitudes. Within the last quarter century that distance was refined to 42 light-years. More recently, the Hipparcos satellite determined that Beta Trianguli Australis lies 40 light-years from Earth. The Milky Way runs through the constellation, enriching it with faint stars.

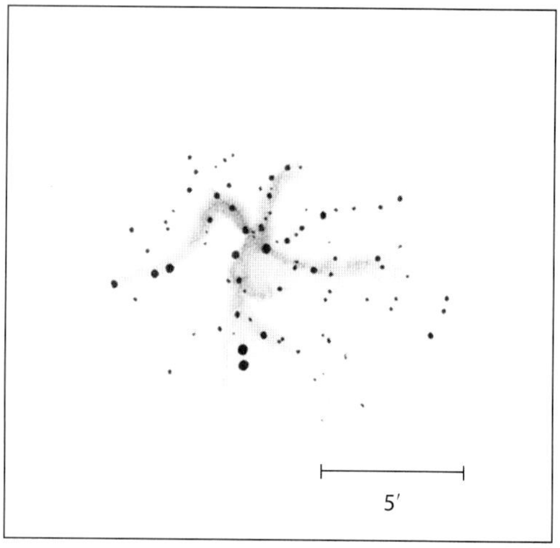

From a dark-sky site, NGC 6025 stands out to the naked eye as a puff of soft light. It lies 1° due east of a tight arc of three 6th-magnitude suns oriented north-south. The cluster's brightest star shines at magnitude 7.1, so it is a good naked-eye challenge. Note that in the table on page 381, my naked-eye estimate of the cluster's apparent magnitude is 0.4 magnitude fainter than Brent Archinal's published value. But the cluster culminates only about 10° above the horizon from Hawaii, so I encourage observers in the Southern Hemisphere to estimate its naked-eye brightness under more reliable conditions.

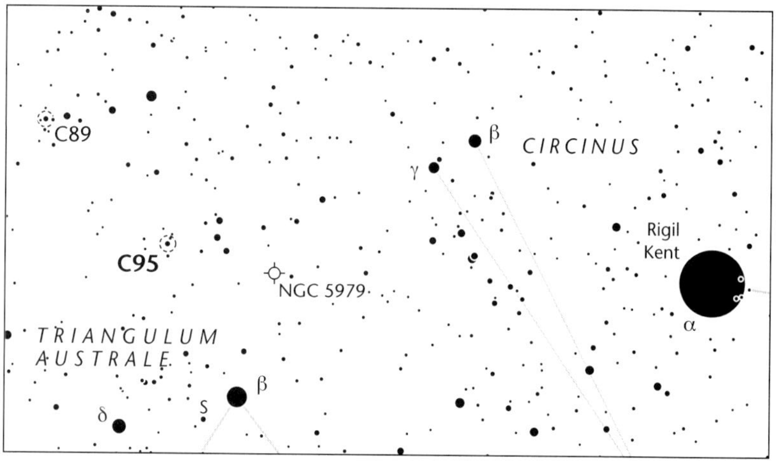

In 7×35 binoculars NGC 6025 looks like a dagger, with a row of three stars surrounded by an elongated halo of unresolved suns. And this is exactly what Nicolas Louis de Lacaille saw when he discovered the cluster with his ½-inch 8× telescope. In his 1755 catalog he describes NGC 6025 as "three faint stars in line with nebulosity." (Seen this way, the cluster resembles Carolyn Shoemaker's famous Jupiter-smashing "squashed comet," Comet Shoemaker-Levy 9, for it looks like a fragmented comet nucleus shrouded in a coma of scintillating dust.) James Dunlop observed the cluster on five occasions and saw no nebulosity. He described it as a "group of about twenty stars of mixt magnitudes, forming an irregular figure, about 5′ or 6′ long . . . and there is no nebula in the group of stars except what is common in the neighbourhood." (Lacaille's nebulosity, of course, was just the haze of unresolved starlight.) The Rev. T. W. Webb simply called the object a "brilliant cluster."

With north up at 23× in the 4-inch, the cluster's dozen or so brightest stars form the Greek letter χ. The brightest star resides in a very wide double of near-equal intensity, and it punctuates the cluster's southeastern arm. This very wide double also forms the eastern half of a trapezoid along with two slightly dimmer suns, the southwestern of which is a much tighter double. The cluster contains no obvious central condensation, though the χ asterism lies on a blanket of fainter suns. Since NGC 6025 is large and scattered, the view is best at low power, but increasing the magnification to 72× or higher will bring out some nice star pairings and a Southern Cross asterism just north of the cluster's visual "center" — which I see as the center of a C asterism. It is hard to determine the cluster's full extent because it is seen against such a rich Milky Way background. Although most observers see 30 cluster stars out to about 5′, Archinal has tallied 139 members out to 15′ (11 light-years). The cluster is about 90 million years old, making it a contemporary (loosely speaking) of the Pleiades (M45) in Taurus.

If you return to the 1°-long arc of three 6th-magnitude suns and move 1° southwest of its southernmost star, you'll encounter the small, pale glow of NGC 5979, a magnitude-11.5 planetary nebula less than 25″ across. And if you return to Beta Trianguli Australis and look about 35′ to the southeast, you'll find the copper-colored classical Cepheid S Trianguli Australis, whose light varies between magnitude 6.1 and 6.8 every 6.32 days.

96

96

Southern Beehive
NGC 2516
Type: Open Cluster
Con: Carina

RA: 07h 58.0m
Dec: –60° 45′
Mag: 3.8; 3.5 (O'Meara)
Diam: 22′
Dist: 1,300 light-years
Disc: Abbé Nicolas Louis de Lacaille, included in his 1755 catalog

J. Herschel: An orange-coloured star 8th magnitude, in middle of a large and magnificent cluster of perhaps 200 to 250 stars [of] 8[th to] 16th mag[nitude]. Many of the [brighter] magnitudes, and really a superb object. Very visible to the naked eye. (h 3111)

GC/NGC: Cluster, very bright, very large, pretty rich, stars of magnitude 7 to 13.

Like rivets, bright open clusters seem to fasten the Milky Way to Carina, the Keel of the Ship Argo. And among the most prominent is NGC 2516. This 4th-magnitude cluster lies just 3¼° southwest of Epsilon (ε) Carinae (Avior), the 2nd-magnitude star marking the foot of the False Cross; the long axis of the Cross points almost right to it. Abbé Nicolas Louis de Lacaille discovered the cluster during his pioneering southern-sky survey. It is the third Class II object (nebulous clusters) in his 1755 catalog, and he recorded it as a "very close group of 10-12 stars." (The "nebulosity" in this case was not dust or gas but unresolved starlight.) John Herschel saw "perhaps 200 to 250 stars" centered on an orange 8th-magnitude star.

NGC 2516 is an easy naked-eye target. It is sometimes mistakenly referred to as the Southern Pleiades, but that name has already been claimed. IC 2602 (Caldwell 102) — also in Carina, nearly 20° to the east-southeast — has been known as the Southern Pleiades ever since Lacaille commented on its similarity to that northern wonder.

When I surveyed the literature on NGC 2516, I discovered a trend: the cluster's reported angular size has been diminishing over the years. In his *Celestial Handbook,* Robert Burnham Jr. describes NGC 2516 as a "brilliant group, easily visible to the naked eye, with more than 100 stars scattered over a field 1° in diameter." The first edition of *Sky Atlas 2000.0* (some versions of which mislabel the cluster as NGC 2561) represents this view fairly well, showing the cluster as a 1°-wide circle centered on a dense core of starlight; the circle extends toward a row of three stars to the northeast but includes only the two southernmost members of that row. Indeed, when I first

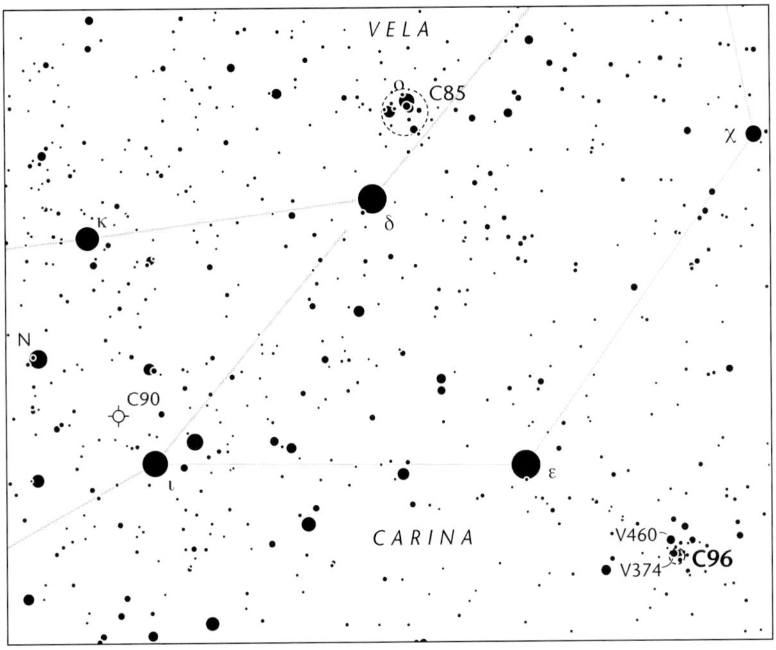

went out and looked at the cluster with the naked eye, I saw a 1°-wide, wishbone-shaped grouping. Three 6th-magnitude stars in a gentle northeast-to-southwest curve formed the thin northwestern branch of the wishbone, while a fuzzy wedge of knitted starlight formed the thicker southeastern branch. The two segments are separated by about ½°.

Imagine my surprise, then, when I noticed that *Uranometria 2000.0,* the second edition of *Sky Atlas 2000.0,* and the *Millennium Star Atlas* all show the cluster as if it were only ½° across. All three of these atlases center NGC 2516 on the dense core of stars in the southeastern part of the wishbone. They also place the region's four brightest naked-eye stars — the three stars in the northwestern branch of the wishbone and the northernmost star in the southeastern branch — outside the cluster.

More recently, Steven Hynes and Brent Archinal have moved the cluster's center 7′ to the north (from –60° 52′ to –60° 45′) and shrunk the cluster's width by another 8′ (from 30′ to 22′). So in the last quarter century, NGC 2516's diameter has diminished by 63 percent. Hynes and Archinal present another problem. They report that the cluster's brightest star shines at magnitude 7.0. But when I center a 22′-wide circle at their position for the cluster I find the variable star V374 Carinae (which fluctuates between magnitude 5.72 and 5.84) within the circle. Is this star a member of the cluster?

Puzzled by this apparent dilemma, I referred to astrophysical journals and checked the Hipparcos data as well as the second edition of *Sky Catalogue 2000.0.* The Hipparcos satellite revealed that the middle star of the trio in the wishbone's northwestern branch is not a cluster member; it is only 53 light-years distant, while the cluster lies 1,300 light-years away. But the other two bright stars in that branch are at the cluster's distance, so they probably are members. Therefore, we've come full circle (to use a pun), because NGC 2516's members do extend beyond the cluster's putative edge. Perhaps this exercise simply demonstrates just how difficult it is to determine the extent of an open cluster in a rich Milky Way region. It also highlights the imprecision of supposedly "hard" astronomical data. Archinal elaborates: "First of all, there appears to be a basic definition problem here regarding what the diameter of a cluster should be. This is another one of those cases where the *NGC* designation really applies to something that's not quite the same thing as the modern astrophysically accepted cluster. Most of the stars are concentrated in a 20-30′ diameter area, but possibly with outlier stars going out to 1°. So, depending on how one wants to define diameter, all the values cited are correct! It all just boils down to deciding whether you prefer the 'original' diameter, the approximate obvious visual diameter (my 22′ seems reasonable), or the maximum diameter, to include all possible cluster members (1° or more)."

Our astronomical predecessors had no way of determining the distances to the stars, so their perception of a star cluster was simply that — stars clustered together on the plane of the sky. When Lacaille discovered NGC 2516, he used a ½-inch telescope at 8×, so his view would have been similar to mine through 7×35 binoculars — a wishbone of stars with a fuzzy "bottom." Through my 7×35s about a dozen stars can be seen to 9th magnitude. So Lacaille probably did include the northwestern branch of the wishbone in his own "very close group of 10-12 stars."

NGC 2516 has 103 known members, several of which are red giants: an orange, magnitude-6.7 *K*1III star (R.A. 7^h 58^m 09.8^s; Dec. –60° 49′ 00″) near the core of the cluster's densest part; the 5.2-magnitude *M*0II irregular variable V460 Carinae (R.A. 7^h 59^m 37.6^s; Dec. –60° 35′ 13″); and a magnitude-5.7 *K*4II star (R.A. 7^h 56^m

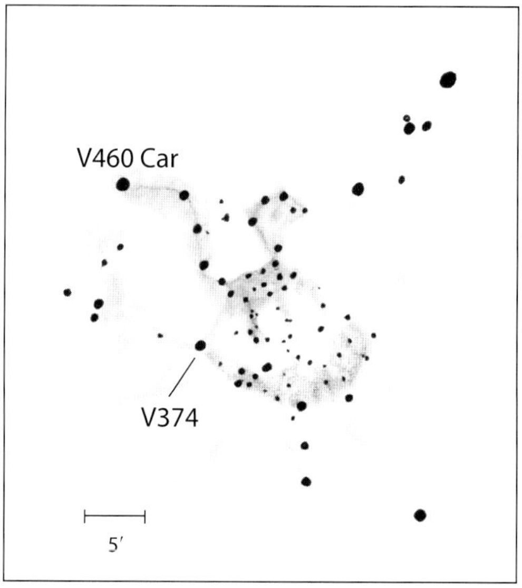

18.6^s; Dec. –60° 31′ 35″). Given the historical discrepancies regarding the cluster's size, one can only wonder which of these reddish stars was Herschel's "orange-coloured star . . . in [the] middle of [this] large and magnificent cluster."

At 23× in the 4-inch NGC 2516's densest part appears boxy, like the core of the Beehive Cluster (M44 in Cancer), and many streams of irregularly bright stars radiate from it. The cluster also contains several dark voids, and with a little imagination I can envision a cat getting ready to sleep on a cushion of starlight whose creases are the shadowy recesses of darkness. This cushion's brightest part is a wide crescent of mixed and colorful suns. Try defocusing the image ever so slightly, and look for a dark rosette centered on this cushion of stars. At 72× the "cat" is surrounded by three prominent pools of blackness, each of which looks like crushed black velvet. These pools are outlined by sparkling jewels. Yet they are not ink-black, for they shimmer softly, as if they were reflecting the starlight surrounding them. Use high magnification to look for a "comet tail" of dim suns flowing to the southwest.

NGC 2516 is 60 million years young and is one of seven similarly aged clusters that move through space together. The Pleiades, IC 2602 (the Southern Pleiades), and the Alpha Persei Cluster also belong to this group. NGC 2516's total mass is somewhere between 800 and 1,500 solar masses and its most populous region spans 9 light-years. The cluster lies 52 light-years below the galactic plane at roughly the same distance as the neighboring (and much younger) Gum Nebula. The latter is famous for the Vela Pulsar and Vela X, a supernova remnant that emits copious radio waves. NGC 2516 is a breeding place for chemically peculiar, binary, and variable stars. Its multiple passages through Vela's dense molecular clouds appear to be triggering repeated bursts of star formation.

97

97

NGC 3766
Type: Open Cluster
Con: Centaurus

RA: $11^h36.2^m$
Dec: –61° 37′
Mag: 5.3
Diam: 15.0′
Dist: 5,800 light-years
Disc: Abbé Nicolas Louis de Lacaille, included in his 1755 catalog

J. HERSCHEL: The preceding of two chief stars of a fine, large, loose, round cluster of stars [of] 8[th to] 12th mag[nitude]; gradually pretty much brighter in the middle, fills field; 150 [to] 200 stars. (h 3352)

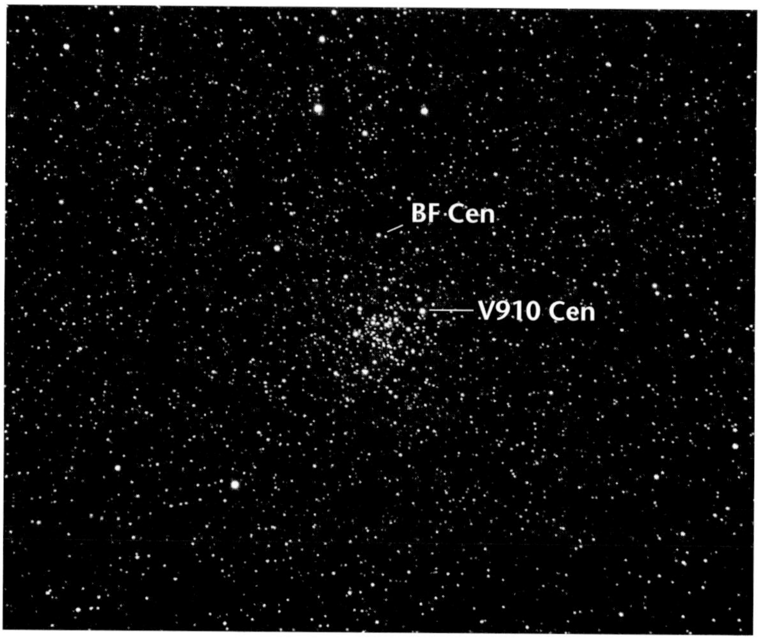

GC/NGC: Cluster, pretty large, pretty rich, pretty compressed, stars of magnitude 8 to 13.

HALFWAY BETWEEN ALPHA (α) CRUCIS, OR Acrux (the foot of the Southern Cross) and the famous Eta (η) Carinae Nebula (Caldwell 92), just 1½° due north of the Lambda (λ) Centauri Cluster (Caldwell 100), and smack-dab on the galactic equator, is yet another young naked-eye open star cluster: NGC 3766 (Caldwell 97). This cluster resides in the Carina molecular complex, a giant collection of molecular-hydrogen clouds that is associated with some of the most massive stars in our galaxy (Eta Carinae among them). Molecular-cloud complexes are hotbeds of star formation, and the Carina complex is no exception. It teems with hot, young bodies, including numerous *B*e-type stars (see the Caldwell 10 installment starting on page 49). And NGC 3766 contains the largest number of *B*e stars known in any single star cluster. A *B*e star loses mass to an expanding shell, and its prodigious rotation draws some of this cast-off matter into a disk surrounding the star's equator. Like their less-massive cousins, the T Tauri stars, *B*e stars are believed to represent the last phase in the pre-main-sequence evolution of young, hot protostars.

In brightness, apparent size, and shape, NGC 3766 appears strikingly similar to M37 in Auriga. Each cluster shines at roughly 5th magnitude, measures 15′ in diameter, and displays an elongated body flecked with bright suns. Spanning 25 light-years of space, NGC 3766 is 20 percent larger than M37, and though its distance is also greater, by about 1,400 light-years, NGC 3766 would outshine M37 were it not for interstellar dust, which diminishes its light by a half magnitude. The main differences between the two clusters lie with their ages and stellar populations. M37 is 200 million years old

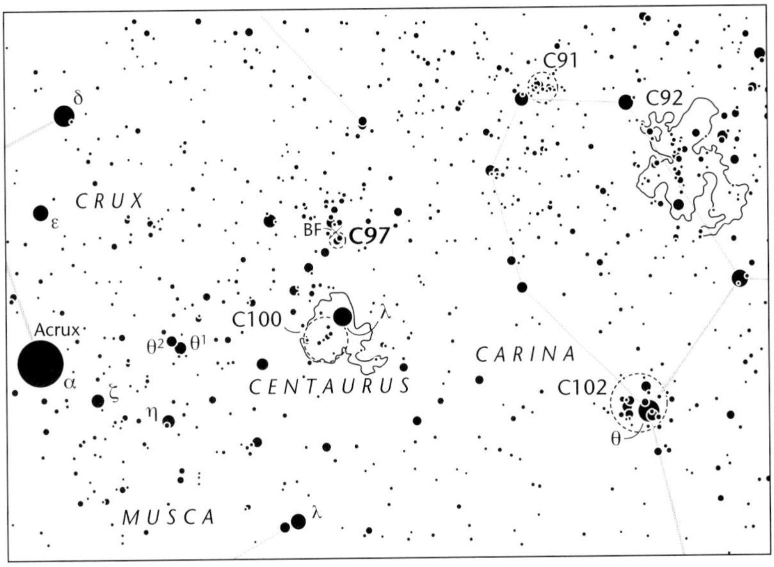

and contains some 1,890 members, while NGC 3766 is a mere 24 million years young with 137 known members (though the larger number of M37 stars may simply be due to that cluster having been studied more often).

NGC 3766 is another one of Abbé Nicolas Louis de Lacaille's finds and is listed in his 1755 catalog as the seventh of his Class III objects (stars accompanied by nebulosity). Through his ½-inch 8× telescope Lacaille saw "three faint neighbored stars surrounded by nebulosity." But, as with many of Lacaille's finds, the nebulosity does not exist; it is merely unresolved starlight. Although NGC 3766 is visible to the naked eye under a dark sky, it is not obvious. It lies in a patchy, star-rich section of the Milky Way, and its 5th-magnitude glow requires one to know precisely where to look for it. Particularly deceptive is a 30′-long rectangular asterism oriented north-south; its southern base lies about 15′ from NGC 3766's northern fringe. This clusterlike grouping looks larger and fuzzier to the naked eye than does NGC 3766, and it could be mistaken for our tiny target. It's best first to identify NGC 3766 in binoculars, then to look with the naked eye. In 7×35 binoculars its half-dozen brightest suns form a sideways T that stands out at the northwestern end of a roughly 1⅓°-long chain of four similarly bright, equally spaced stars. This chain starts about 1° northeast of Lambda Centauri and continues northwestward toward NGC 3766.

At 23× in the 4-inch Genesis the cluster's brightest stars are colorful 8th- to 10th-magnitude jewels strewn across a carpet of dim, white suns. James Dunlop first noted the star's colors in the 1820s, writing: "The greater number of the stars are of a pale white colour. There is a red star near the preceding [western] side; another of the same size and colour near the following [eastern] side; another small red star near the centre; and a yellow star near the south following [southeastern] extremity, all in the cluster." And modern observers still comment on them: in Volume 7 of the *Webb Society Deep-Sky Observer's Handbook,* NGC 3766 is described as having some "orange and yellow

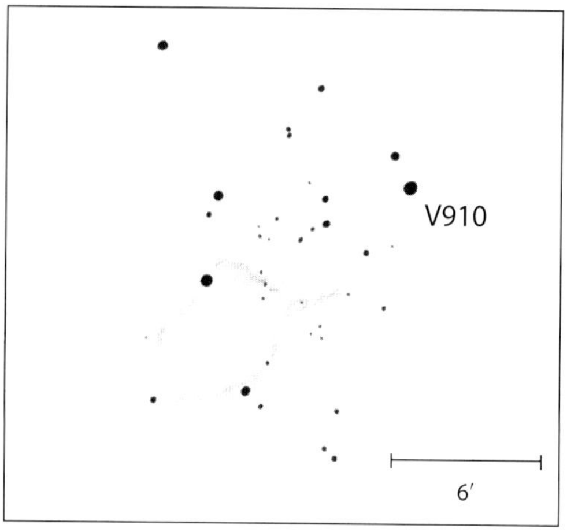

stars." Ernst Hartung calls it a "fine scattered cluster . . . containing orange, yellow, white and bluish stars." And Philip Harrington saw "some appearing golden, others bluish." The most obvious color can be seen in a pair of stars at the cluster's northwestern edge. This wide pair marks the foot of the T. The brighter of this pair's two stars is V910 Centauri, which shines at magnitude 7.2 and has a ruddy spectral class of *M*0Ib.

Now relax your gaze, and see if the southeastern side of the T doesn't widen into a triangle or wedge of starlight. A pair of red stars appears at the triangle's northern tip. The arms of the T are marked by a 6′-long stream of stars flowing to the north and a fainter one flowing an equal distance to the south. The wedge of starlight then can be seen as a body — using my imagination it looks like a great frigate bird with outstretched wings and a long forked tail.

With 72× and averted vision, look for a fantastic 1°-wide pool of darkness abutting the southeastern corner of the triangle. This dark pool has a slightly elliptical shape that manifests itself more clearly the more you concentrate on it. If you suspect that you may have seen this dark lagoon, but aren't sure, try defocusing the image ever so slightly. This will blend the bright stars surrounding the pool, helping the latter to stand out. Focus the image again and look closely at the multitude of stars in the triangle. How many doubles do you see? There should be at least a half dozen bright pairs and an equal number of fainter ones. Immediately east of the triangle, a roughly 9th-magnitude star forms part of a minicluster of dim 12th- to 13th-magnitude suns. Sweep your eyes east and west, then north and south, across NGC 3766, and see if you don't pick up all manner of loops and star chains. At one moment I saw a ring of 11th-magnitude stars within the triangle; it had a string of equally bright stars dropping to the south. At high power the cluster begins to lose its luster, though the extra magnification does help separate the many fine doubles. My drawing shows only the cluster's brightest stars, and serves mainly to show the location and extent of the dark pool mentioned earlier. Use the photograph on page 388 as your guide to the many star patterns visible in this wonderful cluster.

The eclipsing binary star BF Centauri lies at NGC 3766's extreme northern edge. Its light oscillates between magnitude 8.5 and 9.4 every 3.7 days; the star spends 20 percent of this period in eclipse. Following this star's light curve would make a good project for an eager observer with a keen eye, a photometer, or a CCD.

98

Coalsack Cluster
NGC 4609
Type: Open Cluster
Con: Crux

RA: 12^h 42.3^m
Dec: –63° 00′
Mag: 6.9
Diam: 6.0
Dist: 4,200 light-years
Disc: James Dunlop, included in his 1827 catalog

J. HERSCHEL: Cluster class VII. Stars [of] 11[th to] 13th mag[nitude]; about 6′ long and 4′ broad; has 10 stars [of] 11th mag[nitude], and some 20 or 30 smaller. It occurs in the midst of the black space following Alpha Crucis, which is by no means void of stars. (h 3407)

GC: Cluster, pretty large, pretty compressed, extremely extended, star[s] of 10th magnitude and fainter.

NGC: Cluster, pretty large, pretty compressed, considerably extended, star of 10th magnitude.

99

Coalsack; Soot Bag; Shadow of the Cross
Type: Dark Nebula
Con: Crux; Centaurus; Musca

RA: 12^h 53^m
Dec: –63°
Mag: —

98 & 99

Dim: 7° × 5°
Dist: 610 light-years and 790 light-years (two overlapping nebulae)
Disc: Known since antiquity

J. HERSCHEL: None.

GC/NGC: None.

THE COALSACK (CALDWELL 99), A RIVAL OF Crux though devoid of its splendor, is the Milky Way's poster child for visual mystery. The northern tip of the kite-shaped dark nebula lies due south of the Jewel Box (Caldwell 94), while its western flank washes against Acrux, the foot of the Cross. In photographs the Coalsack's size and shape mimic that of the Cross's interior, so I like to call it the Shadow of the Cross. Australian aborigines have probably known of this dark cloud for more than 40,000 years; it appears, on tribal bark paintings from Groote Eylandt, as a speared fish being held up by two brothers over two fires (the brothers and their fires representing the four stars of the Cross). Another Australian legend refers to the Coalsack as "the embodiment of evil in the shape of an Emu." The Polynesians knew it as Te Paniwi a Taewa, a black fish. And ancient Peruvians perceived it as the celestial Doe suckling its fawn. In more modern times the Coalsack was Amerigo Vespucci's *Il Canopo fosco* (the dark Canopus) and Adm. William Henry Smyth's Black Magellanic Cloud. In his *Oceana*, the 19th-century historian James Anthony Froude called it "the inky spot — an opening into the awful solitude of unoccupied space."

Froude's remark reflects William Herschel's pronouncement that each starless void was a *Loch im Himmel*, an opening in the heavens, a curious vacancy "through which we seem to gaze into an uninterrupted infinity." In his sweeps of the night sky, Herschel found a remarkable coincidence: the blackest voids appeared near the richest star clusters, which "would almost authorise a suspicion that the stars, of which it is composed, were collected from that place, and had left the vacancy." The elder Herschel specifically cited the black void near Rho (ρ) Ophiuchi and the globular cluster M80 as a magnificent example. Had he spied the Coalsack and its proximity to the Jewel Box — not to mention the open cluster NGC 4609 (Caldwell 98) within it — his suspicions would have been magnified.

From South Africa in 1834 Herschel's son, John, turned his attention to his father's "holes." At first he presumed them to be "tubular vacancies" penetrating the depths of space. He later retracted this view. But scholars like Froude and authors like Thomas Hardy carried William Herschel's fantastic view of these voids into the late 19th century. In his novel *Two on a Tower* — a work Hardy created "to set the emotional history of two infinitesimal lives against the stupendous background of the stellar universe" — we find astronomer Swithin St. Cleeve showing the sky to his imminent lover, Lady Constantine, to whom he tries to explain the unfathomable size of the universe:

> . . . horrid monsters lie up there waiting to be discovered by any moderately penetrating mind. . . . Impersonal monsters, namely, Immensities. Until a person has thought out the stars and their inter-spaces, he has hardly learnt that there are things much more terrible than monsters of shape, namely, monsters of magnitude without known shape. Such monsters are the voids and waste places of the sky. Look, for instance, at those pieces of darkness in the Milky Way. . . . You see that dark opening in

> it near the Swan? There is a still more remarkable one south of the equator, called the Coal Sack, as a sort of nickname that has a farcical force from its very inadequacy. In these our sight plunges quite beyond any twinkler we have yet visited. Those are deep wells for the human mind to let itself down into, leave alone the human body! and think of the side caverns and secondary abysses to right and left as you pass on!

As early as 1889 Arthur Cowper Raynard, the editor of the London scientific magazine *Knowledge,* believed that these dark clouds were not Herschelian holes but "opaque matter, dust clouds or fog-filled space." Raynard spent years trying to convince Edward Emerson Barnard, who had been photographing these dark regions, of his beliefs. But Barnard strongly promoted the notion of Herschel's vacancies — and he continued to do so until, as William Sheehan explains in his Barnard biography, *The Immortal Fire Within,* he changed his mind on "one beautiful and transparent moonless night" in 1913:

> I was struck with the presence of a group of tiny cumulous clouds scattered over the rich star-clouds of Sagittarius. . . . Against the bright background they appeared as conspicuous and black as drops of ink. They were in every way like the black spots shown on photographs of the Milky Way, some of which I was at that moment photographing. The phenomenon was impressive and full of suggestion. One could not resist the impression that many of the small spots in the Milky Way are due to a cause similar to that of the small black clouds mentioned above.

Barnard died seven years before he could have learned the truth. Not until 1930, when Robert J. Trumpler published his "Preliminary Results on the Distances, Dimensions and Space Distributions of Open Star Clusters," did astronomers have conclusive proof that the galaxy was full of diffuse, obscuring matter. Today we recognize dark nebulae as cold, dusty clouds of molecular hydrogen gas that cloak or dim the light of whatever lurks behind them. Trumpler also discovered that interstellar dust reddens starlight; it scatters blue light more efficiently than red, and the effect points to dust particles ranging between 1/1000 and 1/100 of an inch in diameter — making them something akin to smoke particles encased in ice. The coldest of these dark clouds are only a few degrees above absolute zero.

Nearly all investigations to date find the Coalsack to be a stable, inactive nebula without any signs of hidden star formation. Ultraviolet spectrometers aboard the Voyager 1 and Voyager 2 spacecraft stared toward the Coalsack on four occasions. On each occasion they recorded intense diffuse emission with a spectrum characteristic of a *B*-type star. Present in all four observations, this spectral signature may be the result of starlight scattered by interstellar dust in the foreground of the main mass of the Coalsack. Dust-dimming (*or extinction*) data and distance determinations for 280 stars in the Coalsack region suggest that the Coalsack is not a single sheet of dust, but a conglomerate of two overlapping dark clouds at distances of 610 and 790 light-years, respectively. (These two clouds are among the closest to Earth.) The Coalsack's closest parts span about 54 light-years. The Coalsack may be part of a larger interstellar structure that includes the Chamaeleon-Musca dark clouds some 15° away. If it is, the larger structure may extend like a sheet at right angles to the galactic plane.

To the naked eye the Coalsack recalls H. P. Lovecraft's "unsealed wells of night," or Edgar

98 & 99

Allan Poe's horrifying Pit, where "the blackness of darkness" supervenes and all sensations appear "swallowed up in a mad rushing descent as of the soul into Hades." It is the bleak black heart of the Milky Way. I first saw the Coalsack in 1982 from the cockpit of an Air New Zealand Boeing 747 flying south at an altitude of 40,000 feet. From that darkened cabin the Coalsack needed no introduction. Through the thin high-altitude air it looked like an immense pool of haunting darkness and misty vapors, a black fog hugging the foot of the Cross — the antithesis of Dante's four principal virtues (see Caldwell 94, page 376). The Coalsack was by no means the blackest cloud I'd seen, but the Milky Way in this region is so intensely bright that the enormous dark nebula stood out immediately. The object's size (7° × 5°) also called to mind Smyth's Black Magellanic Cloud. Not only are the Coalsack and the Large Magellanic Cloud roughly the same angular size; the Coalsack is about as conspicuous against the bright Milky Way as the Large Magellanic Cloud is against the relatively dark skies of Dorado and Mensa. It's unfortunate that the brightest objects in the heavens have been given negative magnitude numbers; if it were otherwise, we could create a "negative brightness" scale for dark nebulae. For example, since the Large Magellanic Cloud shines at magnitude +0.4, the Coalsack's "dark magnitude" could've been –0.4. It wouldn't have been a perfect scale, but it would have given observers some idea of these objects' visibility.

The Coalsack has a magical quality to it, and its appearance can change suddenly and eerily under averted vision. I observed it one evening from the nearly 14,000-foot-high summit of Mauna Kea. My 7×35 binoculars revealed a vast smoky clot, a blotch of negative space, an erased section of the Milky Way. Closer examination showed the dark nebula's main body to be square and sharply ribbed. Through the 4-inch at 23× these ribs were thin trails of black smoke, which made me wonder if dark nebulae had anything to do with the ancient notion that stars were flames. Bill Harby, the editor of *Island Scene* magazine, was with me that night, and he compared the Coalsack to black ice with a multitude of streamers. Indeed, under a relaxed gaze the cloud transforms into a series of black satin veils seemingly frayed by age. One prominent veil starts immediately southeast of Acrux, where you'll

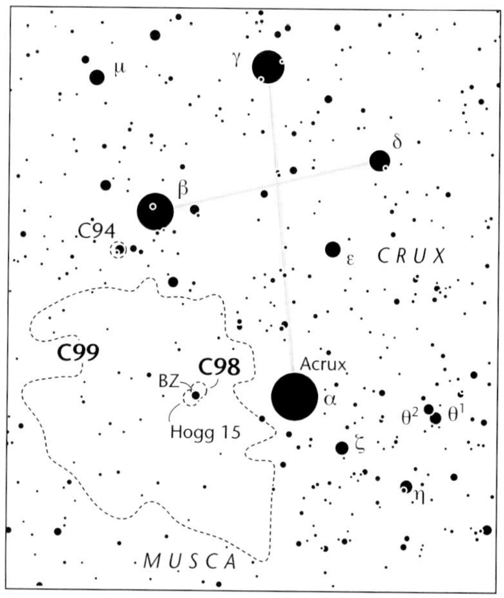

find a conspicuous ink spot with two parallel tails of darkness flowing to the east. The northernmost of these tails passes just south of the sole naked-eye star in the Coalsack, the *Be*-type variable BZ Crucis, whose light fluctuates subtly around magnitude 5.2. Another prominent veil marks the sharp northernmost border of the Coalsack, just south of the Jewel Box. And an isolated black puddle forms the southern "tip" of the kite-shaped nebula. Seeing this

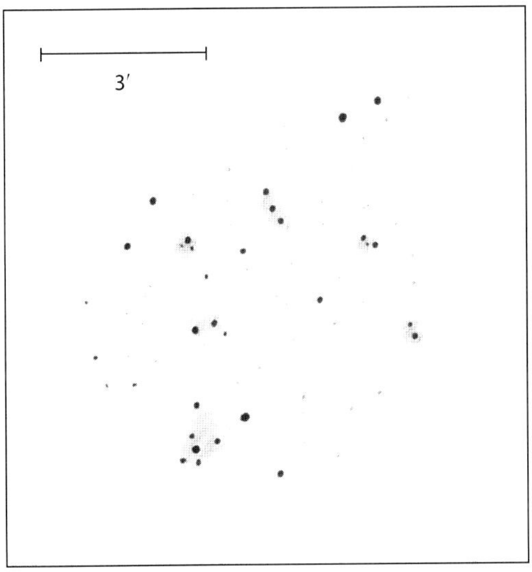

with the naked eye is a challenge. See if you can spot it. In photographs, a thin black tail flows several degrees to the southwest from the kite's dark bottom. Can you see this delicate feature?

Now return to 5th-magnitude BZ Crucis. See anything unusual? Just 5′ northwest is the tiny 6′-long open star cluster NGC 4609 (Caldwell 98), and directly opposite the star, on the southeast, is an even smaller open cluster, Hogg 15. James Dunlop discovered NGC 4609 from Parramatta, New South Wales; it is the 272nd object listed in his catalog. He described it as "a group of five stars of the 8th or 9th magnitude, with a great number of extremely [faint] stars resembling faint nebulae." Brent Archinal says the 7th-magnitude cluster contains 52 stars; if we accept a distance of 4,200 light-years for the cluster, these stars are spread across 7 light-years of space. Of course, that distance also implies that the cluster lies behind the Coalsack, and photometry confirms that its starlight has been considerably reddened by dust. Also, despite their apparent proximity to NGC 4609, the stars of Hogg 15 — four of which can be seen in the 4-inch, forming a small triangular asterism about 1′ across — are not associated with it. NGC 4609 is a young cluster, its age being about 60 million years, which makes it a bit younger than the Pleiades.

A few of NGC 4609's stars can be made out with averted vision through binoculars, and the dimmer ones seem to flare off of BZ Crucis like particles in a dim comet tail. At 23× in the 4-inch the cluster appears as a tiny splash of about a half-dozen suns oriented north-northwest to south-southeast. Sweep your telescope around this area and the cluster may look like a boat sailing across long, rolling waves of darkness. At 72× NGC 4609 is a loose gathering of about 20 stars in a rectangle, many in pairs; under averted vision a dim wedge of faint starlight can be seen to the northwest. Increasing the magnification widens the view, making the cluster appear like a skeleton seen through a dusting of dirt. How utterly different this cluster is from the Jewel Box, which resides just outside the black fog of the Coalsack. The contrast between the two is amazing to behold.

By the way, if you've ever wondered where the Coalsack got its name, we can thank Sir

John Herschel for informing us. In his 1847 "Results of Astronomical Observations" he mentions that the Eta Carinae Nebula (Caldwell 92) was "in the midst of one of those rich and brilliant masses, a succession of which curiously contrasted with dark adjacent spaces (called by the old navigators coal-sacks)."

100

Lambda Centauri Cluster
Collinder 249
Type: Open Cluster
Con: Centaurus

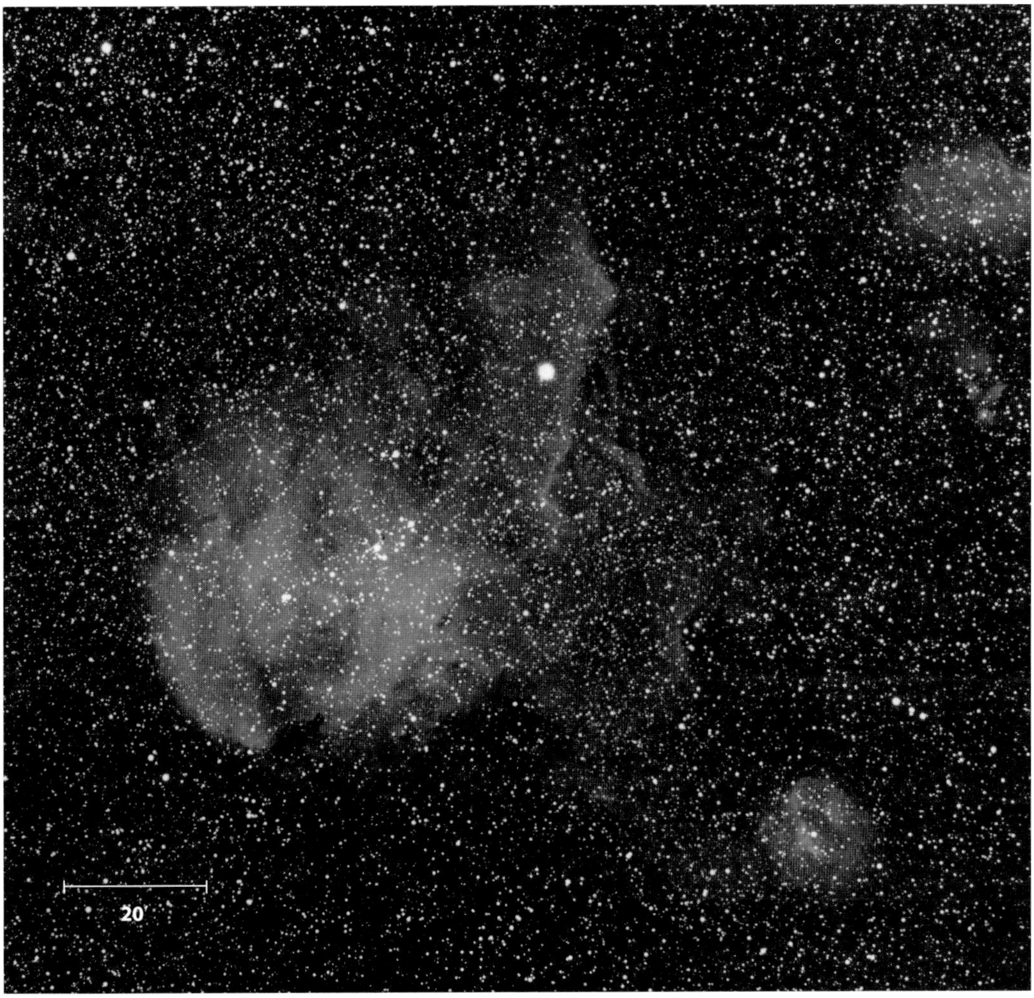

RA: 11^h 38.3^m
Dec: –63° 22′
Mag: 2.9
Dim: 65′ × 40′
Dist: ~6,000 light-years
Disc: Per Arne Collinder, listed in his 1931 treatise, *On structural properties of open galactic clusters and their spatial distribution*

HERSCHEL: None.

GC / NGC: None.

100

THOSE WHO OWN COPIES OF THE ORIGINAL Caldwell Catalog will find the data table accompanying this text somewhat perplexing. In the original catalog Caldwell 100 is listed as the magnitude-4.5 open cluster IC 2944, and it is also called the Gamma Centauri Cluster. But this cannot be. First of all, there is no star cluster associated with Gamma Centauri. The cluster's Caldwell Catalog position makes that point clear, for it is centered on a spot just 5′ east of Lambda Centauri — which does appear to have a large, loose cluster associated with it. Patrick Moore simply mistook Lambda (λ) for Gamma (γ). Okay. But the problem does not end there. According to Brian Skiff and Brent Archinal, IC 2944 is not the Lambda Centauri Cluster. Rather, it is the nebula surrounding the Lambda Centauri Cluster. And the cluster's proper name is Collinder 249.

Mystery solved? Not exactly. After receiving this information I looked at these objects on several star atlases and in catalogs and discovered the following curiosities:

- The 1962 edition of the Skalnate Pleso *Atlas of the Heavens* and the 1981 edition of Wil Tirion's *Sky Atlas 2000.0* plot IC 2944 as a small nebula immediately surrounding Lambda Centauri, with a slight elongation to the northwest. No cluster is shown with Lambda Centauri, though an open cluster labeled IC 2948 is plotted about ½° to the southeast of the star.

- *Uranometria 2000.0* depicts IC 2944 as a cluster of stars immediately to the east of (and including) Lambda Centauri. IC 2948 is depicted as a nebula that surrounds Lambda Centauri and flows to the south and southeast, as well as a bit to the northwest, of that star.

- The *Deep Sky Field Guide* lists IC 2944 as a magnitude-4.5 open cluster southeast of Lambda Centauri and IC 2948 as a nebula surrounding this entire region.

- *Sky Catalogue 2000.0* lists IC 2944 as a 15′-wide cluster centered on the original Caldwell Catalog position, roughly 5′ east of Lambda Centauri.

- The *Millennium Star Atlas* shows IC 2948 as a nebula surrounding Lambda Centauri and extending nearly a full degree to the south and southeast. It does not show any star clusters in the area.

- The second (1998) edition of *Sky Atlas 2000.0* also depicts IC 2948 as a nebula surrounding Lambda Centauri and the regions to the south and southeast. But it also shows IC 2944 as a tiny cluster that seems to include Lambda Centauri, though it is centered on a point just southeast of that star.

What's going on? If the cluster's proper name is Collinder 249, why isn't this designation mentioned in catalogs or plotted on star charts? Better yet, exactly where is the Lambda Centauri Cluster, since so many references are in disagreement? And finally, just what is the difference between IC 2944 and IC 2948, if indeed there is one?

I turned to Archinal for help, and he solved the mystery by checking the sources for the original data for IC 2944, IC 2948, and Collinder 249. Here's what he reports:

> **IC 2944.** The *IC* lists this as Frost 789 with the following description: "star of magnitude 3.4 in an extremely large nebula." The

position updates to (2000.0) R.A. 11^h 35.7^m and [Dec.] –63° 01′; the range updates to approximately (2000.0) from 11^h 35^m to 11^h 36^m, and from –62° 47′ to –63° 13′. The [Digitized] Sky Survey shows at this position a very bright star (Lambda Centauri), and faint nebulosity around it that is mostly overpowered by the star. There is also a very rich background of stars here, which may be a cluster.

IC 2948. The IC lists this as Frost 790 with the following description: "Most extremely large." In the Harvard Annals, Frost 790 is described as a "Neb. patch ext. from 11^h 30.6^m to 11^h 38.1^m, and from –62° 28′ to –63° 14′." The position updates to (2000.0) [R.A.] 11^h 38.8^m, [Dec.] –63° 31′ and the range updates to approximately (2000.0) from 11^h 35^m to 11^h 43^m and from –63° 01′ to –63° 47′. At the precisely given position, the DSS shows nothing unusual. However, with histogram equalization [an image-processing technique], there are concentrations of nebulosity to the northwest, north-northeast, and northeast. A 1° DSS image shows nearly the full 1° square filled with nebulosity, with the patches mentioned above being the brightest portions. Clearly, [Yerkes Observatory astronomer Edwin Brant] Frost [(1866–1935)] had the whole area in mind and not just the brighter concentrations. It is also clear that this is the (larger) nebulosity to the southeast of IC 2944 and is not IC 2944 itself.

Cr 249. Collinder 249 is listed by Collinder as IC 2944 (more comments on this follow). He describes it as 40′ × 15′ in size, 2.7 total magnitude, and 25 stars total. He adds that Lundmark lists it as 65′ × 40′ in size, 2.9 total magnitude, and 40 stars total. It is called a "Neb. cl.", so the involved nebulosity was clearly recognized. Collinder calls it "A replica in Southern Sky of Pleiades" and writes, "This is a chain of bright stars with nebulosity. Areal density of fainter stars the same as in surroundings (from star-counts)."

Collinder goes on to present a photograph of this cluster, copied from the Franklin-Adams charts, showing several bright stars in nebulosity. I couldn't easily match this photo to a DSS image of the area, so I checked the Franklin-Adams chart (no. 18). I was surprised to find that Cr 249 in fact is a very large object, consisting of all of the stars around IC 2944 and IC 2948, and then some! Generally it is the whole group of stars extending from Lambda Centauri (and IC 2944) on the northwest, down to a pair of stars at 11^h 41.8^m and –63° 49′, thus filling [an] oval about 1 degree in length! Lundmark's size estimate of 65′ × 40′ thus matches well.

Thus, Collinder 249 is *not* IC 2944. Frost mentioned nothing about a star cluster in his descriptions of either IC 2944 or IC 2948. And more importantly, the group of stars is centered on IC 2948 and not IC 2944 as Collinder's identification and (incorrect) position implies. These two errors by Collinder are obviously the source of all the later confusion about the objects here.

In summary, the nebula IC 2944 is pretty much concentric on Lambda Centauri. IC 2948 is the nebulosity to the southeast, and it is distinct from IC 2944. And the entire area's ensemble of stars is the cluster Collinder 249. There apparently is no cluster centered on Lambda Centauri, though it does have a rich background of Milky Way stars. (Incidentally, Lambda Centauri itself is much closer than the cluster; the Hipparcos satellite has pegged its distance at 410 light-years.)

Another Caldwell mystery solved! Despite its tortured history, and the way it has been misrepresented on star charts, we can now engage ourselves in finding the true Lambda Centauri Cluster with confidence. (Any nebulosity seen

100

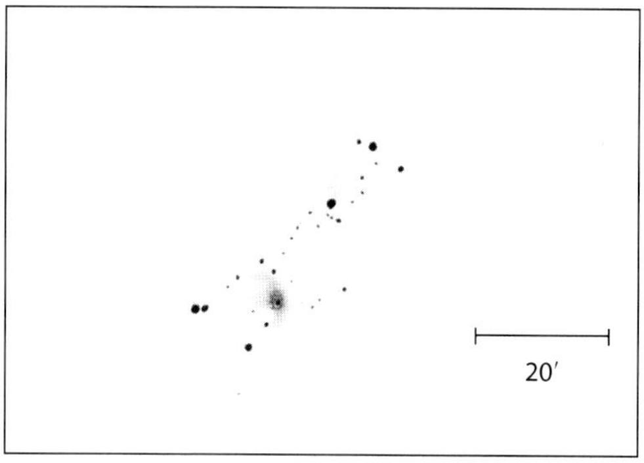

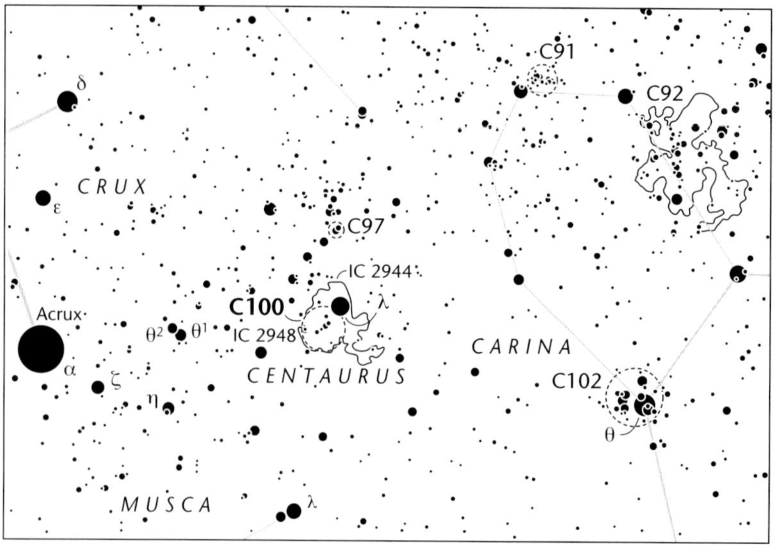

while performing this feat is a bonus.)

From Hawaii 3rd-magnitude Lambda Centauri (a type *B*9II blue giant) is low, achieving an altitude of only 7° at upper culmination. Lambda Centauri has the same declination as Alpha (α) Crucis (Acrux), the southernmost star in the famous Southern Cross, and it can be found about 5½° west of it. Fortunately, the 4,200-foot summit of Kilauea offers a reasonably clear view of the southern Milky Way. Still, my views pale in comparison with those from the Southern Hemisphere, where the Centaurus Milky Way curdles into fluid folds of starlight high overhead.

Lambda Centauri is easy to spot, being the brightest star roughly halfway between Acrux and the great Eta (η) Carinae Nebula (Caldwell 92). Seen on photographs, Lambda Centauri lies on the eastern rim of a conspicuous caldera of darkness that breaches the region's copious starclouds. To be sure you find Lambda Centauri in this star-rich region, start at Acrux and hop two naked-eye stars to the southwest. This will bring you to Eta Crucis, which is the easternmost star in a line of four equally spaced naked-eye stars (oriented southeast to northwest) ending at Lambda Centauri.

To the naked eye, a dusty comet tail of light ½° long flows from Lambda Centauri to the southeast. This glow emanates from an unresolved chain of a half dozen or so 7th- to 8th-magnitude stars in Collinder 249. The view is much the same in 7×50 binoculars, though the star chain is well detached from Lambda Centauri. Under averted vision the chain transforms into a rectangle of scintillating starlight surrounded by a slight fog. Knowing that a nebula (IC 2948) surrounds this region, it is easy for me to assume that the glow comes from interstellar gas, but I could not convince myself of this from my Hawaii site; the glow, I feared, might also come from unresolved stars. The emission nebula does pop out dramatically in many photographs of the region, such as that shown on page 397. In fact, some photographs suggest that IC 2948 rivals the nearby

Eta Carinae Nebula. But remember that photographs of other emission nebulae — Barnard's Loop, say, or the California Nebula — make those nebulae appear much brighter than they do to the eye. I remained unsure of IC 2948's visibility until I observed Collinder 249 with binoculars and David Levy's 3½-inch Questar under dark African skies in June 2001. The nebulosity was apparent in both instruments as a dim fog that swelled into view under averted vision.

From Hawaii my 4-inch refractor reveals a quite obvious circular patch of nebulosity (IC 2944) around Lambda Centauri at 23×. Under averted vision a dim elliptical extension can be followed to the north, where it encounters a roughly 10th-magnitude star 7′ away. This observation agrees with Volume 7 of the *Webb Society Deep-Sky Observer's Handbook*, which describes the nebula as "oval . . . of about 10′ to 15′ extent, surrounding λ Centauri." I also suspected the presence of nebulosity around the brightest stars and star chains to the southeast of Lambda Centauri.

At 72× many pairs of stars can be seen throughout Collinder 249, which is most strongly concentrated about 25′ southeast of Lambda Centauri. An obvious U-shaped asterism marks the chain's southeastern end. (Collinder recorded the cluster as a "chain of bright stars with nebulosity.") The cluster's stars then spread out in two diverging forks to the northwest, each of which ends at a bright pair of binocular suns. The beauty of this star chain notwithstanding, Collinder's suggestion that this cluster is a Southern Pleiades seems unfounded. My eyepiece views certainly did not give such an impression. Perhaps in photographs, where the cluster's stars are embedded in a complex network of nebulosity, one could see how Collinder derived such a name. But I have to wonder. You see, IC 2602 (Caldwell 102) in Carina, which lies about 6° southwest of Lambda Centauri, is popularly known as the Southern Pleiades. Did Collinder mix up these clusters' names?

102

Theta Carinae Cluster;
Southern Pleiades
IC 2602
Type: Open Cluster
Con: Carina

RA: $10^h 42.9^m$
Dec: –64° 24′
Mag: 1.9
Diam: 100′
Dist: 492 light-years
Disc: Abbé Nicolas Louis de Lacaille, included in his 1755 catalog

HERSCHEL: None.

GC/NGC: None.

IC: Cluster, coarse, includes θ Carinae.

DROP A VISUAL PLUMB LINE 4¾° DUE SOUTH of the magnificent Eta (η) Carinae Nebula and it will collide with another brilliant deep-sky wonder, IC 2602, the dazzling Theta (θ) Carinae Cluster (Caldwell 102). Louis de Lacaille discovered this mine of celestial diamonds during his great mid-18th-century exploration of the southern skies. The object is included in his 1755 catalog, where it is the 9th listed under his Class II category ("nebulous clusters"). Lacaille called the object "Theta Argus" (Carina is the keel of the now-defunct constellation Argo Navis) and estimated its brightness as 3rd magnitude. Through his diminutive ½-inch 8× telescope Lacaille saw Theta Carinae "surrounded by many 6/7/8 mag [stars] like the Pleiades." And that is why on modern charts IC 2602 is labeled "The Southern Pleiades."

With a declination of –64° 23.7′, IC 2602 is only 5½° above my Volcano, Hawaii, horizon when at upper culmination. Yet even at that meager altitude it is clearly visible to the unaided eye. The cluster is one of three major glows in the rich Carina Milky Way, the other two being NGC 3532 (Caldwell 91) and the Eta Carinae Nebula (Caldwell 92). Together with the dense starclouds of the attendant Milky Way, the sight of these three objects is without question the most stunning in the entire heavens. Several of IC 2602's stars are visible to the naked eye, but just how many depends upon the cluster's altitude, your site's darkness, and your ability to resolve stars packed tightly together. Photometry of two 15′-wide fields, conducted in 1991 with the 1-meter Swope telescope at Las Campanas Observatory in Chile, revealed that half the stars visible to the telescope in IC 2602 were background stars — not surprising for an open cluster in a rich Milky Way field. Brent Archinal says IC 2602 harbors at least 60 definite members spread across 100′; some 30 of these are brighter than

9th magnitude. Of the latter, 23 have spectral types of *B* and *A* while the others range from *F*0 to *K*5. Theta Carinae, of course, is the brightest, shining at magnitude 2.8 with a spectral class of *B*0V. The nearly 5th-magnitude star 6′ to the southwest of Theta Carinae is V518 Carinae. Its variability was discovered by the Hipparcos satellite, which recorded apparent visual magnitudes ranging from 4.60 to 4.76.

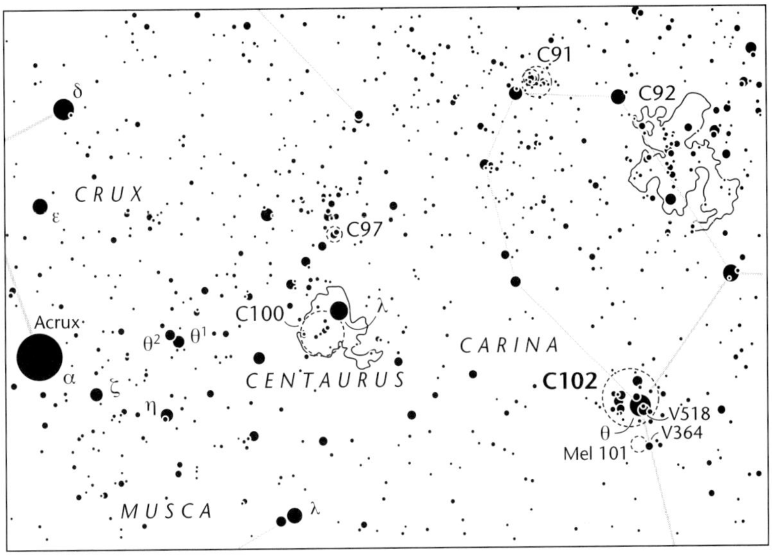

IC 2602 is very young. Its stars burned through their dusty cocoons some 30 million years ago, which makes this cluster less than half the age of the Pleiades. Further evidence of the cluster's relative youth can be found in the spin rates of its stars. Younger stars generally spin faster than older ones. However, there can be a wide range of rotation periods even among a young cluster's stars. Recent studies have found that the rotation periods for 29 of IC 2602's stars range from 0.2 days (one of the shortest known rotation periods of any single open-cluster star) to about 10 days (which is almost twice the longest period previously known for a cluster of this age). Images of IC 2602 from the Rosat X-ray satellite revealed a total of 110 objects within an area of 11 square degrees. Of these, 68 have been identified with at least one visible counterpart; 44 of these are new candidates for cluster membership. In IC 2602 Rosat detected X-rays from stars of all spectral types. The *F*-, *G*-, and early *K*-type stars appear to be more X-ray luminous than their counterparts in the Pleiades, while no significant difference was seen among the two cluster's late-*K*- and *M*-type dwarfs. This may be related to the different rates at which stars of different masses shed rotational energy.

If we accept its distance as 492 light-years, IC 2602 is 85 light-years more distant than the Pleiades and 23 light-years closer than M44, the Beehive in Cancer. It spans 14 light-years, as does the Pleiades. Furthermore, IC 2602 has some 10 stars shining at 6th magnitude or brighter, like the Pleiades, and it is only 0.4 magnitude (39 percent) fainter than the Pleiades. So the two clusters really are twins. Visually, however, IC 2602 resembles M6, the Butterfly Cluster in Scorpius, more than it does the Pleiades. IC 2602 is a member of the Scorpius-Centaurus *OB* association, a widespread group of hot, young stars that appear to have a common origin.

Like the Pleiades, IC 2602 is best observed through binoculars or a rich-field telescope. In the 4-inch at 23× the cluster is a loose aggregation of suns with two distinct groupings. The first grouping includes Theta Carinae and a gentle arc of suns, like a slightly upturned saber glinting with dew. The second group lies to the east, separated from the Theta Carinae grouping by a 20′-wide lagoon of relative darkness, and it looks either like a butterfly, an hourglass, or a bow tie.

Before leaving the field, take the opportunity to view yet another of this stunning region's seemingly countless open clusters. Melotte 101 (R.A. 10^{h} 42.1^{m}; Dec. –65° 06′), an 8th-magnitude fog of stars spanning 13′, lies 45′ due south of Theta Carinae. It is 10′ east of V364 Carinae, a roughly 5th-magnitude variable star. Lacaille, no doubt, saw this star as his southern "Pleione," though unfortunately, it is not a true member of either cluster. Visible in the lower-left corner of my drawing, Melotte 101 contains about 40 faint stars, the brightest of which shine around 11th magnitude.

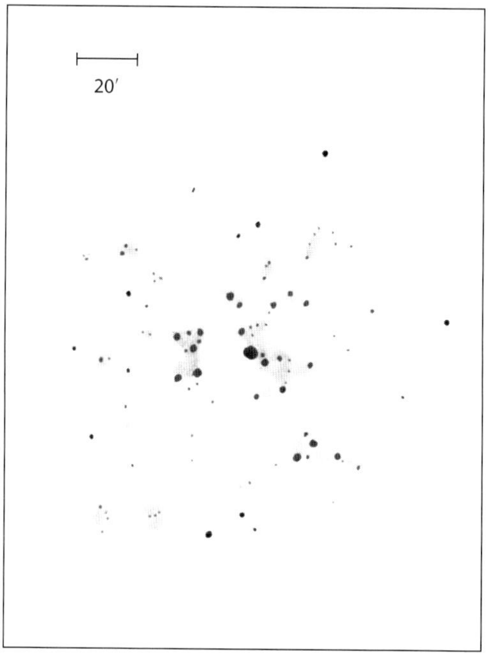

103

Tarantula Nebula;
True Lover's Knot
NGC 2070
Type: Emission Nebula and Star Cluster
Con: Dorado

RA: $05^h 38.6^m$
Dec: –69° 05′
Mag: ~4.0 (O'Meara; nebula)
Dim: 30′ × 20′ (nebula)
Mag: 8.3 (cluster)
Diam: 5′ (cluster)
Dist: ~170,000 light-years
Disc: Abbé Nicolas Louis de Lacaille, included in his 1755 catalog

J. Herschel: An assemblage of loops. [Extensive notes follow.] (h 2941)

GC/NGC: A magnificent object, very bright, very large, looped.

A little more than 60° due south of the Orion Nebula — considered by many to be the heavens' reigning showpiece — lies an equally remarkable southern treasure: NGC 2070 (Caldwell 103), the famous Tarantula Nebula and its associated star cluster. Also known collectively as 30 Doradus, these objects appear to the naked eye as a single irregular glow just north of the eastern tip of the Large Magellanic Cloud's central bar. The Tarantula is the brightest of the 50 or more nebulae populating that neighboring galaxy, and it punctuates the system's visual grandeur.

The Tarantula Nebula is an extraordinarily massive and luminous H II region — a large concentration of (mostly) hydrogen gas ionized by ultraviolet radiation from very hot, luminous stars embedded within it. Indeed, as seen from Earth today, thousands of suns are being born within this gaseous cocoon. With 800,000 Suns' worth of ionized hydrogen, the Tarantula is not only the brightest H II region in the LMC; it is also the most massive H II region known in the Local Group of galaxies. The nebula is truly enormous, spanning about 3,000 light-years; by comparison, the Orion Nebula stretches across a mere 40 light-years of space. The Tarantula's bright inner nebulosity — the part of the nebula most commonly detected by eye — spans about 1,500 light-years, and it contains at its core the star cluster responsible for ionizing the entire nebula. The cluster measures about 250 light-years across.

Astronomers have shown that the Tarantula contains a spectacularly high number of blue supergiants, many of which are of spectral type *O3* (the most massive stars known, each about 100 times as massive as our Sun).

The cluster's total mass is an estimated 200,000 Suns. With that much mass, it's conceivable not only that the stars are gravitationally bound to one another but also that NGC 2070 may harbor a globular cluster in the making. According to a 1998 *Sky & Telescope* article by Georges Meylan and Bernhard Brandl, NGC 2070 shows us how galaxy's old globular clusters may have looked when they formed 10 to 15 billion years ago. At the cluster's center lies a "star," Radcliffe 136 (R136), that recent Hubble Space Telescope and ground-based adaptive-optics systems have resolved into a dense, rich cluster just 10″ (8 light-years) across.

In 1996 one of the youngest known pulsars was discovered on the edge of the Tarantula Nebula. A team led by Francis E. Marshall (NASA/Goddard Space Flight Center) found a signal repeating itself 62 times each second in data from the Rossi X-Ray Timing Explorer. Other observations from RXTE and the Advanced Satellite for Cosmology and Astrophysics then were used to confirm the discovery and to estimate the pulsar's age (4,000 to 5,000 years). The Tarantula's pulsar is in its infancy, just like the Crab Nebula's 944-year-old pulsar, and it appears to be powering an X-ray-luminous supernova remnant on the edge of the nebula.

Visually, the Tarantula Nebula is not an overwhelming sight without optical aid. It shines at roughly 4th magnitude and spans about ½° (though some sources have it extending twice as far). To the naked eye the nebula is conspicuous under dark skies, and many southern observers have detected its strong inner glow even under the light of the full Moon. The Tarantula Nebula's overall naked-eye appearance is that of a slightly smaller, though equally bright, Lagoon Nebula (M8 in Sagittarius).

The 18th-century French astronomer and

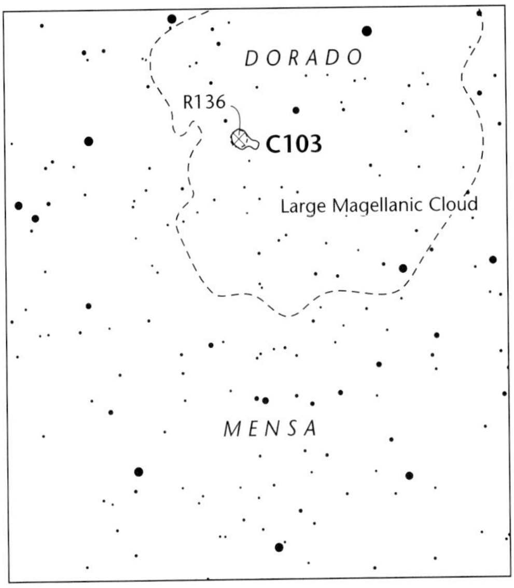

southern-sky explorer Abbé Nicolas Louis de Lacaille discovered the Tarantula Nebula from the Cape of Good Hope in either 1751 or 1752 (no discovery date was recorded). In his catalog of southern-sky objects, published in 1755, he writes simply that the object, when viewed through his modest ½-inch refractor, "resembles [47 Tucanae], but . . . is fainter." He listed the object under the heading "nebulae without stars." Seventy-two years later John Herschel described the nebula as an assemblage of loops, "the centre of a great spiral." He discussed the Tarantula at length:

> This is one of the most singular and extraordinary objects which the heavens present, and derives no small addition to its intrinsic interest from its situation, which is among the thickest of the nebulae and clustering groups of the greater Nubecula [the LMC], of whose total area it occupies one-five hundredth part. . . . [I]ts real nature has been completely misunderstood, and its magnified appearance so strangely misrepresented in the only figure which I am aware to have been made of it as to convey an entirely erroneous impression both of its form and structure. . . . The stars . . . scattered over the area occupied by this nebula may or may not be systematically connected with it, either as an individual object, or as part of the vast and complex system which constitutes the Nubecula. In respect of their arrangement there is nothing to distinguish them from those which occupy the rest of the area covered by the Nubecula, in which every variety of condensation and mode of distribution is to be met with. The nebula itself (as seen in the 20-feet reflector) is of the milky or irresolvable kind — quite as free from any mottling or incipient stellar appearance as any other nebula which I can remember to have examined with that instrument.

The magnitude-8.3 cluster (which is also embraced by the designation NGC 2070) spans 5′ and contains about 100 suns visible through small-aperture instruments; its principal member shines at magnitude 11.8, and most of the others are in the 12th- to 14th-magnitude range. Some catalogs list the magnitude of the nebula *and* the cluster as 8.3, but this certainly is in error, for that value pertains only to the cluster.

Through a telescope the Tarantula Nebula and its associated cluster together stand out as one of the most profound sights in the night sky. Apertures of 2 inches and larger will reveal the object's complex tangle of gaseous loops and sprays of starlight. A moderate-size amateur instrument like an 8-inch Schmidt-Cassegrain will reveal Herschel's assemblage of loops dramatically. Indeed, the Tarantula's visible structure rivals its photographic appearance. The most obvious feature is the nebula's strong central bar, which extends from north to south and is centered on 30 Doradus. From it sprawl several crablike arms and legs, all extending in seemingly animated ways. A distinct loop, like

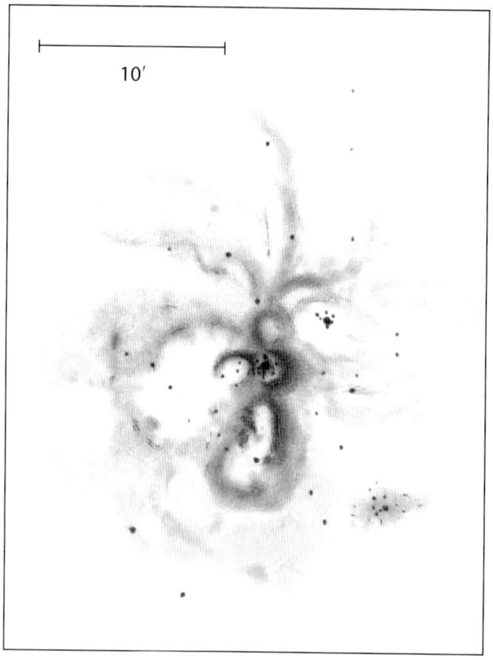

a hangman's noose, can be seen at the bar's southern end. Another nebulous arm juts from the bar's northern end, then heads northwest before curving slightly south to all but touch a slightly swollen cluster of stars. Together the noose, the arm, and the cluster look remarkably like a wide-eyed celestial seahorse navigating an ocean of nebulosity.

Due east of the Tarantula's central star cluster, a small but bright coil of nebulosity encompasses a dark bay, east of which lies another larger, though less conspicuous, bay encircled by a faint prominence of feathery nebulosity. The southern extension of this latter feature flows into the eastern edge of the "hangman's noose," forming the True Lover's Knot (a bowknot used as a symbol of love). This name was assigned to the nebula by the imaginative and romantic 19th-century explorer of the skies, Adm. William Henry Smyth.

The Tarantula's grandeur does not end with these loops. Bright nebulous extensions, each laced with wisps of fine detail, can be traced to the north and west, and several stellar groupings lie in and around the glowing gaseous loops. The fact that amateur astronomers can see so much detail in a galaxy other than our Milky Way is a marvel unto itself. If the Tarantula Nebula were as close as the Orion Nebula, it would illuminate a patch of sky nearly 60° across, outshining Venus. Keep that in mind when looking through the eyepiece at the Tarantula Nebula, and the object will transcend its ghostly appearance and stand before you as a rare vision of celestial magnificence.

By the way, despite the Tarantula's association with the LMC, which is summarily considered a Southern Hemisphere object, the nebula *can* be seen from the Northern Hemisphere during upper culmination. Kevin Krisciunas observed Supernova 1987A after it appeared about 18′ west and 10′ south of the Tarantula's core. He was standing on the nearly 14,000-foot summit of Mauna Kea on the Big Island of Hawaii (19° 50′ N latitude).

104 & 106

104

NGC 362
Type: Globular Cluster
Con: Tucana

RA: $01^h\ 03^m\ 14.3^s$
Dec: –70° 50′ 54″
Mag: 6.4
Diam: 14′
Dist: 27,700 light-years
Disc: James Dunlop, included in his 1827 catalog, though probably noted throughout history as a star

J. HERSCHEL: Fine, highly condensed globular cluster, pretty suddenly brighter in the middle, diameter 4′. (h 2375)

GC/NGC: Globular cluster of stars, very bright, very large, very compressed, very much brighter in the middle, stars of magnitude 13 to 14.

106

47 Tucanae; Xi Tucanae
NGC 104
Type: Globular Cluster
Con: Tucana

RA: $00^h\ 24^m\ 05.2^s$
Dec: –72° 04′ 51″
Mag: 3.95
Diam: 50′
Dist: 14,700 light-years
Disc: Abbé Nicolas Louis de Lacaille, included in his 1755 catalog, though probably noted throughout history as a star

J. HERSCHEL: The great cluster preceding the Nubecula Minor [Small Magellanic Cloud]. Estimated diameter of the denser portion 5′; of the whole (not, however, including those stragglers) 8′. Stars 14[th to] 16th mag[nitude] and one of 12th mag[nitude] north preceding [to the northwest of] the center. Excessively compressed. (h 2322)

GC/NGC: Globular cluster of stars, *remarkable*, very bright, very large, very much compressed in the middle.

104 & 106

A FIST'S WIDTH SOUTHWEST OF FIRST-MAGnitude Achernar, the end of the celestial River Eridanus, is the body of one of the sky's most "flighty" constellations: Tucana, the Toucan. With an area of 295 square degrees, Tucana is the sky's 48th-largest constellation; it contains some 80 stars between magnitudes 2.9 and 7. Johann Bayer first introduced this colorful bird simply as "Toucan" in his 1603 *Uranometria*; the name was later latinized. The Toucan flies around the South Celestial Pole with three of the heavens' greatest wonders dangling from its claws: the Small Magellanic Cloud (a dwarf galaxy that orbits our Milky Way) and the globular star clusters NGC 362 and 47 Tucanae (Caldwell 104 and 106, respectively).*

NGC 362 and 47 Tucanae bracket the Small Magellanic Cloud to the north and west, respectively, so it is only natural that we discuss these globulars together. Recent proper-motion studies of these clusters' stars confirm that they belong to our Milky Way system and are merely projected against the Small Magellanic Cloud's halo some 176,000 light-years away. NGC 362 is the more distant of the two clusters. Indeed, 47 Tucanae is one of the closest globulars to our Sun; it is as far from the Sun as is M10 in Ophiuchus, though M10 is about 9,000 light-years closer to the galactic center.

James Dunlop discovered NGC 362 in the 1820s. It is the 62nd object in his *Catalogue of Nebulae and Clusters of Stars in the Southern Hemisphere, Observed at Parramatta in New South Wales*. Of it he writes, "A beautiful bright round nebula, about 4′ diameter, exceedingly condensed. This is a good representation of the 2nd of the *Connaissance des Tems* [sic] in figure, colour, and distance; it is but a very little easier resolved, rather a brighter white, and perhaps more compact and globular. This is a beautiful globe of white light; resolvable; the stars are very little scattered." Dunlop's reference to "the 2nd of the Connaissance des Tems" was to none other than Messier 2, the second object listed in Charles Messier's catalog, which was published in the French almanac *Connaissance des Temps*.

NGC 362 lies about 2¼° southwest of golden 4.9-magnitude Kappa (κ) Tucanae and 1½° north-northeast of NGC 346 and NGC 371 — two large balls of gas and stars on the northern edge of the brighter part of the Small Magellanic Cloud. In 10×50 binoculars NGC 362 looks like a slightly out-of-focus star, especially with averted vision. The cluster shines with an apparent magnitude of 6.4, and no similarly bright object competes with it in its immediate vicinity. Under a dark sky I have seen it with the naked eye. Through the telescope its apparent size and magnitude recall those of the globular star cluster M2 in Aquarius (as Dunlop himself noted so long ago).

NGC 362 is a relatively metal-rich cluster. On average, each of its stars has about 1⁄14 as much iron as does the Sun. This is a higher metallicity than that found in most globulars. Several generations of stars must have preceded the cluster's formation in order to enrich it thus. This suggests that the metal-rich globulars, though old, are younger than metal-poor ones. NGC 362 also has an overabundance of binary stars and an exceptionally tight core 1.6′ (13 light-years) in diameter. Apparently this tight core has saved the cluster from certain destruction. NGC 362 is in a highly eccentric orbit that takes it within 1 kiloparsec (3,260 light-years) of the galactic center. Had its core not been so tight, the cluster probably wouldn't have survived the 100 or so encoun-

*Caldwell-number users beware: 47 Tucanae (NGC 104) is Caldwell 106, while NGC 362 is Caldwell 104.

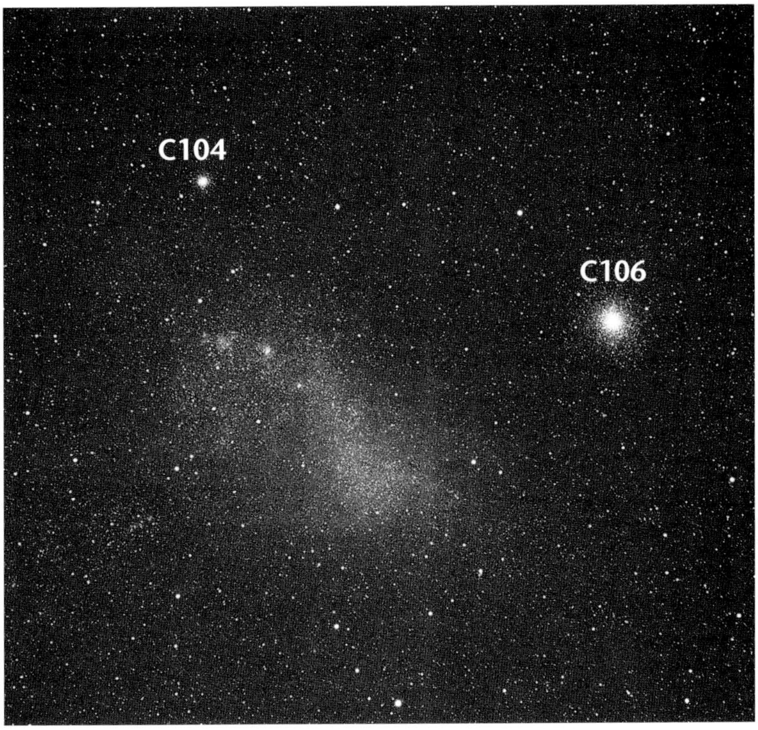

ters it has had thus far with the galactic center. Only the concentrated clusters like NGC 362 can survive repeated close-range encounters with our galactic bulge. A looser cluster would have been ripped apart by tidal forces, its members tossed helter-skelter into the galactic halo. Presently NGC 362 is approaching us at 220 km per second and is 30,000 light-years from the galactic center, so we are not witnessing one of its close encounters with the galactic bulge, nor will we in our lifetimes.

In photographs taken with large telescopes it is difficult to appreciate NGC 362's dense core because it is usually overexposed. The cluster's faint outliers appear to be fizzing off that burned-out ball like gas bubbles from an Alka-Seltzer tablet in water. The only way to truly appreciate the cluster's dynamic beauty is to go out and see it through the eyepiece of a telescope. My view of NGC 362, from Wellington, New Zealand, was spectacular. I used an 8-inch Celestron Schmidt-Cassegrain telescope that Rob and Lesley Hall had set up for me in their Wellington backyard. I almost didn't get to see NGC 362. New Zealand is noted for its clouds, and my Caldwell-hunting trip was nearly at an end. In fact, I was returning to Hawaii the next morning. But I still hadn't studied this cluster in any detail. Fortunately, the sky cleared in the wee hours of the morning, just hours before my flight. Rob woke me up, pointed to the sky, and I was outside in a flash.

The Small Magellanic Cloud is dimly visible to the naked eye even from Wellington. I had no trouble spotting the cluster in the telescope's finder. I simply directed it 1½° due north of NGC 346 and *voilá!* At low power, the cluster displayed a perfectly round sphere of stars packed into a tight ball that became increasingly denser toward the center, until it burned with what appeared to be the light of a single intense star. NGC 362 has a relatively faint horizontal-branch magnitude of 15.4, so it is not easily resolved even with the 8-inch. I saw a suggestion of resolution toward the center, but no more. Trying to resolve stars in the cluster's outer halo was like trying to track a piece of dust in a sunbeam; it's all but impossible from a suburban site with a moderately sized telescope. But I bet if I had had the time, I could have eked out more and more of the cluster's members, especially if the skies were darker. Still, what a glorious sight to see a core so tight, to realize that the bonds at this cluster's heart had defied the gravitational tug of our galaxy. I looked upon this timeworn celestial warrior and felt honored to have made its acquaintance.

Barbara Wilson observed NGC 362 on November 4, 1994, while observing at an altitude of 9,700 feet (2,900 meters) from Tacna, Peru. Her view of it, through a 7-inch f/5.1 telescope, was equally impressive: "Much smaller than 47 Tucanae with a bright core, very bright, stars are fainter. . . . Cluster is fuzzy; with a 20-mm Nagler [eyepiece] it is about 3′ in size but the core is very bright, not resolved in the 7-inch. Core is near stellar, but I can see about 20 or 30 stars near the core." Ernst Hartung found that he could definitely resolve the cluster with a 12-inch telescope: "This beautiful globular cluster is well resolved to a very bright compressed centre, the main part 2′ wide and the scattered outliers reaching to 4′." He also claims that NGC 362 can be resolved with a 6-inch and that it "looks granular" in a 3-inch.

Hartung observed NGC 362 from much darker skies than I had in Wellington. So maybe that's the key. But why, then, didn't Wilson resolve the cluster in her 7-inch under the clear, dark skies of the Peruvian Andes? Wilson's response to this question raises an interesting point, one that I haven't addressed elsewhere: that resolving a globular cluster is highly subjective. "Regarding [the] resolution of NGC 362," Wilson writes, "I guess it depends on one's definition of resolution. I don't know what Hartung used for his criteria. Does one mean clarity, crispness, focus[,] smoothness of image, visibility, distinctness as all synonyms for resolution? My definition of resolution of a globular cluster is when one is

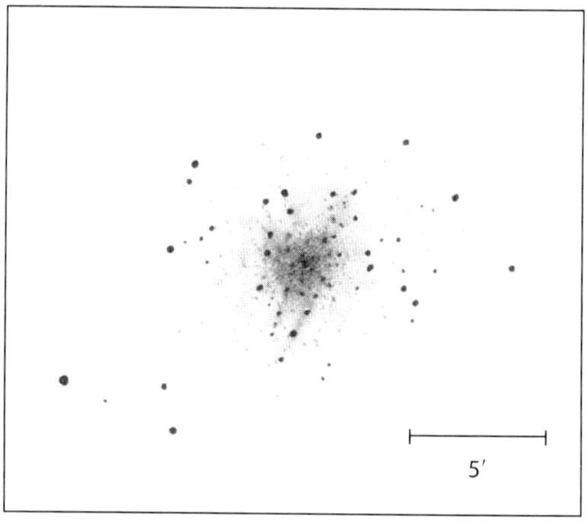

resolved to the core. When I can see stars all across the face of the core of the cluster, I call it totally resolved. The outer parts are a given, meaning these stars usually are resolved in a bright cluster. If I can see most of the cluster's stars from its outer periphery to its mysterious center, it is 80 to 90 percent resolved. I called NGC 1261 in Horologium [(Caldwell 87)] semi-resolved because its bright little non-stellar nucleus did not show stars, and had fuzzy edges. I saw only a few stars clearly. I remember NGC 1851 in Columba from the Winter Star Party appearing mottled in the center using 500×, otherwise it resolved in its outer portions. But I would not call it resolved, but nearly so."

CCD images of NGC 362 have revealed dark patches distributed asymmetrically around the cluster's center. Argentine astronomer Mariano Mendez and his colleagues estimate that the cluster contains about 0.1 solar mass of dust. Patches of dust have been imaged in other globulars, but have they ever been detected visually? Walter Scott Houston thought they were worth pursuing because "it seems reasonable that some dust clouds could be visible to the eye." So give it a try.

Before we move on to 47 Tucanae, I have a challenge for you. Moving 1° north and slight-

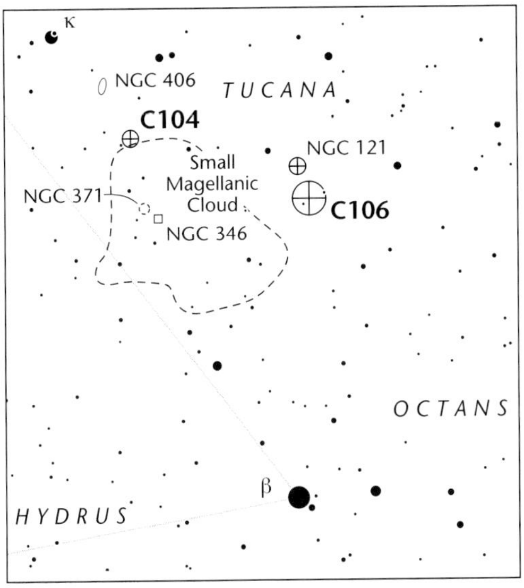

ly east of NGC 362 we find the magnitude-12.3 galaxy NGC 406, a tiny spindle just 2.9′ long. In photographs it displays two peculiar spiral arms that are festooned with bright knots. How large a telescope is needed to see the knots?

Next to Omega Centauri (Caldwell 80), 47 Tucanae is the brightest globular in the heavens. But Omega Centauri's top ranking seems insignificant once you realize that it is only 0.05 magnitude (5 percent) brighter than 47 Tucanae. The magnificent 47 Tucanae is also 1.25 magnitudes (3 times) brighter than the brightest Messier globular, M22 in Sagittarius, and its diameter is 30 percent larger than M22's. The cluster's true physical size is 214 light-years — nearly twice that of NGC 362 — and its apparent diameter is some 60 percent larger than the full Moon's. Like NGC 362, 47 Tucanae is a relatively metal-rich cluster, with ⅙ as much iron in each of its stars as there is in our Sun. Its integrated spectral type is a yellowish *G*4. Unlike NGC 362, 47 Tucanae is receding from us, at a paltry 19 km per second.

Many works say that 47 Tucanae was not discovered by *astronomers* before 1751, when Abbé Nicolas Louis de Lacaille swept up this spectacular object and kicked off his 1755 catalog with it. After seeing it through a ½-inch 8× telescope he said 47 Tucanae was "like the nucleus of a fairly bright comet." (Ironically he placed it in his Class I, "nebulae without stars.") But 47 Tucanae does have a history that predates Lacaille's telescopic observation. The cluster appears as a "star" just west of Nubecula Minor (the Small Magellanic Cloud) on Bayer's *Uranometria*, which was first published in 1603. Bayer never did see that "star" himself, and his information on far-southern stars came not from astronomers but from explorers who journeyed south of the equator. Indeed, the first catalogs of southern stars were those of the Dutch navigators Pieter Dirckszoon Keyser (published around 1595, and now missing) and Frederick de Houtman (published in 1603). Had these explorers possessed a keener understanding of the heavens, they undoubtedly would have noticed that the "star" just west of the Small Magellanic Cloud might appear swollen to the attentive naked eye. (I have seen it this way from Africa, with the "star" only 10° above the horizon, and I'm certain many aboriginal skywatchers would have done the same.) Astronomers did not study the southern sky in any detail until a 20-year-old Edmond Halley went to St. Helena in 1676 with the sole purpose of cataloging the southern stars. Halley, who had "poor eyesight," spent 18 months on that remote South Atlantic island, plagued by inclement weather, an arrogant governor, and harsh living conditions. In the end Halley cataloged only 341 stars. He did discover the diffuse nature of Omega Centauri (Caldwell 80), however, and given more clear nights he undoubtedly would have discovered the fuzzy nature of 47 Tucanae before Lacaille did.

In the 19th century Benjamin Apthorp Gould (1824–1896), the founder and first director of Cordoba Observatory in Argentina, bestowed the Greek letter Xi (ξ) to the cluster in his 1879 *Uranometria Argentina*. This moniker was not universally adopted, though — which is unfortunate, since "Xi Tucanae" would complement "Omega Centauri" nicely. To this day Tucana does not have a star labeled Xi. The name 47 Tucanae refers to the Flamsteed number of the "star", and NGC 104 is the only nonstellar naked-eye object to bear such a number.

It's a shame Lacaille didn't spy 47 Tucanae with an instrument larger than his ½-inch 8× telescope. What a missed opportunity to look into the eye of God. Dunlop observed the cluster with a 9-inch f/12 reflector and described it as "a beautiful large round nebula, about 8′ diameter, very gradually condensed to the centre. This beautiful globe of light is easily resolvable into stars of a dusky colour. The compression to the centre is very great, and the stars are considerably scattered south preceding and north following [to the southwest and northeast]."

John Herschel called 47 Tucanae "the great cluster preceding the Nubecula Minor [the SMC]." With his 18¼-inch f/13 speculum-mirror telescope, he estimated the diameter of the cluster's denser portion as 5′ and called it "excessively compressed." Herschel also saw the cluster's innermost stars shining with a rose-colored tint while the outliers were pure white. Agnes Clerke, in *The Sphere of the Stars* (1905), deemed Herschel's color perception rather subjective. "To the present writer, in 1888," she penned, "the sheeny radiance of this exquisite object appeared of uniform quality from centre to circumference. . . . Perhaps no other cluster exhibits an equal degree of compression. . . . it was, indeed, for several nights after his arrival in Peru, mistaken by [Alexander von] Humboldt for a comet."

Is it only in modern times that we are able to comprehend the beauty of this object? In my estimation, no other globular in the heavens, not even Omega Centauri, can compare. Seeing 47 Tucanae through a telescope is like cracking open a geode and finding it filled with gold dust. It calls to mind Robert Frost's poem, "All Revelation":

> But the impervious geode
> Was entered, and its inner crust
> Of crystals with a ray cathode
> At every point and facet glowed.

To the naked eye 47 Tucanae shines singularly at 4th magnitude, like the head of a tailless comet sailing 2½° west of the SMC. And there is nothing in the sky background to interfere with the view of these two splendid objects. They are like two vessels alone in the void. Under dark skies there is almost no taking your gaze off them, so splendid they are together. Seen through 10×50 binoculars, 47 Tucanae immediately captures the imagination. Here is a starlike nucleus surrounded by a halo of soft, pearly light. In no other globular do I recall having seen such an opalescent hue through so small an instrument. My first telescopic view of 47 Tucanae was through the Carter Observatory's 9-inch Cook refractor in 1981; it stole my breath. The cluster literally burned from within. A tight 2.8′-wide bead of topaz light sizzled like an electric flame while thousands of sparks illuminated sphere upon sphere of starlight, which diffused outward from the cluster's center. Photographs reveal several million stars — as many as can be found in some minor galaxies — across the cluster's 50′-wide disk, but it seemed to me that there were millions of stars in the tiny core itself. (Of course, this begs the question: are all globular clusters the products of merging minor galaxies?) Omega Centauri lacks 47 Tucanae's awesome central blaze, which draws

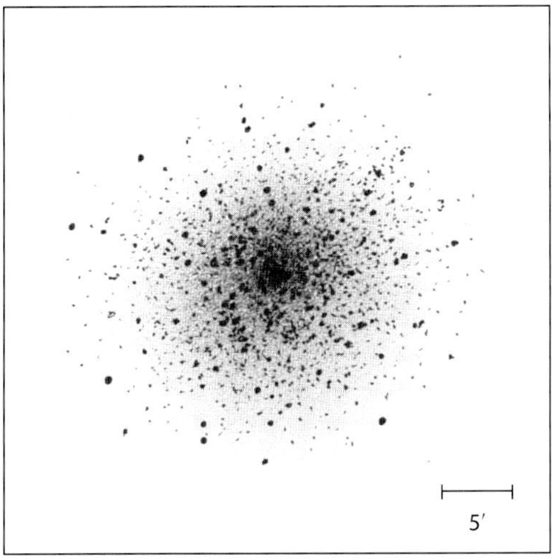

the eye ever inward and takes your mind on a fantastic journey. With the help of the telescope I felt like I was *in* the cluster.

The horizontal-branch magnitude of 47 Tucanae is a reasonable 14.1, so the cluster can be resolved in a 4-inch telescope. When I studied 47 Tucanae in an 8-inch from Wellington in 1997, I noticed something peculiar, something I have not read much about in the observing literature: the cluster is not perfectly round but elliptical, extending from north to south. Furthermore, its interior is not as uniform as it appears at first glance. With quiet scrutiny I detected at least eight clearly defined spidery arms. Barbara Wilson had a similar impression when she observed 47 Tucanae on November 4, 1994, from Tacna, Peru, under the darkest, steadiest, and most transparent skies she had experienced thus far. "The core is a brilliant crowded mass of stars," she penned, describing the view through her 7-inch telescope, "and the cluster has an almost delicate appearance, with spiraling lines of stars emanating in all directions. In a 20-mm Nagler [eyepiece], the cluster resolves to the core, some stars are reddish in appearance in [the] center, and the cluster's center is sharply brighter than the outer portions. This is the finest globular I've seen."

While living in New Zealand in 1999, William Sheehan had the opportunity to view 47 Tucanae with the 1-meter telescope at the Mount John Observatory. "At 48×," he says, "one immediately notices the intensely concentrated nucleus with a large spreading field of stars decreasing in density away from the center. It is extremely bright, in fact the brightest globular available to the eye, inasmuch as Omega [Centauri] is diffusely spread and has [a] low surface brightness. At higher power the brilliance of the globular seems to increase as it fills the whole field of view and the core becomes ever more condensed."

We often hear about the colors of globular clusters. To my eyes, none makes a stronger impression of color than 47 Tucanae. To most observers 47 Tucanae displays a yellowish cast, as its spectral type would suggest. But can that color really be perceived at the eyepiece? The cluster's concentrated core, it seems, should be bright enough to stimulate the eye's color-sensitive cones. If so, what color should we see? David Malin, of the Anglo-Australian Observatory, writes: "Though the light of globular clusters is dominated by so-called 'red' giant stars, their colour is no redder than a domestic tungsten lamp, so the true colour of 47 [Tucanae] is close to . . . pale yellow."

Globular clusters harbor some of the oldest stars known in our galaxy. But when the Hubble Space Telescope peered into the heart of 47 Tucanae, it revealed that the cluster's entire core is crowded with mysterious "blue stragglers" — stars that are bluer and brighter than the other cluster members, as if somehow they were born more recently than their siblings. Blue stragglers were discovered about a half century ago and were enigmatic until recently. However, in recent years astronomers have become convinced that these stellar nonconformists are formed either when the stars in a double-star system slowly merge or when two unrelated stars collide. Data from HST's Faint Object Spectrograph established the temperature, size, and rotation rate of one massive blue straggler (BSS-19) in 47 Tucanae. And for BSS-19, at least, astronomers favor the slow-merger scenario. BSS-19 has a mass of 1.7 Suns; it is the first globular-cluster blue straggler whose mass has been measured directly. The cluster's crowded center also contains rapidly spinning pulsars discovered recently by radio astronomers.

Finally, here's a little oddity. How many of us have heard of, or given much thought to, NGC 121, a pathetic little globular cluster about 30′ north-northeast of 47 Tucanae's core and just 10′ from the edge of its outer halo? NGC 121 measures a mere 1.5′ and shines with the light of a magnitude-11.2 star. It is not plotted in the first edition of Wil Tirion's *Sky Atlas 2000.0,* though it is in the second edition. The cluster also is not included in the *Catalog of Parameters for Milky Way Globular Clusters* compiled by expert William E. Harris (McMaster University). Why? Because NGC 121 is an old globular cluster in the Small Magellanic Cloud. By studying Hubble Space Telescope images of NGC 121, astronomers detected the first blue stragglers known in an extragalactic globular cluster. The globular's true size is 75 light-years — only one-third as large as 47 Tucanae. Still, seeing the two together offers observers incredible depth perception! How many observers have overlooked NGC 121 for the glory of 47 Tucanae?

105

Southern Butterfly
NGC 4833
Type: Globular Cluster
Con: Musca

RA: $12^h\ 59^m\ 35.0^s$
Dec: –70° 52′ 29″
Mag: 6.9
Diam: 13.5′
Dist: 19,600 light-years
Disc: Abbé Nicolas Louis de Lacaille, included in his 1755 catalog

J. HERSCHEL: Globular, bright, large, round, gradually brighter in the middle, stars [of] 14th mag[nitude], and one [of] 7th mag[nitude to the northwest of] the centre; a fine object. (h 3444)

GC/NGC: Globular cluster, bright, large, round, gradually, then very suddenly brighter in the middle, stars of magnitude 12.

108

NGC 4372
Type: Globular Cluster
Con: Musca

RA: $12^h\ 25^m\ 45.4^s$
Dec: –72° 39′ 33″
Mag: 7.2
Diam: 18.6′
Dist: 18,900 light-years
Disc: James Dunlop, included in his 1827 catalog

J. HERSCHEL: Globular cluster, very faint, large, very gradually brighter in the middle, 6′ diameter, resolved into stars of 15th magnitude, rich in stars; a delicate and faint object; has a star [at a position angle of] 45

105 & 108

[degrees to the northwest], distance 5′ from centre. Almost perfectly insulated in a very large space almost entirely devoid of stars, being the smaller and southern lacuna below the great "coal sack."

GC/NGC: Globular cluster, pretty faint, large, round, stars of magnitude 12 to 16.

BUZZING AROUND THE FOOT OF THE SOUTHERN Cross, a tongue's length north of Chamaeleon, is one of the sky's tiniest constellations: Musca, the Fly. Originally the constellation was known as Apis, the Bee, a name Johann Bayer introduced in 1603. Abbé Nicolas Louis de Lacaille substituted Musca for Apis in or around 1752. But, as Richard Hinckley Allen points out in his popular book *Star Names: Their Lore and Meaning,* Edmond Halley had called the constellation Musca Apis in 1679, and before him Giovanni Battista Riccioli cataloged it as Apis seu Musca. Actually, the constellation's proper full name is Musca Australis (Southern Fly), because a Musca Borealis also once buzzed around the northern sky, over the back of Aries, the Ram. You also may find some star charts calling Musca Australis by its more taxonomical name, Musca Indica (Indian Fly). What a busy history for such a tiny bee (or fly, as the case may be)! Actually, with an area of 138 square degrees, Musca is deceivingly large, with slightly more than twice as much celestial real estate as far more famous Crux.

Shining at magnitudes 2.7 and 3.0, respectively, Alpha (α) and Beta (β) Muscae skirt the horizon at my observing site, the 4,200-foot-high summit of Kilauea (latitude 19° 25′ N). Alpha Muscae is the more southerly of the two stars and culminates just 1½° above my southern horizon. Fortunately the sky is very dark and transparent at the summit and Alpha Muscae is easy to see. In fact, on paper I should be able to dip down to declinations as far south as my colatitude of –70° 35′. Now here's the rub. The globular cluster NGC 4833 (Caldwell 105) is only 17′ farther south; it lies 2½° southeast of Alpha Muscae at a declination of –70° 52′. But wait. Atmospheric refraction can lift starlight by some 30′ at the horizon. So, on paper at least, it is possible to see NGC 4833 from Hawaii, though it will peek just 13′ above the horizon. (Actually it should climb a bit higher for me because I observe several thousand feet above sea level.) Of course, atmospheric conditions at the horizon would have to be superb — no vog (volcanic smog), ocean haze, or orographic clouds — to achieve this feat. Do perfect conditions like these occur on Kilauea volcano? Yes, occasionally. Have I succeeded in spotting NGC 4833 from there? After several tries, yes. NGC 4833, then, is the southernmost Caldwell object visible from the United States. Only four Caldwell objects require a U.S. resident to depart the Fifty States for more southerly locales.

NGC 4833 is one of two Caldwell globulars in Musca. The other, NGC 4372 (Caldwell 108), lies 3° to the southwest of NGC 4833. At a declination of –72° 39′, it is too far south to be spied from Hawaii. To see NGC 4372 I had to travel to New Zealand. In fact, I observed both clusters with the 4½-inch f/7 finderscope on the 20-inch f/13.5 Zeiss reflector at Auckland Observatory (as well as with the 20-inch itself). So let's look at both clusters in more detail.

The Abbé Nicolas Louis de Lacaille (1713–62) discovered NGC 4833 during his two-year exploration of the southern skies from the Cape of Good Hope. The cluster is included in his 1755 list of 42 southern nebulae and clusters as a Class I object (a nebula "without stars"). Lacaille also likened NGC 4833's appearance to that of a small, faint comet. John Herschel placed NGC 4833 in his Class VII, deeming it a very loose and sparse cluster.

NGC 4833 is easy to find. It lies ¾° northwest of 3.6-magnitude Delta (δ) Muscae, and there is no mistaking its glow. But how big and bright is that glow? Well, that's a matter of some debate, and I wish I'd known as much *before* I went to New Zealand! Here's the present — and still unresolved (no pun intended!) — situation:

The *Deep Sky Field Guide to Uranometria 2000.0* lists NGC 4833's magnitude as 7.0 and its diameter as 13.5′. Ernst Hartung fails to record the cluster's magnitude but says it's a "fairly compact globular cluster with small outliers to a diameter of about 4′." The original Caldwell Catalog lists a magnitude of 7.3 and a diameter of 14.0′. *Burnham's Celestial Handbook* offers a magnitude of 8.5 and a diameter of 6′. In his globular-cluster database Brian A. Skiff gives a magnitude of 8.5 but no diameter. The professional literature is just as confusing. "The [cluster's] total magnitude," Brent Archinal explains, "apparently is poorly determined, given the range of values in the professional papers of 6.80 to 8.35. However, this is mostly dependent on what diameter was chosen for this total magnitude. The cluster has a fairly obvious 'half-light' core that extends to a diameter of about 5′ with outer halo stars out to at least about 10′ on the Digital Sky Survey. Thus the 13.5′ *Deep Sky Field Guide* value and 14.0′ Caldwell Catalog values seem a little large, especially for visual observers." For better or worse, the values I've tabulated on page 417 come from the *Catalog of Parameters for Milky Way Globular Clusters* that McMaster University astronomer William E. Harris compiled in 1999, and from the *Deep Sky Field Guide to Uranometria 2000.0*.

Adopting an angular diameter of 13.5′ and a distance of 19,600 light-years implies that the cluster's physical diameter is 77 light-years, making it 1.7 times smaller than M13 and 1.3 times smaller than M92. NGC 4833's color-magnitude diagram places it in a class of extremely metal-deficient globular clusters, like M92. The cluster's stars are metal-poor because they formed about 15 billion years ago from gas that contained only a small sprinkling of heavy chemical elements. The typical star in NGC 4833 contains only about 1/60 as much iron (per unit of hydrogen) as there is in our Sun.

NGC 4833's brightest star shines at magnitude 12.4, and its horizontal-branch magnitude is 15.5, meaning the cluster can be resolved in moderate-sized amateur telescopes but challenges smaller instruments like my Genesis. I spotted NGC 4833 in 10×50 binoculars from city-bound Auckland Observatory, under the light of a waxing gibbous Moon. It lies just 15′ south of an 8th-magnitude star. In the binoculars the cluster's core appeared to possess a stellar radiance. But through the 4½-inch finderscope on the observatory's 20-inch reflector I discovered that this was an illusion created by a 9th-magnitude star superposed on the cluster's envelope, only about 2.5′ north of the compressed center. In the 4½-inch the cluster's core looks layered. The brightest inner part measures about 2′ and appears distinctly square. The square is lined by moderately bright stars, making the core look rather hollow (save for a small gathering of about four tightly packed "stars" — really bunches of suns — in the very center). A dark lane running east-west separates the box from a thin, similarly sized bar of stars to the north. The rest of the inner core sweeps out from this boxy body in four magnificent loops, to the northeast, northwest, southwest, and southeast, like the wings of a butterfly (thus my nickname for the cluster). The "butterfly" appears to be resting in a field of stars, which may or may not be related to the cluster. The drawing below reflects the view through the 4½-inch finderscope (at powers of 19×, 75×, and 150×) and through the Auckland Observatory 20-inch reflector.

The cluster's fainter, roughly 5′-wide outer halo was not noticeable as such in the 20-inch. Perhaps if I had seen the cluster at low power

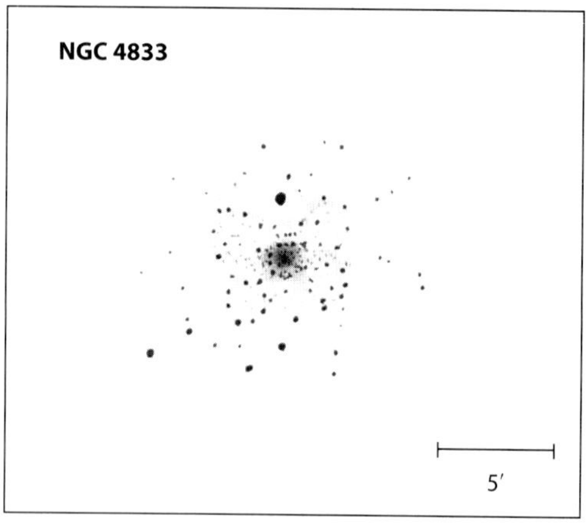

from a dark site, with no Moon, I would have realized that there was more to the cluster than first met my eye. In any case, the outer shell is obvious in photographs, though it is hard to tell where the cluster ends and the star fields of the southern Milky Way begin. (No wonder astronomers have had trouble determining the cluster's angular size and total apparent magnitude.)

NGC 4372 is another extremely metal-poor globular cluster. Its 7th-magnitude form lies 40′ southwest of 4th-magnitude Gamma (γ) Muscae and 5′ southeast of a 6.6-magnitude star. The cluster also lies about ¼° due east of the southern tip of a narrow, 3°-long river of obscuring matter called the Dark Doodad. The Doodad starts just 1¼° due south of Alpha Muscae and flows southwest from there. (The presence of this dark cloud also suggests that the cluster is being dimmed by obscuring matter, and indeed, Harris's data suggest that the cluster loses a full magnitude of its brightness to interstellar extinction.) Telescope users under dark skies should try to sail "down" this river, because it is almost free of bright starlight, though two roughly 9th-magnitude stars (separated by about 15′) lie nearly at the halfway mark. (The brighter of

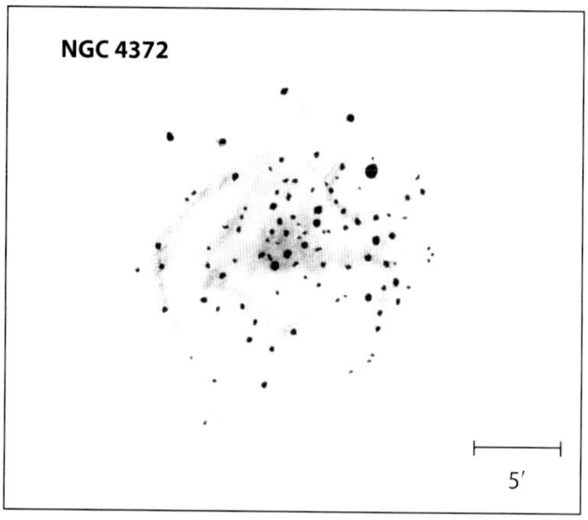

these two suns is 168 light-years distant). Can you see the Dark Doodad with binoculars? With the naked eye?

The 19th-century "gentleman scientist" James Dunlop discovered NGC 4372 while surveying the southern skies from Australia with a 9-inch f/12 reflector. It is the 67th of the 629 objects he recorded in *A Catalogue of Nebulae and Clusters of Stars in the Southern Hemisphere, Observed at Parramatta in New South Wales*. Here is his description:

> A star of the 6th magnitude, with a beautiful well-defined milky ray proceeding from it south following [to the southeast]; the ray is conical, and the star appears in the point of the cone, and the broad or south following extremity is circular, or rounded off. The ray

> is about 7′ in length, and nearly 2′ in breadth at the broadest part, near the southern extremity. With the sweeping power this appears like a star with a very faint milky ray south following, the ray gradually spreading in breadth from the star, and rounded off at the broader end. But with a higher power it is not a star with a ray, but a very faint nebula, and the star is not involved or connected with it: I should call it a very faint nebula of a long oval shape, the smaller end towards the star; this is easily resolvable into extremely minute points or stars, but I cannot discover the slightest indications of attraction or condensation towards any part of it. I certainly had not the least suspicion of this object being resolvable when I discovered it with the sweeping power, nor even when I examined it a second time; it is a beautiful object, of a uniform faint light.

To honor Dunlop for his achievements, John Herschel, on behalf of the Royal Society, presented him with a gold medal on February 8, 1828. Herschel praised the man for his observational prowess and his work, describing the latter as "zealous, active, ready, but above all, industrious and methodical." Ironically, when Herschel made his own expedition to survey the southern skies in 1834, he discovered that about two-thirds of the objects in Dunlop's catalog could not be reidentified. "A want of sufficient light or defining power in the instrument used by Mr. Dunlop has been the cause of his setting down objects as nebulae where none really exist." Of course, NGC 4372 is not one of those nonexistent objects, as Herschel learned firsthand.

NGC 4372 is indeed a very open globular cluster, though, as is the case with its neighbor, NGC 4833, astronomers do not agree on its size or magnitude. The *Deep Sky Field Guide* lists its magnitude as 7.3 and its diameter as 18.6′. Robert Burnham Jr. lists its magnitude as 8 and its diameter as 18′. Hartung doesn't offer a magnitude but measures the cluster out to 10′, and Skiff asserts values of 7.2 and 5′, respectively. Archinal summarizes the problem:

> This cluster is open enough, and has a bright enough star in the field, that it's no wonder the diameters and total magnitudes are discordant. The Skiff value may correctly indicate that the 25 magnitude per square arcsecond profile [a standard surface-brightness threshold] lies at a diameter of 5′, but this does not seem to represent the overall size of the cluster very well. The half light diameter of 7.79′ matches the obvious size of the cluster on the Digital Sky Survey, and there are outliers to at least 10′. On the other hand, the *Deep Sky Field Guide* size of 18.6′ would appear to be far too large.
>
> So as usual with deep-sky information (as in the cases with NGC 4833 and 4372), it apparently just depends on what your definition of "diameter" and "total magnitude" is. The diameter is going to depend quite a lot on the limiting magnitude and how one determines where the cluster merges into the background stars. And then given various diameters, of course the total magnitude enclosed varies quite a bit. So there appears to be no definitive sizes and magni-

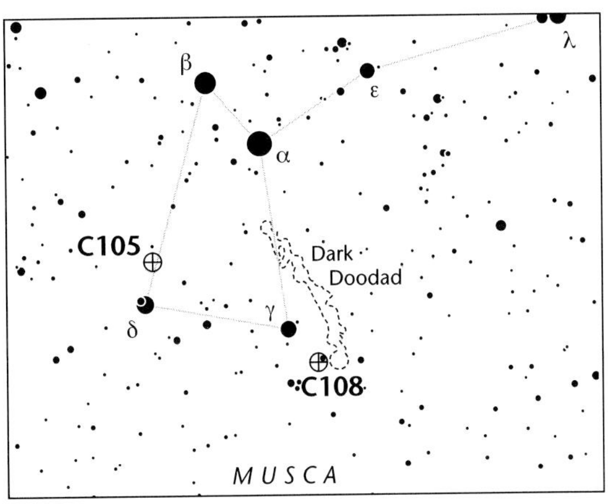

> tudes for these objects. At least that will give you something to write about in your book!

These observations reveal one overarching truth about deep-sky objects: we still have a lot to learn!

Through binoculars NGC 4372's tiny puff of light is overpowered by the neighboring 7th-magnitude star, but see if you can't pick it out with averted vision. The view through large binoculars or a 60-mm refractor gives the best impression of Dunlop's conical ray of light following the star like a comet's flowing tail. I could barely see NGC 4372 in the 4½-inch finder at 19× through the moonlight, so weak is its glow. At high power the cluster displayed loose ribbons of starlight lying atop a wrinkled, semitransparent sheet. I could also imagine a slightly denser sheet punctuated by two roughly 12th-magnitude stars. Seen through the 20-inch, each ribbon was part of a wavy row of 13th- to 14th-magnitude stars; the rows flowed more or less parallel to one another in a northeast-southwest direction. Dark troughs separated these waves, though they were not completely black, as they were lightly peppered with dim stars. These wavy features and dark troughs reminded me of the "rippled dunes" of starlight I saw in the open cluster M25 in Sagittarius through the 4-inch Genesis. If you defocus the cluster slightly and use averted vision, can you locate some very thin "dark doodads" marring the wrinkled sheet? My drawing of NGC 4372 is a composite of low-power views with the Auckland Observatory 4½-inch finderscope and higher-power views with the 20-inch.

Over the years astronomers have characterized the light curves of 19 variable stars in NGC 4372. Eight of these variables belong to the SX Phoenicis class (metal-poor analogs of the pulsating Delta Scuti stars), and eight are contact binaries. The variables range in brightness from 14th to 19th magnitude, so all are within the reach of modest-sized telescopes when they are at maximum light. CCD owners should be able to monitor all of these stars throughout their variability cycles. All the SX Phoenicis variables are concentrated toward the center of the globular and are blue stragglers. The contact binaries are distributed throughout the cluster. Two more variables, possibly binaries, have also been detected. The cluster's total lack of RR Lyrae variables confirms that it is extremely metal-poor, like NGC 4833. Each of NGC 4372's stars has about 1/20 the iron content of our Sun. The cluster is approaching us at a speed of 72 km per second and has an estimated age of 15 billion years, plus or minus 4 billion years.

107

NGC 6101
Type: Globular Cluster
Con: Apus

RA: $16^h\ 25^m\ 49.6^s$
Dec: –72° 12′ 06″
Mag: 9.2
Diam: 10.7′
Dist: 49,900 light-years
Disc: James Dunlop, included in his 1827 catalog

J. HERSCHEL: Globular cluster, large, faint, round, very gradually a little brighter in the middle, all resolved into stars 15th to 18th magnitude, 4′ [in] diam[eter], with stragglers. A delicate and beautiful object. (h 3623)

GC: Globular cluster, large, faint, round, very gradually a little brighter in the middle, all resolved into stars 15..18th mag, 4′ diam, with stragglers. A delicate and beautiful object.

NGC: Globular cluster, pretty faint, large, irregularly round, very gradually brighter in the middle, partially resolved, stars of 14th magnitude.

APUS, THE BIRD OF PARADISE, HAS AN intriguing history. Its name is derived from the Greek *apous*, meaning "without feet." It refers to the fabled legless sparrows of Greek mythology, which Keats later glorified in his "Eve of Saint Mark" as "legless birds of paradise." However, it is not the sparrow that Apus represents but the large feathered curiosities brought aboard the European sailing vessels of the 16th century. These birds the explorers called *Apus* (also *Avis*) *Indica*, the Indian Bird. The bird was first introduced to the people of Europe in 1522, when some skins were found on the sole surviving ship of Magellan's fateful expedition. It is from these and subsequent remains that the Spaniards, and later other Europeans, created their own myths about the bird. So it is of little wonder, then, that Apus ultimately found its way into a section of the southern sky dedicated to the tools and treasures of these great explorers. Ironically, when the great bird finally appeared in Johann Bayer's original 1603 atlas (and its corresponding page of text) it did so under the name "Apis Indica" (the Indian Bee). But this, notes Richard Hinckley Allen, is obviously a typographical error. Remember, Bayer did have a constellation called Apis, which Abbé Nicolas Louis de Lacaille later renamed Musca, the Fly (see page 418). The typographers and engravers obviously misread Apus for Apis. (So book reviewers take note: to err has long been human.) The error was fixed on later editions of Bayer's atlas.

Apus is conveniently placed just south of Triangulum Australe, and its two brightest stars, both of 4th magnitude, lie about a fist's width

107

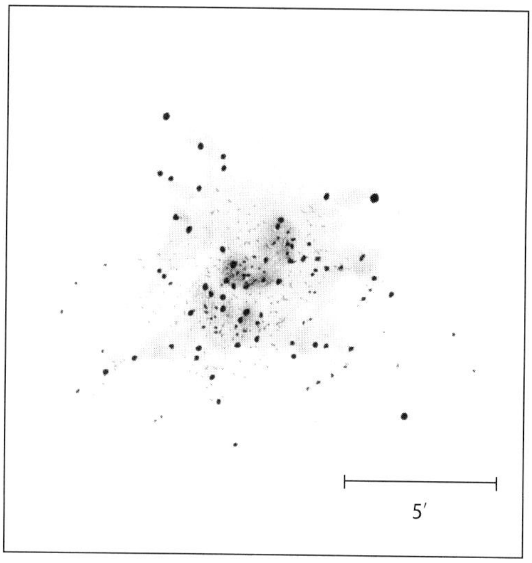

from the south celestial pole. With 206 square degrees, the constellation ranks 67th in area. Two globular clusters lie within its boundaries: NGC 6101 (Caldwell 107) and dim IC 4499, the closest globular star cluster to the South Celestial Pole.

James Dunlop discovered NGC 6101 in the 1820s and cataloged it as his 68th object. He found it to be a "pretty large rather faint round nebula, about 3½′ or 4′ diameter, a little brighter in the middle. There is a very small nebula on the north preceding [northwestern] side joining the margin of the large nebula." Dunlop's last statement alarmed me at first; did he chance upon a comet and not know it? But I quickly recalled John Herschel's dismay with Dunlop's catalog. After he set out for South Africa and conducted his own southern-sky survey, Herschel concluded that a "want of sufficient light or defining power in the instrument used by Mr. Dunlop has been the cause of his setting down objects as nebulae where none really exist." On photographs of NGC 6101 (like that on page 423) I have identified a distinct oval of roughly 14th-magnitude stars in the cluster's northwestern halo, and I believe this could have manifested itself to Dunlop as a nebulous extension, especially if his optics were poor.

NGC 6101's brightest stars shine at magnitude 13.5, but the cluster's horizontal-branch magnitude, which determines the cluster's resolvability, is a dim 16.6 — a challenge for a 12-inch telescope. NGC 6101 is intrinsically a rather large globular, measuring 160 light-years, which places it roughly halfway between NGC 362 and 47 Tucanae (Caldwell 104 and 106, respectively) in terms of size. NGC 6101 appears small and faint to us because it is so distant (about 3½ times farther away than 47 Tucanae). It is about 13,000 light-years closer to the galactic center than to our Sun. The cluster's iron-to-hydrogen ratio is slightly below normal for a globular; each cluster member has only 1/66 as much iron as does our Sun. According to William Harris (McMaster University, Canada) its integrated spectral type is *F*5. A 1991 *Astronomical Journal* article by Ata Sarajedini (Yale University) described CCD photometry that

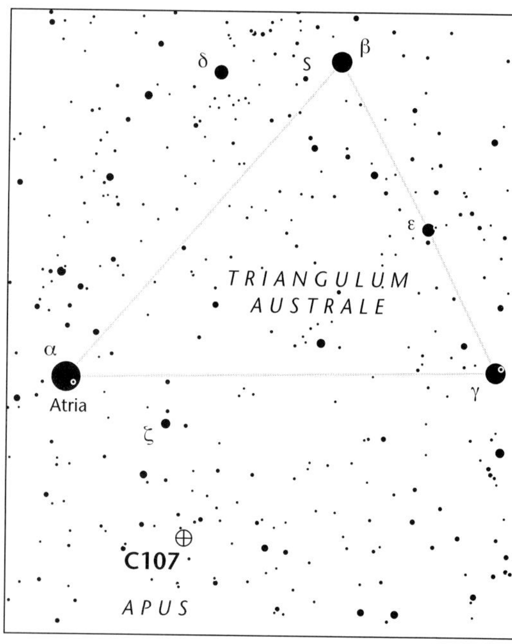

revealed a significant blue-straggler population in NGC 6101. It was found that these stars are more centrally concentrated than the cluster's similarly bright subgiants. This indicates that NGC 6101's blue stragglers have larger masses than their ostensible counterparts, supporting the theory that they either are binary systems or have been formed by the merging of stars. Fifteen variable stars have been discovered in the region of NGC 6101. Eleven of these appear to be RR Lyrae stars, one of the types of variables used to size up our galaxy.

To find NGC 6101 start with magnitude-1.9 Atria, Alpha (α) Trianguli Australis. The cluster lies 3½° to the southwest of it. If that's too far a jump, first locate 5th-magnitude Zeta (ζ) Trianguli Australis 2° to the southwest of Atria; the globular is another 2° south of Zeta Trianguli Australis. NGC 6101 can be seen in binoculars. Look for a tight triangle of 7th- to 8th-magnitude stars whose westernmost, 7th-magnitude member appears double. The cluster should be just south of the triangle and will appear as a dim glow with a slight central condensation. It's best to have some magnification handy if you want to confirm your sighting.

With an 8-inch telescope sited in Wellington, New Zealand, I saw the cluster as a moderately bright sphere surrounded by a dim gray halo. The roughly 3′-wide core is fractured into three parts. The innermost section is an uneven wedge of suns. A dark lane separates this mass from a tiny cap of stars to the northwest (Dunlop's "very small" nebula?), while a chevron of starlight to the south appears to have calved off the core. Through Auckland Observatory's 20-inch Zeiss reflector, the core of NGC 6101 looks herniated. A splash of 13th- to 15th-magnitude suns crosses the cluster's entire face, as if some disgruntled artist shook a wet paint brush at it. The drawing here is a composite of all these views. Through his 12-inch telescope, Ernst Hartung discerned a "rather faint but very rich globular cluster . . . irregularly round, rising broadly to the centre, about 3′ across with rays of faint stars emerging."

109

109

NGC 3195
Type: Planetary Nebula
Con: Chamaeleon

RA: $10^h\ 09^m\ 21^s$
Dec: –80° 51′ 31″
Mag: 11.6
Diam: 38″
Dist: 5,400 light-years
Disc: John Herschel, 1835

J. HERSCHEL: [Observed 24 February 1835] Planetary nebula, pretty bright, not quite uniform in its light, having two brighter patches, slightly elongated towards a star A; slightly hazy; diameter = 15 or 18 arcsec (in RA 13 seconds of time). Pos[ition angle] of star A = 265.7[°], dist[ance] = 0.7 diam[eter] from edge, 11th mag[nitude]; star C pos[ition angle] = 210.7[°], dist[ance] = 1.75 diam[eters] from edge (NB 13 seconds in time = 33.75 arcsec in arc).

GC/NGC: Remarkable, planetary nebula, pretty bright, small, little extended, 13″ diameter, 3 small stars nearby.

THE LAST CALDWELL OBJECT, NGC 3195, IS the southernmost planetary nebula known and one of the most beautiful in the southern skies — at least in photographs. It resides in Chamaeleon, a "small and unimportant constellation" (so says Richard Hinckley Allen) south of Carina. Chamaeleon is separated from the South Celestial Pole by Octans. In 1603 the Bavarian lawyer and astronomy aficionado Johann Bayer introduced Chamaeleon, along with 11 other south-circumpolar constellations, in his *Uranometria* atlas. The constellation's pattern is based on 16th-century observations from the Dutch mariner Pieter Dirckszoon Keyser, who, like other explorers sailing south of the equator, visually decorated the night sky with the tools of his trade as well as with the birds and animals discovered during the Age of Exploration. Famous for its ability to camouflage itself, the chameleon befits this dim and elusive region, because its eponymous constellation clings to the skirt of the southern Milky Way virtually unnoticed.

Once found, Chamaeleon offers a dash of visual solace to those who explore the night sky's otherwise bland south-polar region. Its

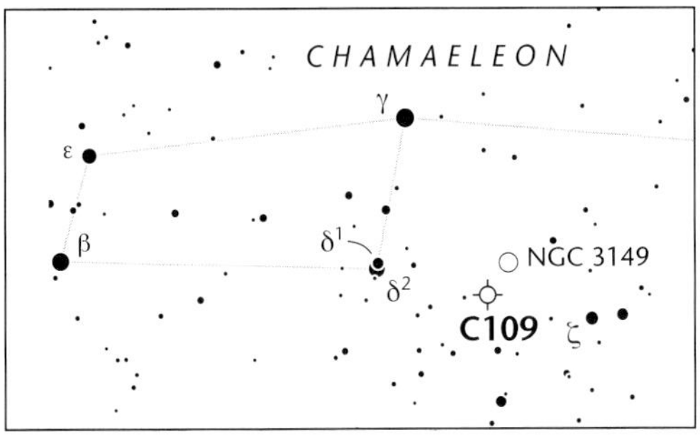

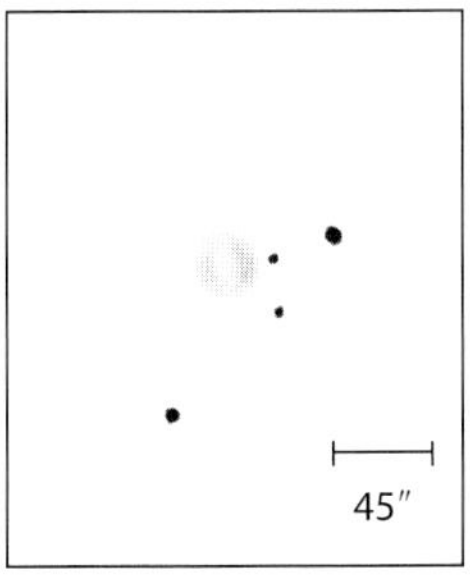

brightest stars shine around 4th magnitude and form a nearly flattened diamond. Its central and easternmost stars also can be seen as a dipper — a comforting sight for us northerners, who have grown fond of dippers in the polar regions of the sky. Two of the dipper's bowl stars, Gamma (γ) and Delta^{1,2} ($\delta^{1,2}$) Chamaeleontis, even point to a spot within 2° of the South Celestial Pole. The dipper also harbors our target, NGC 3195 (Caldwell 109), which lies halfway between 5th-magnitude Zeta (ζ) Chamaeleontis, the closest star in the dipper's handle to the bowl, and the slightly brighter pair of stars that constitute Delta^{1,2} Chamaeleontis, the bowl's southernmost "star."

Small-telescope users may have some difficulty picking out the magnitude-11.6 glow of NGC 3195 from the background sky at low power. Indeed, it is the second-faintest planetary in the Caldwell Catalog. And though NGC 40 (Caldwell 2) in Cepheus is nearly a full magnitude fainter, it is easier to see because it has a magnitude-11.6 central star. NGC 3195 does not share this visual luxury; its central star shines at a feeble magnitude 15.3. Fortunately NGC 3195 has a high surface brightness (10.6 magnitudes per square arcminute). So all a small-telescope user needs to do is sweep the region between Zeta and Delta Chamaeleontis with powers between 75× and 100× and look for a swollen star. As with many planetaries, magnification is needed to differentiate the object from field stars. Averted vision helps, too, since it captures the nebula's outer shell and causes the planetary to swell. Kent Wallace of western Australia writes that through an 8-inch Schmidt-Cassegrain telescope at 62.5× the planetary displays a "nice small disk" under averted vision. He uses an Oxygen III (O III) filter, however, to find the nebula. Wallace concurs that the planetary is easier to see at 100× and up.

Large-instrument users will have no problem sighting the nebula, but seeing detail in it will still be a challenge. Australian amateur Andrew Murrell observed NGC 3195 in February 1999 with a 20-inch f/5 Dobsonian. He saw it as a slightly flattened ring and noted a brightening on the ring's inner eastern edge. The planetary responded well to an O III filter, which made it appear slightly larger and its annular shape easier to pick out. But Murrell saw little else. In August 1997 I studied NGC 3195 under the light of a waxing gibbous Moon with the 20-inch f/13.5 Zeiss reflector at Auckland Observatory and with its 4½-inch finderscope. The planetary was virtually stellar in the finderscope, though it swelled slightly with an increase in magnification. The view was different through the 20-inch. Despite the moonlight the planetary displayed a robust, elliptical inner ring with a swollen eastern lip. The ellipse is oriented roughly north-south. It has a pale aqua inner annulus that floats in a fainter sea of light whose visual fabric appears rippled, like leathery skin; but this detail was highly suggestive. I had no problem seeing the 15th-magnitude central star at 686×, but I am not aware of anyone else who has. Many amateurs fail to see the central stars of planetaries because they fail to use sufficiently high magnification. A 10-inch telescope should show NGC 3195's central star at magnifications greater than 350×. NGC 3195 has a high surface brightness, so don't be afraid to crank up the power to more than 50× for each inch of your telescope's aperture. The way to detect the faint central stars of planetary nebulae is to use power, lots of power. The drawing on the previous page shows the nebula as I saw it under moderate magnification through the 20-inch, without the central star.

The Hubble Space Telescope transformed NGC 3195's seemingly uniform glow into an irregular array of coiled nebulosity. Under HST's gaze the inner annulus appears to be made of two loops seen at an oblique angle; it looks more like a tilted hourglass than a ring. The outer envelope may be two expanding bubbles traveling in opposite directions. Spectroscopy has revealed that the planetary is approaching us at a speed of about 16 km per second while the nebulosity expands at a speed of about 40 km per second. The expanding shell has about 9 percent as much mass as our Sun and extends about one light-year, which makes it similar in size to NGC 2392 (Caldwell 39), the famous Eskimo Nebula in Gemini.

Just 30′ northwest of NGC 3195 is the little enigmatic galaxy NGC 3149. Its *NGC* description reads: "Faint, small, little extended, very little brighter in the middle, star of 15th magnitude involved." But what's enigmatic is that some star charts and field guides fail to recognize the galaxy's very existence. Not that it isn't there. It is. But the second edition of *Sky Catalogue 2000.0* does not include it, nor does the *Deep Sky Field Guide*. It is not plotted in the first edition of Wil Tirion's *Sky Atlas 2000.0*, nor does it appear in *Uranometria 2000.0*.

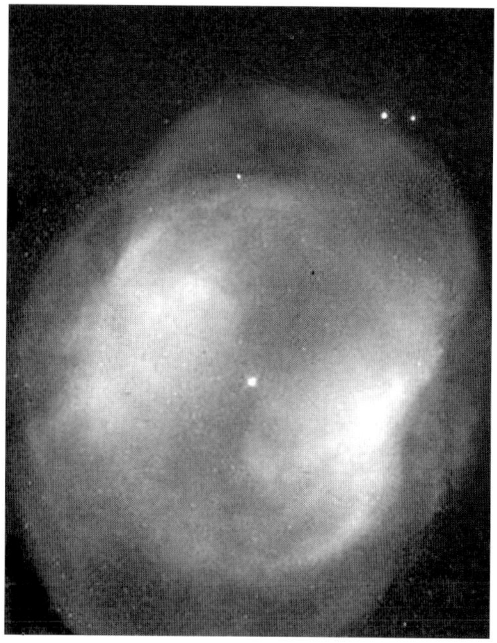

Modern atlases, such as the *Millennium Star Atlas* and the second edition of *Sky Atlas 2000.0*, do include it. The galaxy shines at magnitude 13.1 and is visible in an 8-inch telescope as a timid circular patch of light 2′ across.

Before leaving Chamaeleon, take a moment to contemplate the South Celestial Pole a mere fist's width away. Where exactly is the Pole? Sadly, the southern sky lacks a conspicuous pole star. The closest reasonably bright star to it is Sigma (σ) Octantis, which lay just 1° 03′ from the pole as the year 2000 began. This magnitude-5.5 star certainly is not a retina burner, however, nor a guiding light. No wonder so many people throughout the ages have sung the praises of Polaris; with a glance voyagers in the Northern Hemisphere can sense its comforting light and orient themselves. But this sense of security vanishes when one ventures south of the equator. Looking toward the South Celestial Pole, I feel as if I'm peering down a dark alley, and though I know where I'm looking, I see no guiding light, only a void. The region looks cold and disturbing, not warm and comforting. But what beauty and wonder surrounds the pole — the Large and Small Magellanic Clouds; the Coalsack and its glittering Jewel Box; the vast Eta Carinae Nebula. If I sum up the parts, the far-southern skies attain a visual balance. Their attraction, as in the north, is a sense of unity among the stars. "Attraction is therefore the supreme law among the worlds, among atoms, and among beings," astronomy popularizer Camille Flammarion penned in his 1923 book *Dreams of an Astronomer*. "[T]he bird which flies from branch to branch seeking a place for its nest, the nightingale which with incomparable song charms the sweet mistress of the night, the man whose heart is troubled at the appearance of a beloved being, the sound of a beloved voice, or a fond memory — all these beings, all these things obey the same law, that of universal attraction which in diverse forms governs all nature and guides it — whither? Towards yet another attraction, to the attraction of the unknown!"

CHAPTER 3

Twenty Spectacular Non-Caldwell Objects

Globular cluster NGC 6723

ANYONE WHO HAS USED HIS OR HER TELESCOPE to sweep the night sky knows the thrill of finding bright but little-observed nebulae, galaxies, and star clusters. Indeed, the sky is host to hundreds of "hidden" celestial treasures, some of which easily rival many of the Messier and Caldwell objects. While hunting for comets with my 4-inch Genesis during this project, I encountered many deep-sky objects not listed in the Messier or Caldwell Catalogs. Of course, most of them were dim or featureless, but a handful were surprisingly obvious or had some visual charm, like an edge-on galaxy's sharp needle of light or an open cluster's provocative arrangement of stars. A sampling of 20 such hidden treasures is listed below in order of declining declination. All are easily seen under a dark sky from a latitude of 20° N and are fine objects in a 4-inch or larger telescope.

1

NGC 4605
Type: Barred Spiral Galaxy (SBc pec)
Con: Ursa Major

RA: $12^h\ 40.0^m$
Dec: +61° 36′
Mag: 10.3
SB: 13.1
Dim: 5.7′ × 2.5′
Dist: 13 million light-years

NGC 4605 is a peculiar barred spiral galaxy (seen only 21° from edge on) about 5½° northeast of Delta (δ) Ursae Majoris (Megrez) in the Big Dipper's Bowl. Despite its proximity to the Bowl, amateurs do not commonly target the galaxy, probably because it has tough competition from the better known Ursa Major galaxies M51, M101, and M108. Yet NGC 4605 rivals M108 in brightness, apparent size, shape, and the level of detail it presents at the eyepiece.

The galaxy received some incidental recognition after January 1996, when the public first saw the now-famous Hubble Deep Field — a 10-day-long Hubble Space Telescope exposure revealing at least 1,500 new galaxies, some as faint as 30th magnitude, in an area of sky only 2′ across. The Deep Field lies only about 42′ to the northwest of NGC 4605, which often appeared in photographs and finder charts that were published to help amateurs locate the tiny sky patch scrutinized by Hubble. Through the Genesis NGC 4605 (whose true linear diameter is a modest 22,000 light-years) appears as a warped bar, oriented northwest to southeast, surrounded by a spindle of light. Its disk displays a mottled texture and has a large and obvious patch of dust in its northwestern quadrant. Long-exposure photographs reveal a larger envelope surrounding the spindle, several H II regions, and what appears to be a tiny galaxy protruding from NGC 4605's southeastern tip. NGC 4605 resides in the Coma-Sculptor Cloud of galaxies and is receding from us at 140 km per second.

2

NGC 7789
Type: Open Cluster
Con: Cassiopeia

RA: 23^h 57.0^m
Dec: +56° 43′
Mag: 6.7
Diam: 25′
Dist: 6,000 light-years

Just 3° southwest of Beta (β) Cassiopeiae, or Caph (the westernmost star in Cassiopeia's famous W asterism), and midway between Rho (ρ) and Sigma (σ) Cassiopeiae, lies the large, uniform glow of NGC 7789. This star cluster is a magnificent spectacle in 7×35 binoculars, appearing nearly as large as the full Moon and just as round. In fact, when seen through binoculars the cluster looks like a tailless comet crossing the rich star fields of the northern Milky Way. NGC 7789 lies at or just beyond the limit of unaided vision, making it a great naked-eye challenge under dark skies. The Rev. T. W. Webb, in his *Celestial Objects for Common Telescopes,* calls it a "[b]eautiful large faint cloud of minute stars," while John Herschel called it a "most superb" cluster. Piazzi Smyth, though, saw NGC 7789 as a "mere condensed patch in a vast region of inexpressible splendour, spreading over many fields; including the whole Galaxy through this and the adjacent constellations." At its estimated distance of 6,000 light-years, the cluster's angular diameter corresponds to a respectable physical size of 44 light-years. There are 583 stars in the cluster, with the brightest shining at magnitude 10.0. Many of NGC 7789's stars can be spied at 23×, and higher magnification shows a distinct background of unresolved suns. Equally impressive in wide-field telescopes is the complex of dark nebulae in and around the cluster. These impressive veils are not outlined on any star atlas that I'm aware of.

3

NGC 281 and IC 1590
Type: Emission Nebula and Open Cluster
Con: Cassiopeia

RA: 00^h 52.8^m
Dec: +56° 38′
Mag: 8.0 (nebula); 7.4 (cluster)
Dim: 35′ × 30′ (nebula)
Diam: 4.0′ (cluster)
Dist: ~7,500 light-years

The third object in this short list, NGC 281 in Cassiopeia, is a bit of an enigma. *Sky Atlas*

2000.0 (second edition), *Uranometria 2000.0,* and the *Millennium Star Atlas* all depict it as a nebula and a cluster, and it is described as such in *NGC 2000.0* and in the *Observing Handbook and Catalogue of Deep-Sky Objects*. But, as Brent Archinal of the U.S. Naval Observatory notes, NGC 281 is not a star cluster. Rather, he explains, it is the nebulosity associated with the star cluster IC 1590. NGC 281 was an Edward Emerson Barnard discovery, while IC 1590 was discovered later by Guillaume Bigourdan. "In any case," Archinal writes, "Bigourdan . . . clearly stated these two were not the same object. Therefore the correct name for the cluster is IC 1590."

Archinal notes a further complication — namely, that the *Deep Sky Field Guide to Uranometria 2000.0* lists NGC 281 and IC 1590 as two distinct open clusters at different positions, while *Sky Catalogue 2000.0* asserts, in its Open Cluster Cross Index, that NGC 281 and IC 1590 are the same object. That confusion aside, the nebula (NGC 281) is quite an obvious irregular patch of glowing gas a mere 1¾° east of Alpha (α) Cassiopeiae (Shedir); its Moon-sized glow forms the apex of a triangle with Shedir and Eta (η) Cassiopeiae and is visible in 7×35 binoculars under a dark sky. It's easily picked up in a sweep with the 4-inch at 23×. At first glance the nebula has a cometlike quality to it, but once you focus your attention on it, the small cluster (IC 1590) embedded within it draws in your attention. The cluster sports 90 stars within its tiny disk, with the brightest being the magnitude-7.8 triple star HD 5005, which contributes most of the ultraviolet light that powers the nebula. Archinal suspects many more of the stars that appear within the nebula may be associated with the cluster.

4

NGC 5866
Type: Lenticular Galaxy (S0)
Con: Draco

RA: $15^h\ 06.5^m$
Dec: +55° 46′
Mag: 9.9
SB: 12.6
Dim: 7.3′ × 3.5′
Dist: 50 million light-years

Just 4° southwest of Iota (ι) Draconis, or Edasich, lies a remarkable lenticular galaxy seen almost exactly edge on. In long-exposure photographs NGC 5866 sports a very bright ring and a prominent dust lane, which are slightly tilted with respect to one another and to the equatorial plane of the galaxy's central lens — making each side of the galaxy look like a pair of slightly opened scissors. X-ray emission has been observed along the galaxy's major axis, as has ionized gas. Early photographs overexposed the dust lane, leading astronomers to misclassify this lenticular system as an elliptical galaxy. NGC 5866 may contain a considerable 1 trillion solar masses or thereabouts. Assuming its estimated distance of 50 million light-years is correct, the galaxy spans some 90,000 light-years.

Until the late 1990s even the reliable NASA Extragalactic Database boldly listed NGC 5866 as Messier 102. So the history behind the mysterious M102, and its questionable relationship to NGC 5866, deserves repeating. In Messier's catalog, M102 is described in the following way:

> Nebula between the stars (ο) [Omicron] Boötis and (ι) [Iota] Draconis. It is very faint. Close to it is a sixth-magnitude star.

This entry is obviously in error, because Omicron Boötis and Iota Draconis lie nearly 40° apart — making them far too imprecise a marker for visual stalwarts like Messier or his contemporary, Pierre Méchain, who purportedly discovered this nebulous object. So what *is* M102?

One could reasonably argue that the star Theta (θ) Boötis was misread as Omicron (ο) Boötis, since an observer can find several faint galaxies between Iota Draconis and Theta Boötis, which lie about 11° apart, and the brightest and most conspicuous of these galaxies is NGC 5866. This explanation is enticing, but there is no need for it, nor for any other theories on M102 (though cooking up such theories remains popular to this day, probably because everyone enjoys a good mystery). For Méchain explained the mystery of M102 in a letter that he published in the 1786 *Berliner Astronomisches Jahrbuch:*

> On page 267 of the *Connaissance des Temps* for 1784 M. Messier lists under No. 102 a nebula which I have discovered between omicron *Boötis* and iota *Draconis.* This is nothing but an error. This nebula is the same as the preceding No. 101. In the list of my nebulous stars communicated to him M. Messier was confused due to an error in the sky-chart.

Simply put, M102 is not a mystery unless you want it to be; in light of the historical evidence, NGC 5866 should lose any association with the name "M102."

Certainly one can wonder why neither Messier nor Méchain discovered NGC 5866, because it is a wonderful blaze of light. In the Genesis at 23× the galaxy is immediately obvious as a condensed elliptical glow with white winglike extensions. At higher powers the galaxy shows a prominent ring with a tight core that gradually focuses to a pinpoint. Overall, NGC 5866 is a much more dramatic object than M99 or M100, or M101, for that matter. So it is understandable why some amateur organizations include NGC 5866 in their Messier Catalogs and Marathons. We like to think of Messier's catalog as being chock full of the sky's most magnificent objects, so why not insert the meritorious NGC 5866 into that gap between M101 and M103?

Unfortunately, creating a "hit list" of the sky's most wonderful deep-sky objects was not Messier's intent. His catalog was drafted to document objects that masqueraded as comets, objects that for the most part had been swept up during comet hunts. Indeed, NGC 5866 cannot be singled out, because Messier and Méchain did not include other conspicuous deep-sky objects, as the Caldwell Catalog amply demonstrates. Not until 1789, when William Herschel was methodically sweeping the heavens with his telescopes, did anyone unambiguously document an encounter with the object now known as NGC 5866. We should be careful not to let our modern desires interfere with history.

5

NGC 1528
Type: Open Cluster
Con: Perseus

RA: 04^h 15.3^m
Dec: +51° 13′
Mag: 6.4
Diam: 18′
Dist: 2,400 light-years

During the cool winter nights of the Northern Hemisphere, our galaxy's outer arms arc gracefully over the North Celestial Pole like a garden trestle covered in snow. Night after night, the stars of this Milky Way region sail up and over the pole until, by February, the Perseus Arm of our home spiral galaxy dominates the view high in the north after sunset. And it is here that many amateurs will train their telescopes on some of the sky's most prominent star clusters and stellar associations, including the famous Double Cluster (Caldwell 14), M34, and the Alpha Persei Association. But there is a rather distinctive cluster, NGC 1528, which is casually overlooked in favor of its more popular neighbors. Located only about 1¾° east-northeast of 4th-magnitude Lambda (λ) Persei in the Champion's knee, NGC 1528 shines at magnitude 6.4 and makes a decent naked-eye challenge for anyone under dark skies. It is a fine object whose brightest members will start to appear to the dedicated observer in even the smallest of binoculars.

At 23× in the Genesis the cluster appears large and bright, with a veil of well-resolved suns scintillating in front of a fuzzy backdrop of fainter ones. Brent Archinal catalogs 165 stars in NGC 1528, with the brightest shining at magnitude 8.7. The cluster's brightest stars form a teardrop-shaped asterism that is elongated along a northwest-to-southeast axis; its narrow tip points to the southeast. In *Star-Hopping for Backyard Astronomers,* Alan M. MacRobert notes that through a 6-inch reflector under suburban skies a dark U-shaped bay can be seen at the cluster's western end, where it is outlined by seven bright stars. This dark bay is but one of many wisps of darkness that meander through the region. Under the dark skies of Hawaii the cluster appears to be composed of shards of starlight separated by dark cracks, as if a crystal teardrop had accidentally shattered on a sheath of black ice illuminated by moonlight.

6

NGC 4490
Type: Barred Spiral Galaxy (SBd pec)
Con: Canes Venatici

RA: $12^h 30.6^m$
Dec: +41° 38′
Mag: 9.8
SB: 13.0
Dim: 5.6′ × 2.8′
Dist: 25 million light-years

Canes Venatici harbors a wealth of extragalactic treasures, and NGC 4490 is one of the more prominent ones. This peculiar galaxy lies a mere 40′ northwest of 4th-magnitude Beta (β) Canes Venaticorum (Chara), and it is an easy target for small-telescope users. The 19th-century British observer Rev. T. W. Webb described NGC 4490 as a "[l]ong, easily resolvable" nebula. Indeed, the galaxy's size, shape, and brightness rival those of some of the more fashionable Messier galaxies in neighboring Coma Berenices, such as M88, M99, and M100. At a distance of 25 million light-years, however, NGC 4490 is half as far away as they are and, with a diameter of 36,000 light-years, less than half as big.

A chaotic late-type spiral, NGC 4490 is inclined 65° from face on and is a most fascinating system, with a ringtail core wrapped in a cocoon of stellar filigree and H II regions. The galaxy's northwestern end looks warped, and it is here that bizarre drips of starlight make a dogleg to the north, to yet another pool of tortured nebulosity surrounding the galaxy's companion, NGC 4485. In high-resolution images taken with large telescopes, NGC 4490 looks like an eviscerated galaxy, one whose entrails have been ghoulishly ripped from the carcass and tossed wildly asunder. Through the 4-inch NGC 4490 is a flying-saucer-shaped lens of light with faint extensions.

7

NGC 404
Type: Lenticular Galaxy (S0)
Con: Andromeda

RA: $01^h\ 09.4^m$
Dec: +35° 43′
Mag: 10.3
SB: 14.0
Diam: 6.6′
Dist: 8 million light-years

NGC 404 is truly a hidden treasure. This lenticular galaxy lies only 6′ northwest of the 2nd-magnitude topaz star Beta (β) Andromedae (Mirach), whose light interferes with — but does not drown out — this bright little gem. NGC 404 belongs to a small group of galaxies in the Coma-Sculptor Cloud and is approaching us at a speed of 35 km per second. The galaxy is quite small, measuring only 16,000 light-years in diameter, and images of it through large telescopes reveal it to be a roundish spiral. Its small bulge contrasts well with a surrounding dust ring. The galaxy is just visible in a 60-mm refractor, and it takes magnification remarkably well, so use enough power to remove Beta Andromedae from your field of view. In the 4-inch the circular galaxy has a milky-smooth texture. A faint "star" punctuates its core; high-resolution images show that NGC 404 has a very small, extremely bright nucleus, with a dark crescent and patchy whorls of darkness in its lens. A 1995 Hubble Space Telescope image of this region reveals a bright, unresolved point source (less than 1 parsec across) surrounded by several fainter ones. The latter are presumably star-forming regions responsible for the galaxy's extended nebular emission and for the young stellar features seen in spectra of the nucleus.

8

Hockey Stick; The Hook
NGC 4656
Type: Barred Spiral Galaxy (SBm pec)
Con: Canes Venatici

RA: $12^h\ 44.0^m$
Dec: +32° 10′
Mag: 10.5
SB: 14.8
Dim: 18.8′ × 3.2′
Dist: 23.5 million light-years

Larry Mitchell of the Houston Astronomical Society first introduced me to NGC 4656 during the 1998 Texas Star Party. He calls it the Hockey Stick galaxy, while his observing compatriot Barbara Wilson refers to it as The Hook. After seeing this galaxy through their large reflectors, I understand both monikers. Situated ½° southeast of NGC 4631 (Caldwell 32), NGC 4656 is an irregular barred spiral galaxy, merely 10° from edge on, with a warped appendage on its northeastern end. Known as NGC 4657, this appendage juts out at right angles to the main shaft like the hooked blade of a hockey stick. The most likely explanation for NGC 4656's distorted appearance is a tidal interaction between it and NGC 4631. In his 1994 *Carnegie Atlas of Galaxies*, Allan R. Sandage writes that the pair's atomic-hydrogen gas has provided evidence of a tidal encounter between the two galaxies. The two are 350,000 light-years apart on the plane of the sky. Wilson likens these interacting spirals to "two ships nearly crashing in the night, both listing from the encounter, with [NGC] 4656 definitely the loser, but [NGC] 4631 . . . not unscathed."

NGC 4656 is easily visible in the 4-inch Genesis. Southwest of what may be considered its nuclear region, the galaxy is bright; on the opposite side it is very faint and appears segmented along its major axis. NGC 4656 is a large system, with a physical diameter of 91,000 light-years. In long-exposure photographs the galaxy resolves into a smattering of stars and H II regions, the brightest of which lie approximately 18″ west of the apparent nucleus.

9

NGC 1333
Con: Perseus
Type: Bright Nebula

RA: 03^h 29.3^m
Dec: +31° 25′
Mag: 3.6
Dim: 6′ × 3′
Dist: ~1,100 light-years

NGC 1333 is a simple but beautiful reflection nebula. It lies 3¼° southwest of Omicron (ο) Persei and almost nicks the spot where Perseus, Taurus, and Aries meet. At magnitude 3.6 the nebula is quite bright, but it's also very tiny. Several times I have swept it up during my comet hunts (even in moonlight) with the 4-inch refractor, and Brian Skiff and Christian Luginbuhl say it's not difficult to see in a 60-mm refractor. At first glance, the nebula has a round form with a uniform luster, but closer inspection reveals it to be a mottled kidney bean of light elongated from north-northeast to south-southwest.

On most star charts NGC 1333 is far from eye-catching. In the second edition of *Sky Atlas 2000.0*, for instance, the symbol for NGC 1333 is a tiny box, no bigger than the disk representing a 4th-magnitude star. The box looks like an afterthought, perhaps a move to flesh out that region of sky with pretty symbols. Certainly observers moving their fingers across their star charts while selecting objects for a night's work would not stop there. NGC 1333 is literally dwarfed by the Pleiades star cluster just 8½° to the southeast, and by the challenging but famous California Nebula (NGC 1499) a similar distance to the northeast. Yet surprisingly, at the eyepiece, NGC 1333 is easier to see than the Merope Nebula (IC 349).

I find the very simplicity of NGC 1333 an attraction — the face of a shy specter masked in a white veil. The nebula is illuminated by a magnitude-10.5 star adorning its northeastern corner like a jewel in a crown. Recent near-infrared images also reveal some 80 pre-main-sequence stars embedded in this region of active star formation. The dark nebula Barnard 205 lies just to the south of NGC 1333, and its two fingers of darkness seem to be pinching the reflection nebula's southern tip. This dark nebula stretches for nearly ½° and is a wonderfully complex agglomeration of shadows. Exploring this hidden corner of Perseus through the Genesis at low power is like tiptoeing through an old, forgotten house dusted with cobwebs and littered with keepsakes.

10

NGC 1647
Type: Open Cluster
Con: Taurus

RA: 04^h 45.7^m
Dec: +19° 07′
Mag: 6.4
Diam: 40′
Dist: 1,800 light-years

I thank my wife, Donna, for introducing me to NGC 1647, a wonderful but somewhat neglected open cluster near the V-shaped face of Taurus, the Bull. One night Donna stood in the driveway of our home in Volcano, Hawaii, and pointed her 7×35 binoculars toward the Hyades cluster (Caldwell 41), then in the eastern sky. Suddenly she exclaimed that she had discovered a comet and pointed a finger to a spot about 3½° northeast of Aldebaran. There, to my surprise, I saw a beautiful open cluster that had, for all these years, escaped my gaze. Later, while studying the Hyades (Caldwell 41) for this book, I learned that the magnitude-6.4 glow of NGC 1467 could be seen with the naked eye from a dark sky.

Through binoculars NGC 1467 is a lovely cluster, appearing as a round ghostly glow with an apparent size larger than that of the full Moon. Undoubtedly, since I had lived most of my life in the Boston area, light pollution must have drowned out the cluster's pale glow, making it easy to sweep past without notice. The Rev. T. W. Webb's description of the cluster in his *Celestial Objects for Common Telescopes* is rather perplexing, for he writes: "Stars [of magnitudes] 8.5 to 10, round wide pair" (I believe there should be a comma between "round" and "wide pair"). The cluster is indeed quite round, and most of its 200 stars can be found within ½° of its core, where a fine pair of 9th-magnitude stars dominates the view. Other bright stars spiral out from the core in crooked or disjointed arms. NGC 1647 is not densely packed, however, and the view is best at low power; in fact, Roger Clark suggests 20× for a 6-inch telescope and 27× for an 8-inch. In 1996 Michael Geffert and his colleagues visually inspected NGC 1647 on a *National Geographic Society – Palomar Observatory Sky Survey* print and determined the position of this cluster's center; they also determined that the Cepheid variable SZ Tauri (whose apparent visual magnitude ranges from 6.50 to 6.86 every 3.15 days) is probably not a member of this cluster.

11

Ghost Globular
NGC 5897
Type: Globular Cluster
Con: Libra

RA: $15^h\ 17.4^m$
Dec: –21° 01′
Mag: 8.5
Diam: 11′
Dist: 42,000 light-years

NGC 5897 is an unusually loosely structured globular cluster that haunts the dim and unassuming stars of Libra. Ernst Hartung calls it "irregularly round" and "conspicuous." The cluster can be seen on clear, moonless nights as a large amorphous glow nearly one-third the apparent size of Omega (ω) Centauri (Caldwell 80). Look for what appears to be the detached head of a flaring comet nearly midway between, and just south of, a line joining Delta (δ) Scorpii and the wide double star $Alpha^{1,2}$ ($\alpha^{1,2}$) Librae (Zubenelgenubi). At a glance in the 4-inch at 23×, the cluster's oblate disk appears buffed and featureless, as if its stars had been scrubbed away with bleach. But with a little concentration you can resolve the glow into a patchwork of glittering hazes that lacks a strong central condensation. The cluster's brightest stars shine at magnitude 13.3, and they appear to form slight needles of light that border minute "holes" at the core. I call NGC 5897 the Ghost Globular because it resembles a ghost image of the globular M55 in Sagittarius. Actually NGC 5897 measures nearly 135 light-years across, making it 68 percent larger than M55.

Adm. William Henry Smyth wrote extensively about this "large compressed cluster of minute stars," saying "it is faint and pale, but owing to the fineness of the night, steadiness of gaze, and excellent action of the telescope, [it] was as well seen as so low and so awfully remote an object could be expected to be." William Herschel found the cluster in March 1785 and considered it an important discovery because, in his view, it was "one of the gradations from the palpable congeries of stars . . . towards the distant nebulae."

12

NGC 1360
Type: Planetary Nebula
Con: Fornax

RA: 3^h 33.2^m
Dec: –25° 52′
Mag: 9.4
Diam: 6.5′
Dist: 1,100 light-years

Concealed among the dim suns of Fornax, mixed within the wide fanfare of galaxies that populate the region, lies one of the sky's brightest planetary nebulae, one which richly rewards small-telescope users: NGC 1360. This often-overlooked gem has an immediate "wow, what is that?" appeal, especially when stumbled upon at the eyepiece. Robert Burnham Jr. refers to NGC 1360 as a "peculiar nebula . . . usually classified as a planetary." In 1968 Luboš Kohoutek called it an "uncommon planetary." And Mario Perinotto, who studied the central star's spectrum in the early 1980s, placed it in the Wolf-Rayet category of planetary nebulae, meaning the central star appears to be in a state of flux and is violently ejecting matter into the surrounding shell. Steven J. Hynes lists NGC 1360 in his work *Planetary Nebulae* and classifies it as a Type III (a planetary with an irregular disk).

Not much is heard about NGC 1360's visual appearance from amateurs living at midnorthern latitudes, which is surprising. Then again, as seen from the north, Fornax, the Furnace, is not among the most glamorous constellations, as its 110 4th- to 7th-magnitude stars all lie near the southern horizon in mid-December. Still, NGC 1360 is at nearly the same declination as Antares and the globular cluster M4.

You will find this "peculiar planetary" about 5½° northeast of magnitude-3.8 Alpha (α) Fornacis, less than 15′ southeast of a magnitude-6.4 star. (Note that the second edition of *Sky Atlas 2000.0* curiously fails to show NGC 1360.) NGC 1360 is not only bright as planetaries go, it's huge; while many planetaries measure mere arcseconds across, NGC 1360 spans a respectable 6½ arcminutes, making it about five times larger than M57, the Ring Nebula, on the plane of the sky. Its central star also burns at an unusually bright 11th magnitude, making it a cinch to find in even the smallest of backyard instruments. Christian Luginbuhl and Brian Skiff note that the nebula is a large, smooth, elongated glow with hints of patchiness visible in larger instruments. Through the Genesis NGC 1360 is a simple yet spectacular sight — a conspicuous central star surrounded by a fantastic oval shell that's evident at 23×. At 72× the shell appears delicately mottled, like slightly curdled milk, or an egg made of white marble. The nebula's eastern half is the most noticeably mottled.

13

NGC 288
Type: Globular Cluster
Con: Sculptor

RA: 00^h 52.8^m
Dec: –26° 35′
Mag: 8.1
Diam: 13′
Dist: 27,000 light-years

Globular cluster NGC 288 resides near one of the southern sky's greatest attractions — the oblique spiral galaxy NGC 253 (Caldwell 65). Although the globular lies only 1¾° to the southeast of the galaxy, that's just about ¾° too far to be placed in the same low-power field as NGC 253 in many modern telescopes. But with a field of view nearly 3° wide, the Genesis removes NGC 288 from the shadows of obscurity and thrusts it into the telescopic "spotlight." And there are few sights in the sky as rewarding as this dramatic pairing of deep-sky splendors. Curiously, Patrick Caldwell Moore did not include NGC 288 in his catalog, and Robert Burnham Jr. fails to highlight it in his *Celestial Handbook*, even in passing. The Rev. T. W. Webb acknowledges NGC 288 in his *Celestial Objects for Common Telescopes*, though he simply calls it "bright."

Both NGC 253 and NGC 288 can be easily spotted in 7×35 binoculars, shining at 7th and 8th magnitude, respectively. NGC 288 is less than 9° southeast of Beta (β) Ceti (Deneb Kaitos), 3° northwest of Alpha (α) Sculptoris, and just 40′ north and slightly east of the South Galactic Pole. The cluster looks like a perfect sphere at 23× and starts to resolve into individual stars at 72× (the cluster's brightest member shines at magnitude 12.6), especially at the fringes, which become patchy with clumps of starlight. High power reveals many veins both dark and bright emanating from a hollow central wedge of stars. In larger telescopes the cluster is quite remarkable. Christian Luginbuhl and Brian Skiff compare their view of NGC 288 through a 12-inch reflector to that of the great globular M4 in Scorpius, though they admit it is not as rich.

14

NGC 7793
Type: Spiral Galaxy (Sdm)
Con: Sculptor

RA: 23^h 57.8^m
Dec: –32° 35′
Mag: 9.2
SB: 13.6
Dim: 9.2′ × 6.4′
Dist: 9 million light-years

NGC 7793 is a fine galaxy hidden in the star-poor region of southwestern Sculptor. It lies some 13° east-southeast of brilliant Fomalhaut, Alpha (α) Piscis Austrini, and is near the apex of a nearly perfect isosceles triangle whose other corners are the magnitude-4.4 stars Beta (β) Sculptoris (7¼° to the southwest of the galaxy) and Gamma (γ) Sculptoris (8° to the west). The galaxy is a prototypical late-type spiral, and it is receding from us at a speed of 230 km per second. Through the Genesis at 23× the 9th-magnitude galaxy appears quite large, and it looks very much like the smoothly textured head of a comet. At 72× the galaxy's core appears sharply stellar; the bright haze immediately surrounding the core is dappled with fuzzy clumps that would cause any small-telescope user to believe he or she has chanced upon a globular cluster near the limit of resolution. The galaxy's fainter outer envelope is slightly elliptical and shows less mottling. Although NGC 7793 is the faintest of the five major members of the Sculptor Group, the detail it displays in small instruments makes it a superb, yet visually challenging, target for beginners. If we accept a distance of 9 million light-years for NGC 7793, its true linear diameter is 23,000 light-years. We see its disk inclined 50° from face on, so its tiny nuclear bulge and dominating spiral arms are prominently revealed.

15

NGC 3621
Type: Spiral Galaxy (Sd)
Con: Hydra

RA: 11^h 18.3^m
Dec: –32° 49′
Mag: 8.9
SB: 12.9
Dim: 14.9′ × 7.4′
Dist: 23 million light-years

Hydra is host to innumerable galaxies, many of which can be picked up in small amateur instruments. Its biggest deep-sky prizes, of course, are M83, a magnificent spiral galaxy, and NGC 3242 (Caldwell 59), the Ghost of Jupiter planetary nebula. Spiral galaxy NGC 3621 is another bright denizen of Hydra, but very little attention has been paid to it. The galaxy is not included in many popular handbooks, though it is similar in size to M83 (albeit nearly 1½ magnitudes fainter). In a 1993 *Astrophysical Journal* article NGC 3621 was referred to as "another comparatively poorly studied galaxy despite having a large angular size and abundance of H II regions." Once again, the low southerly declination and a lack of bright nearby stars may explain the neglect.

To find NGC 3621 look 10° south and slightly east of Beta (β) Crateris and 3¼° west and slightly south of Xi (ξ) Hydrae. Telescopically, the galaxy appears as a big diffuse glow with a bright core and a faint, elliptical halo. Its spiral arms are faint and will challenge small-telescope users. The galaxy is reasonably large (100,000 light-years across) and the universe's expansion is carrying it away at a speed of 730 km per second. In long-exposure photographs NGC 3621 is a remarkable pinwheel just 25° from edge on. Its small, bright nucleus is nestled in a bright ellipse of spiral arms (cut by rifts of dust) that gradually unwind into a fainter outer halo. One prominent arm extends southward from the small bulge; it carries some of this galaxy's most luminous H II regions.

16

NGC 134
Type: Mixed Spiral Galaxy (SABbc)
Con: Sculptor

RA: 00^h 30.4^m
Dec: –33° 15′
Mag: 10.4
SB: 13.3
Dim: 7.6′ × 1.5′
Dist: 62 million light-years

NGC 134 is an extraordinary nearly edge-on spiral galaxy in Sculptor, which just happens to be surrounded by, and somewhat lost among, a cast of extragalactic prima donnas — namely NGC 253 (Caldwell 65), NGC 300 (Caldwell 70), and NGC 55 (Caldwell 72) — all of which are savagely beautiful galaxies for small telescopes. Nevertheless, NGC 134 is a treasure that's relatively easy to acquire. First locate 4th-magnitude Alpha (α) Sculptoris, which is some 12° south-southeast of Deneb Kaitos, or Beta (β) Ceti. Next, use binoculars to locate 5th-magnitude Eta (η) Sculptoris about 7½° southwest of Alpha Sculptoris. Now center your telescope on Eta Sculptoris and look for a 10th-magnitude needle of pale light 30′ to the east-southeast. NGC 134 is 7.6′ long and its southwestern tip points to a small but obvious trapezoid of 9.0- to 10.5-magnitude stars. A 13th-magnitude star can be seen north-northwest of the galaxy's nuclear region and could easily be mistaken for a supernova.

Seen through small telescopes, wafer-thin NGC 134 appears to be a near twin of the spectacular edge-on galaxy NGC 4565 (Caldwell 38) in Coma Berenices. Photographs, however, reveal the two galaxies to be worlds apart in appearance. NGC 4565 is a classic example of a nearly edge-on, nonbarred spiral galaxy with an oval hub surrounded by a largely symmetrical disk of dust and starlight. Although NGC 134 is similarly oriented, it is a different class of spiral, with hints of a bar. Its bright, very small nucleus is partly hidden by a series of dark, feathery dust lanes that delineate the galaxy's clumpy, moderately tightly wound arms.

Long exposures also reveal bizarre plumes streaming off of the galaxy's major axis in both directions like the flagella of some extraterrestrial euglena. Such plumes are found on many a galaxy that has an interacting companion. Apparently a smaller galaxy has gravitationally stretched matter from NGC 134, forming the plumes. One likely candidate would seem to be 13th-magnitude NGC 131, which lies only 5′ west of NGC 134, but this galaxy does not seem to be involved; a still fainter galaxy, ESO 350-G21, may be the disturbing culprit.

From our perspective, NGC 134 lies in a haystack of galaxies centered on the impressive Sculptor Group, which, at a distance of 10 million light-years, is the nearest galaxy gathering to our own Local Group. But NGC 134 is a not a true member of the Sculptor Group, according to R. Brent Tully (University of Hawaii). Rather, it belongs to the Telescopium-Grus Cloud of galaxies and lies 62 million light-years from Earth. NGC 134 is a massive system with a diameter of 99,000 light-years, making it nearly as large as NGC 4565.

17

NGC 5102
Type: Lenticular Galaxy (S0 pec)
Con: Centaurus

RA: 13^h 22.0^m
Dec: –36° 38′
Mag: 8.8
SB: 12.6
Diam: 8.3′ × 3.5′
Dist: 11 million light-years

NGC 5102 is a fantastic lenticular galaxy just 20′ east-northeast of 3rd-magnitude Iota (ι) Centauri. The galaxy hides in the bright "skirt" of its stellar "companion," though moderate to high magnification can easily remove the star from the field of view. NGC 5102 is more easily spied than NGC 404 in Andromeda, being both brighter and farther from its neighboring star, and it makes a perfect hidden treasure for Southern Hemisphere observers (or those at relatively low northern latitudes). Interestingly, this galaxy is not mentioned in most handbooks and is not acknowledged in Ernst Hartung's *Astronomical Objects for Southern Telescopes*. Yet the galaxy's size, apparent magnitude, and surface brightness are comparable to those of M32, the elliptical companion to the Great Andromeda Spiral. This makes NGC 5102 a binocular object, albeit a challenging one because of its proximity to Iota Centauri. I first noticed the galaxy during a comet sweep across Centaurus with the Genesis at 23×. It was a nagging glow just beyond the glare of Iota Centauri. I was ready to dismiss it as a ghost image of that star, but the star and the galaxy maintained their relative spatial relationship no matter how I moved the telescope.

NGC 5102 is regarded as a classical example of a type-S01 lenticular galaxy. It spans 24,000 light-years and is receding from us at 460 km per second. We see the galaxy inclined 19° from edge on. Under moderate magnification, the galaxy displays a tack-sharp nucleus in a bright and obvious saucer of light; a larger, uniform oval glow surrounds both these features. Long-exposure photographs show that the lens may be dappled with dust, and one can try to imagine other dusty forms in the outer disk as well. A 1993 *Astrophysical Journal Supplement Series* article classifies NGC 5102 as a very blue low-luminosity galaxy. In addition to an underlying population of old, red, metal-poor stars, the galaxy contains an abundance of hot, young, metal-rich stars. Together with infrared observations gathered in the 1980s, this fact has led astronomers to suggest that a burst of star formation occurred in the galaxy about 400 million years ago.

18

NGC 6723
Type: Globular Cluster
Con: Sagittarius

RA: $18^h\ 59.6^m$
Dec: –36° 38′
Mag: 7.0
Diam: 13′
Dist: 29,000 light-years

In the case of Caldwell 68 (NGC 6729), the R Coronae Australis Nebula, Patrick Caldwell-Moore seems to have missed the forest for a tree, because a mere 30′ northwest of this rather obscure nebula lies the glorious Sagittarius globular cluster NGC 6723 — one of the most spectacular objects of its kind in the night sky. (Of course, the Caldwell Catalog is not a list of the sky's most *spectacular* non-Messier objects but of Moore's *favorite* non-Messier objects, whether he chose them on the basis of visual appeal or astrophysical interest.) NGC 6723 marks the apex of a near-perfect isosceles triangle with Zeta (ζ) and Epsilon (ε) Sagittarii, which lie 6¾° to the north-northeast and 7½° to the northwest of the cluster, respectively. It also just skirts the northern boundary of Coronae Australis, so it is well within the reach of most Northern Hemisphere observers.

Shining at 7th magnitude, the cluster is a tad faint to be detected with the unaided eye, but it is so concentrated that someone with young eyes should take up the challenge, especially if attending one of the more southerly star parties, such as the Texas Star Party. Needless to say the object is a cinch in 7×35 binoculars and a fine sight in the smallest of telescopes. The cluster's brightest star shines at magnitude 12.8 and at low power the remainder form a sizzling haze of tiny suns. Christian Luginbuhl and Brian Skiff say that in a 60-mm refractor the cluster looks similar to M4 in Scorpius. I came to the same conclusion independently after I resolved the needle-like row of stars at the cluster's core. At high power NGC 6723 looks like a blizzard of starlight. Look for dark lanes running through its core and fringes.

19

NGC 6281
Type: Open Cluster
Con: Scorpius

RA: $17^h\ 04.8^m$
Dec: –37° 53′
Mag: 5.4
Diam: 8′
Dist: 1,800 light-years

The tail of Scorpius, the Scorpion, is projected against the hub of our galaxy, a region littered with open clusters and asterisms. Some might argue that once you move beyond the Messier and Caldwell Catalogs, open clusters are all the same — little puffs of concentrated starlight. But one, NGC 6281, stands out from the multitude as a fine target for small telescopes. In fact, at magnitude 5.4, NGC 6281 is the brightest non-Messier/non-Caldwell open cluster in Scorpius, though one hardly hears about it. Brent Archinal lists about 70 stars within its 8′-wide disk, with the brightest shining at magnitude 9.0. The cluster's surface brightness is also quite high, making it a fine naked-eye target.

Alas, the cluster is near a host of other popular objects. It is 4½° north-northeast of NGC 6231 (Caldwell 76), 2° southwest of the Bug Nebula (Caldwell 69), 10° southwest of M7, and 8° northeast of NGC 6124 (Caldwell 75). At least NGC 6281 is conveniently located. It simply lies 2½° east-northeast of $Mu^{1,2}$ ($\mu^{1,2}$) Scorpii, about one-third of the way toward the paired Stinger stars at the end of the Scorpion's tail. Through the Genesis NGC 6281 is best seen at 23×, when its stars form a marvelous dipper of bright suns superimposed on a fog of Milky Way. Ernst Hartung concurs, saying the cluster is "very beautiful with a pyramidal group of stars . . . standing out prominently in the centre; it contains several pairs and two bright orange stars, and the linear pattern is most striking." Christian Luginbuhl and Brian Skiff add that the cluster has "20 stars arranged over haze like Christmas tree lights," making it "a pretty cluster" for a 60-mm telescope.

20

SL 7
Type: Dark Nebula
Con: Lupus

RA: $16^h\ 01.8^m$
Dec: –41° 52′
Mag: —
Dim: 160′ × 10′
Dist: unknown

Our final hidden treasure is an incredible veil of darkness in southeastern Lupus that dips ever so slightly into Norma and Scorpius. On a scale of opacity ranging from 1 to 6, SL 7 is classified as a 6, meaning "highly opaque." The nebula is an elongated swath of darkness running from east-northeast to west-southwest. Though it is about 4 Moon diameters long, only about a dozen stars between 9th and 11th magnitude lie within its confines! The darkest section, to the southwest, is an empty 1°-wide corridor that begins about 3¾° due south of Eta (η) Lupi, just south of a 5th-magnitude star. Compared to SL 7, the Coalsack (Caldwell 99) appears very dark to the unaided eye. Yet through a telescope the Coalsack is littered with stars; its clouds do not appear black (like those of SL 7) but gray through the 4-inch.

Appendix A
The Caldwell Catalog

C	Name	Type	Con.	R.A.	Dec.	Mag.	Size (′)	Notes	Page
1	NGC 188	OC	Cep	00^h 48^m	+85° 14′	8.1	15		19
2	NGC 40	PN	Cep	00^h 13.0^m	+72° 31′	12.3	0.6	Bow-Tie Nebula	22
3	NGC 4236	SG	Dra	12^h 16.7^m	+69° 28′	9.6	20 × 8		24
4	NGC 7023	BN	Cep	21^h 01.6^m	+68° 10′	7.7	10 × 8	reflection nebula	27
5	IC 342	SG	Cam	03^h 46.8^m	+68° 06′	8.4	16		30
6	NGC 6543	PN	Dra	17^h 58.6^m	+66° 38′	8.1	0.4 × 0.3	Cat's Eye Nebula	33
7	NGC 2403	SG	Cam	07^h 36.9^m	+65° 36′	8.4	24 × 13		37
8	NGC 559	OC	Cas	01^h 29.5^m	+63° 18′	9.5	7.0	Ghost's Goblet	41
9	Sh2-155	BN	Cep	22^h 56.8^m	+62° 37′	10.0	50 × 30	Cave Nebula	44
10	NGC 663	OC	Cas	01^h 46.3^m	+61° 13′	7.1	15	Horseshoe Cluster	49
11	NGC 7635	BN	Cas	23^h 20.7^m	+61° 12′	10.0	15 × 8	Bubble Nebula	52
12	NGC 6946	SG	Cep	20^h 34.9^m	+60° 09′	8.8	11 × 9		55
13	NGC 457	OC	Cas	01^h 19.5^m	+58° 17′	6.4	20	E.T. Cluster	58
14	NGC 869/84	OC	Per	02^h 20.5^m	+57° 08′	5/6	18/18	Double Cluster	62
15	NGC 6826	PN	Cyg	19^h 44.8^m	+50° 31′	8.8	0.5 × 0.4	Blinking Planetary	68
16	NGC 7243	OC	Lac	22^h 15.0^m	+49° 54′	6.4	30		72
17	NGC 147	EG	Cas	00^h 33.2^m	+48° 30′	9.5	18 × 11	M31 companion	75
18	NGC 185	EG	Cas	00^h 39.0^m	+48° 20′	9.2	17 × 14	M31 companion	75
19	IC 5146	BN	Cyg	21^h 53.4^m	+47° 16′	9.3	10	Cocoon Nebula	79
20	NGC 7000	BN	Cyg	20^h 58.8^m	+44° 20′	5.0	100 × 60	No. America Neb.	83
21	NGC 4449	IG	CVn	12^h 28.2^m	+44° 06′	9.6	5 × 4	Box Galaxy	89
22	NGC 7662	PN	And	23^h 25.9^m	+42° 32′	8.3	0.5	Lt. Blue Snowball	92
23	NGC 891	SG	And	02^h 22.6^m	+42° 21′	9.9	12 × 3	Outer Limits Gal.	96

C	Name	Type	Con.	R.A.	Dec.	Mag.	Size (′)	Notes	Page
24	NGC 1275	EG	Per	$03^h\ 19.8^m$	+41° 31′	11.9	3 × 2	Perseus A	100
25	NGC 2419	GC	Lyn	$07^h\ 38.1^m$	+38° 53′	10.4	4.6	Intergalactic Tramp	104
26	NGC 4244	SG	CVn	$12^h\ 17.5^m$	+37° 48′	10.4	16 × 2	Silver Needle Gal.	107
27	NGC 6888	BN	Cyg	$20^h\ 12.0^m$	+38° 21′	8.8	18 × 13	Crescent Nebula	109
28	NGC 752	OC	And	$01^h\ 57.6^m$	+37° 50′	5.7	75		112
29	NGC 5005	SG	CVn	$13^h\ 10.9^m$	+37° 03′	9.8	6 × 3		115
30	NGC 7331	SG	Peg	$22^h\ 37.1^m$	+34° 25′	9.5	9.7 × 4.5		118
31	IC 405	BN	Aur	$05^h\ 16.2^m$	+34° 16′	9.2	30 × 20	Flaming Star Neb.	122
32	NGC 4631	SG	CVn	$12^h\ 42.1^m$	+32° 32′	9.2	15 × 4		126
33	NGC 6992/5	SN	Cyg	$20^h\ 56.4^m$	+31° 43′	7.5	60 × 8	Eastern Veil Neb.	129
34	NGC 6960	SN	Cyg	$20^h\ 45.7^m$	+30° 43′	7.9	70 × 6	Western Veil Neb.	129
35	NGC 4889	EG	Com	$13^h\ 00.1^m$	+27° 58′	11.5	3 × 2	in Coma Cluster	135
36	NGC 4559	SG	Com	$12^h\ 36.0^m$	+27° 58′	10.0	11 × 5		139
37	NGC 6885	OC	Vul	$20^h\ 11.6^m$	+26° 28′	8.1	20		143
38	NGC 4565	SG	Com	$12^h\ 36.3^m$	+25° 59′	9.6	16 × 2		150
39	NGC 2392	PN	Gem	$07^h\ 29.2^m$	+20° 55′	9.2	0.8 × 0.7	Eskimo Nebula	154
40	NGC 3626	SG	Leo	$11^h\ 20.1^m$	+18° 21′	11.0	3 × 2		159
41	Mel 25	OC	Tau	$04^h\ 26.9^m$	+15° 52′	0.5	330	Hyades	162
42	NGC 7006	GC	Del	$21^h\ 01.5^m$	+16° 11′	10.6	3.6		169
43	NGC 7814	SG	Peg	$00^h\ 03.2^m$	+16° 09′	10.6	6 × 2	Electric Arc	173
44	NGC 7479	SG	Peg	$23^h\ 05.0^m$	+12° 19′	10.8	4 × 3		173
45	NGC 5248	SG	Boo	$13^h\ 37.5^m$	+08° 53′	10.3	6 × 5		179
46	NGC 2261	BN	Mon	$06^h\ 39.2^m$	+08° 44′	~10 var.	4 × 2	Hubble's Var. Neb.	182

C	Name	Type	Con.	R.A.	Dec.	Mag.	Size (′)	Notes	Page
47	NGC 6934	GC	Del	20^h 34.2^m	+07° 24′	8.8	7.1		187
48	NGC 2775	SG	Cnc	09^h 10.3^m	+07° 02′	10.1	5 × 4		191
49	2237-8/46	BN	Mon	06^h 32.3^m	+05° 03′	~5.0	80 × 60	Rosette Nebula	194
50	NGC 2244	OC	Mon	06^h 32.3^m	+04° 51′	4.8	30	assoc. w/Rosette	194
51	IC 1613	IG	Cet	01^h 04.8^m	+02° 07′	9.2	19 × 17	The Scarecrow	202
52	NGC 4697	EG	Vir	12^h 48.6^m	–05° 48′	9.2	7 × 5	in Virgo Cluster	207
53	NGC 3115	SG	Sex	10^h 05.2^m	–07° 43′	8.9	7 × 3	The Spindle	211
54	NGC 2506	OC	Mon	08^h 00.0^m	–10° 46′	7.6	12		215
55	NGC 7009	PN	Aqr	21^h 04.2^m	–11° 22′	8.0	0.7 × 0.4	Saturn Nebula	218
56	NGC 246	PN	Cet	00^h 47.1^m	–11° 52′	10.9	5 × 4	Pac-Man Nebula	223
57	NGC 6822	IG	Sgr	19^h 44.9^m	–14° 48′	8.8	15	Barnard's Galaxy	227
58	NGC 2360	OC	CMa	07^h 17.7^m	–15° 38′	7.2	14		231
59	NGC 3242	PN	Hya	10^h 24.8^m	–18° 39′	7.8	0.8 × 0.6	Ghost of Jupiter	234
60	NGC 4038	IG	Crv	12^h 01.9^m	–18° 52′	10.5	11 × 6	Ringtail Gal. NW	239
61	NGC 4039	IG	Crv	12^h 01.9^m	–18° 53′	10.3	10 × 4	Ringtail Gal. SE	239
62	NGC 247	SG	Cet	00^h 47.1^m	–20° 46′	9.2	22 × 7	Milkweed Seed Gal.	245
63	NGC 7293	PN	Aqr	22^h 29.6^m	–20° 50′	7.3	12 × 10	Helix Nebula	248
64	NGC 2362	OC	CMa	07^h 18.7^m	–24° 57′	4.1	6	Tau CMa Cluster	254
65	NGC 253	SG	Scl	00^h 47.6^m	–25° 17′	7.6	26 × 6	Sculptor Galaxy	258
66	NGC 5694	GC	Hya	14^h 39.6^m	–26° 32′	10.2	4.3		263
67	NGC 1097	SG	For	02^h 46.3^m	–30° 16′	9.2	10 × 6		266
68	NGC 6729	BN	CrA	19^h 01.9^m	–36° 57′	~9.5 var.	1.3 var.	R CrA Nebula	270
69	NGC 6302	PN	Sco	17^h 13.7^m	–37° 06′	9.6	1.4 × 0.4	Bug Nebula	274

C	Name	Type	Con.	R.A.	Dec.	Mag.	Size (′)	Notes	Page
70	NGC 300	SG	Scl	$00^h\ 54.9^m$	–37° 41′	8.1	20 × 15	Southern Pinwheel	277
71	NGC 2477	OC	Pup	$07^h\ 52.2^m$	–38° 32′	5.8	20		284
72	NGC 55	SG	Scl	$00^h\ 15.1^m$	–39° 13′	8.1	34 × 6		277
73	NGC 1851	GC	Col	$05^h\ 14.1^m$	–40° 03′	7.1	12		288
74	NGC 3132	PN	Vel	$10^h\ 07.0^m$	–40° 26′	9.4	1.4 × 0.9	So. Ring Neb.	292
75	NGC 6124	OC	Sco	$16^h\ 25.3^m$	–40° 40′	5.8	40		296
76	NGC 6231	OC	Sco	$16^h\ 54.2^m$	–41° 50′	2.6	14	False Comet Cluster	300
77	NGC 5128	EG	Cen	$13^h\ 25.5^m$	–43° 01′	6.7	17 × 13	Centaurus A	305
78	NGC 6541	GC	CrA	$18^h\ 08.0^m$	–43° 43′	6.3	15		311
79	NGC 3201	GC	Vel	$10^h\ 17.6^m$	–46° 25′	6.7	20		314
80	NGC 5139	GC	Cen	$13^h\ 26.8^m$	–47° 29′	3.7	55	Omega Centauri	318
81	NGC 6352	GC	Ara	$17^h\ 25.5^m$	–48° 25′	8.0	7.1		324
82	NGC 6193	OC	Ara	$16^h\ 41.3^m$	–48° 46′	5.2	14		328
83	NGC 4945	SG	Cen	$13^h\ 05.4^m$	–49° 28′	8.8	19 × 4	Tweezers Galaxy	332
84	NGC 5286	GC	Cen	$13^h\ 46.4^m$	–51° 22′	7.3	11		336
85	IC 2391	OC	Vel	$08^h\ 40.3^m$	–52° 55′	2.5	60	Omicron Vel Cl.	339
86	NGC 6397	GC	Ara	$17^h\ 40.7^m$	–53° 40′	5.7	31		342
87	NGC 1261	GC	Hor	$03^h\ 12.3^m$	–55° 13′	8.3	6.8		345
88	NGC 5823	OC	Cir	$15^h\ 05.4^m$	–55° 36′	7.9	12		348
89	NGC 6087	OC	Nor	$16^h\ 18.9^m$	–57° 54′	5.4	15	S Normae Cluster	351
90	NGC 2867	PN	Car	$09^h\ 21.4^m$	–58° 19′	9.7	0.2		354
91	NGC 3532	OC	Car	$11^h\ 05.5^m$	–58° 44′	3.0	50	Pincushion Cluster	358
92	NGC 3372	BN	Car	$10^h\ 43.8^m$	–59° 52′	4.8 var.	120 var.	Eta Carinae Nebula	361

C	Name	Type	Con.	R.A.	Dec.	Mag.	Size (′)	Notes	Page
93	NGC 6752	GC	Pav	19^h 10.9^m	–59° 59′	5.4	29	Starfish Cluster	370
94	NGC 4755	OC	Cru	12^h 53.6^m	–60° 21′	4.2	10	Jewel Box	376
95	NGC 6025	OC	TrA	16^h 03.3^m	–60° 25′	5.1	15		381
96	NGC 2516	OC	Car	07^h 58.0^m	–60° 45′	3.8	22	Southern Beehive	384
97	NGC 3766	OC	Cen	11^h 36.2^m	–61° 37′	5.3	15		388
98	NGC 4609	OC	Cru	12^h 42.3^m	–63° 00′	6.9	6.0	in Coalsack	391
99	—	DN	Cru	12^h 53^m	–63°	—	420 × 300	Coalsack	391
100	Cr 249	OC	Cen	11^h 38.3^m	–63° 22′	2.9	65 × 40	Lambda Cen Cl.	397
101	NGC 6744	SG	Pav	19^h 09.7^m	–63° 51′	8.6	21 × 13		371
102	IC 2602	OC	Car	10^h 42.9^m	–64° 24′	1.9	100	Theta Car Cl.	402
103	NGC 2070	BN	Dor	05^h 38.6^m	–69° 05′	~4.0	30 × 20	Tarantula Neb.	405
104	NGC 362	GC	Tuc	01^h 03.2^m	–70° 51′	6.4	14		409
105	NGC 4833	GC	Mus	12^h 59.6^m	–70° 52′	6.9	14	Southern Butterfly	417
106	NGC 104	GC	Tuc	00^h 24.1^m	–72° 05′	4.0	50	47 Tucanae	409
107	NGC 6101	GC	Aps	16^h 25.8^m	–72° 12′	9.2	11		423
108	NGC 4372	GC	Mus	12^h 25.8^m	–72° 40′	7.2	19		417
109	NGC 3195	PN	Cha	10^h 09.4^m	–80° 52′	11.6	0.6		426

Key:

BN = bright nebula
DN = dark nebula
PN = planetary nebula
SN = supernova remnant
OC = open cluster
GC = globular cluster
EG = elliptical galaxy
SG = spiral galaxy
IG = irregular galaxy

Appendix B Why Messier Did Not Include the Double Cluster

NGC 884 and NGC 869

ONE MYSTERY THAT STILL HAUNTS THE HALLS of astronomy is why the 18th-century French comet hunter Charles Messier did not include the Double Cluster in his famous catalog. "This splendid object . . . was certainly known in his day," Robert Burnham Jr. comments in his *Celestial Handbook,* "and he included other bright clusters such as Praesepe and the Pleiades." So why not the Double Cluster?

I believe Messier had a very logical reason to omit it. But first a powerful myth must be dispelled.

The Messier Catalog is not a list of the best or the brightest deep-sky objects in the heavens. This cannot be said enough. It is not a manifest of celestial superlatives. But who hasn't loosely described the Messier objects as the finest in the heavens? On the jacket of *The Messier Album,* by John H. Mallas and Evered Kreimer, no less an authority than Harvard University astronomer Owen Gingerich encourages readers to "find a telescope and a clear night, and . . . begin the chase of the *most spectacular* sidereal objects of the northern skies" (emphasis mine).

Without question, the Messier Catalog contains some of the most spectacular deep-sky objects visible in small telescopes. But their inclusion in the catalog is largely circumstantial. Messier never intended to create a list of the most visually pleasing objects in the heavens. His was a catalog of "comet masqueraders," as the late comet discoverer Leslie Peltier called them. Messier started his list in 1758, after he encountered a stationary fuzzy patch (M1) near Zeta (ζ) Tauri while looking for the Comet of 1758. As comet hunter David Levy writes in the first chapter of *Deep-Sky Companions: The Messier Objects,* "Realizing he had been fooled by the sky's version of a practical joke, Messier began to build a catalogue of what he called these 'embarrassing objects.'"

Messier's first catalog, published in 1774, contained not 110 but 45 "embarrassing" objects. Actually, his original list, compiled in 1765, had only 41 objects. "Before submitting the list for publication," Levy explains, "he decided to round it out with a few more objects." In order of their inclusion, these are the two components of the Orion Nebula (M42 and M43), the Beehive Cluster (M44) in Cancer, and the Pleiades star cluster (M45) in Taurus. But why not the Double Cluster?

To understand Messier's intent, we have to forget what we now know about the nature of the objects in his list, forget how spectacular these objects look in photographs and CCD images. Instead, we have to journey back in time some two and a half centuries and think like Messier, who was a comet hunter, not a celestial tourist.

Like Messier, I have spent several years hunting for comets. Unlike Messier, I have never discovered one that was not already known. But I have had the privilege to meet many "embarrassing objects." Identifying them has been relatively easy for me because most are plotted on the star charts I use in the field. Messier and his contemporaries did not have the luxury of star charts with reams of nebulae plotted on them. And nothing is more unproductive to a comet hunter than sitting behind an eyepiece, waiting to see if an uncharted object will move; the agony increases as twilight advances, or as the object nears the western horizon. Understandably, then, Messier wanted to "cage" these celestial pests, once and for all, by cataloging their positions.

When Messier systematically began searching for comets in 1758 competition was scarce. "For about 15 years nearly all comet discoveries were made by Messier," Gingerich writes in *The Messier Album,* "so that he almost considered them his own property." These comets were not the dim and distant objects that (mostly professional) astronomers are accustomed to discovering today (mostly on films or CCD images, well before amateurs have a chance to spy the comets in their eyepieces). In Messier's day all comets were swept up visually, and all were bright and relatively near the Sun. Of the 15 comets discovered from 1758 to 1774 (when his first catalog was published), 7 were discovered by Messier (and he independently found one other first discovered in 1759 by observers in Lisbon); seven were discovered with the unaided eye; and more than half were sighted within two weeks of perihelion (each comet's closest approach to the Sun). No comets were found more than about 2 months before or after perihelion, and the faintest ones shone near the limit of naked-eye visibility at magnitude 6.5. These statistics provide us with some valuable insight into where Messier and other comet hunters concentrated their searches.

It has long been known that to maximize their chances of discovery, comet hunters search a "haystack" of sky centered on the ecliptic, near the position of sunset or sunrise. In the *Guide to Observing Comets* (edited by Daniel W. E. Green and published as a special issue of the *International Comet Quarterly*), veteran comet observer John Bortle explains how these searches are conducted:

> A typical pattern for evening hunting is to begin in late twilight by choosing a point near the horizon and about 45° north or south of the sunset point and sweeping parallel to the horizon until reaching an azimuthal point around 45° on the other side of the sun. . . . After completing the first sweep, the instrument is raised about one-third to one-half of a field diameter and the observer proceeds to sweep backward to the original starting point. This procedure should be continued until one reaches at least 45° altitude above the horizon.

The search pattern is reversed when scanning the morning sky.

Of the 41 deep-sky objects in Messier's original list, 17 clearly were discovered first by Messier; and of the 109 objects in the final catalog only 37 were Messier discoveries. Just as we might expect from a comet hunter, all of his original finds were within 45° of the point where the ecliptic met the horizon at the time of discovery. In fact, all but one (M40) of the first 41 objects lies reasonably within that limit; 60 percent of them were within about 25° of that point. Since it was Messier's intent to catalog as many of these "embarrassing" objects as possible, he thoroughly searched the literature on previously reported nebulae and, after observing and obtaining accurate positions for them, added these objects to his list. Messier did not care whether these objects were his discoveries, because they were not comets. He meticulously acknowledged each discoverer in his catalog. In fact, he was probably quite happy to do so — their discoveries only benefited him.

One question immediately surfaces, however: why did Messier include in his catalog the bright and obvious objects M31 in Andromeda, M6 and M7 in Scorpius, and M8 in Sagittarius? All were well known at the time, some since antiquity. The answer would be obvious to an 18th-century comet hunter. All of these objects lie in the comet "haystack" at various times of the year, and all look nebulous to the naked eye. Obviously, no intelligent observer would confuse, say, M31, for a comet when it was in the dark of the sky, but what about in a twilight sky?

In Messier's day, twilight comets were not infrequent. The first and only comet discovered in 1758 (when Messier started looking), was sighted from Bourbon Island in the Indian Ocean just 15 days before perihelion; the comet shone at magnitude 2.5 and had a 1.5°-long tail. Messier's fourth comet was sighted (on March 8, 1766) in the twilight; the 6th-magnitude comet, which was approaching conjunction with the Sun, never escaped the twilight and was last seen by Messier with difficulty low over the horizon. On October 13, 1773, one year before Messier's first list was published, Messier discovered another twilight comet; this one shone at magnitude 4.5, about the same brightness as both M6 and M8. In 1780 Messier would find yet another twilight comet, a telescopic 7th-magnitude one. So it was not only reasonable but also prudent that Messier include in his catalog the positions of *all* cometlike objects, however bright, as long as they were in the comet haystack.

It's easy to imagine how the compact glows of M6, M7, or M8 could be confused as comets when hiding in the twilight, many would agree, but M31? Interestingly, in the last 20 years, I know of two cases where respected comet hunters have reported the discovery of a binocular comet in the twilight sky only to be "embarrassed" when they learned that they had simply rediscovered M31. Of course, the small cometlike form of M32 is almost sitting on top of M31; how strange it would have been for Messier to include M32 in a catalog of pseudocomets while neglecting the conspicuous, elongated mass beside it.

As noted above, of the remaining objects in Messier's first list of 41 noncometary objects, all but one (the double star M40) lie within the magical comet haystack at various times of the year. But even M40 appears to be an explainable exception. After searching through the lists of previously reported nebulae to make his list as complete as possible, Messier set out to locate each of these objects, some of which he could not find. "He therefore published as part of his memoir," Gingerich writes, "a number of objects previously reported, but which he could not find. He realized that many of them were nebulous to the naked eye but were mere asterisms when examined with a telescope." This left Messier with a list of 39 cometlike objects. Given Messier's penchant for rounding out his lists, he included in his "final" listing of 1764 the "nebula" that had fooled not him, but Johannes Hevelius in 1660 (namely, M40). The "nebula," as Messier pointed out, is merely "two stars very close to one another and very faint." As many comet hunters today know, close pairs of stars seen at low magnifications can look remarkably like dim comets. Messier's choice to include M40 was not only logical but educational — a warning, if you will, to comet hunters that there are things other than nebulae and star clusters that can masquerade as comets.

In January 1765, just as Messier had finished rounding out his list, he discovered, by chance (not in the course of a systematic comet search) a cluster of stars (M41) just south of brilliant Sirius. And that is why, four years later, he decided once again to round off his list. The inclusion of M42 seems logical because of M43, the small nebula north of M42 that has a round cometlike head, starlike nucleus, and "tail" of nebulosity. Both the Praesepe and the Pleiades star clusters lie nearly on the ecliptic. Although they would not necessarily be problematical to telescopic comet hunters (being bright and easily resolvable into stars), they still might confuse someone looking in the comet haystack with the unaided eye in the twilight.

The Double Cluster, then, is the only celestial object visible from Messier's latitude and known since classical times that is not included in the comet hunter's original list of 45 objects. The remaining 65 objects in Messier's final catalog are not part of the mystery. "[The] first group of [45] nebulae was the only part of the catalogue prepared for its own sake," Gingerich explains; "most of the subsequent objects were found and reported in comet searches, so that the remainder of the catalogue is less systematic than the first part."

So why not include the Double Cluster? From Messier's latitude of nearly 50° N, the Double Cluster never sets. Even at lower culmination it would still be almost 20° above Messier's northern horizon. Its far northerly declination of nearly +60° places it well outside of the comet haystack during most of the year. Only in the late spring and early fall does it lie on the far northern fringe of the comet haystack. Furthermore, in the twilight, not only is the Double Cluster obviously a pair of close objects (a dead giveaway), its appearance coincides with that of Cassiopeia's brightest stars. Even in the unlikely event that Messier were to spot its glow in the early evening twilight, by nightfall he certainly would have recognized it. Nor is the Double Cluster a problem when it lies close to the comet haystack in the morning sky. Predawn comet searches begin before the onset of twilight. The observer starts by looking 45° above the horizon and works his way toward the Sun. Since the Double Cluster is circumpolar, its presence and identity would be obvious in a dark sky. Furthermore, as the

Earth turns, the cluster would climb ever higher in the sky, well away from the comet haystack. Meanwhile the comet hunter's sweeps would take him (or her; Caroline Herschel was a near-contemporary of Messier) lower and lower as he (or she) searched for brighter and brighter comets near the Sun. So only someone who did not know the sky or was unskilled in the comet-hunting trade would have been fooled by this bright circumpolar glow; it would be folly to list the Double Cluster as an object that could be mistaken for a comet by a skilled observer. And though Messier's sky-sweeping technique is not known with certainty, Peter Lancaster Brown, in his book *Comets, Meteorites, and Men,* writes that "from sketchy accounts [Messier] left, there is no reason to suppose that it differed much from those in use by present generation amateur-visual observers dedicated to making a persistent search of the skies night after night, month after month and year after year."

Appendix C
William Herschel:

The Greatest Visual Observer of All Time

Based on The Scientific Papers of Sir William Herschel (London: The Royal Society and the Royal Astronomical Society, 1912).

Larry Mitchell

Fig. 533. William Herschel (1738–1822).

AMONG THE ASTRONOMICAL LUMinaries of the 18th century, none shone as brightly as William Herschel (1738–1822). His contributions to astronomy were unrivaled in his day, and they greatly expanded humanity's perception of the universe. Herschel was passionately devoted to astronomy. He tirelessly pursued his endeavors both day and night, often becoming thoroughly consumed with each project. He had to engineer, and then build, the great instruments he used at night — often all night. Such devotion is almost unheard of today. Yet Herschel pursued his passion in this way for about 40 years.

The Caldwell Catalog is replete with Herschelian discoveries, which is not surprising. After Charles Messier (1730–1817) published his famous catalog of comet masqueraders, Herschel undertook the first systematic visual survey of the night sky with a large-aperture telescope, discovering 2,508 additional deep-sky objects. Herschel, however, was much more than a deep-sky devotee. He was a prolific observer of everything in the heavens. Indeed, his views of deep-sky objects were influenced in part by his observations of the stars and planets, as well as his belief in the plurality of worlds.

BEGINNINGS AND FAMILY

Friedrich Wilhelm (later William) Herschel was born on November 15, 1738, in Hanover, Germany. He was one of 10 children born to Isaac and Anna and is said to have inherited his energy and perseverance from his father, a musician in the Hanoverian Guard. William followed in his father's footsteps, though not necessarily by choice. When he was four his father sat him on a table and commanded him to play the fiddle. By his late teens William was an excellent musician and eminent composer, and in 1756 he toured England for several months with his father and the Guard. The country and culture left a lasting impression on the young composer, and in the autumn of 1757 William left Germany for England, where he lived for the rest of his life.

Initially Herschel made his living in England by performing and copying music. In 1766 he became the organist at the Octagon Chapel in Bath and, within a few years, was promoted to Director of Public Concerts. But behind this veneer of musical aptitude lay a keen interest in scientific and mathematical studies, which magnified after his equally talented and devoted sister Caroline joined him in Bath in 1772. In time Caroline gave up her own musical interests to support her brother's growing passion for science, not only because William wished it, but because she knew she would be indispensable to him. Her own devotion would win her praise from astronomers and other scientists. She became an astute telescope maker and observer in her own right, as well as the discoverer of seven comets and several deep-sky objects. She also produced a revised version of Flamsteed's star catalog. Caroline Herschel was awarded the Gold Medal of the Royal Astronomical Society in 1828 for carefully and thoroughly cataloging 2,500 of the nebulae and clusters found by her brother.

On May 8, 1788, at the age of 50, Herschel married Mary Pitt, a wealthy widow and daughter of Mr. Adee Balwin, a merchant. She brought him financial security that alleviated him of the necessity of building telescopes to supplement his income. Her friendly disposition and gentle character made Herschel's home life thoroughly happy, and she did not distract him from his work. Equally important, if not more so, she and Caroline became the best of friends. William and Mary's only child, John Frederick William Herschel, became a renowned scientist and astronomer who extended his father's studies to the southern celestial hemisphere.

THE TURNING POINT

Herschel penned his earliest known astronomical note on May 10, 1773: "Bought a book of astronomy and one of astronomical tables." He had begun to read about astronomy that spring, and the subject was to consume him from that point on. During the day, Herschel would study Newton's works; at night he would retire to bed with a glass of milk or water and read Smith's *Optics* or Ferguson's *Astronomy*. These books turned his life around, for they aroused in him a passionate desire to see the wonders in the heavens for himself.

After fussing with lenses, tin tubes, and tools for making refracting telescopes, he quickly turned his interest to reflectors. His first purchase, a Gregorian reflector with a focal length of 2 feet, was so convenient to use that he decided to build his own — though he knew nothing about telescope making. Come October, Herschel had some mirrors cast for a 2-foot focal-length reflector. These were made of metal alloys called speculum, which contained copper and tin and "produced a very good sound white metal." By March 1774, the young telescope maker had turned a 5.5-foot focal-length reflector to the sky. His first astronomical journal entry, dated March 1, 1774, describes observations of Saturn at 40×, then of the "Lucid Spot" (M42) in Orion's Sword. This marked the modest beginning to his exceptional astronomical career.

In the summer of 1774 William and Caroline moved to 19 New King Street, also in Bath. The house had more room for telescope building, a roof for observing, and a grass plot that could support a large telescope. Soon, Caroline noted, the house was turned upside down as William churned out numerous mountings and mirrors with focal lengths as large as 20 feet. By January 1781 William had developed a good case of aperture fever. He built a furnace and melting oven on the ground floor of his house, which he used in August of that year to cast a speculum-metal mirror with a focal length of

30 feet. The mirror weighed 469 pounds and had to be lifted by a crane. It cracked while cooling. Herschel recast the mirror, which now weighed 538 pounds, but this time the metal fell through the bottom of the furnace, damaging the floor and causing the Herschels to run for cover. At the end of this episode, Caroline later wrote, her "poor brother fell, exhausted with heat and exertion, on a heap of brickbats."

A NEW WORLD

Between these mirror-making escapades, Herschel made his greatest discovery, that of the planet Uranus. He found it on Tuesday, March 13, 1781, from the backyard of his new residence at Bath. The discovery was no accident. That night Herschel was using a telescope (with a focal length of 7 feet, an aperture of 6.2 inches, and a magnification of 227×) to systematically study, for the purpose of registering double stars, all stars brighter than 8th magnitude, when he recorded a new object in his journal:

> In the quartile near ζ Tauri the lowest of two is a curious either nebulous star or perhaps a comet. A small star follows the comet at ⅔ of the fields distance.

After verifying that the "comet" had moved, Herschel notified the Astronomer Royal, Nevil Maskelyne, at Greenwich, who first suggested that the object might be a new planet. He also contacted his fellow astronomer and correspondent, Rev. T. Hornsby at Oxford. Subsequent observations proved that Herschel had indeed become the first person in history to discover a planet beyond those visible to the naked eye, and this granted him immediate and lasting fame. In November 1781 he was awarded the Royal Society's Copley Medal for the discovery.

In response to a request by Maskelyne the following year, Herschel named his new object Georgium Sidus (the star of George) after his king and mentor, King George III. "I cannot but wish to take this opportunity of expressing my sense of gratitude," Hershel wrote, "by giving the name Georgium Sidus to a star, which with respect to us first began to shine under his auspicious reign." (The name Uranus, proposed by Johann Elert Bode, was not universally accepted until after Herschel's death). Herschel's request did not go unnoticed by the king, who suggested that Herschel concentrate on astronomy full time. To this end, he granted Herschel an annual stipend of £200 (not a large sum), in exchange for which the astronomer was to occasionally show members of the royal family something of interest in the sky.

Herschel's career as a professional astronomer commenced in August of 1782. To fulfill his duties to the king, Herschel (with Caroline) moved to a house in Datchet, about one and a half miles from Windsor Castle, on the left bank of the Thames. It had a convenient garden, in which he placed a "small 20-foot reflector of 12 inches aperture." By 1783, Herschel had built a "large 20 feet" (focal length) telescope with an aperture of 18.7 inches, and he began observing the heavens with it on October 23rd of that year. All of his "sweeps" were made with this instrument, as were most of his other important observations.

Upon completing the large 20-foot telescope, Herschel did not miss a single hour of "Star-Light weather from which I used either to watch myself or to keep somebody to watch. My leisure hours of the day were spent in preparing and improving telescopes." Herschel was to build numerous telescopes for sale as well as for his own use.

SWEEPING THE HEAVENS

Five days after he deployed his large 20-foot reflector, Herschel began a systematic sweep of the heavens with it. For his sweeps he usually used a magnification of 157×, which offered a small 15′ field of view. The eyepiece was mounted on the upper side of the telescope's octagonal tube, where it made a 45° angle with the vertical when the telescope was horizontal. Herschel preferred this observing position, looking down as if at a reading desk, and used this arrangement until September 22, 1786, in Sweep 600, when he changed to the "front view," as he called it. By removing the small diagonal mirror and placing the eyepiece at the front end of the tube, he found the images to be just as good and well defined as at the Newtonian focus and the light "incomparably more brilliant." He found this position very convenient.

Before starting a sweep, Herschel pointed the telescope toward the meridian and observed from a 9-foot-long movable gallery. The telescope hung freely in the center of its mount, and the observer moved it with a handle fastened near the eyepiece. Drawing the telescope along by hand in a lateral motion, Herschel would walk backward and forward on the movable gallery. This enabled him to perform slow oscillations, 12° to 14° in breadth, each taking generally 4 to 5 minutes. At the end of each oscillation he took short notes on what he had seen. If he had found a new nebula or star cluster, he carefully noted the stars in the field of view of both the finderscope and the main telescope, so as to be able to locate the object again. After this procedure, Herschel would then raise or lower the instrument about 8′ or 10′ along a north-south line and then perform another oscillation. He then continued for 10, 20, or 30 oscillations according to circumstances, and the whole sweep was numbered and registered in his journal.

At first, Herschel took his own notes, but after 41 sweeps the disadvantages of doing so became obvious. Because he needed light to write, his eye could never achieve the dark adaptation required for his "delicate observations." Furthermore, the gallery was not well placed, and holding down the heavy telescope at the ends of the oscillations, when it tended to arch upward, led him to become fatigued. Herschel therefore began to sweep by elevating and lowering the telescope manually with a vertical motion. This method increased the labor of moving the telescope, so he hired workmen to do the job. Caroline wrote down his observations, repeating everything back to him so he could verify the "picture before me" while preserving his night vision. A few sweeps (numbers 42 to 45) were made experimentally, and with Sweep 46 on December 18, 1783, Herschel's long survey of the heavens commenced. This series was not closed until September 30, 1802, with Sweep No. 1112.

Each object's description and position relative to the nearest Flamsteed star was copied on single sheets of Foolscap (the "Register Sheets"). A great deal of effort was made to obtain accurate coordinates, and Herschel experimented with various techniques, using pulleys, knotted ropes, bells, and iron plates to tell him where he was in the heavens. Eventually there was a total of 2,508 new objects. Caroline also had a similar set of sheets for Messier's nebulae and clusters. She was very careful and accurate and made only two or three positional mistakes of 1° in all of her entries — and these she made in old age. A total of seven "Sweep Books" was eventually presented to the Royal Society.

THE 40-FOOT MONSTER

In 1785 Herschel petitioned the king for a large instrument with a 30- or 40-foot focal length. The king chose the 40-foot option and granted £2,000 toward its construction, and in Sep-

mirror eventually was discarded. However, first light for the 40-foot telescope was achieved with this mirror on February 19, 1787, when Herschel crawled inside the tube and hand-held the eyepiece to find the focal point. The view of the "Lucid Spot" in Orion was far from perfect, but it was better than expected and extremely bright. The Herschels immediately began trying to refine the mirror.

In August 1787, King George granted Herschel yet another £2,000. A second mirror cracked during the casting process, but in February 1788 a third mirror was cast successfully. This mirror weighed 2,118 pounds, had a diameter of 48 inches, and was 3½ inches thick throughout (its back was not flat). An iron ring surrounded the mirror, and an iron cross supported it from behind. The mirror was successfully polished by machine.

The telescope's optical tube was 39 feet, 4 inches long, with a diameter of 4 feet, 10 inches.

Herschel did not deem the Great Telescope completed until August 28, 1789 — the same night he discovered the sixth satellite of Saturn (Enceladus) with it. (This actually proved to be a rediscovery, as Herschel had observed the satellite two years earlier with the 20-foot telescope but did not recognize it as such at the time.) Herschel observed with the 40-foot at the "front view," from an enclosed, adjustable seat that was mounted on the lower side of the tube's front end. No secondary mirror was used. The primary mirror's focal point was placed 4 inches above the lower side of the

tember work commenced. The king also granted a yearly salary of £50 to Caroline Herschel. However, frequent flooding at Datchet and consequent ill health slowed William's progress. Thus he and his sister moved in 1786 to a house at Slough, also in the neighborhood of Windsor Castle. There William's health improved.

The telescope, its massive mount, and its mirror all were designed exclusively by William Herschel. The first of its two functioning mirrors was cast in London; it spanned 49½ inches in diameter and was 2½ inches thick. Ten men polished it, giving the speculum "[a] very white surface." The mirror was very brittle, however, and came out of the mold thinner than intended. The resulting weakness would never permit it to give a good image, and the tube's front end. This meant that the observer's head would not block incoming light. A "slider" with an adjustable base was located at the lower end of the tube, where it pointed directly at the center of the great mirror. It carried a brass tube, into which eyepieces or micrometers were inserted.

Due to its massive size, Herschel seldom used the great telescope, even though it was his most celebrated instrument (and, at the time, the world's largest). The telescope required much time to ready for use and two workmen to move. Because of its weight, the mirror had to be left inside the telescope, so it tarnished quickly. (By contrast, Herschel's two 18¾-inch mirrors remained highly polished, as they were stored in protective boxes and placed in their telescope tube only when observations were about to be made.) Herschel used the 40-foot reflector sparingly and wrote that it should be used only to examine objects that other instruments could not reach.

Interestingly, Herschel never reported seeing spiral structure in any "nebula" through the 40-foot telescope, even though such structure should have been obvious in some galaxies with so large an instrument. Equally strange is that Herschel did not distinctly mention the fifth or sixth stars of the Trapezium in M42, an object that he observed repeatedly. On one occasion the gigantic telescope even failed to show Saturn's rings. Obviously it gave poor images, even with its second mirror.

THE MOON AND PLANETS

Like many of his contemporaries, Herschel observed the Moon and, among other tasks, he carefully measured shadows cast upon its surface by its mountains. He deduced a lunar diameter of 2,180 miles, surprisingly close to today's more accurate figure (2,160 miles), but he also made some very controversial lunar observations. On May 28, 1776, he used a 10-foot reflector at 240× to view what he believed were forests in Mare Humorum, complete with shadows cast by trees at the edge of the woods. The next night he could not see any "woods" and seemed to have dropped the idea completely. However, in a 1780 letter to Maskelyne he wrote of his absolute certainty of the Moon being inhabited, and if given the chance, Herschel vowed,

> I would not hesitate a moment to fix upon the Moon for my habitation. What a glorious view of the heavens from the Moon. Do not all the elements seem at war when we compare the Earth with the Moon? Air, water, fire, clouds, tempests, volcanoes; all these are either not on the Moon or at least in much greater subjection than here.

In 1783 and again in 1787, Herschel reported a volcano in the Earthlit part of the new Moon, in the crater Aristarchus, stating that it resembled a red 4th-magnitude star. However, other astronomers in Paris thought that the Aristarchus region, which was relatively bright, might well have produced this volcano-like apparition by reflecting earthshine. Herschel may have been persuaded of this also, as the whole issue was quietly dropped.

Herschel's observations of Mars include many references to "luminous spots" (ice caps) that appeared to project beyond the edge of the planet's disk. He deduced that the polar spots were frozen, covered with mountains of ice and snow that melted only partway when alternately exposed to the Sun. He also noted dark spots on the surface and determined a rotation period of $24^h\ 37^m\ 26.3^s$ — close to recent values. And he was aware that Mars had an atmosphere, noting an observation by Cassini that "a star in the water of Aquarius, at the distance of six minutes from the disk of

Mars, became so faint before its occultation, that it could not be seen by the naked eye, nor with a 3-feet telescope." To Herschel, Cassini's observation indicated that Mars had "a considerable but moderate atmosphere" and "its inhabitants probably enjoy a situation in many respects similar to ours."

Herschel's observations of Jupiter were focused mainly on the Galilean satellites. He noted considerable changes in their brightnesses and correctly reasoned that they have unequally tinted regions. As the moons rotated, he suspected, they would not present us with a constant quantity of reflected light. He wrote that "the satellites have a rotatory motion upon their axes, of the same duration with their periodical revolutions about the primary planet." He was correct.

Saturn particularly interested Herschel, and he was known to follow it continuously for up to six hours at a time. This dedication led to his discoveries of the two Saturnian satellites mentioned previously. He also observed that the Encke Division upon the A ring is not in the middle of that ring's breadth, and noted:

> That this black belt is not the shadow of a chain of mountains [which some astronomers believed] may be gathered from its being visible all around on the ring. It is evident that this black zone is contained between two concentric circles.

When Saturn's rings turned edgewise, Herschel correctly noted that they were thinner than the disks of its satellites. He surmised that the planet's poles were flattened, like those of Jupiter, and that its atmosphere was dense. He also determined the ringed planet's rotation rate and noted that its rings rotated as well.

On January 11, 1787, Herschel continued his observations of "Georgium Sidus," noting the position of some very faint stars near the planet. On the next night, two of those "stars" were missing. By February 5th, after a half dozen more observations, he knew he had discovered a new satellite of the planet. Before making the announcement, however, he wanted to see it move. So, on the evening of February 7th he kept the satellite in view for a total of nine hours, watching "this planet faithfully attend its primary planet." Herschel called this new satellite Oberon. During these studies he also discovered a second satellite, which he called Titania.

Herschel was a great believer in using high magnification or power:

> When you want to practice seeing, apply a power something higher than what you can see well with and go on increasing it after you have used it some time. These practices I have acquired and I can now see with powers that I used to reject for a long time.

To search for satellites around Uranus, he would use magnifications up to 7,200× and a "Field Bar" to hide the planet. Both his satellite discoveries, however, were made at the relatively low magnification of 157×, the satellites being "very nearly the dimmest objects that can be seen." From observations made in 1787 and 1792 Herschel discovered that very "small" (faint) stars dimmed when they approached Uranus, and that the satellites became regularly invisible when they arrived at certain distances from the planet. He did not realize he was detecting evidence of Uranus's ring system, which would not be discovered for another 200 years.

TO THE STARS

Herschel was fascinated with variable stars and spent much time trying to understand them. On February 2, 1781, he wrote "A Memorandum of the Uncommon Lustre of the Periodical

Star in Collo Ceti" about the star we know as Mira in Cetus.

> A star (a Sun I should say) perhaps surrounded with a system of Planets depending upon it, undergoes a change, which, were it to happen to our Sun would probably be the total destruction of every living creature. What an amazing alteration from the first magnitude down to the 6th, 7th or 8th.

Regarding the star Algol, Herschel penned the following comments in a paper presented to the Royal Society and read on May 8, 1783:

> The most extraordinary Phenomenon of the occultation of Algol is so interesting a subject, that we cannot too soon collect every observation that we may serve to ascertain its period, or permit us to find what quantity of light the star [loses] on these occasions. That stars are Suns has long been inferred from the intensity of their light at such great distances, and that these suns may have Planets around them. It has even been surmised that the change in the appearance of periodical stars might be owing either to spots revolving on their surfaces or to dark clouds in their atmospheres. The likes of a small sun revolving around a large opaque body has also been mentioned in the list of such conjectures.

Today we know Algol as one of the best-studied eclipsing binary stars in the heavens.

While undertaking his variable-star studies in January 1778, Herschel wondered if the annual parallax of stars could be used to determine their distances, as Galileo had suggested. He reasoned that if one were to look at double stars that appeared to be connected but in reality were widely separated in distance from the Earth, that the motion of the Earth around the Sun would cause the closer of the two stars to appear to move back and forth relative to the more distant star. This led him to search for, and measure, hundreds of double stars, which he called his "First Stellar Review." This review covered all stars down to 4th magnitude and was made with a 7-foot Newtonian telescope (4½ inches in aperture) at 222×. His "Second Review" was made with a far superior 7-foot instrument (with an aperture of 6.2 inches) at 227× and included all stars down to 8th magnitude. This work resulted in his first *Catalogue of Double Stars,* published in January 1782. The catalog contained 269 double stars, 227 of which, to Herschel's knowledge, had not been noticed by anyone before. Herschel's catalog was submitted to the Royal Society, and the discovery of Georgium Sidus (Uranus) was a result of this survey.

At the end of December 1781 Herschel started his third sky review using the same instrument. He measured the relative magnitudes, colors, and positions of the paired stars at 460× and reviewed all of John Flamsteed's stars for companions as faint as 12th magnitude. These observations were published on December 9, 1784, as the *Second Double Star Catalogue,* and it included 434 additional double stars. Many years later, on June 8, 1821, he published his last listing, "On the places of 145 new Double Stars."

Suspecting that many stars had changed their brightnesses since Flamsteed had published his stellar catalog, Herschel began a series of naked-eye observations to compare the brightnesses of stars within each constellation. He soon discovered that Flamsteed's magnitudes were inconsistent and of little use. Herschel's first catalog of comparative stellar brightnesses was published in 1796 and was followed by five more such publications. (Incidentally, Flamsteed's observations of the "star" he called "34th Tauri" proved to actually be of the planet Uranus.)

REALM OF THE NEBULAE

Herschel is perhaps best known for his observations of nebulae, which in his day were anything appearing fuzzy and extended. Today we know a few of these objects really are nebulae (clouds of gas and/or dust), but most are actually galaxies, which Herschel termed "milky" or "detached" nebulae. The publication of Charles Messier's short list of nebulae and clusters in the *Connaissance des Temps* for the year 1783 at once attracted Herschel's attention and he soon began to examine them with his "small" 20-foot reflector of 12 inches aperture. Herschel was not a skilled artist and never attempted to make elaborate sketches of his objects. His efforts were solely intended to show the positions of nebulous objects with respect to stars.

After observing Messier's objects, Herschel came to believe that all "nebulae" could be resolved into clusters of stars, if only they could be viewed with a sufficiently large telescope. He noted that many of Messier's nebulae yielded to the superior force of his larger telescope and were resolved into individual stars. Messier had deemed 27 of his catalog's entries to be "nebula[e] without stars," but Herschel found that these objects "either plainly appeared to be nothing but stars, or at least contained stars and gave every other indication of consisting of stars entirely." He concluded that Messier and Pierre Méchain (1744–1805) — who also discovered several objects in Messier's catalog — "saw only the more luminous parts of the Nebula in their small telescopes and the feeble shape of the remainder for want of light escaped their notice."

Herschel decided to carry out a systematic search for nebulae with his great telescopes. In the next 20 years he discovered, cataloged, and added to the science of astronomy no fewer than 2,508 new objects to the 100 or so that had been known previously.

On June 17, 1784, Herschel wrote that he had already discovered 466 new nebulae and clusters, none of which had been seen before and most of which were not within reach of the best common telescopes then in use. His first *Catalogue of One Thousand new Nebulae and Clusters of Stars,* published on April 27, 1786, was extracted from a series of sweeps with the large 20-foot telescope that began on December 19, 1783, with an observation of NGC 1055 in Cetus. He carefully arranged his finds into eight classes according to appearance and size. Using the same 20-foot reflector as in the first survey, Herschel began his second series of sweeps on April 28, 1785, with an observation of NGC 3813 in Ursa Major. Four years later, on June 11, 1789, his *Catalogue of a Second Thousand of new Nebulae and Clusters of Stars* was published. His final series of sweeps commenced on December 3, 1788, with an observation of NGC 2787, and on July 1, 1802, he published his *Catalogue of 500 new Nebulae, nebulous Stars, planetary Nebulae, and Clusters of Stars.*

Upon completing the sweeps in 1802, Herschel went back to his double-star studies and reobserved many double stars for proper motion. A colleague, John Michell, had a theory that double stars must be physically connected binary systems rather than chance, or optical, double systems. Herschel was able to visually confirm this in 1802 when he reexamined some of his double stars and noted that their relative positions had shifted. This observation provided the first direct evidence that an attractive force (namely, gravity) actually operated beyond the confines of the solar system. From time to time during the next 13 years, Herschel would observe some of the more interesting of the bright Messier objects with different instruments, from the 40-foot reflector downward. His advancing age gradually put a stop to the marvelous activity he had shown as an observer.

COSMIC INTERPRETATIONS

Herschel carefully analyzed everything he saw in the night sky, and he tried to understand what all astronomical objects were composed of, and how and why they acquired their diverse forms.

Herschel believed that stars were all about as bright as the Sun, and that first-magnitude stars were the nearest to us, while 2nd-magnitude ones were twice as far away. Mathematically he figured that his large 20-foot telescope showed stars up to 497 times the distance of Sirius, and that a larger telescope would reach even farther.

With a telescope, he reasoned, he could not only see objects at great distances; he could also look back in time:

> A telescope with a power of penetrating into space like my 40 feet has also a power of penetrating into time past. To explain this we must consider that from the known velocity of light it may be proved that when we look at Sirius the rays that enter the eye cannot have been less than 6 years and 4½ months coming from that star to the observer. Hence it follows that when we see one of these remote nebulae the rays of light which convey its image to the eye must have been almost 2 million years on the way and that so many years ago this object must already have had an existence in the sidereal heavens in order to send out those rays by which we now perceive it.

William Herschel's thinking was limited to multiples of millions of miles, and therefore it is quite obvious that he had no idea how large the universe really was. Nevertheless, he remarkably proposed a nearly correct distance for Sirius, and he concluded correctly that the Andromeda Nebula (M31) was "the farthest object that can make an impression on the eye when not assisted by telescopes."

Herschel saw the Milky Way as an extensive stratum of scattered stars, not unlike the many objects he had discovered through his telescope, and concluded:

> It is probable that the Sun is placed inside the Milky Way though perhaps not in the very center of its thickness. We gather this from the appearance of the galaxy which seems to encompass the whole heavens as it must do if the Sun is within the same.

Nebulae, he long reasoned, were composed of bodies very much like our planetary system, with the Sun, planets, moons, and comets. Planets like the Earth could not be seen around other stars since "bodies shining only with borrowed light can never be seen at very great distances." Since nebulae were often seen in beds of stars — and since the stars must have been nearer to us than the nebulae — Herschel concluded that the stars we see one by one in the heavens belonged to a detached system of which we are a part, while the Milky Way itself was composed of an unresolved background of planetary systems at much greater distances.

As stated previously, Herschel initially thought that all nebulae could eventually be resolved into individual stars if only a large enough telescope could be applied to the task. However, a series of detailed observations, from 1774 to 1811, of the Orion Nebula — which Herschel considered the nearest in the heavens — led him to reconsider this belief. His observations led him to believe (rightly or wrongly) that the nebulosity around the 7th-magnitude star in the heart of nearby M43 had steadily decreased in intensity and size until nothing could be seen around it in 1811, even with the 40-foot telescope. This led him to believe that some of the nebulous matter must be in motion with respect to the star, which was behind it, and that the nebulous matter scattered the starlight.

To Herschel, these subtle changes in the nebula's shape and luster meant that the nebula was composed not of stars but of an unknown type of luminous matter. In 1802 he wrote, "To attempt even a guess at what this light may be would be presumptuous. If it is surmised to be of the nature of the zodiacal light, what is the cause?" Other observations of the "Lucid Spot in Orion" convinced him of its intrinsic faintness; he figured it could be only two to three times as far as Sirius, while the dim 6th- or 7th-magnitude stars had to be far more distant. These nebulous objects, Herschel reasoned, had to be closer than the stars in order to be seen at all. This was a complete reversal of his previous ideas regarding the distances to, and the composition of, nebulae.

After many years and thousands of observations Herschel gradually settled on the idea of a universe in transition. Nebulae and star systems slowly developed over time under the constant action of gravity, and the source of their luminosity was unknown. A nebula that was a "little brighter in the middle" than on its periphery had not undergone much central attraction and therefore was not very advanced. A nebula "gradually brighter in the middle" was in a more advanced evolutionary state, while one that appeared "gradually much brighter in the middle" was even more evolved. Here gravity had had a longer time, possibly millions of years, to act.

Herschel's "Round Nebula[e]" with nuclei were special cases:

> When round nebulae have a nucleus, it is an indication that they have already undergone a high degree of condensation. I suppose there are not less than one thousand of these round objects. These clusters of stars are more condensed and brighter in the middle and the central brightness must be the result of central powers. [A] Nebula that [is] composed of a thousand stars must arrive at the perfection of its form sooner than another which takes in a range of a million years. A cluster or nebula [that] is very gradually more compressed and brighter to the middle may be in the perfection of its growth while another with a more equal compression, which I have called Planetary, may be looked upon as very aged.

Nebulae that enveloped stars within diffuse material represented a further stage of stellar development and became an important analytical tool. On October 16, 1784, Herschel discovered the nebula in Monoceros now known as NGC 2170 and described it thus:

> A star of about 9th magnitude, surrounded by a milky nebulosity or chevelure [luminous atmosphere], of about 3 minutes in diameter. The nebulosity is very faint and a little extended or elliptical. The chevelure involves a small star, which is about 1½ minute north of the cloudy star; other stars of equal magnitude are perfectly free from this appearance.

It seemed reasonable that the nebulosity belonged to the star in its center (or vice versa).

Herschel's discovery of "Nebulous Stars," or stars with atmospheres — which previously had been mentioned by other astronomers — compelled him, again, to conclude that there must be some kind of luminous fluid in space, the nature of which was not yet understood. He surmised the star was condensing out of the fluid, but added that ages of observations probably would be required to test the notion. The object we know as NGC 6543 (Caldwell 6) was one example.

Herschel reasoned that if there really were a connection between a star and its surrounding nebulosity, then the star would have to be very distant and uncommonly luminous if the surrounding nebulosity were in fact a multitude of unresolved stars. On the other hand, the star could be ordinary in its properties, but the surrounding luminosity then had to be "a shining fluid of a nature totally unknown to us. I can adopt no other sentiment than the latter," he decided, "since the probability is certainly not for the existence of so enormous a body as would be required to shine like a star of the 8th magnitude at a distance sufficiently great to cause a vast system of stars to put on the appearance of a very diluted milky nebulosity."

Planetary nebulae genuinely puzzled Herschel, as they were nebulous in appearance yet seemed to have planetlike disks.

> Some look so much like planets that [I] can hardly suppose them to be nebula[e]. If they were singly large stars with large diameters then they should be brighter. They might be comets at aphelion, however they must consist of stars that are compressed and accumulated in the highest degree.

Herschel eventually decided that planetaries were highly condensed, highly evolved nebulae, and he surmised that there might be a connection between planetary nebulae, nebulous stars, and stellar regeneration.

Herschel considered globular clusters to be the most interesting objects in the heavens. He noted that some of them bordered what we call dark nebulae and speculated that the stars in the globular clusters somehow had been gathered from adjacent parts of the heavens, which thereafter were completely devoid of matter.

> The 80th object of the *Connaisance des Temps* [M80, a globular cluster in Scorpius] is one of the richest clusters of small stars and is located on the western border of an opening or hole [today called a dark nebula] as if the stars were collected from that place and had left a vacancy.

While he did not understand the true nature of the objects we now call galaxies, many of Herschel's thoughts about them were remarkably prescient. For example, he was the first to note their tendency to cluster. When he came to one, he noted, generally he would find several others nearby on the dome of the sky.

Herschel finally put together a construction plan of sorts for the heavens by stating that stars condensed out of a surrounding fluid, the true nature of which was unknown. Diffuse clouds of nebulosity would collect into smaller concentrated clouds, which then condensed into scattered stars. These individual stars then were attracted into clusters, and the light given off from these clusters eventually would collect into diffuse nebulous clouds, which then would begin the cycle all over again. Perhaps, Herschel thought, he had too hastily surmised that all milky nebulosity was made up of unresolved gatherings of stars. In fact, it has been postulated that one reason Herschel quit using the 40-foot telescope was his realization that larger telescopes would not resolve all nebulae into individual stars.

THE GENIUS OF WILLIAM HERSCHEL

While nobody has ever questioned William Herschel's observational skills or telescope-building abilities, he has been criticized by some people for his resistance to advice and commitment to speculation. His common-sense approach to the science occasionally led him astray, but his ideas were original and many have been proved to be correct. Nearly all of his published papers were destined to influence the trend of European scientific progress. Many years later, Johann Louis Emil Dreyer, author of *A New General Catalogue of Nebulae and Clusters of Stars,* wrote, "The genius of Herschel, which acknowledged no boundaries to its explorations, planned work on original lines and found no untrodden paths." Herschel carefully preserved all the letters he received from his colleagues, and they show the universal admiration and respect in which he was held. Gradually all the learned societies and academies of Europe enrolled William Herschel as a member. In 1816 Herschel was knighted by the Prince Regent for his scientific accomplishments, and in 1820 he was nominated as the first President of The Royal Astronomical Society (originally the Astronomical Society of London).

William Herschel died on August 25, 1822, at Observatory House in Slough and was buried in the nearby church of Saint Lawrence at Upton. He had been very weak for a long time, yet his death was unexpected. Interestingly, his lifetime of just under 84 years spanned almost precisely the time it takes the planet he discovered to orbit the Sun.

Caroline Herschel at once left England and returned to Hanover, where she spent the remaining 25 years of her life. She prepared a catalog of 2,500 nebulae and clusters found by her brother and herself entitled *A Catalogue of the Nebulae which have been observed by Wm. Herschel in a Series of Sweeps.* The catalog was primarily intended for the use of his son, Sir John Herschel, who was engaged in the revision and extension of his father's work, using the old "large" 20-foot telescope. Sir John never published his father's catalog, in part because he shared the opinion, universal at the time, that very few of his father's nebulae could be seen with any but the largest telescopes. Yet today we know this work as the very foundation upon which modern astronomy rests.

William Herschel devoted his life to astronomy and was a tireless observer and theoretician whose ideas were way ahead of his time. Those of us lucky enough to be astronomers today will be forever grateful to him. As Dreyer wrote in 1912:

> No earthly monument is wanted to perpetuate the memory of the man who, solely guided by his genius and his indefatigable energy, rose to the greatest eminence in a hitherto almost unexplored field of science, and widened the horizon of man to an extent no one had imagined possible before his time.

Larry Mitchell
Houston, Texas
November 29, 2001

Author's Note: In 1987 I started to observe all 2,508 objects in William and Caroline Herschel's catalog. At first I used a 10-inch telescope but soon graduated to a 24-inch f/5 telescope I built myself. After seven years I achieved my goal. What's amazing is that William Herschel used a smaller telescope (an 18¾-inch f/12.8) than I, and one with inferior reflectivity. Some of the objects he discovered I found exceedingly difficult to see, even though I had the advantages of superior optics and knowing exactly where to look. In a few cases I had to return on a second night, under different atmospheric conditions, to see one of his discoveries. Yet Herschel was able to detect this same object — without prior knowledge of its existence — the first time it moved through his field of view. After completing this project, I came away with a feeling of awe, admiration, and reverence for this man's dedication, abilities, and accomplishments. He truly was the best. — L.M.

Appendix D
Photo Credits *in order of appearance*

FOREWORD
Patrick Moore

PREFACE
Luke Dodd

ACKNOWLEDGMENTS
Akira Fujii

CHAPTER 1
Akira Fujii
Donna Donovan O'Meara
VdS-Fachsgruppe Astrophotographie

CHAPTER 2
Caldwell 1 Ivan King / Harvard College Observatory
Caldwell 2 Martin Germano
Caldwell 3 Martin Germano
Caldwell 4 George Greaney / Dean Easton
Caldwell 5 Kim Zussman
Caldwell 6 Sean Walker / John Boudreau
J. P. Harrington / K. J. Borkowski / NASA
Caldwell 7 Preston Scott Justis
Caldwell 8 Lee Coombs
Caldwell 9 Digitized Sky Survey[1]
Caldwell 10 Preston Scott Justis
Caldwell 11 Dennis Sørensen
Caldwell 12 Kim Zussman
Caldwell 13 Preston Scott Justis
Caldwell 14 Akira Fujii (2)
Caldwell 15 Sean Walker / John Boudreau
Bruce Balick / Jason Alexander / Arsen Hajian / Yervant Terzian / Mario Perinotto / Patrizio Patriarchi / NASA
Caldwell 16 Digitized Sky Survey[1]
Caldwell 17/18 Martin Germano
Caldwell 19 George Greaney
Caldwell 20 Ethan Van Matre
Dennis di Cicco
Caldwell 21 Adam Block / NOAO / AURA / NSF
Caldwell 22 Sean Walker
Howard Bond / NASA
Caldwell 23 Dale E. Mais
Caldwell 24 Digitized Sky Survey[1]
Caldwell 25 Preston Scott Justis
Caldwell 26 Harvey Freed
Caldwell 27 Preston Scott Justis
Caldwell 28 Lee Coombs
Caldwell 29 Preston Scott Justis
Caldwell 30 William McLaughlin
Caldwell 31 George Greaney
Caldwell 32 Robert Bickel
Caldwell 33/34 Jim Riffle
Caldwell 35 Digitized Sky Survey[1]
Caldwell 36 Digitized Sky Survey[1]
Edwin P. Hubble, *The Realm of the Nebulae*, Yale University Press
Caldwell 37 Preston Scott Justis
Caldwell 38 Robert and Janice Fera
Caldwell 39 Sean Walker
NASA / Andrew Fruchter / ERO Team
Caldwell 40 Digitized Sky Survey[1]
Caldwell 41 Akira Fujii
Stephen James O'Meara
Caldwell 42 Digitized Sky Survey[1]
Caldwell 43/44 Preston Scott Justis
Chuck Vaughn
Caldwell 45 Martin Germano
Caldwell 46 Preston Scott Justis
Eric R. Craine (3)
Caldwell 47 Digitized Sky Survey[1]
Caldwell 48 Digitized Sky Survey[1]
Caldwell 49/50 Hans Vehrenberg
Caldwell 51 Martin Germano
Caldwell 52 Digitized Sky Survey[2]
Caldwell 53 Ray Gralak
Caldwell 54 Laurence A. Marschall / Liang-Tai George Chiu / William van Altena
Caldwell 55 C. R. Lynds / Kitt Peak National Observatory
Bruce Balick / Jason Alexander / Arsen Hajian / Yervant Terzian /Mario Perinotto / Patrizio Patriarchi / NASA
Caldwell 56 Kim Zussman
Caldwell 57 Martin Germano
Caldwell 58 Akira Fujii
Caldwell 59 Martin Germano
Caldwell 60/61 Kim Zussman

	Brad Whitmore / NASA
Caldwell 62	Kim Zussman
Caldwell 63	Chuck Vaughn
	C. Robert O'Dell / Kerry P. Handron / NASA
	Mount Wilson and Palomar Observatories
Caldwell 64	Preston Scott Justis
Caldwell 65	Preston Scott Justis
	Hubble Heritage Team (AURA/STScI/NASA)
Caldwell 66	Jeffrey L. Jones
Caldwell 67	Preston Scott Justis
Caldwell 68	Steven Quirk
	Mount Wilson Observatory
Caldwell 69	VdS-Fachsgruppe Astrophotographie
Caldwell 70/72	Martin Germano
	Andrew Peters
Caldwell 71	Akira Fujii
Caldwell 73	Akira Fujii
Caldwell 74	*Sky & Telescope* archives
	Hubble Heritage Team (AURA/STScI/NASA)
Caldwell 75	Digitized Sky Survey[2]
Caldwell 76	Martin Germano
	Akira Fujii (2)
Caldwell 77	Luke Dodd
	Ethan J. Schreier / NASA
	Reginald Dufour / Cerro Tololo Inter-American Observatory
Caldwell 78	Digitized Sky Survey[2]
Caldwell 79	Digitized Sky Survey[2]
Caldwell 80	Luke Dodd
Caldwell 81	Steven Lee
Caldwell 82	European Southern Observatory
Caldwell 83	Steven Quirk
Caldwell 84	Digitized Sky Survey[2]
Caldwell 85	Steven Lee
Caldwell 86	Akira Fujii
Caldwell 87	Digitized Sky Survey[2]
Caldwell 88	Digitized Sky Survey[2]
Caldwell 89	Akira Fujii
Caldwell 90	Digitized Sky Survey
	Howard Bond / NASA
Caldwell 91	Akira Fujii
Caldwell 92	Luke Dodd
	Pablo Prado
	Jon Morse / NASA
Caldwell 93/101	Steven Quirk
	Reginald Dufour / Raymond J. Talbot / Eric B. Jensen
Caldwell 94	Cerro Tololo Inter-American Observatory
	Akira Fujii
Caldwell 95	Digitized Sky Survey[2]
Caldwell 96	Akira Fujii
Caldwell 97	Akira Fujii
Caldwell 98/99	Akira Fujii (2)
Caldwell 100	VdS-Fachsgruppe Astrophotographie
Caldwell 102	Akira Fujii
Caldwell 103	Andrew Peters
	Luke Dodd
Caldwell 104/106	Steven Lee
	Andrew Peters
	Akira Fujii
Caldwell 105/108	Akira Fujii (2)
Caldwell 107	Digitized Sky Survey[2]
Caldwell 109	Digitized Sky Survey[2]
	Howard Bond / NASA

CHAPTER 3

Steven Quirk

APPENDIX B

Akira Fujii

APPENDIX C

Wolbach Library, Harvard-Smithsonian Center for Astrophysics (2)

[1]From the Digitized Sky Survey, Northern Hemisphere, courtesy Palomar Observatory and NASA/AURA/STScI.

[2]From the Digitized Sky Survey, Southern Hemisphere, courtesy the U.K. Schmidt Telescope Unit and NASA/AURA/STScI.

Index

A Abell 426, 101
Acrux, 388, 392, 394, 400
adaptive-optics camera, 180, 184, 188
Advanced Satellite for Cosmology and Astrophysics, 406
AE Aurigae, **122**–125
Albireo, **87**, 148, 172
Aldebaran, 164–166
Algol, 461
Allen, David A., 93
Allen, Richard Hinckley, 10, 63, 288, 363, 418, 423
Alpha Persei Moving Cluster, 339
al-Sufi, Abd al-Rahman, 340
American Association of Variable Star Observers, 181
Ames, Adelaide, 307
Andromeda Galaxy. *See* M31
Anglo-Australian Observatory, 259, 266, 357
Antlia-Hydra Cloud, 192
Apis, 423
Apis Indica, 423
apparent magnitudes, 8–10
Apus, 423
Archinal, Brent, 8
  and catalog/atlas discrepancies, 27, 145–146, 198–199, 432
  and galaxies, 202
  and globular clusters, 105, 172, 419
  and nebulae, 80
  and open clusters, 20, 59–60, 86, 113, 286, 298, 382, 385, 402–403
Arcturus, **1**
Arecibo, 192
Arnett, Bill, 257
Arp, Halton C., 76, 267, 332
Ashbrook, Joseph, 318–319
astronomical atlases
  discrepancies between, 20, 86, 144–146, 286, 384–385, 398, 432
  Millennium Star Atlas, 14, 52, 67, 79, 86, 95, 103, 136, 144, 146, 153, 165, 172, 192, 206, 269, 286, 347, 374, 385, 398, 428
  Skalnate Pleso *Atlas of the Heavens,* 143, 397
  Sky Atlas 2000.0, 14, 20, 46, 52, 79, 86, 143–144, 153, 159, 198, 240, 286, 384–385, 398, 416, 428, 431–432, 436, 438
  Uranometria, 318–319, 381, 413, 418, 423, 426
  Uranometria 2000.0, 14, 46, 52, 86, 144, 153, 198, 269, 286, 385, 398, 428, 432
astronomical catalogs
  discrepancies between, 144–146, 197–199, 228, 294, 398–399, 419, 432
  of James Dunlop, 274, 278, 285, 288, 312, 315, 320, 337, 343, 346, 351, 372, 395, 410, 420, 424
  General Catalogue of Nebulae (GC), xiii, 7, 50, 92, 145, 198
  of William Herschel, 7–8, 461–462, 466
  of Nicolas Louis de Lacaille, 284–285, 320, 341, 343, 358, 378, 382, 384, 389, 413, 418
  Messier Catalog, vii, 19, 30, 305, 336, 432–433, 449–453
  New General Catalogue of Nebulae and Clusters of Stars (NGC), 7, 86, 145, 152, 223, 279, 466
  prefixes in, 12
  Revised New General Catalogue of Nonstellar Astronomical Objects (RNGC), 86
Athens Observatory, 182
Auckland Observatory, 5, 418, 422, 425, 427
Australia Telescope Compact Array, 374

B Baade, Walter, 101, 104, 255, 307
Bailey, Solon, 169–170
Balick, Bruce, 34, 69, 93, 115, 220–221
Barker, Edmond S., 145
Barlow lens, 4, 226
Barnard 48, 303, **304**
Barnard 205, 436
Barnard, Edward Emerson, 94, 183–184, 197–198, 227, 274, 393, 432
Barnard's Galaxy (Caldwell 57), **227**–230
Barrado y Navascués, David, 340
Baume, Gustavo, 301

Bayer, Johann
constellations created by, 418, 423, 426
and Greek letters for stars, 318–319
Uranometria. See astronomical atlases
Bennett, Jack, 337
Bettoni, Daniela, 160
BF Centauri, **388**, 390
Bigourdan, Guillaume, 319–320, 432
binary stars
in Caldwell 56, 226
and core collapse in globular clusters, 372–373
double stars and, 462
Eta Carinae as, 365
evolution of, 343
Black Forest Star Party, 113
black holes, 213, 279, 308, 333
Blinking Planetary Nebula (Caldwell 15), **68**–71
Blow, Graham, 368
blue stragglers, 189, 216, 289, 372, 415–416, 425
Bode, Johann Elert, 456
Bok, Bart J., 195, 369
Bok globules, 195
Bond, George P., 9, 306
Bond, Howard, 226
Bond, William Cranch, 9
Bortle, John, 450–451
Bottari, Claudio, 117
Bow-Tie Nebula (Caldwell 2), 9, **22**–23
Box Galaxy (Caldwell 21), **89**–91
Brandl, Bernhard, 406
Braun, W. G., 74
Brown, Peter Lancaster, 453
brown dwarfs, 246, 340
Bruck, Mary T., 349
Bubble Nebula (Caldwell 11), **52**–54
Bug Nebula (Caldwell 69), **274–**276
Burbidge Chain, 247
Burnham, Jr., Robert, 124, 179, 229, 235, 438
Burnham's Celestial Handbook, 6, 31, 39, 94, 99, 133, 143, 155, 212, 384, 449
BZ Crucis, **391**, 394–395
Caldwell 1, **19**–21

C Caldwell 2, 9, **22**–23
Caldwell 3, **24**–26
Caldwell 4, 7, **27**–29
Caldwell 5, 7, **30**–32, 150
Caldwell 6, **33**–36
Caldwell 7, **37**–40
Caldwell 8, **41**–43, **64**
Caldwell 9, **44**–48
Caldwell 10, **49**–51, **64**
Caldwell 11, **52**–54
Caldwell 12, **55**–57
Caldwell 13, **58**–61, **64**
Caldwell 14, **62**–67, **64, 449**–453
Caldwell 15, **68**–71, **87**
Caldwell 16, **72**–74
Caldwell 17, **75**–78
Caldwell 18, **75**–78
Caldwell 19, **79**–92
Caldwell 20, **83**–88, **87**
Caldwell 21, **89**–91
Caldwell 22, **92**–95, 121
Caldwell 23, **96**–99, 150–151, 153
Caldwell 24, **100**–103
Caldwell 25, **104**–106
Caldwell 26, 13, **107**–108
Caldwell 27, **87, 109**–111
Caldwell 28, **112**–114
Caldwell 29, **115**–117
Caldwell 30, **118**–121
Caldwell 31, **122**–125
Caldwell 32, **126**–128, 435
Caldwell 33, **87, 129**–134
Caldwell 34, **87, 129**–134
Caldwell 35, **135**–138
Caldwell 36, **139**–142
Caldwell 37, 8, 13, **87,143**–149
Caldwell 38, **150**–153, 441
Caldwell 39, **154**–158, **155**
Caldwell 40, **159**–161
Caldwell 41, 13–14, **162**–168, **165**, 339, 437
Caldwell 42, **169**–172
Caldwell 43, **173**–176

Caldwell 44, **173**, 176–178
Caldwell 45, **179**–181
Caldwell 46, **182**–186, **183**
Caldwell 47, **187**–190
Caldwell 48, **191**–193
Caldwell 49, 14, **194**–201
Caldwell 50, **194**–201
Caldwell 51, **202**–206
Caldwell 52, **207**–210
Caldwell 53, **211**–214
Caldwell 54, **215**–217
Caldwell 55, **218**–222, **220**
Caldwell 56, **223**–226
Caldwell 57, **227**–230
Caldwell 58, **231**–233
Caldwell 59, **234**–238, 440
Caldwell 60, 102, **239**–244, **241**
Caldwell 61, 102, **239**–244, **241**
Caldwell 62, **245**–247
Caldwell 63, **248**–253, **250, 253**
Caldwell 64, **254**–257
Caldwell 65, **258**–262, **260**, 439
Caldwell 66, **263**–265
Caldwell 67, **266**–269
Caldwell 68, **270**–273, **271**, 442
Caldwell 69, **274**–276
Caldwell 70, **277**–283
Caldwell 71, **284**–287
Caldwell 72, **277**–283
Caldwell 73, **288**–291
Caldwell 74, **292**–295, **293**
Caldwell 75, **296**–299
Caldwell 76, **300**–304, **302, 304**
Caldwell 77, **305**–310, **307, 309**
Caldwell 78, **311**–313
Caldwell 79, **314**–317, **377**
Caldwell 80, **318**–323, 413
Caldwell 81, **324**–327
Caldwell 82, **328**–331
Caldwell 83, **332**–335
Caldwell 84, **336**–338
Caldwell 85, **339**–341
Caldwell 86, 274, **342**–344
Caldwell 87, **345**–347, 412
Caldwell 88, **348**–350
Caldwell 89, 14, **351**–353
Caldwell 90, **354**–357, **356**
Caldwell 91, **358**–360, **377**
Caldwell 92, **ix**, **361**–369, **363, 377**
Caldwell 93, **370**–374
Caldwell 94, **376**–380, **377**
Caldwell 95, **381**–383
Caldwell 96, **377, 384**–387
Caldwell 97, **377, 388**–390
Caldwell 98, **377, 391**, 395–396
Caldwell 99, **377, 391**–396
Caldwell 100, 8, **13**, 14, **377, 397**–401
Caldwell 101, **371**, 374–375
Caldwell 102, 339, **377, 402**–404
Caldwell 103, **405**–408, **406**
Caldwell 104, **409**–413, **411**
Caldwell 105, 3, 5, **377, 417**–420
Caldwell 106, 321, **409**, **411,** 413–416
Caldwell 107, 5, **423**–425
Caldwell 108, 5, **377, 417**, 420–422
Caldwell 109, 5, **426**–429, **428**
Caldwell Catalog
  creation of, vii–viii
  errors in original publication of, 12–14, 398–399
Camelopardalis, 40
Canada-France-Hawaii Telescope (CFHT), 180, 184, 204
Canes Venatici Spur, 115
Capa, Cristina E., 110
Cape of Good Hope, 7, 236, 258, 278, 284, 297, 300, 345, 372, 407, 418
Carter, David, 266
Carter Observatory, 368, 414
Cat's Eye Nebula (Caldwell 6), **33**–36, **34**
Cave Nebula (Caldwell 9), **44**–48
Centaurus A (Caldwell 77), **305**–310, **307, 309**
Cepheid variable stars, 29, 205, 351
CFHT. *See* Canada-France-Hawaii Telescope

challenging targets
galaxies, 26, 91, 98, 107, 152–153, 202–206, 247, 252–253, 412–413
nebulae, 35–36, 47–48, 81, 111, 195, 234
star clusters, 51, 60, 104–106, 256
stars, 94–95, 201
Chamaeleon, 426
Christmas Tree, 185
Clark, Roger N., 25–26, 31, 91, 95, 113, 124, 437
Clark, Tom and Jeannie, 54, 133, 368
Clerke, Agnes, 414
Close, Laird M., 184
Clown-Face Nebula (Caldwell 39), **154**–158, **155**
Coalsack (Caldwell 99), **377**, **391**–396
Cocoon Nebula (Caldwell 19), **79**–92
Coe, Steve, 235–236, 280, 295
Cogshall, Wilbur A., 134
Collinder, Per Arne, 27, 145–146, 397
Collinder 228, 368
Collinder 249 (Caldwell 100), 8, **13**, 14, **397**–401
Collinder 416, 146
Collinder 417, 146
Collinder 428, 88
Collinder 429, 27
Collins, Peter, 68
Coma Cluster, 136
Coma-Sculptor Cloud, 32, 38, 56, 77, 151
comets
Halley's, 309
methods of searching for, 450–453
Shoemaker-Levy 9, 282
Cone Nebula, 185
constellations
boundaries redefined, 20, 191
defined by Bayer, 418, 423, 426
defined by Keyser and de Houtman, 370
defined by Lacaille, 345, 418
Cooper, Ian, 360
Copeland, Leland, 92, 185
Cor Caroli, 126
Corona Australis, 311
Corvus Cloud, 243
Corwin, Harold G., 147–148, 202, 281
Crab Nebula (M1), 131
Crescent Nebula (Caldwell 27), **109**–111
Crux, 376
Curtis, Heber D., 22, 306
Cygnus Loop, 130, 132

D Dark Doodad, 420
dark matter, 246
dark nebulae, nature of, 392–393
data sources, 10–12
Davidson, Kris, 364
Dawes, William, 9
de Bruijpe, Jos, 166
Deep-Sky Wonders, 6, 236
de Houtman, Frederick, 370, 413
Dejonghe, Herwig, 208
Delgado, Rosa M. Gonzalez, 178
Delporte, Eugene, 20
Delta Geminorum, 156–157
Delta Scuti variable stars, 256
Deneb, 84, **87**–88
Denning, William E., 30
density waves
and star formation in spiral galaxies, 328–329
de Vaucouleurs, Gérard, 126–127, 212, 281
Digitized Sky Survey (DSS), 16, 399, 467–468
Double Cluster (Caldwell 14), **62**–67, **449**–453
double stars
binary stars and, 462
in Caldwell 41,167–168
Herschel's catalogs of, 461
drawing deep-sky objects, 16–18
Dreyer, Johann Louis Emil, 7, 145, 279
on William Herschel, 466
DSS. See Digitized Sky Survey
Dunlop, James
catalog of southern deep-sky objects by, 278
and galaxies, 278, 305–306, 335
and globular clusters, 288–289, 312, 320, 337, 372, 410, 420–421, 424
John Herschel and, 278–279, 424

and nebulae, 271, 274, 362
and open clusters, 297, 303, 330, 351, 358, 389
dust
in Caldwell 99, 392–393
in Caldwell 104, 412
dust lanes
in Caldwell 20, 84
in galaxies, 97–98, 115, 150–153, 175, 177, 306–310
in the nucleus of Caldwell 52, 208
dwarf galaxies
accompanying M31, 75–78
dark matter in, 246
globular clusters as nuclei of, 324–325

E EGGs. See Evaporating gaseous globules
18 Vulpeculae, 146
80 Tauri, 167
81 Tauri, 167
Einstein satellite, 110, 266
Electric Arc (Caldwell 43), **173**–176
elliptical galaxies
dust lanes and stellar disks within, 208
and the Hubble sequence, 139–140, 207
as X-ray emitters, 101
Elmegreen, Debra M., 180
Eridanus Cloud, 266
Eskimo Nebula (Caldwell 39), **154**–158, **155**
Espin, Thomas E., 79
E.T. Cluster (Caldwell 13), **58**–61
Eta Carinae, **363–365**
Eta Carinae Nebula (Caldwell 92), **ix, 361**–369, **363, 377**
European Southern Observatory
New Technology Telescope, 208, 346
Very Large Telescope, 274
Evans, Rev. Robert, 309
evaporating gaseous globules (EGGs), 53
eye, anatomy of, 68–69
Eye Nebula (Caldwell 59), **234**–238, 440

F False Comet, **302**
False Comet Cluster (Caldwell 76), **300**–304, **302, 304**
Fan, Zuhui, 56
fast low-ionization emission regions (FLIERs), 69–71, 94, 237
Ferris, Timothy, 102–103, 179
Filamentary Nebula (Caldwell 34), **129**–134
Fisher, J. Richard, 115
Flaming Star Nebula (Caldwell 31), **122**–125
Flammarion, Camille, 429
Flamsteed, John, 196
FLIERs. *See* fast low-ionization emission regions
Fornax Cluster, 266
47 Tucanae (Caldwell 106), 321, **409**, 413–416
FP Virginis, 181
Fremont Park Observatory, 257
Frost, Robert, 2, 414

G galactic structure
globular clusters and, 169–170
galaxies. *See also* dwarf galaxies, elliptical galaxies, irregular galaxies, lenticular galaxies, spiral galaxies, and starburst galaxies
bars in, 24–25
classification of, 8, 139–140, 207–208
collision of or interaction between, 90, 101–102, 127, 137, 160–161, 192, 240, 243, 307
dust lanes in, 97–98, 115, 150–153, 175, 177, 267, 306–310
and 18th- and 19th-century perceptions of universe, 8, 89, 119, 178
hot spots in, 180
Hubble sequence of, 139–**140**
orbital motions within, 160–161
Seyfert, 101–102, 115, 266, 305
velocities of, 76
Gamma Centauri, 398
Gamma Delphini, 172
Geffert, Michael, 437
Geisler, Douglas, 149
Gemini North Telescope, 189

General Catalogue of Nebulae (GC). See astronomical catalogs
George III, 456, 458
Georgium Sidus. *See* Uranus
Ghost Globular. *See* NGC 5897
Ghost's Goblet (Caldwell 8), **41**–43
Ghost of Jupiter (Caldwell 59), **234**–238, 440
Gingerich, Owen, 285, 449, 450
globular clusters
ages of, 289, 324
blue stragglers in, 189, 289, 372, 415–416, 424–425
core collapse of, 312, 372–373
evolution of, 289, 324, 343, 372, 406
extragalactic, 102, 151, 416
and galactic structure, 169–170, 410–411
William Herschel and, 465
orbital motions of, 263–264, 314–315
variable stars in, 169–170, 336, 422, 425
Glyn Jones, Kenneth, 380
Gómez, Yolanda, 275
Gonzales, Guillermo, 315
Gould, Benjamin Apthorp, 413–414
Graham, John A., 271
Great Sculptor Galaxy (Caldwell 65), **258**–262, **260**, 439
Great Square of Pegasus, 174
Greek letters
in star names, 318–319
Green, Daniel W. E., 450
Green, David A., 110–111
Gum Nebula, 387

H Hall, Rob and Lesley, 5, 411
Halley, Edmond, 319, 362, 413
Halley's Comet, 309
Handler, Gerald, 70
Hanover (Germany), 454
Harby, Bill, 394
Hardy, Thomas, 392–393
Harris, William E., 104, 264, 289, 373, 419, 424
Hartung, Ernst J.
and galaxies, 181, 208, 212
and globular clusters, 325, 347, 412, 437
and nebulae, 295, 356
and open clusters, 166, 232–233, 254, 299, 303, 331
Harvard College Observatory, 169
Oak Ridge Station, 151
refractors of, 9, 132, 249, 307
Harvard-Smithsonian Center for Astrophysics, 68
Hawaii
light pollution in, 4–5
volcanic smog (vog) above, 2–3
Hawaii Volcanoes National Park, 2
Hazen, Martha, 264
HD 111973, **376**, 379–380
HD 112887, **135,** 137
HD 150136, 329–331
HD 192163, 109–110
HD 192182, 111
HD 199579, 84–85
HD 200775, 28–29
Head, Marilyn, 374
Heckman, Timothy, 115
Helix Nebula (Caldwell 63), **248**–253, **250, 253**
Herbig, George, 183
Herschel, Caroline, 96, 231–232, 258, 455–458, 466
Herschel, John, 7, 9–10, 145–147, 177, 188, 198, 320, 354, 364, 407
James Dunlop and, 278–279, 424
Herschel, William, 6, 27, 119, 188, 223, 234–235, **454**–466
catalogs published by, 461–462, 466
categories of deep-sky objects, 8
and 18th-century perception of universe, 8, 89, 119, 178, 392
family and musical career, 454–455
and Immanuel Kant's hypothesis, 178
planetary observations by, 458–460
telescopes of, 455–459, **458**
discovery of Uranus by, 456

Hevelius, Johannes, 72, 81, 452
Hipparcos satellite, 123, 165, 181, 261, 380, 385, 403
Hockey Stick (NGC 4656/7), **126**, 127, 435–436
Hodge, Paul, 77
Hodierna, Giovanni Batista, 254–255, 300
Hogg 15, **391**, 395
Hogg, David E., 192
Hogg, Helen Sawyer, 105, 336
Holmberg, Erik B., 107
Hoogerwerf, Ronnie, 123
Hoopes, Charles, 108
Horologium, 345
Horseshoe Cluster (Caldwell 10), **49**–51
Houston, Walter Scott, 6, 19, 199, 212, 234–236, 242, 322
Hydra Hysteria, 263
HST. *See* Hubble Space Telescope
H II regions, 45–46, 255–256, 362, 405
in galaxies, 38–39, 178, 259, 267, 374–375
star formation in, 32
Hubble, Edwin P., 30, 108, 182–183, 192, 227
galaxy classification and, 139–140, 207
Hubble Space Telescope (HST)
and black holes, 213, 308
and galaxies, 90, 102, 136, 180, 259, 308, 435
and nebulae, 69, 93, 134, 220, 292–293, 428
and star clusters, 151, 189, 286, 416
Hubble's Variable Nebula (Caldwell 46), **182**–186
Huggins, William, 33, 236
Humason, Milton L., 30
Hutchings, John B., 228
Hyades (Caldwell 41), 13–14, **162**–168, 339, 437
stereogram of, **165**
Hynes, Steven J., 11, 69, 385, 438

I IAU. See International Astronomical Union
IC 342 (Caldwell 5), 7, **30**–32, 150
IC 405 (Caldwell 31), **122**–125
IC 410, **122**, 124
IC 1340, 133
IC 1496 (Caldwell 19), **79**–92
IC 1590, 431–432
IC 1613 (Caldwell 51), **202**–206
IC 2391 (Caldwell 85), **339**–341
IC 2488, 357
IC 2602 (Caldwell 102), 339, **377, 402**–404
IC 2944, 8, **397**–401
IC 2948, **397**–401
IC 4628, 303–**304**
IC 4637, 303
IC 4651, 326–327
IC 4677, 35–36
IC 4812, **270**, 272
IC 4836, 374
IC 4842, 374
IC 4845, 374
IC 5067, 84
IC 5070, 84
Iddings, Joseph P., 134, 154
Infrared Astronomical Satellite (IRAS), 110, 196, 208
Intergalactic Tramp (Caldwell 25), **104**–106
Intergalactic Wanderer. *See* Intergalactic Tramp
International Astronomical Union (IAU), 191
International Ultraviolet Explorer (IUE), 357
Iota Orionis, 123
IRAS. *See* Infrared Astronomical Satellite
irregular galaxies, 204

J Janes, Kenneth A., 349
Jensen, Peter, 363
Jewel Box (Caldwell 94), **376**–380
Johnson, Harold, 255
Jupiter, 460

K Kahanpää, Jere, 50, 181
Kant, Immanuel, 154, 178
Kappa Crucis (Caldwell 94), **376**–380, **377**
Kappa Tauri, 167
Keeler, James, 175
Kelly, Al, 128
Keyhole Nebula, **363**, 368
Keyser, Pieter Dirkszoon, 370, 413, 426

Kilauea Volcano, 2
Kitt Peak National Observatory, 241
Knapen, Johan H., 180
Koester, Detlev, 353, 359
Kohle, Sven, 90
Kohoutek, Luboš, 52, 292
Kreimer, Evered, 449
Krisciunas, Kevin, 408
Kwok, Sun, 155–156, 275

L Lacaille, Abbé Nicolas Louis de
catalog of, 284–285
constellations defined by, 345, 418
and globular clusters, 320, 413, 418–419
and nebulae, 362, 407
and open clusters, 297, 300, 341, 358, 378, 384, 389
telescopes used by, 297, 378
Lacerta, 72
Lada, Charles, 255
Lambda Centauri, **13, 397**–401
Lambda Centauri Cluster (Caldwell 100), 8, **13**–14, **377, 397**–401
Lampland, Carl Otto, 182–183, 263
Laplace, Pierre Simon, 89
Large Magellanic Cloud (LMC), 90, 204, 228, 360, 394, 405–**406**
Lassell, William, 92, 183, 271
Laustsen, Svend, 292
Leander McCormick Observatory, 228
Leavitt, Henrietta S., 352
Lee, Young-Wook, 321
lenticular galaxies, 139–140, 159–160, 175, 207–208, 213, 432
Leo Cloud, 159
Leuschner Observatory Supernova Search, 192
Levy, David H., ix, 401, 450
Lick Observatory, 175, 183
Light Blue Snowball (Caldwell 22), **92**–95
light pollution, 4–5, 199
Liller, William, 336
LINERs. *See* low-ionization nuclear emission-line regions
Ling, Alister, 71, 198
Lion Nebula (Caldwell 39), **154**–158, **155**
Liquid Pyramid (Caldwell 64), **254**–257
Little Jewel Box. *See* NGC 3293
LMC. *See* Large Magellanic Cloud
Local Group, 76, 204
Lorenzin, Tom, 59
Lossing, Fred, 199
Lou, Yu-Qing, 56
Low, Frank, 183
low-ionization nuclear emission-line regions (LINERs), 115–116, 119
Luginbuhl, Christian B., 11
and galaxies, 40, 98, 128, 141, 181, 280
and nebulae, 35, 236
and star clusters, 114, 217
Lynch, Richard, 133

M M1 (Crab Nebula), 131
M15, 121
M20 (Trifid Nebula), 81
M27 (Dumbbell Nebula), 148
M31 (Andromeda Galaxy), 119, 174, 451, 463
dwarf companions to, 75–78
M32, 77
M33, 113, 279
M37, 388–389
M39, 80
M40, 81, 452
M42 (Orion Nebula), 80, 122–123, 195
M44 (Beehive), 216
M50, 196
M52, 53
M57 (Ring Nebula), 438
M63, 119
M65, 160
M74, 30
M81, 97
M83, 440
M101, 30
M102, 432–433

M103, 49–50, 58–59
M110, 77
MACHOs. *See* massive compact halo objects
MacRobert, Alan M., 58, 124, 185, 434
Magellanic Clouds, 263. *See also* Large Magellanic Cloud, Small Magellanic Cloud
magnitudes
  discrepancies and variations in, 8–10, 64–65, 94, 223–224, 294, 419, 421–422
  methods for determining, 39
Malin, David, 259, 415
Mallas, John H., 449
Marconi, Alessandro, 334
Mars, 459–460
Marshall, Francis E., 406
Marth, Albert, 197, 271
Martin, Martha Evans, 2
massive compact halo objects (MACHOs), 246
Mayall, Nicholas U., 123
Mayor, Michel, 233
McClure, Robert, 215
McNeil, Jay, 33, 35–36, 70, 156
Meaburn, John, 250–251
Méchain, Pierre, 218, 433, 462
Meier, Rolf, 199
Melotte 25 (Caldwell 41), 13–14, **162**–168, 339, 437
Melotte 101, 404
Mendez, Mariano, 412
Mermilliod, Jean-Claude, 233
Messier Catalog. *See* astronomical catalogs
Messier, Charles, 63, 81, 112, 131, 157, 196, 211, 300
  William Herschel and, 462
Mexican Jumping Star, 257
Meylan, Georges, 406
Mice, The, 243
Michell, John, 462
Milkweed Seed Galaxy (Caldwell 62), **245**–247
Milky Way, **xiii**, 99, 121
  and cannibalized dwarf galaxies, 324–325
  globular clusters and structure of, 169–170
  orbital motions within, 314–315
Minkowski, Rudolph L., 101, 134, 307
Mitchell, Larry, x, 119, 128, 243–244, 435
molecular hydrogen, 98, 388
Moon
  William Herchel and, 459
Moore, Patrick, **vii**, ix, 442
  Caldwell catalog creation and, vii–viii
Morales, Ronald, 223
Morris, Charles, 39
Moulton, Forest Ray, 134
Mount John Observatory, 415
Mount Wilson
  60-inch reflector, 212, 221–222
  100-inch reflector, 89, 213, 227
Mu Columbae, 123
Murrell, Andrew, 356, 427
Musca, 418
mythologies
  Australian, 392
  Chinese, 363
  Greek, 163–164, 370, 423

N

Nagler, Al, 3
National Geographic Society–Palomar Observatory Sky Survey, 298, 437
nebulae. *See also* dark nebulae, planetary nebulae, reflection nebulae, and variable nebulae
  classification of, 284–285
  18th-century perception of universe and, 8
  gas knots in, 251
  illumination of, 45, 84–85, 93, 182–184, 219, 271–272, 292, 405
  as imagined groupings of stars, 462
  William Herschel and, 462–465
nebular hypothesis, 154
Neckel, Thorston, 85
Network Nebula (Caldwell 33), **129**–134
New General Catalogue of Nebulae and Clusters of Stars (NGC). *See* astronomical catalogs
New Technology Telescope, 208, 346

New Zealand, 5, 360, 368
NGC 14, 174
NGC 40 (Caldwell 2), 9, **22**–23
NGC 55 (Caldwell 72), **277**–283
NGC 57, 175
NGC 104 (Caldwell 106), 321, **409**, 413–416
NGC 121, 416
NGC 134, 440–441
NGC 147 (Caldwell 17), **75**–78
NGC 185 (Caldwell 18), **75**–78
NGC 188 (Caldwell 1), **19**–21
NGC 246 (Caldwell 56), **223**–226
NGC 247 (Caldwell 62), **245**–247
NGC 253 (Caldwell 65), **258**–262, **260**, 439
NGC 255, 225
NGC 281, 431–432
NGC 288, 260, 439
NGC 300 (Caldwell 70), **277**–283
NGC 346, 410, 411
NGC 362 (Caldwell 104), **409**–413, **411**
NGC 371, 410, 411
NGC 404, 435
NGC 406, 413
NGC 436, 61
NGC 457 (Caldwell 13), **58**–61
NGC 559 (Caldwell 8), **41**–43
NGC 654, **49**
NGC 659, 49, 51
NGC 663 (Caldwell 10), **49**–51
NGC 708, 114
NGC 752 (Caldwell 28), **112**–114
NGC 753, 114
NGC 869. *See* Caldwell 14
NGC 884. *See* Caldwell 14
NGC 891 (Caldwell 23), **96**–99, 150–151, 153
NGC 1097 (Caldwell 67), **266**–269
NGC 1097A, **266**, 269
NGC 1261 (Caldwell 87), **345**–347, 412
NGC 1270, **100,** 103
NGC 1272, **100,** 103
NGC 1273, **100,** 103
NGC 1275 (Caldwell 24), **100**–103
NGC 1277, **100,** 103
NGC 1278, **100,** 103
NGC 1281, **100**, 103
NGC 1333, 436
NGC 1360, 438
NGC 1528, 433–434
NGC 1647, 168, 436–437
NGC 1792, 291
NGC 1808, 291
NGC 1851 (Caldwell 73), **288**–291
NGC 1893, 124
NGC 2070 (Caldwell 103), **405**–408, **406**
NGC 2237. *See* Caldwell 49
NGC 2238. *See* Caldwell 49
NGC 2244 (Caldwell 50), **194**–201
NGC 2246. *See* Caldwell 49
NGC 2261 (Caldwell 46), **182**–186
NGC 2264, 185
NGC 2354, 257
NGC 2360 (Caldwell 58), **231**–233
NGC 2362 (Caldwell 64), **254**–257
NGC 2392 (Caldwell 39), **154**–158, **155**
NGC 2403 (Caldwell 7), **37**–40
NGC 2419 (Caldwell 25), **104**–106
NGC 2420, 158
NGC 2451, 287
NGC 2477 (Caldwell 71), **305**–310, **308, 309**
NGC 2506 (Caldwell 54), **215**–217
NGC 2516 (Caldwell 96), **377, 384**–387
NGC 2539, 216
NGC 2669, 341
NGC 2775 (Caldwell 48), **191**–193
NGC 2777, 192
NGC 2867 (Caldwell 90), **354**–357, **356**
NGC 3115 (Caldwell 53), **211**–214
NGC 3132 (Caldwell 74), **292**–295, **293**
NGC 3149, 428
NGC 3195 (Caldwell 109), 5, **426**–429, **428**
NGC 3200, 238
NGC 3201 (Caldwell 79), **314**–317, **377**
NGC 3242 (Caldwell 59), **234**–238, 440
NGC 3293, 369

NGC 3324, 369
NGC 3372 (Caldwell 92), ix, **361**–369, **363, 377**
NGC 3532 (Caldwell 91), **358**–360
NGC 3605, **159**
NGC 3607, **159**
NGC 3608, **159**
NGC 3621, 440
NGC 3626 (Caldwell 40), **155**–161
NGC 3766 (Caldwell 97), **388**–390
NGC 4027, 242
NGC 4038 (Caldwell 60), 102, **239**–244, **241**
NGC 4039 (Caldwell 61), 102, **239**–244, **241**
NGC 4128, 26
NGC 4190, 108
NGC 4214, 108
NGC 4236 (Caldwell 3), **24**–26
NGC 4244 (Caldwell 26), 13, **107**–108
NGC 4372 (Caldwell 108), 5, **377**, **417**, 420–422
NGC 4449 (Caldwell 21), **89**–91
NGC 4460, 91
NGC 4485, 434
NGC 4490, 434
NGC 4494, 153
NGC 4550, 160
NGC 4559 (Caldwell 36), **139**–142
NGC 4562, 152
NGC 4565 (Caldwell 38), **150**–153, 441
NGC 4605, 430–431
NGC 4609 (Caldwell 98), **377, 391**, 395—396
NGC 4627, 127
NGC 4631 (Caldwell 32), **126**–128, 435
NGC 4656. *See* Hockey Stick
NGC 4657. *See* Hockey Stick
NGC 4697 (Caldwell 52), **207**–210
NGC 4755 (Caldwell 94), **376**–380, **377**
NGC 4833 (Caldwell 105), 3, 5, **377, 417**–420
NGC 4874, 136–137
NGC 4889 (Caldwell 35), **135**–138
NGC 4898AB, **135,** 137
NGC 4921, 136–137
NGC 4945 (Caldwell 83), **332**–335
NGC 4976, 335
NGC 5005 (Caldwell 29), **115**–117
NGC 5033, 116
NGC 5082, 310
NGC 5086, 310
NGC 5090, 310
NGC 5090A, 310
NGC 5090B, 310
NGC 5091, 310
NGC 5102, 441–442
NGC 5128 (Caldwell 77), **305**–310, **307, 309**
NGC 5139 (Caldwell 80), **318**–323, 413
NGC 5248 (Caldwell 45), **179**–181
NGC 5286 (Caldwell 84), **336**–338
NGC 5307, 338
NGC 5694 (Caldwell 66), **263**–265
NGC 5749, 350
NGC 5823 (Caldwell 88), **348**–350
NGC 5866, 160, 432–433
NGC 5897, 437–438
NGC 5979, 383
NGC 6025 (Caldwell 95), **381**–383
NGC 6087 (Caldwell 89), 14, **351**–353
NGC 6101 (Caldwell 107), 5, **423**–425
NGC 6124 (Caldwell 75), **296**–299
NGC 6139, 299
NGC 6153, 299
NGC 6188, 329
NGC 6193 (Caldwell 82), **328**–331
NGC 6231 (Caldwell 76), **300**–304, **302, 304**
NGC 6242, 303, **304**
NGC 6268, 303, **304**
NGC 6281, 443
NGC 6302 (Caldwell 69), **274**–276
NGC 6337, 276
NGC 6352 (Caldwell 81), **324**–327
NGC 6397 (Caldwell 86), 274, **342**–344
NGC 6496, 312
NGC 6541 (Caldwell 78), **311**–313
NGC 6543 (Caldwell 6), **33**–36
NGC 6723, **270, 430**, 442
NGC 6726, **270**, 272–273, **430**
NGC 6727, **270**, 272–273, **430**

NGC 6729 (Caldwell 68), **270**–273, **271**, 442
NGC 6744 (Caldwell 101), **371**, 374–375
NGC 6752 (Caldwell 93), **370**–374
NGC 6818, 230
NGC 6822 (Caldwell 57), **227**–230
NGC 6826 (Caldwell 15), **68**–71
NGC 6882, 143–149
NGC 6885 (Caldwell 37), 8, 13, **143**–149
NGC 6888 (Caldwell 27), **109**–111
NGC 6934 (Caldwell 47), **187**–190
NGC 6939, 56–57
NGC 6946 (Caldwell 12), **55**–57
NGC 6960 (Caldwell 34), **129**–134
NGC 6992. *See* Caldwell 33
NGC 6995. *See* Caldwell 33
NGC 6996, 86–88
NGC 6997, 86–88
NGC 7000 (Caldwell 20), **83**–88, **87**
NGC 7006 (Caldwell 42), **169**–172
NGC 7009 (Caldwell 55), **218**–222, **220**
NGC 7023 (Caldwell 4), 7, **27**–29
NGC 7209, 72, 82
NGC 7243 (Caldwell 16), **72**–74
NGC 7293 (Caldwell 63), **248**–253, **250, 253**
NGC 7331 (Caldwell 30), **118**–120
NGC 7335, **118**, 120
NGC 7336, **118**, 120
NGC 7337, **118**, 120
NGC 7340, 120
NGC 7354, 48
NGC 7479 (Caldwell 44), **173**, 176–178
NGC 7635 (Caldwell 11), **52**–54
NGC 7640, 95
NGC 7662 (Caldwell 22), **92**–95, 121
NGC 7789, 431
NGC 7793, 439–440
NGC 7814 (Caldwell 43), **173**–176
19 Vulpeculae, 146
Noctua, 264
North America Nebula (Caldwell 20), **83**–88, **87**
Nubecula Minor, 413
nuclear dust lanes, 208

O O'Connor, John J., 231
O'Dell, C. Robert, 251
O'Meara, Stephen James, **3**
  celestial play by, 62–63
  Deep-Sky Companions: The Messier Objects, 2
  eyepieces used by, 3–4
  method of magnitude determination, 39
  and TeleVue Genesis refractor, **3**
  Texas Star Party and, 128, 133, 158, 243–244, 303
Omega Centauri (Caldwell 80), **318**–323, 413
Omicron Velorum Cluster (Caldwell 85), **339**–341
open clusters
  blue stragglers in, 216
  formation of, 215
  planetary nebula in, 353
Orion Nebula (M42), 122–123, 195
Outer Limits Galaxy (Caldwell 23), **96**–99, 150–151, 153
Owens Valley Radio Observatory, 116

P Pac-Man Nebula (Caldwell 56), **223**–226
Palitzsch, Johann Georg, 197
Palomar Mountain
  Schmidt telescope, 160
  200-inch reflector, 38, 101, 186, 213
Paraskevopoulos, John S., 307
Parramatta, New South Wales, 278, 285, 288, 305, 312, 320, 372, 378, 395, 410, 420
Parsons, William (3rd Earl of Rosse), 150, 218
Pavo, 370–371
Pease, Frank G., 212
Pegasus, 118, 121, 174
Pegasus Cloud, 176
Pegasus Spur, 119, 175
Pelican Nebula, **83**–85
Pellepoix, Antoine Darquier de, 33
Peltier, Leslie C., 2, 450
Perek, Luboš, 52
Perinotto, Mario, 438
Perseid meteor shower, 67
Perseus A (Caldwell 24), **100**–103

Perseus Cluster, 101
Phi Cas Cluster (Caldwell 13), **58**–61
photographs, selection criteria for, 15–16
Piccolomini, Alessandro, 319
Pickering, William, 358
Pincushion (Caldwell 91), **358**–360
Pitt, Mary, 455
planetary nebulae
  formation of, 22–23, 155–156
  gas jets in, 34
  motions within, 93–94, 219–220
  in an open cluster, 353
Pleiades, 166, 339
  Greek mythology of, 163–164
Pogson, Rev. Norman R., 9–10
Polakis, Tom, 107, 268
Polaris, 19, 428
Ptolemy, 319
pulsars, 406

Q quasars, 266–267, 332

R Radcliffe 136 (R136), 406
Raymo, Chet, 84
Raynard, Arthur Cowper, 393
R Canis Majoris, 233
R Coronae Australis, **270**–273, **271**
R Coronae Australis Nebula (Caldwell 68), **270**–273, **271**, 442
red-dwarf stars, 343
Reed, Darren, 34
reflection nebulae, 27–29, 182–186, 270–273, 436
Reimers, Dieter, 359
Revised New General Catalogue of Nonstellar Objects (RNGC). See astronomical catalogs
Ring Nebula (M57), 237, 438
Ringtail Galaxy. *See* Caldwell 60/61
R Monocerotis, 182–184
Roberts, Issac, 29, 306
Roberts, Morton S., 192
Robertson, Edmond F., 231
Robichon, Noël, 149
Rosat satellite, 45–46, 110, 119, 140, 241, 246, 279, 403
Rosette Nebula (Caldwell 49), 14, **194**–201
Rossi X-Ray Timing Explorer (RXTE), 406
Roth, Joshua, 133, 153
Roy, Jean-René, 127
Royal Astronomical Society, 466
Royal Society, 456, 461
RR Lyrae-type variable stars, 169–170, 289, 336
RS Persei, 67
Rubin, Vera, 160
Rudaux, Lucien, 175, 212
runaway stars, 122–123
RXTE. See Rossi X-Ray Timing Explorer

S Sagar, Ram, 379
Samus, Nikolai, 336
Sandage, Allan R., 435
SAO 20575, 53
SAO 228708, 311
SAO 252073, **376**, 380
Sarajedini, Ata, 424
Saturn, 4, 460
Saturn Nebula (Caldwell 55), **218**–222, **220**
Scarecrow, The (Caldwell 51), **202**–206
Schaeberle, J. M., 123
Schmidt, J. F. Julius, 182, 271
Sculptor Galaxy (Caldwell 65), **258**–262, **260**, 439
Sculptor Group, 246, 259, 278
Seagrave Observatory, 133
Secchi, Angelo, 221, 236
Sechler, Tim, 59
Serviss, Garrett P., 166
Sextans, 211
Seyfert, Carl K., 98, 101
Seyfert galaxies, 100–103, 115, 266, 305, 333
Shain, C. A., 307
Shapley, Harlow, 104, 169–170, 286, 307, 359
Sharpless, Stewart, 45
Sharpless 2-155 (Caldwell 9), **44**–48
Sharpless 310, 255–256

Sheehan, William, 17, 415
E. E. Barnard biography by, 393
Shoemaker, Carolyn, 282, 382
Shoemaker, Eugene, 282
Shoemaker-Levy 9, 282, 382
Sigma Octantis, 428
Sigma Tauri, 167
Silver Coin (Caldwell 65), **258**–262, **260**
Silver Needle Galaxy (Caldwell 26), **107**–108
Sinnott, Roger W., 10–11, 14, 165
Sirius supercluster, 339
Skiff, Brian A., 11
and galaxies, 98, 128, 141, 181, 280
and nebulae, 35, 195, 236
and star clusters, 40, 114, 217
Skyscrapers Astronomy Club, 133
SL 7, 443
Small Magellanic Cloud (SMC), 228, 409–**411**, 413
Smith, Bruce J., 183
Smyth, Adm. William Henry, 9–10, 167, 177, 188, 211, 219, 232, 234–236, 255, 437–438
star colors and double stars, 156, 167, 172, 185
Smyth, Piazzi, 431
S Normae, 351–353
S Normae Cluster (Caldwell 89), 14, **351**–353
Sofue, Yoshiaki, 259
solar activity
effect on airglow, 4
solar system, 314
South Celestial Pole, 428
South Galactic Pole Group, 246
Southern Beehive (Caldwell 96), **384**–387
Southern Butterfly (Caldwell 105), **417**–420
Southern Cross, 315, 318, 348, 358, 362, 376–378, **377**, 388, 418
Southern Pinwheel (Caldwell 70), **277**–283
Southern Pleiades (Caldwell 102), **402**–404
Southern Ring (Caldwell 74), **292**–295, **293**
Space Telescope Science Institute, 33
spectra and spectroscopy, 33, 166, 224–225, 357
Spindle, The (Caldwell 53), **211**–214
spiral galaxies
barred, 24, 177
classification of, 139–140
magnetic fields in, 56
mixed, 139–140
molecular gas in, 116
unbarred, 139–140
Staal, Julius, 345
starburst galaxies, 241–242, 259, 334
star clusters. *See also* globular clusters, open clusters
in Caldwell 65, 259
Starfish, The (Caldwell 93), **370**–374
star formation, 57, 90, 119, 196, 388
and density waves, 328–329
in H II regions, 32, 38–39, 45–46, 178, 204, 259, 267, 374
in interacting galaxies, 240–242
star names
Greek letters in, 318–319
stars, binary. *See* binary stars
stellar winds
and nebula formation, 22–23, 52–53, 196
Stephan's Quintet, **118**, 121
Stone, Edward James, 380
Straizys, Vytantas, 84
String of Pearls (Caldwell 72), **277**–283
Struve 953, 186
Struve 1645, 91
Struve 2890, 73–74
Struves, 9–10
Sullivan, III, Woodruff T., 121
supergiant stars
extragalactic, 205, 279, 405–406
and Type II supernovae, 130–131
supernovae, 117, 178, 269, 309
remnants of, 25, 110–111, 129–134, 196, 241, 259, 279
star formation and, 25, 334
Type Ia, 55, 193
Type Ic, 178
Type II, 55–56, 130–131
Swift, Lewis, 197

T Tarantula Nebula (Caldwell 103), **405**–408, **406**
Tau Canis Majoris Cluster (Caldwell 64), **254**–257
Tau Canis Majoris, 256
Taurus, 162–163
T Cephei, 29
T Coronae Australis, 273
telescopes
built by William Herschel, 455–456, 457–459
Hubble Space Telescope. *See* Hubble Space Telescope
limiting magnitudes of, 9–10
Palomar Observatory. *See* Palomar Mountain
used by author, 3–6
Telescopium-Grus Cloud, 441
Tele Vue Optics, 3–4
Texas Star Party, 128, 133, 158, 243, 303, 435
Theis, Christian, 90
Theodor, Pieter, 381
Theta Carinae Cluster (Caldwell 102), **377**, **402**–404
Thomson, Malcomb J., 70
3C 84, 101
Tirion, Wil, 14, 143, 398, 416, 428
Tombaugh, Clyde, 156, 263
Townes, Charles, 164
Trapezium, 9, 122
Triangulum Australe, 381
True Lover's Knot, 408
Trumpler, Robert J., 145–146, 255, 393
Trumpler 1, 49
Trumpler 14, 367
Trumpler 16, 367
Trumpler 24, 300, 303, **304**
T Tauri stars, 45
Tucana, 410
Tully, R. Brent, 96, 115
Tweezers Galaxy (Caldwell 83), **332**–335
20 Vulpeculae, **143**–146
TY Coronae Australis, 272

U U3, 192
Ultra-high contrast (UHC) filter, 199
Uranus, 4, 460
William Herschel's discovery of, 456
2 Ursae Minoris, 20

V V364 Carinae, 404
V374 Carinae, 385
V382 Carinae, **358**, 360
V460 Carinae, 386
V518 Carinae, 403
V910 Centauri, **388**, 390
V973 Scorpii, 302
Valentijn, Edwin A., 98
Vanderbilt University Observatory, 227
van der Werf, Paul P., 98
variable nebulae, 182–186, 270–273, 366
variable stars
within the *Be* spectral class, 51, 388
Cepheid type of, 205, 351
Delta Scuti type of, 256
William Herschel's observations of, 460–461
in nebulae, 272, 363
RR Lyrae type of, 170, 289, 336
in star clusters, 169–170, 336, 380, 422, 425
vdB 96, 256
vdB-Ha 99, 367
Vehrenberg, Hans, 249
Veil Nebula. *See* Caldwell 33/34
Vela Pulsar, 387
Vela X, 387
Very Large Array, 192
Virgo Cluster, 207
visual acuity, 68–69
volcanic smog (vog), 2–3
Volcano, Hawaii, 2
von Auwers, Georg Fredrich Julius Arthur, 248
von Gothard, E., 123
VY Canis Majoris, 256

W Walker, Alistair R., 289
Wallace, Kent, 427
Warren Rupp Observatory, 146–147, 172

Webb, Rev. T. W., 157, 197, 233, 267, 434
Webb Society Deep-Sky Observer's Handbook, 145, 223–224, 280, 380, 389, 401
white-dwarf stars, 343, 359
William Herschel Telescope, 205
Wilner, David, 255
Wilson, Barbara, 6, 8
  and elliptical galaxies, 103, 138, 210
  and globular clusters, 265, 291, 313, 412, 415
  and irregular galaxies, 91
  and nebulae, 54, 95, 221
  and open clusters, 42, 50, 144, 201
  and spiral galaxies, 40, 57, 99, 116, 127–128, 160, 214, 282
Winnecke 4 (M40), 81
Winter Star Party, 54, 234, 284, 291, 412
Wolf, Max, 29, 45, 79–80, 84, 123, 202, 228
Wolf-Rayet stars, 109–110, 301
  extragalactic, 228, 279
  in planetary nebulae, 22, 357, 438
WZ Velorum, 317

X X-ray sources, 101, 110, 119, 140, 246, 266–267, 289

Y Y Lupi, 349

Z Zeta Scorpii, 300–301, **304**
Zwicky, Fritz, 136

Bold type indicates where the subject is illustrated.

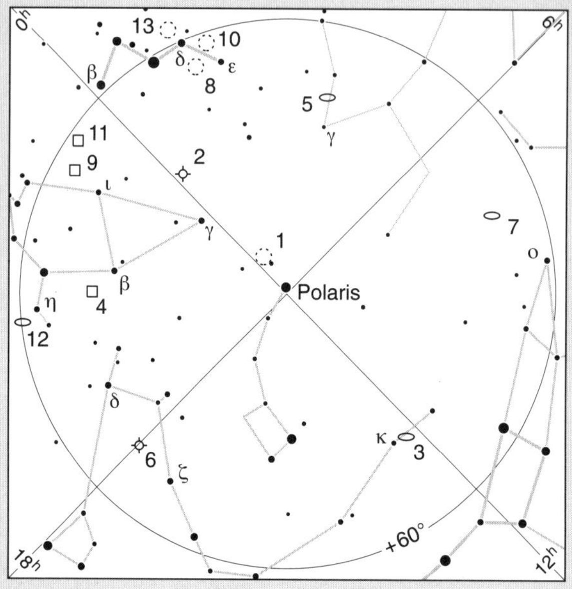

0h
6h
12h
18h
+60°
Polaris
1
2
3
4
5
6
7
8
9
10
11
12
13
β
γ
δ
ε
ζ
η
ι
κ
ο

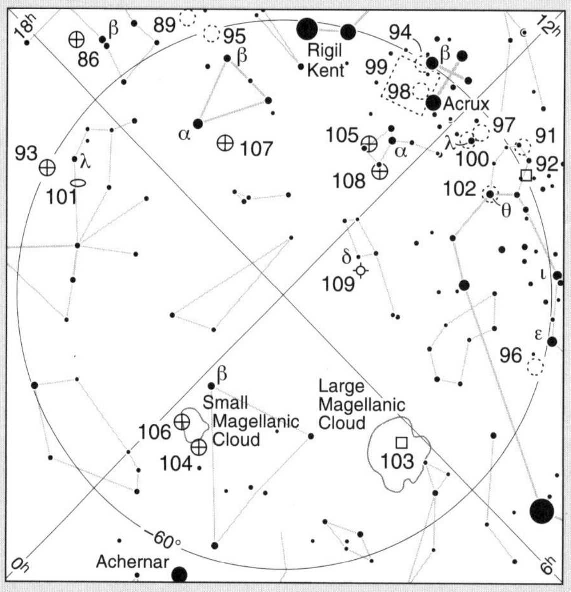

18h
12h
6h
0h
−60°
Rigil Kent
Acrux
Achernar
Small Magellanic Cloud
Large Magellanic Cloud
86
89
91
92
93
94
95
96
97
98
99
100
101
102
103
104
105
106
107
108
109
α
β
δ
ε
θ
ι
λ